**Handbook of RNA
Biochemistry**
*Edited by R. K. Hartmann,
A. Bindereif, A. Schön,
E. Westhof*

Further Titles of Interest

U. Schepers

RNA Interference in Practice

2005
ISBN 3-527-31020-7

G. Cesareni, M. Gimona, M. Sudol, M. Yaffe (Eds.)

Modular Protein Domains

2005
ISBN 3-527-30813-X

G. Kahl

The Dictionary of Gene Technology, 3rd edition

2004
ISBN 3-527-30765-6

S. Brakmann, A. Schwienhorst (Eds.)

Evolutionary Methods in Biotechnology

2004
ISBN 3-527-30799-0

S. Lorkowski, P. Cullen (Eds.)

Analysing Gene Expression

2002
ISBN 3-527-30488-6

Handbook of RNA Biochemistry

*Edited by Roland K. Hartmann, Albrecht Bindereif,
Astrid Schön, Eric Westhof*

WILEY-VCH Verlag GmbH & Co. KGaA

Editors

Prof. Dr. Roland K. Hartmann
Philipps-Universität Marburg
Institut für Pharmazeutische Chemie
Marbacher Weg 6
35037 Marburg
Germany
roland.hartmann@staff.uni-marburg.de

Prof. Dr. Albrecht Bindereif
Justus-Liebig-Universität Giessen
Institut für Biochemie
Heinrich-Buff-Ring 58
35392 Giessen
Germany
albrecht.bindereif@chemie.bio.uni-giessen.de

Dr. Astrid Schön
Universität Leipzig
Institut für Biochemie
Brüderstr. 34
04103 Leipzig
Germany
schoena@uni-leipzig.de

Prof. Dr. Eric Westhof
CNRS – UPR 9002
Institut de Biologie Moléculaire et Cellulaire
15 rue René Descartes
67084 Strasbourg
France
e.westhof@ibmc.u-strasbg.fr

All books published by Wiley-VCH are carefully produced. Nevertheless, authors, editors, and publisher do not warrant the information contained in these books, including this book, to be free of errors. Readers are advised to keep in mind that statements, data, illustrations, procedural details or other items may inadvertently be inaccurate.

Library of Congress Card No.: applied for
A catalogue record for this book is available from the British Library.
Bibliographic information published by Die Deutsche Bibliothek
Die Deutsche Bibliothek lists this publication in the Deutsche Nationalbibliografie; detailed bibliographic data is available in the Internet at http://dnb.ddb.de

© 2005 WILEY-VCH Verlag GmbH & Co. KGaA, Weinheim
All rights reserved (including those of translation in other languages). No part of this book may be reproduced in any form – by photoprinting, microfilm, or any other means – nor transmitted or translated into machine language without written permission from the publishers. Registered names, trademarks, etc. used in this book, even when not specifically marked as such, are not to be considered unprotected by law.

Printed in the Federal Republic of Germany.
Printed on acid-free paper.

Typesetting Asco Typesetters, Hong Kong
Printing Strauss GmbH, Mörlenbach
Bookbinding Litges & Dopf Buchbinderei GmbH, Heppenheim
Cover Design Matthes + Traut Werbeagentur GmbH, Darmstadt

ISBN-13 978-3-527-30826-2
ISBN-10 3-527-30826-1

Short Contents

Volume 1
Part I   RNA Synthesis   *1*
I.1   Enzymatic RNA Synthesis, Ligation and Modification   *3*
I.2   Chemical RNA Synthesis   *95*

Part II   Structure Determination   *131*
II.1   Molecular Biology Methods   *133*
II.2   Biophysical Methods   *385*
II.3   Fluorescence and Single Molecule Studies   *453*

Volume 2
Part III   RNA Genomics and Bioinformatics   *489*

Part IV   Analysis of RNA Function   *665*
IV.1   RNA–Protein Interactions *in vitro*   *667*
IV.2   RNA–Protein Interactions *in vivo*   *729*
IV.3   SELEX   *783*

Part V   RNAi   *895*

Appendix: UV Spectroscopy for the Quantitation of RNA   *910*

Handbook of RNA Biochemistry. Edited by R. K. Hartmann, A. Bindereif, A. Schön, E. Westhof
Copyright © 2005 WILEY-VCH Verlag GmbH & Co. KGaA, Weinheim
ISBN: 3-527-30826-1

Contents

Preface *XXXI*

List of Contributors *XXXIV*

Volume 1

Part I **RNA Synthesis** *1*

I.1 **Enzymatic RNA Synthesis, Ligation and Modification** *3*

1 **Enzymatic RNA Synthesis using Bacteriophage T7 RNA Polymerase** *3*
Heike Gruegelsiepe, Astrid Schön, Leif A. Kirsebom and Roland K. Hartmann
1.1 Introduction 3
1.2 Description of Method – T7 Transcription *in vitro* 4
1.2.1 Templates 5
1.2.2 Special Demands on the RNA Product 6
1.2.2.1 Homogeneous 5′ and 3′ Ends, Small RNAs, Functional Groups at the 5′ End 6
1.2.2.2 Modified Substrates 7
1.3 Transcription Protocols 8
1.3.1 Transcription with Unmodified Nucleotides 8
1.3.2 Transcription with 2′-Fluoro-modified Pyrimidine Nucleotides 14
1.3.3 Purification 15
1.4 Troubleshooting 17
1.4.1 Low or No Product Yield 17
1.4.2 Side-products and RNA Quality 17
1.5 Rapid Preparation of T7 RNA Polymerase 17
1.5.1 Required Material 18
1.5.2 Procedure 18
1.5.2.1 Cell Growth, Induction and Test for Expression of T7 RNAP 18
1.5.2.2 Purification of T7 RNAP 19
1.5.3 Notes and Troubleshooting 20

Handbook of RNA Biochemistry. Edited by R. K. Hartmann, A. Bindereif, A. Schön, E. Westhof
Copyright © 2005 WILEY-VCH Verlag GmbH & Co. KGaA, Weinheim
ISBN: 3-527-30826-1

Acknowledgement 21
References 21

2 Production of RNAs with Homogeneous 5′ and 3′ Ends 22
Mario Mörl, Esther Lizano, Dagmar K. Willkomm and Roland K. Hartmann

2.1 Introduction 22
2.2 Description of Approach 23
2.2.1 *Cis*-cleaving Autocatalytic Ribozyme Cassettes 23
2.2.1.1 The 5′ Cassette 23
2.2.1.2 The 3′ Cassette 23
2.2.1.3 Purification of Released RNA Product and Conversion of End Groups 26
2.2.2 *Trans*-cleaving Ribozymes for the Generation of Homogeneous 3′ Ends 26
2.2.3 Further Strategies Toward Homogeneous Ends 29
2.3 Critical Experimental Steps, Changeable Parameters, Troubleshooting 29
2.3.1 Construction of *Cis*-cleaving 5′ and 3′ Cassettes 29
2.3.2 Dephosphorylation Protocols 33
2.3.3 Protocols for RNase P Cleavage 34
2.3.4 Potential Problems 34
References 35

3 RNA Ligation using T4 DNA Ligase 36
Mikko J. Frilander and Janne J. Turunen

3.1 Introduction 36
3.2 Overview of the RNA Ligation Method using the T4 DNA Ligase 37
3.3 Large-scale Transcription and Purification of RNAs 38
3.4 Generating Homogeneous Acceptor 3′ Ends for Ligation 40
3.5 Site-directed Cleavage with RNase H 42
3.6 Dephosphorylation and Phosphorylation of RNAs 43
3.7 RNA Ligation 44
3.8 Troubleshooting 45
3.9 Protocols 46
Acknowledgments 51
References 51

4 T4 RNA Ligase 53
Tina Persson, Dagmar K. Willkomm and Roland K. Hartmann

4.1 Introduction 53
4.2 Mechanism and Substrate Specificity 54
4.2.1 Reaction Mechanism 54
4.2.2 Early Studies 56
4.2.3 Substrate Specificity and Reaction Conditions 57
4.3 Applications of T4 RNA Ligase 58

4.3.1	End-labeling 58
4.3.2	Circularization 59
4.3.3	Intermolecular Ligation of Polynucleotides 59
4.4	T4 RNA Ligation of Large RNA Molecules 61
4.5	Application Examples and Protocols 64
4.5.1	Production of Full-length tRNAs 64
4.5.2	Specific Protocols 65
4.5.3	General Methods (GM) 69
4.5.4	Chemicals and Enzymes 70
4.5.4.1	Chemical Synthesis and Purification of Oligoribonucleotides 70
4.5.4.2	Chemicals 71
4.5.4.3	Enzymes 72
4.6	Troubleshooting 72
	Acknowledgments 72
	References 72

5 Co- and Post-Transcriptional Incorporation of Specific Modifications Including Photoreactive Groups into RNA Molecules 75
Nathan H. Zahler and Michael E. Harris

5.1	Introduction 75
5.1.1	Applications of RNA Modifications 75
5.1.2	Techniques for Incorporation of Modified Nucleotides 77
5.2	Description 79
5.2.1	5′-End Modification by Transcription Priming 79
5.2.2	Chemical Phosphorylation of Nucleosides to Generate 5′-Monophosphate or 5′-Monophosphorothioate Derivatives 80
5.2.3	Attachment of an Arylazide Photo-crosslinking Agent to a 5′-Terminal Phosphorothioate 82
5.2.4	3′-Addition of an Arylazide Photo-crosslinking Agent 83
5.3	Troubleshooting 84
	References 84

6 3′-Terminal Attachment of Fluorescent Dyes and Biotin 86
Dagmar K. Willkomm and Roland K. Hartmann

6.1	Introduction 86
6.2	Description of Method 87
6.3	Protocols 88
6.3.1	3′ Labeling 88
6.3.1.1	Biotin Attachment [12] 88
6.3.1.2	Fluorescence Labeling [5] 89
6.3.2	Preparatory Procedures: Dephosphorylation of RNA Produced with 3′ Hammerheads 89
6.3.3	RNA Downstream Purifications 90
6.3.3.1	Gel Chromatography 90
6.3.3.2	Purification on Denaturing Polyacrylamide Gels 90

6.3.4	Quality Control 91
6.4	Troubleshooting 91
6.4.1	Problems Caused Prior to the Labeling Reaction 91
6.4.2	Problems with the Labeling Reaction Itself 92
6.4.3	Post-labeling Problems 93
	References 93

I.2 Chemical RNA Synthesis 95

7 Chemical RNA Synthesis, Purification and Analysis 95
Brian S. Sproat

7.1	Introduction 95
7.2	Description 97
7.2.1	The Solid-phase Synthesis of RNA 97
7.2.1.1	Manual RNA Synthesis 99
7.2.1.2	Automated RNA Synthesis 100
7.2.2	Deprotection 101
7.2.2.1	Deprotection of Base Labile Protecting Groups 101
7.2.2.2	Desilylation of Trityl-off RNA 102
7.2.2.3	Desilylation of Trityl-on RNA 102
7.2.3	Purification 103
7.2.3.1	Anion-exchange HPLC Purification 103
7.2.3.2	Reversed-phase HPLC Purification of Trityl-on RNA 104
7.2.3.3	Detritylation of Trityl-on RNA 105
7.2.3.4	Desalting by HPLC 106
7.2.4	Analysis of the Purified RNA 107
7.3	Troubleshooting 107
	References 110

8 Modified RNAs as Tools in RNA Biochemistry 112
Thomas E. Edwards and Snorri Th. Sigurdsson

8.1	Introduction 112
8.1.1	Modification Strategy: The Phosphoramidite Method 113
8.1.2	Modification Strategy: Post-synthetic Labeling 115
8.2	Description of Methods 116
8.2.1	Post-synthetic Modification: The 2′-Amino Approach 116
8.2.1.1	Reaction of 2′-Amino Groups with Succinimidyl Esters 119
8.2.1.2	Reaction of 2′-Amino Groups with Aromatic Isothiocyanates 119
8.2.1.3	Reaction of 2′-Amino Groups with Aliphatic Isocyanates 120
8.3	Experimental Protocols 120
8.3.1	Synthesis of Aromatic Isothiocyanates and Aliphatic Isocyanates 120
8.3.2	Post-synthetic Labeling of 2′-Amino-modified RNA 122
8.3.3	Post-synthetic Labeling of 4-Thiouridine-modified RNA 125
8.3.4	Verification of Label Incorporation 125
8.3.5	Potential Problems and Troubleshooting 126
	References 127

Part II	Structure Determination *131*
II.1	Molecular Biology Methods *133*

9 Direct Determination of RNA Sequence and Modification by Radiolabeling Methods *133*
Olaf Gimple and Astrid Schön

9.1	Introduction *133*
9.2	Methods *133*
9.2.1	Isolation of Pure RNA Species from Biological Material *134*
9.2.1.1	Preparation of Size-fractionated RNA *134*
9.2.1.2	Isolation of Single Unknown RNA Species Following a Functional Assay *134*
9.2.1.3	Isolation of Single RNA Species with Partially Known Sequence *135*
9.2.2	Radioactive Labeling of RNA Termini *137*
9.2.2.1	5' Labeling of RNAs *137*
9.2.2.2	3' Labeling of RNAs *138*
9.2.3	Sequencing of End-labeled RNA *140*
9.2.3.1	Sequencing by Base-specific Enzymatic Hydrolysis of End-labeled RNA *141*
9.2.3.2	Sequencing by Base-specific Chemical Modification and Cleavage *144*
9.2.4	Determination of Modified Nucleotides by Post-labeling Methods *146*
9.2.4.1	Analysis of Total Nucleotide Content *146*
9.2.4.2	Determination of Position and Identity of Modified Nucleotides *148*
9.3	Conclusions and Outlook *149*
	Acknowledgments *149*
	References *149*

10 Probing RNA Structures with Enzymes and Chemicals *In Vitro* and *In Vivo* *151*
Eric Huntzinger, Maria Possedko, Flore Winter, Hervé Moine, Chantal Ehresmann and Pascale Romby

10.1	Introduction *151*
10.2	The Probes *153*
10.2.1	Enzymes *153*
10.2.2	Chemical Probes *153*
10.2.3	Lead(II) *155*
10.3	Methods *155*
10.3.1	Equipment and Reagents *155*
10.3.2	RNA Preparation and Renaturation Step *156*
10.3.3	Enzymatic and Lead(II)-induced Cleavage Using End-labeled RNA *157*
10.3.4	Chemical Modifications *160*
10.3.5	Primer Extension Analysis *161*
10.3.6	*In Vivo* RNA Structure Mapping *163*

10.3.6.1	*In Vivo* DMS Modification	*163*
10.3.6.2	*In Vivo* Lead(II)-induced RNA Cleavages	*165*
10.4	Commentary	*166*
10.4.1	Critical Parameters	*166*
10.4.2	*In Vivo* Mapping	*168*
10.5	Troubleshooting	*168*
10.5.1	*In Vitro* Mapping	*168*
10.5.2	*In Vivo* Probing	*169*
	Acknowledgments	*169*
	References	*170*
11	**Study of RNA–Protein Interactions and RNA Structure in Ribonucleoprotein Particles** *172*	
	Virginie Marchand, Annie Mougin, Agnès Méreau and Christiane Branlant	
11.1	Introduction	*172*
11.2	Methods	*175*
11.2.1	RNP Purification	*175*
11.2.2	RNP Reconstitution	*176*
11.2.2.1	Equipment, Materials and Reagents	*176*
11.2.2.2	RNA Preparation and Renaturation Step	*177*
11.2.3	EMSA	*178*
11.2.3.1	EMSA Method	*179*
11.2.3.2	Supershift Method	*180*
11.2.3.3	Identification of Proteins Contained in RNP by EMSA Experiments Coupled to a Second Gel Electrophoresis and Western Blot Analysis	*184*
11.2.4	Probing of RNA Structure	*185*
11.2.4.1	Properties of the Probes Used	*185*
11.2.4.2	Equipment, Material and Reagents	*186*
11.2.4.3	Probing Method	*187*
11.2.5	UV Crosslinking and Immunoselection	*195*
11.2.5.1	Equipment, Materials and Reagents	*195*
11.2.5.2	UV Crosslinking Method	*196*
11.3	Commentaries and Pitfalls	*196*
11.3.1	RNP Purification and Reconstitution	*198*
11.3.1.1	RNA Purification and Renaturation	*198*
11.3.1.2	EMSA	*199*
11.3.2	Probing Conditions	*199*
11.3.2.1	Choice of the Probes Used	*199*
11.3.2.2	Ratio of RNA/Probes	*200*
11.3.3	UV Crosslinking	*200*
11.3.3.1	Photoreactivity of Individual Amino Acids and Nucleotide Bases	*200*
11.3.3.2	Labeled Nucleotide in RNA	*201*
11.3.4	Immunoprecipitations	*201*
11.3.4.1	Efficiency of Immunoadsorbents for Antibody Binding	*201*

11.4	Troubleshooting *201*
11.4.1	RNP Reconstitution *201*
11.4.2	RNA Probing *201*
11.4.3	UV Crosslinking *202*
11.4.4	Immunoprecipitations *202*
	Acknowledgments *202*
	References *202*

12	**Terbium(III) Footprinting as a Probe of RNA Structure and Metal-binding Sites** *205*
	Dinari A. Harris and Nils G. Walter
12.1	Introduction *205*
12.2	Protocol Description *206*
12.2.1	Materials *206*
12.3	Application Example *210*
12.4	Troubleshooting *212*
	References *213*

13	**Pb^{2+}-induced Cleavage of RNA** *214*
	Leif A. Kirsebom and Jerzy Ciesiolka
13.1	Introduction *214*
13.2	Pb^{2+}-induced Cleavage to Probe Metal Ion Binding Sites, RNA Structure and RNA–Ligand Interactions *216*
13.2.1	Probing High-affinity Metal Ion Binding Sites *216*
13.2.2	Pb^{2+}-induced Cleavage and RNA Structure *220*
13.2.3	Pb^{2+}-induced Cleavage to Study RNA–Ligand Interactions *221*
13.3	Protocols for Metal Ion-induced Cleavage of RNA *222*
13.4	Troubleshooting *225*
13.4.1	No Pb^{2+}-induced Cleavage Detected *225*
13.4.2	Complete Degradation of the RNA *225*
	Acknowledgments *226*
	References *226*

14	***In Vivo* Determination of RNA Structure by Dimethylsulfate** *229*
	Christina Waldsich and Renée Schroeder
14.1	Introduction *229*
14.2	Description of Method *230*
14.2.1	Cell Growth and *In Vivo* DMS Modification *230*
14.2.2	RNA Preparation *231*
14.2.3	Reverse Transcription *232*
14.3	Evidence for Protein-induced Conformational Changes within RNA *In Vivo* *233*
14.4	Troubleshooting *235*
	References *237*

15	**Probing Structure and Binding Sites on RNA by Fenton Cleavage** 238
	Gesine Bauer and Christian Berens
15.1	Introduction 238
15.2	Description of Methods 240
15.2.1	Fe^{2+}-mediated Cleavage of Native Group I Intron RNA 240
15.2.2	Fe^{2+}-mediated Tetracycline-directed Hydroxyl Radical Cleavage Reactions 241
15.3	Comments and Troubleshooting 245
	References 247
16	**Measuring the Stoichiometry of Magnesium Ions Bound to RNA** 250
	A. J. Andrews and Carol Fierke
16.1	Introduction 250
16.2	Separation of Free Magnesium from RNA-bound Magnesium 251
16.3	Forced Dialysis is the Preferred Method for Separating Bound and Free Magnesium Ions 252
16.4	Alternative Methods for Separating Free and Bound Magnesium Ions 254
16.5	Determining the Concentration of Free Magnesium in the Flow-through 255
16.6	How to Determine the Concentration of Magnesium Bound to the RNA and the Number of Binding Sites on the RNA 256
16.7	Conclusion 257
16.8	Troubleshooting 258
	References 258
17	**Nucleotide Analog Interference Mapping and Suppression: Specific Applications in Studies of RNA Tertiary Structure, Dynamic Helicase Mechanism and RNA–Protein Interactions** 259
	Olga Fedorova, Marc Boudvillain, Jane Kawaoka and Anna Marie Pyle
17.1	Background 259
17.1.1	The Role of Biochemical Methods in Structural Studies 259
17.1.2	NAIM: A Combinatorial Approach for RNA Structure–Function Analysis 262
17.1.2.1	Description of the Method 262
17.1.2.2	Applications 265
17.1.3	NAIS: A Chemogenetic Tool for Identifying RNA Tertiary Contacts and Interaction Interfaces 268
17.1.3.1	General Concepts 268
17.1.3.2	Applications: Elucidating Tertiary Contacts in Group I and Group II Ribozymes 269
17.2	Experimental Protocols for NAIM 271
17.2.1	Nucleoside Analog Thiotriphosphates 271
17.2.2	Preparation of Transcripts Containing Phosphorothioate Analogs 271
17.2.3	Radioactive Labeling of the RNA Pool 273

17.2.4	The Selection Step of NAIM: Three Applications for Studies of RNA Function *273*
17.2.4.1	Group II Intron Ribozyme Activity: Selection through Transesterification *273*
17.2.4.2	Reactivity of RNA Helicases: Selection by RNA Unwinding *277*
17.2.4.3	RNA–Protein Interactions: A One-pot Reaction for Studying Transcription Termination *279*
17.2.5	Iodine Cleavage of RNA Pools *283*
17.2.6	Analysis and Interpretation of NAIM Results *284*
17.2.6.1	Quantification of Interference Effects *284*
17.3	Experimental Protocols for NAIS *287*
17.3.1	Design and Creation of Mutant Constructs *287*
17.3.2	Functional Analysis of Mutants for NAIS Experiments *289*
17.3.3	The Selection Step for NAIS *289*
17.3.4	Data Analysis and Presentation *290*
	Acknowledgments *291*
	References *291*

18	**Nucleotide Analog Interference Mapping: Application to the RNase P System** *294*
	Simona Cuzic and Roland K. Hartmann
18.1	Introduction *294*
18.1.1	Nucleotide Analog Interference Mapping (NAIM) – The Approach *294*
18.1.2	Critical Aspects of the Method *296*
18.1.2.1	Analog Incorporation *296*
18.1.2.2	Functional Assays *297*
18.1.2.3	Factors Influencing the Outcome of NAIM Studies *297*
18.1.3	Interpretation of Results *298*
18.1.4	Nucleotide Analog Interference Suppression (NAIS) *300*
18.2	NAIM Analysis of *Cis*-cleaving RNase P RNA–tRNA Conjugates *300*
18.2.1	Characterization of a *Cis*-cleaving *E. coli* RNase P RNA–tRNA Conjugate *300*
18.2.2	Application Example *301*
18.2.3	Materials *305*
18.2.4	Protocols *306*
18.2.5	Data Evaluation *311*
18.3	Troubleshooting *313*
	References *317*

19	**Identification and Characterization of Metal Ion Binding by Thiophilic Metal Ion Rescue** *319*
	Eric L. Christian
19.1	Introduction *319*
19.2	General Considerations of Experimental Conditions *323*
19.2.1	Metal Ion Stocks and Conditions *323*

19.2.2	Consideration of Buffers and Monovalent Salt	*324*
19.2.3	Incorporation of Phosphorothioate Analogs	*325*
19.2.4	Enzyme–Substrate Concentration	*327*
19.2.5	General Kinetic Methods	*328*
19.2.6	Measurement of Apparent Metal Ion Affinity	*329*
19.2.7	Characterization of Metal Ion Binding	*333*
19.2.8	Further Tests of Metal Ion Cooperativity	*336*
19.3	Additional Considerations	*337*
19.3.1	Verification of k_{rel}	*337*
19.3.2	Contributions to Complexity of Reaction Kinetics	*338*
19.3.3	Size and Significance of Observed Effects	*339*
19.4	Conclusion	*340*
	Acknowledgments	*341*
	References	*341*

20 Identification of Divalent Metal Ion Binding Sites in RNA/DNA-metabolizing Enzymes by Fe(II)-mediated Hydroxyl Radical Cleavage *345*

Yan-Guo Ren, Niklas Henriksson and Anders Virtanen

20.1	Introduction	*345*
20.2	Probing Divalent Metal Ion Binding Sites	*346*
20.2.1	Fe(II)-mediated Hydroxyl Radical Cleavage	*346*
20.2.2	How to Map Divalent Metal Ion Binding Sites	*347*
20.2.3	How to Use Aminoglycosides as Functional and Structural Probes	*349*
20.3	Protocols	*350*
20.4	Notes and Troubleshooting	*351*
	References	*352*

21 Protein–RNA Crosslinking in Native Ribonucleoprotein Particles *354*

Henning Urlaub, Klaus Hartmuth and Reinhard Lührmann

21.1	Introduction	*354*
21.2	Overall Strategy	*354*
21.3	UV Crosslinking	*355*
21.4	Identification of UV-induced Protein–RNA Crosslinking Sites by Primer Extension Analysis	*357*
21.5	Identification of Crosslinked Proteins	*361*
21.6	Troubleshooting	*364*
21.7	Protocols	*367*
	Acknowledgments	*372*
	References	*372*

22 Probing RNA Structure by Photoaffinity Crosslinking with 4-Thiouridine and 6-Thioguanosine *374*

Michael E. Harris and Eric L. Christian

22.1	Introduction	*374*
22.2	Description	*377*

22.2.1	General Considerations: Reaction Conditions and Concentrations of Interacting Species *377*
22.2.2	Generation and Isolation of Crosslinked RNAs *380*
22.2.3	Primer Extension Mapping of Crosslinked Nucleotides *381*
22.3	Troubleshooting *382*
	References *384*
II.2	**Biophysical Methods** *385*
23	**Structural Analysis of RNA and RNA–Protein Complexes by Small-angle X-ray Scattering** *385*
	Tao Pan and Tobin R. Sosnick
23.1	Introduction *385*
23.2	Description of the Method *387*
23.2.1	General Requirements *387*
23.2.2	An Example for the Application of SAXS *389*
23.3	General Information *389*
23.4	Question 1: The Oligomerization State of P RNA and the RNase P Holoenzyme *390*
23.5	Question 2: The Overall Shape *392*
23.6	Question 3: The Holoenzyme–Substrate Complexes *392*
23.7	Troubleshooting *395*
23.7.1	Problem 1: Radiation Damage and Aggregation *395*
23.7.2	Problem 2: High Scattering Background *395*
23.7.3	Problem 3: Scattering Results cannot be Fit to Simple Models *396*
23.8	Conclusions/Outlook *396*
	Acknowledgments *396*
	References *397*
24	**Temperature-Gradient Gel Electrophoresis of RNA** *398*
	Detlev Riesner and Gerhard Steger
24.1	Introduction *398*
24.2	Method *399*
24.2.1	Principle *399*
24.2.2	Instruments *400*
24.2.3	Handling *400*
24.3	Optimization of Experimental Conditions *401*
24.3.1	Attribution of Secondary Structures to Transition Curves in TGGE *401*
24.3.2	Pore Size of the Gel Matrix *402*
24.3.3	Electric Field *402*
24.3.4	Ionic Strength and Urea *402*
24.4	Examples *402*
24.4.1	Analysis of Different RNA Molecules in a Single TGGE *403*
24.4.2	Analysis of Structure Distributions of a Single RNA – Detection of Specific Structures by Oligonucleotide Labeling *405*

24.4.3	Analysis of Mutants *409*	
24.4.4	Retardation Gel Electrophoresis in a Temperature Gradient for Detection of Protein–RNA Complexes *409*	
24.4.5	Outlook *413*	
	References *414*	
25	**UV Melting, Native Gels and RNA Conformation** *415*	
	Andreas Werner	
25.1	Monitoring RNA Folding in Solution *415*	
25.2	Methods *417*	
25.3	Data Analysis *420*	
25.4	Energy Calculations and Limitations *422*	
25.5	RNA Concentration *424*	
25.6	Salt and pH Dependence *424*	
25.7	Native Gels *426*	
	References *427*	
26	**Sedimentation Analysis of Ribonucleoprotein Complexes** *428*	
	Jan Medenbach, Andrey Damianov, Silke Schreiner and Albrecht Bindereif	
26.1	Introduction *428*	
26.2	Glycerol Gradient Centrifugation *429*	
26.2.1	Equipment *429*	
26.2.2	Reagents *429*	
26.2.3	Method *430*	
26.2.3.1	Preparation of the Glycerol Gradient *430*	
26.2.3.2	Sample Preparation and Centrifugation *430*	
26.2.3.3	Preparation of RNA from Gradient Fractions *431*	
26.2.3.4	Simultaneous Preparation of RNA and Proteins *431*	
26.2.3.5	Control Gradient with Sedimentation Markers *432*	
26.2.3.6	Notes and Troubleshooting *433*	
26.3	Fractionation of RNPs by Cesium Chloride Density Gradient Centrifugation *434*	
26.3.1	Equipment *434*	
26.3.2	Reagents *434*	
26.3.3	Method *435*	
26.3.3.1	Preparation of the Gradient and Ultracentrifugation *435*	
26.3.3.2	Preparation of RNA from the Gradient Fractions *435*	
26.3.3.3	Control Gradient for Density Calculation *435*	
26.3.3.4	Notes and Troubleshooting *435*	
	Acknowledgments *437*	
	References *437*	
27	**Preparation and Handling of RNA Crystals** *438*	
	Boris François, Aurélie Lescoute-Phillips, Andreas Werner and Benoît Masquida	
27.1	Introduction *438*	

27.2	Design of Short RNA Constructs	439
27.3	RNA Purification	439
27.3.1	HPLC Purification	439
27.3.2	Gel Electrophoresis	440
27.3.3	RNA Recovery	441
27.3.3.1	Elution of the RNA from the Gel	441
27.3.3.2	Concentration and Desalting	441
27.4	Setting Crystal Screens for RNA	442
27.4.1	Renaturing the RNA	447
27.4.2	Setting-up Crystal Screens	447
27.4.3	Forming Complexes with Organic Ligands: The Example of Aminoglycosides	447
27.4.4	Evaluate Screening Results	449
27.4.5	The Optimization Process	449
27.5	Conclusions	451
	References	452

II.3 Fluorescence and Single Molecule Studies 453

28 Fluorescence Labeling of RNA for Single Molecule Studies 453
Filipp Oesterhelt, Enno Schweinberger and Claus Seidel

28.1	Introduction	453
28.2	Fluorescence Resonance Energy Transfer (FRET)	456
28.2.1	Measurement of Distances via FRET	456
28.3	Questions that can be Addressed by Single Molecule Fluorescence	458
28.3.1	RNA Structure and Dynamics	459
28.3.2	Single Molecule Fluorescence in Cells	460
28.3.2.1	Techniques used for Fluorescent Labeling RNA in Cells	460
28.3.2.2	Intracellular Mobility	462
28.3.3	Single Molecule Detection in Nucleic Acid Analysis	462
28.3.3.1	Fragment Sizing	462
28.3.3.2	Single Molecule Sequencing	462
28.4	Equipment for Single Molecule FRET Measurements	463
28.4.1.1	Excitation of the Fluorophores	463
28.4.1.2	Fluorescence Detection	464
28.4.1.3	Data Analysis	465
28.5	Sample Preparation	466
28.5.1	Fluorophore–Nucleic Acid Interaction	466
28.5.2	RNA Labeling	466
28.5.2.1	Fluorophores for Single Molecule Fluorescence Detection	466
28.5.2.2	Fluorophores used for FRET Experiments	467
28.5.2.3	Attaching Fluorophores to RNA	467
28.5.2.4	Linkers	468
28.5.3	Fluorescence Background	468
28.5.3.1	Raman Scattered Light	468

28.5.3.2	Cleaning Buffers	468
28.5.3.3	Clean Surfaces	469
28.5.4	Surface Modification	469
28.5.4.1	Coupling Single Molecules to Surfaces	469
28.5.4.2	Surface Passivation	470
28.5.5	Preventing Photodestruction	470
28.6	Troubleshooting	470
28.6.1.1	Orientation Effects	470
28.6.1.2	Dissociation of Molecular Complexes	471
28.6.1.3	Adsorption to the Surface	471
28.6.1.4	Diffusion Limited Observation Times	471
28.6.1.5	Intensity Fluctuations	472
	References	472
29	**Scanning Force Microscopy and Scanning Force Spectroscopy of RNA**	**475**
	Wolfgang Nellen	
29.1	Introduction	475
29.2	Questions that could be Addressed by SFM	477
29.3	Statistics	481
29.4	Scanning Force Spectroscopy (SFS)	481
29.5	Questions that may be Addressed by SFS	483
29.6	Protocols	483
29.7	Troubleshooting	485
29.8	Conclusions	486
	Acknowledgments	487
	References	487

Volume 2

Part III	**RNA Genomics and Bioinformatics**	**489**
30	**Comparative Analysis of RNA Secondary Structure: 6S RNA**	**491**
	James W. Brown and J. Christopher Ellis	
30.1	Introduction	491
30.1.1	RNA Secondary Structure	492
30.1.2	Comparative Sequence Analysis	492
30.1.3	Strengths and Weakness of Comparative Analysis	493
30.1.4	Comparison with Other Methods	494
30.2	Description	495
30.2.1	Collecting Sequence Data	495
30.2.2	Thermodynamic Predictions	498
30.2.3	Initial Alignment	500
30.2.4	Terminal Helix (P1a)	502
30.2.5	Subterminal Helix (P1b)	506
30.2.6	Apical Helix (P2a)	506

30.2.7	Subapical Helices (P2b and P2c)	507
30.2.8	Potential Interior Stem–loop (P3)	508
30.2.9	Is There Anything Else?	508
30.2.10	Where To Go From Here	509
30.3	Troubleshooting	510
	Acknowledgments	511
	References	511

31	**Secondary Structure Prediction**	**513**
	Gerhard Steger	
31.1	Introduction	513
31.2	Thermodynamics	513
31.3	Formal Background	516
31.4	mfold	518
31.4.1	Input to the mfold Server	518
31.4.1.1	Sequence Name	518
31.4.1.2	Sequence	518
31.4.1.3	Constraints	519
31.4.1.4	Further Parameters	521
31.4.1.5	Immediate versus Batch Jobs	522
31.4.2	Output from the mfold Server	524
31.4.2.1	Energy Dot Plot	524
31.4.2.2	RNAML (RNA Markup Language) Syntax	524
31.4.2.3	Extra Files	524
31.4.2.4	Download All Foldings	525
31.4.2.5	View ss-count Information	525
31.4.2.6	View Individual Structures	525
31.4.2.7	Dot Plot Folding Comparisons	527
31.5	RNAfold	527
31.5.1	Input to the RNAfold Server	528
31.5.1.1	Sequence and Constraints	528
31.5.1.2	Further Parameters	529
31.5.1.3	Immediate versus Batch Jobs	530
31.5.2	Output from the RNAfold Server	531
31.5.2.1	Probability Dot Plot	531
31.5.2.2	Text Output of Secondary Structure	531
31.5.2.3	Graphical Output of Secondary Structure	531
31.5.2.4	Mountain Plot	533
31.6	Troubleshooting	533
	References	534

32	**Modeling the Architecture of Structured RNAs within a Modular and Hierarchical Framework**	**536**
	Benoît Masquida and Eric Westhof	
32.1	Introduction	536

32.2	Modeling Large RNA Assemblies	537
32.2.1	The Modeling Process	538
32.2.1.1	Getting the Right Secondary Structure	539
32.2.1.2	Extrusion of the Secondary Structure in 3-D	540
32.2.1.3	Interactive Molecular Modeling	540
32.2.1.4	Refinement of the Model	542
32.3	Conclusions	543
	References	544
33	**Modeling Large RNA Assemblies using a Reduced Representation**	**546**
	Jason A. Mears, Scott M. Stagg and Stephen C. Harvey	
33.1	Introduction	546
33.2	Basic Modeling Principles	547
33.2.1	Pseudo-atoms and Reduced Representation	549
33.2.2	Implementing RNA Secondary Structure	550
33.2.3	Protein Components	551
33.2.4	Implementing Tertiary Structural Information	551
33.2.5	Modeling Protocol	552
33.3	Application of Modeling Large RNA Assemblies	554
33.3.1	Modeling the Ribosome Structure at Low Resolution	554
33.3.2	Modeling Dynamic Assembly of the Ribosome with Reduced Representation	556
33.4	Conclusion	557
33.5	Troubleshooting	557
	References	559
34	**Molecular Dynamics Simulations of RNA Systems**	**560**
	Pascal Auffinger and Andrea C. Vaiana	
34.1	Introduction	560
34.2	MD Methods	560
34.3	Simulation Setups	562
34.3.1	Choosing the Starting Structure	562
34.3.1.1	Model Built Structures	563
34.3.1.2	X-ray Structures	563
34.3.1.3	NMR Structures	563
34.3.2	Checking the Starting Structure	563
34.3.2.1	Conformational Checks	563
34.3.2.2	Protonation Issues	564
34.3.2.3	Solvent	564
34.3.3	Adding Hydrogen Atoms	564
34.3.4	Choosing the Environment (Crystal, Liquid) and Ions	564
34.3.5	Setting the Box Size and Placing the Ions	565
34.3.5.1	Box Size	565
34.3.5.2	Monovalent Ions	565
34.3.5.3	Divalent Ions	565

34.3.6	Choosing the Program and Force Field	565
34.3.6.1	Programs	565
34.3.6.2	Force Fields	566
34.3.6.3	Parameterization of Modified Nucleotides and Ligands	566
34.3.6.4	Water Models	567
34.3.7	Treatment of Electrostatic Interactions	567
34.3.8	Other Simulation Parameters	568
34.3.8.1	Thermodynamic Ensemble	568
34.3.8.2	Temperature and Pressure	568
34.3.8.3	Shake, Time Steps and Update of the Non-bonded Pair List	568
34.3.8.4	The Flying Ice Cube Problem	568
34.3.9	Equilibration	569
34.3.10	Sampling	569
34.3.10.1	How Long Should a Simulation Be?	569
34.3.10.2	When to Stop a Simulation	570
34.3.10.3	Multiple MD (MMD) Simulations	570
34.4	Analysis	570
34.4.1	Evaluating the Quality of the Trajectories	570
34.4.1.1	Consistency Checks	571
34.4.1.2	Comparison with Experimental Data	571
34.4.1.3	Visualization	571
34.4.2	Convergence Issues	571
34.4.3	Conformational Parameters	572
34.4.4	Solvent Analysis	572
34.5	Perspectives	572
	Acknowledgments	573
	References	573
35	**Seeking RNA Motifs in Genomic Sequences**	**577**
	Matthieu Legendre and Daniel Gautheret	
35.1	Introduction	577
35.2	Choosing the Right Search Software: Limitations and Caveats	578
35.3	Retrieving Programs and Sequence Databases	581
35.4	Organizing RNA Motif Information	581
35.5	Evaluating Search Results	583
35.6	Using the RNAMOTIF Program	585
35.7	Using the ERPIN Program	589
35.8	Troubleshooting	592
35.8.1	RNAMOTIF	592
35.8.1.1	Too Many Solutions	592
35.8.1.2	Program Too Slow	592
35.8.2	ERPIN	592
35.8.2.1	Too Many Solutions	592
35.8.2.2	Program Too Slow	593
	Acknowledgments	593
	References	593

36	**Approaches to Identify Novel Non-messenger RNAs in Bacteria and to Investigate their Biological Functions: RNA Mining** *595*
	Jörg Vogel and E. Gerhart H. Wagner
36.1	Introduction *595*
36.2	Searching for Small, Untranslated RNAs *597*
36.2.1	Introduction *597*
36.2.2	Direct Labeling and Direct Cloning *598*
36.2.3	Functional Screens *599*
36.2.4	Biocomputational Screens *602*
36.2.5	Microarray Detection *605*
36.2.6	Shotgun Cloning (RNomics) *606*
36.2.7	Co-purification with Proteins or Target RNAs *609*
36.2.8	Screens for *Cis*-encoded Antisense RNAs *610*
36.3	Conclusions *610*
	Acknowledgments *611*
	References *611*
37	**Approaches to Identify Novel Non-messenger RNAs in Bacteria and to Investigate their Biological Functions: Functional Analysis of Identified Non-mRNAs** *614*
	E. Gerhart H. Wagner and Jörg Vogel
37.1	Introduction *614*
37.2	Approaches for Elucidation of Bacterial sRNA Function *615*
37.2.1	Large-scale Screening for Function *615*
37.2.2	Preparing for Subsequent Experiments: Strains and Plasmids *615*
37.2.3	Experimental Approaches *618*
37.2.4	Physiological Phenotypes (Lethality, Growth Defects, etc.) *618*
37.2.5	Analyzing sRNA Effects on Specific mRNA Levels by Microarrays *619*
37.2.6	Analyzing sRNA Effects by Proteomics *620*
37.2.7	Analyzing sRNA Effects by Metabolomics *621*
37.2.8	Finding Targets by Reporter Gene Approaches *621*
37.2.9	Bioinformatics-aided Approaches *623*
37.2.10	Prediction of Regulatory Sequences in the Vicinity of sRNA Gene Promoters *623*
37.2.11	Finding Interacting Sites (Complementarity/Antisense) *624*
37.3	Additional Methods Towards Functional and Mechanistic Characterizations *625*
37.3.1	Finding sRNA-associated Proteins *625*
37.3.2	Regulation of the Target RNA – Use of Reporter Gene Fusions *626*
37.3.3	Northern Analyses *627*
37.3.4	Analysis of sRNAs – RACE and Primer Extensions *627*
37.3.5	Structures of sRNAs and Target RNAs *628*
37.4	Conclusions *629*
37.5	Protocols *629*
	Acknowledgments *639*
	References *640*

38	**Experimental RNomics: A Global Approach to Identify Non-coding RNAs in Model Organisms** *643*	
	Alexander Hüttenhofer	
38.1	Introduction *643*	
38.2	Materials *644*	
38.2.1	Oligonucleotide Primers *644*	
38.2.2	Enzymes *644*	
38.2.3	Buffers *644*	
38.2.4	Reagents, Kits, Vectors and Bacterial Cells *645*	
38.3	Protocols for Library Construction and Analysis *645*	
38.4	Computational Analysis of ncRNA Sequences *652*	
38.5	Troubleshooting *653*	
	Acknowledgments *653*	
	References *653*	
39	**Large-scale Analysis of mRNA Splice Variants by Microarray** *655*	
	Young-Soo Kwon, Hai-Ri Li and Xiang-Dong Fu	
39.1	Introduction *655*	
39.2	Overview of RASL Technology *655*	
39.3	Description of Methods *657*	
39.3.1	Preparation of Index Arrays *657*	
39.3.2	Annotation of Alternative Splicing *658*	
39.3.3	Target Design *658*	
39.3.4	Preparation of Target Pool *658*	
39.3.5	The RASL Assay Protocol *659*	
39.3.6	PCR Amplification *659*	
39.3.7	Hybridization on Index Array *660*	
39.3.8	Data Analysis *661*	
39.4	Troubleshooting *661*	
39.4.1	System Limitation and Pitfalls *661*	
39.4.2	Potential Experimental Problems *662*	
	References *663*	
IV	**Analysis of RNA Function** *665*	
IV.1	**RNA–Protein Interactions in vitro** *667*	
40	**Use of RNA Affinity Matrices for the Isolation of RNA-binding Proteins** *667*	
	Steffen Schiffer, Sylvia Rösch, Bettina Späth, Markus Englert, Hildburg Beier and Anita Marchfelder	
40.1	Introduction *667*	
40.2	Materials *668*	
40.2.1	CNBr-activated Sepharose 4B Affinity Column *668*	
40.2.2	NHS-activated HiTrap Columns *669*	
40.3	Methods *669*	

40.3.1	Coupling of tRNAs to CNBr-activated Sepharose 4B	669
40.3.2	Coupling of tRNAs to a 5-ml NHS-activated HiTrap Column	671
40.4	Application	672
40.4.1	Purification of the Nuclear RNase Z from Wheat Germ	672
40.5	Notes	674
	References	674

41   Biotin-based Affinity Purification of RNA–Protein Complexes   676

Zsofia Palfi, Jingyi Hui and Albrecht Bindereif

41.1	Introduction	676
41.2	Materials	677
41.2.1	Oligonucleotides	677
41.2.2	Affinity Matrices	678
41.2.3	Cell Extracts	678
41.2.4	Buffers and Solutions	679
41.2.5	Additional Materials	679
41.3	Methods	680
41.3.1	Affinity Purification of RNPs	680
41.3.1.1	Depletion of Total Cell Lysate from SAg-binding Material (Pre-clearing)	680
41.3.1.2	Pre-blocking SAg Beads	681
41.3.1.3	Affinity Selection of RNPs for Structural Studies	681
41.3.1.4	Affinity Selection of RNPs for Functional Studies by Displacement Strategy	685
41.3.2	Affinity Purification of Specific RNA-binding Proteins by Biotinylated RNAs	689
41.4	Troubleshooting	691
41.4.1	Biotinylated 2′OMe RNA Oligonucleotides	691
41.4.2	Extracts and Buffers	691
41.4.3	Optimization of the Experimental Conditions: When Yields are Low	691
41.4.4	Optimization of the Experimental Conditions: When Non-specific Background is Too High	692
	References	692

42   Immunoaffinity Purification of Spliceosomal and Small Nuclear Ribonucleoprotein Complexes   694

Cindy L. Will, Evgeny M. Makarov, Olga V. Makarova and Reinhard Lührmann

42.1	Introduction	694
42.2	Generation of Anti-peptide Antibodies: Peptide Selection Criteria	694
42.3	Immunoaffinity Selection of U4/U6.U5 Tri-snRNPs	697
42.4	Immunoaffinity Purification of 17S U2 snRNPs	699
42.5	Approaches for the Isolation of Native, Human Spliceosomal Complexes	702
42.6	Isolation of Activated Spliceosomes by Immunoaffinity Selection with Anti-peptide Antibodies against the SKIP Protein	703

Acknowledgments 709
References 709

43 Northwestern Techniques for the Identification of RNA-binding Proteins from cDNA Expression Libraries and the Analysis of RNA–Protein Interactions 710

Ángel Emilio Martínez de Alba, Michela Alessandra Denti and Martin Tabler

43.1 Introduction 710
43.2 Methods 712
43.2.1 Preparation of Probes and Buffers 712
43.2.1.1 Preparation of ^{32}P-labeled RNA Probes 712
43.2.1.2 Preparation of Blocking RNA 712
43.2.1.3 Preparation of the Northwestern Buffer 713
43.2.2 Protocol 1: Northwestern Screening for Identification of RNA-binding Proteins from cDNA Expression Libraries 713
43.2.2.1 Preparation of the Host Plating Culture 714
43.2.2.2 Plating of the cDNA Phage Expression Library 714
43.2.2.3 Adsorbing Recombinant Proteins to Nitrocellulose Membranes 715
43.2.2.4 Incubation with an RNA Ligand 716
43.2.2.5 Washing of Membranes 717
43.2.2.6 Identification of True Positives 717
43.2.3 Protocol 2: Northwestern Techniques to Detect and Analyze RNA–Protein Interactions 719
43.2.3.1 Protein Sample Preparation 719
43.2.3.2 Protein Electrophoresis and Transfer 721
43.2.3.3 Incubation of the Membranes with an RNA Probe 722
43.2.3.4 Washing of Membranes and Autoradiography 722
43.3 Troubleshooting 723
43.3.1 Probe Quality 723
43.3.2 Background Signals 724
43.3.3 Signal-to-Background Ratio 724
43.3.4 Protein Conformation 726
43.3.5 Weak Binding Signals 726
43.3.6 False Positives 726
43.3.7 Quality of the cDNA Library 727
43.3.8 Fading Signals 727
43.3.9 Supplementary 727
References 727

IV.2 RNA–Protein Interactions *in vivo* 729

44 Fluorescent Detection of Nascent Transcripts and RNA-binding Proteins in Cell Nuclei 729

Jennifer A. Geiger and Karla M. Neugebauer

44.1 Introduction 729
44.2 Description of the Methods 730

44.2.1	Overview 730
44.2.2	Preparation of Fluorescent DNA Probes for *In Situ* Hybridization 731
44.2.2.1	Method 1: Nick Translation of Plasmid DNA 731
44.2.2.2	Method 2: PCR Amplification and DNase I Digestion 732
44.2.3	Performing Combined Immunocytochemistry and FISH 733
44.2.4	Troubleshooting 735
	Acknowledgments 735
	References 735

45 Identification and Characterization of RNA-binding Proteins through Three-hybrid Analysis   737

Felicia Scott and David R. Engelke

45.1	Introduction 737
45.2	Basic Strategy of the Method 738
45.3	Detailed Components 739
45.3.1	Yeast Reporter Strain 740
45.3.2	Plasmids 740
45.3.3	Hybrid RNA 740
45.3.3.1	Technical Considerations for the Hybrid RNA 742
45.3.4	Activation Domain FP2 743
45.3.4.1	Technical Considerations for the Activation Domain FP2 743
45.3.5	Positive Controls 744
45.4	Protocols 745
45.4.1	Transformation of Yeast 745
45.4.2	Assaying for *HIS3* Expression 748
45.4.3	Assaying for β-Galactosidase Activity 748
45.5	Troubleshooting 749
45.6	Additional Applications 751
45.7	Summary 752
	Acknowledgments 752
	References 752

46 Analysis of Alternative Splicing *In Vivo* using Minigenes   755

Yesheng Tang, Tatyana Novoyatleva, Natalya Benderska, Shivendra Kishore, Alphonse Thanaraj and Stefan Stamm

46.1	Introduction 755
46.2	Overview of the Method 755
46.3	Methods 757
46.3.1	Construction of Minigenes 757
46.3.2	Transfection of Cells 757
46.3.3	Analysis 771
46.3.3.1	RT-PCR 771
46.3.3.2	Other Analysis Methods 771
46.3.4	Necessary Controls 773
46.3.5	Advantages and Disadvantages of the Method 773
46.3.6	Related Methods 774

46.4	Troubleshooting 774
46.5	Bioinformatic Resources 775
46.6	Protocols 775
	References 780

IV.3 SELEX 783

47 Artificial Selection: Finding Function amongst Randomized Sequences 783
Ico de Zwart, Catherine Lozupone, Rob Knight, Amanda Birmingham, Mali Illangasekare, Vasant Jadhav, Michal Legiewicz, Irene Majerfeld, Jeremy Widmann and Michael Yarus

47.1	The SELEX Method 783
47.2	Understanding a Selection 784
47.2.1	Sequence Motif Representation and Abundance 786
47.2.2	The Recovery Efficiency of Different RNAs 787
47.2.3	Stringency 789
47.2.4	Amplification and Transcription Biases 789
47.3	Isolation of RNAs that Bind Small Molecules 790
47.3.1	Stringency and K_D 792
47.3.2	Selection for Multiple Targets in One Column 793
47.3.3	Characterizing Motif Activity 793
47.4	Techniques for Selecting Ribozymes 794
47.4.1	Making the RNA a Substrate for the Reaction 795
47.4.2	The Inherent Reactivity of RNA 795
47.4.3	Selecting Active RNAs 797
47.4.4	Negative Selections 798
47.4.5	Stringency 799
47.4.6	Analysis of the Product 799
47.4.7	Determining the Scope of the Reaction 799
47.5	Sequence Analysis 800
47.5.1	Identifying Related Sequences 800
47.5.2	Predicting Structure 802
47.5.3	Chemical and Enzymatic Mapping 803
47.5.4	Finding Minimal Requirements 804
47.5.5	Three-dimensional Structural Modeling 804
	Acknowledgments 805
	References 805

48 Aptamer Selection against Biological Macromolecules: Proteins and Carbohydrates 807
C. Stefan Vörtler and Maria Milovnikova

48.1	Introduction 807
48.2	General Strategy 808
48.2.1	Choosing a Suitable Target 810
48.2.1.1	Protein Targets 810
48.2.1.2	Carbohydrate Targets 811

48.2.2	Immobilization of the Target	*812*
48.2.3	Selection Assays	*812*
48.2.4	Design and Preparation of the Library	*813*
48.3	Running the *In Vitro* Selection Cycle	*814*
48.4	Analysis of the Selection Outcome	*815*
48.5	Troubleshooting	*816*
48.6	Protocols	*817*
	Acknowledgments	*836*
	References	*836*
49	***In Vivo* SELEX Strategies**	*840*
	Thomas A. Cooper	
49.1	Introduction	*840*
49.2	Procedure Overview	*841*
49.2.1	Design of the Randomized Exon Cassette	*843*
49.2.2	Design of the Minigene	*844*
49.2.3	RT-PCR Amplification	*846*
49.2.4	Monitoring for Enrichment of Exon Sequences that Function as Splicing Enhancers	*846*
49.2.5	Troubleshooting	*847*
49.3	Protocols	*848*
	Acknowledgments	*852*
	References	*852*
50	***In Vitro* Selection against Small Targets**	*853*
	Dirk Eulberg, Christian Maasch, Werner G. Purschke and Sven Klussmann	
50.1	Introduction	*853*
50.2	Target Immobilization	*856*
50.2.1	Covalent Immobilization	*857*
50.2.1.1	Epoxy-activated Matrices	*857*
50.2.1.2	NHS-activated Matrices	*859*
50.2.1.3	Pyridyl Disulfide-activated Matrices	*860*
50.2.2	Non-covalent Immobilization	*861*
50.3	Nucleic Acid Libraries	*862*
50.3.1	Library Design	*862*
50.3.2	Starting Pool Preparation	*863*
50.4	Enzymatics	*865*
50.4.1	Reverse Transcription	*865*
50.4.2	PCR	*866*
50.4.3	*In Vitro* Transcription	*867*
50.5	Partitioning	*868*
50.6	Binding Assays	*873*
50.6.1	Equilibrium Dialysis	*873*
50.6.2	Equilibrium Filtration Analysis	*874*
50.6.3	Isocratic Competitive Affinity Chromatography	*875*
	References	*877*

51	SELEX Strategies to Identify Antisense and Protein Target Sites in RNA or Heterogeneous Nuclear Ribonucleoprotein Complexes 878
	Martin Lützelberger, Martin R. Jakobsen and Jørgen Kjems
51.1	Introduction 878
51.1.1	Applications for Antisense 879
51.1.2	Selecting Protein-binding Sites 879
51.2	Construction of the Library 879
51.2.1	Generation of Random DNA Fragments from Genomic or Plasmid DNA 881
51.2.2	Preparing RNA Libraries from Plasmid, cDNA or Genomic DNA 881
51.3	Identification of Optimal Antisense Annealing Sites in RNAs 882
51.4	Identification of Natural RNA Substrates for Proteins and Other Ligands 884
51.5	Cloning, Sequencing and Validating the Selected Inserts 884
51.6	Troubleshooting 885
51.6.1	Sonication of Plasmid DNA does not Yield Shorter Fragments 885
51.6.2	Inefficient Ligation 885
51.6.3	Inefficient *MmeI* Digestion 885
51.6.4	The Amplification of the Unselected Library is Inefficient 886
51.6.5	The Library Appears to be Non-random in the Unselected Pool 886
51.6.6	The Selected RNAs do not Bind Native Protein 886
51.7	Protocols 886
	References 894
V	**RNAi** 895
52	**Gene Silencing Methods for Mammalian Cells: Application of Synthetic Short Interfering RNAs** 897
	Matthias John, Anke Geick, Philipp Hadwiger, Hans-Peter Vornlocher and Olaf Heidenreich
52.1	Introduction 897
52.2	Background Information 898
52.3	Ways to Induce RNAi in Mammalian Cells 900
52.3.1	Important Parameters 901
52.3.1.1	siRNA Design 901
52.3.1.2	Target Site Selection 901
52.3.1.3	Preparation of siRNA Samples 902
52.3.2	Transfection of Mammalian Cells with siRNA 902
52.3.3	Electroporation of Mammalian Cells with siRNA 904
52.3.4	Induction of RNAi by Intracellular siRNA Expression 905
52.4	Troubleshooting 908
	References 908
	Appendix: UV Spectroscopy for the Quantitation of RNA 910
	Index 915

Preface

The field of RNA research has experienced an incredible boom in recent years, with no calming down in sight. Stimulated by a common fascination for RNA, we became convinced about two years ago that something is missing in the RNA field: a comprehensive handbook of RNA biochemistry that combines protocols of methods and techniques in RNA chemistry, biochemistry and bioinformatics – a handbook that is not a guide for specialists only, but also for graduate and PhD students as well as experienced scientists who wish to embark on RNA research projects. We had in mind to merge several core features into our book concept, making it unique among related publications. These include a thorough introduction to the individual approach or method, providing the necessary background for non-specialists and addressing the scientific potential as well as the limitations in terms of applicability, resolution and interpretation. A second feature is the detailed description of experimental protocols, such that the reader should be able to apply the technique(s) directly, without extensive further reading of the original literature. Related to this is the incorporation of troubleshooting sections, describing pitfalls and discussing critical experimental steps as well as ideas for adequate problem solving in cases of failure.

A substantial fraction of the handbook (Part I) describes basic approaches and methods of RNA synthesis and modification (e. g. T7 transcription, co- and post-transcriptional modification, enzymatic RNA ligation, chemical RNA synthesis, co- and post-synthetic modifications), many of which may be considered timeless as they have been continuously applied and developed during the last four decades. Several of these basic techniques are routinely used in many RNA laboratories, but in most cases PhD students just apply a protocol inherited from former lab members. This fulfils the purpose as long as the method works smoothly. In cases of failure, though, the devil is usually in the details, and a deeper insight into the procedure or enzymatic reaction is then required. Likewise, techniques have often been tailored to a specific task, but variation of application for other purposes is not always straightforward. An example is T4 RNA ligase primarily used for [^{32}P]pCp end-labeling. When [^{32}P]pCp is replaced with longer donor substrates, the reaction is often inefficient and more liable to disturbance, and unwanted ligation products tend to be the rule rather than the exception. The handbook will be a valuable guide in such cases.

The biological insight gained from any RNA or RNA-protein structure as determined, for example, by X-ray crystallography is thoroughly enhanced when flanked by biochemical experiments that investigate RNA structure in solution, or even in its cellular environment. Thus, in Part II, a string of chapters is dedicated to the investigation of RNA structure in solution, naked as well as in the context of RNA-protein complexes. Both *in vitro* and *in vivo* approaches are presented that use enzymatic, chemical or metal ion probes. Elaborate crosslinking methods to investigate higher-order RNA structure or RNA-protein interactions are covered as well, and two contributions outline approaches that are based on complex nucleic acid libraries and allow to screen RNA molecules for important functional groups (NAIM, NAIS). In general, the techniques described in this handbook have been developed to a prodigious level of detail and sophistication by expert laboratories in recent years, and timely RNA research depends on the accuracy and reproducibility of experiments conducted on the basis of such protocols.

The integral role of metal ions in RNA architecture and catalysis was first put in a nutshell by Mike Yarus in his *Cheshire Cat* metaphor (*FASEB J.*, 1993). Any RNA researcher will consequently devote at least some experiments to the role of metal ions. Thus, a cornucopia of approaches and protocols in Part II of the handbook is dedicated to the study of metal ion binding to RNA molecules and, as an extension, to RNA/DNA-metabolizing enzymes. This includes the characterization of high affinity metal ion binding sites and strategies to unravel the localization and functional role of catalytic metal ions, as well as a method to quantify Mg^{2+} binding to RNA.

Understanding RNA function is unimaginable without biophysical and computer-based approaches. Thus, Part II further contains several chapters that deal with physico-chemical techniques: various types of spectroscopy (e. g. CD, fluorescence, UV, small angle X-ray scattering) as applied to RNA are described; a number of special techniques address the physical properties of naked or protein-complexed RNAs, such as their melting and sedimentation behaviour. Fluorescently labeled RNAs are nowadays indispensable tools for the study of RNA structure and dynamics, and single molecule studies in living cells give detailed insight into molecular distribution and dynamic processes. Single molecule research has been further addressed by a contribution on scanning force microscopy (SFM) and a derived technique, scanning force spectroscopy (SFS), used to determine inter- and intramolecular binding forces. Although X-ray crystallography *per se* is not covered in this book, an introduction to the preparation and handling of RNA crystals is included.

Bioinformatics, traditionally relegated to the application of BLAST-type programs for the search and annotation of unknown genes, more and more extends to the understanding of the relationships between sequences and three-dimensional structures in the context of molecular Darwinian evolution. Part III begins with the powerful applications of phylogenetic and sequence comparisons in the RNA field, especially the determination of secondary structures. Theoretical prediction of secondary structures, the foundations of which are explained extensively in one chapter, can be successful, although on the basis of *in silico* approaches only it is often difficult to decide between several possible solutions. The search for RNA motifs in genomic sequences is exemplified for motifs recognized by protein cofactors in the maturation or translation of messenger RNAs. The identification of such motifs contributes to the proper annotation of genes and the sorting out of proteomics data. In the end, RNA molecules exert their functions because they adopt a specific tertiary fold. Three chapters tackle this question, focusing first on how to assemble complex tertiary structures *in silico* starting from fragments which have been structurally characterized, then giving reasons for the advantages of increasing the coarseness of the representations, and finally describing the calculations of the molecular dynamics simulations in order to apprehend movements, and binding of water molecules, ions and ligands. The very recent field of RNomics has led to the discovery of a plethora of potential regulatory RNAs, whose functional and structural analysis will keep us busy for the next decades. Several pioneers in the field address the strategic as well as the bioinformatic and experimental aspects of RNomics. Part III is completed by a contribution introducing a novel genome-wide approach, the analysis of alternative splicing variants by microarrays. Although still at an early stage of development, this will certainly become an important tool in the future.

Since nakedness of RNA in most cases reflects an artificial *in vitro* state, Part IV of the handbook focuses on RNA-protein interaction. A queue of chapters present different techniques directly related to the analysis of RNA function in conjunction with proteins – both *in vitro* and *in vivo*. Addressed are various affinity purification methods, Northwestern techniques, three-hybrid screening, fluorescent detection of RNA and RNA-binding proteins, and the *in vivo* analysis of alternative splicing. Another series of functional approaches is dedicated to *in vitro* and *in vivo* SELEX strategies, encompassing the different applications of this technique and detailing the experimental steps and associated pitfalls. This provides a wealth of in-depth information in this still developing area of research.

The vast field of RNA interference is represented by a contribution (Part V) written by authors who have pioneered the field of gene silencing in mammalian cells. We have refrained from extending the RNAi part because its specific methodology is currently under rapid evolution and varies greatly between different biological systems. Furthermore, specialized monographs have become available recently.

In the end, we hope that this comprehensive collection of protocols spanning RNA research will be a helpful and useful toolbox for all researchers already working with RNA as well as for those planning to foray in the RNA world.

Last but not least, we are indebted to all authors for their engagement, patience and the high quality of their contributions. Special thanks go to Dagmar K. Willkomm for her continuous assistance in the editing process. The editors would also like to thank Frank Weinreich and the staff of Wiley-VCH for making the production of this handbook possible.

Roland K. Hartmann October 2004
Albrecht Bindereif
Astrid Schön
Eric Westhof

67084 Strasbourg
France

Mikko J. Frilander
University of Helsinki
Institute of Biotechnology
Viikinkaari 9
000014 University of Helsinki
Finland

Xiang-Dong Fu
University of California, San Diego
Department of Cellular and Molecular Medicine
San Diego, CA 92093-0651
USA

Daniel Gautheret
INSERM ERM-206
Université de la Méditerranée
Luminy Case 906
13288 Marseille Cedex 09
France

Anke Geick
Alnylam Europe AG
Fritz-Hornschuch-Str. 9
95326 Kulmbach
Germany

Jennifer A. Geiger
Max-Planck-Institute of Molecular Cell Biology and Genetics
Pfotenhauerstrasse 108
01307 Dresden
Germany

Olaf Gimple
Universität Würzburg
Institut für Biochemie
Am Hubland
97074 Würzburg
Germany

Heike Gruegelsiepe
Philipps-Universität Marburg
Institut für Pharmazeutische Chemie
Marbacher Weg 6
35037 Marburg
Germany

Philipp Hadwiger
Alnylam Europe AG
Fritz-Hornschuch-Str. 9
95326 Kulmbach
Germany

Dinari A. Harris
University of Michigan
Department of Chemistry
Ann Arbor, MI 48109
USA

Michael E. Harris
Case Western Reserve University
Center for RNA Molecular Biology and Department of Molecular Biology and Microbiology
Cleveland OH 44106-4973
USA

Roland K. Hartmann
Philipps-Universität Marburg
Institut für Pharmazeutische Chemie
Marbacher Weg 6
35037 Marburg
Germany

Klaus Hartmuth
Max-Planck-Institute for Biophysical Chemistry
Department of Cellular Biochemistry
Am Fassberg 11
37077 Göttingen
Germany

Stephen C. Harvey
Georgia Institute of Technology
Department of Biology
310 Ferst Dr.
Atlanta, GA 30332
USA

Olaf Heidenreich
University of Tübingen
Department of Molecular Biology
Auf der Morgenstelle 15
72076 Tübingen
Germany

Niklas Henriksson
Uppsala University
Department of Cell and Molecular Biology
BMC Box 596
751 24 Uppsala
Sweden

List of Contributors

Jingyi Hui
Justus-Liebig-Universität Giessen
Institut für Biochemie
Heinrich-Buff-Ring 58
35392 Giessen
Germany

Eric Huntzinger
CNRS – UPR 9002
Institut de Biologie Moléculaire et Cellulaire
15 rue René Descartes
67084 Strasbourg
France

Alexander Hüttenhofer
Innsbruck Medical University
Division of Genomics and RNomics
Peter-Mayr-Str. 4b
6020 Innsbruck
Austria

Mali Illangasekare
University of Colorado
Department of Molecular, Cellular and Developmental Biology
Boulder, CO 80309-0347
USA

Vasant Jadhav
University of Colorado
Department of Molecular, Cellular and Developmental Biology
Boulder, CO 80309-0347
USA

Martin R. Jakobsen
University of Aarhus
Department of Molecular Biology
Møllers Allé
8000 Åarhus C
Denmark

Matthias John
Alnylam Europe AG
Fritz-Hornschuch-Str. 9
95326 Kulmbach
Germany

Jane Kawaoka
Yale University
Department of Molecular Biophysics and Biochemistry
266 Whitney Ave
New Haven, CT 06520
USA

Leif A. Kirsebom
Uppsala University
Department of Cell and Molecular Biology
Biomedical Center
751 24 Uppsala
Sweden

Shivendra Kishore
University of Erlangen
Institute of Biochemistry
Fahrstrasse 17
91054 Erlangen
Germany

Jørgen Kjems
University of Aarhus
Department of Molecular Biology
Møllers Allé
8000 Åarhus C
Denmark

Sven Klussmann
NOXXON Pharma AG
Max-Dohrn-Strasse 8–10
10589 Berlin
Germany

Rob Knight
University of Colorado
Department of Molecular, Cellular and Developmental Biology
Boulder, CO 80309-0347
USA

Young-Soo Kwon
University of California, San Diego
Department of Cellular and Molecular Medicine
San Diego, CA 92093-0651
USA

Matthieu Legendre
INSERM ERM-206
Université de la Méditerranée
Luminy Case 906
13288 Marseille Cedex 09
France

Mihail Legiewicz
University of Colorado
Department of Molecular, Cellular and Developmental Biology
Boulder, CO 80309-0347
USA

Aurélie Lescoute-Philipps
CNRS – UPR 9002
Institut de Biologie Moléculaire et Cellulaire
15 rue René Descartes
67084 Strasbourg
France

Hai-Ri Li
University of California, San Diego
Department of Cellular and Molecular
Medicine
San Diego, CA 92093-0651
USA

Ester Lizano
Max-Planck-Institute for Evolutionary
Anthropology
Deutscher Platz 6
04103 Leipzig
Germany

Catherine Lozupone
University of Colorado
Department of Molecular, Cellular and
Developmental Biology
Boulder, CO 80309-0347
USA

Reinhard Lührmann
Max-Planck-Institute for Biophysical
Chemistry
Department of Cellular Biochemistry
Am Fassberg 11
37077 Göttingen
Germany

Martin Lützelberger
University of Aarhus
Department of Molecular Biology
Møllers Allé
8000 Århus C
Denmark

Christian Maasch
NOXXON Pharma AG
Max-Dohrn-Strasse 8–10
10589 Berlin
Germany

Irene Majerfeld
University of Colorado
Department of Molecular, Cellular and
Developmental Biology
Boulder, CO 80309-0347
USA

Evgeny M. Makarov
Max-Planck-Institute for Biophysical
Chemistry
Department of Cellular Biochemistry
Am Fassberg 11
37077 Göttingen
Germany

Olga V. Makarova
Max-Planck-Institute for Biophysical
Chemistry
Department of Cellular Biochemistry
Am Fassberg 11
37077 Göttingen
Germany

Virginie Marchand
UMR 7567 CNRS-UHP
Université Herni Poincaré Nancy 1
Boulevard des Aiguillettes, BP 239
54506 Vandoeuvre-Lès-Nancy
France

Anita Marchfelder
Universität Ulm
Molekulare Botanik
Albert-Einstein-Allee 11
89069 Ulm
Germany

Ángel Emilio Martínez de Alba
Universidad Politécnica de Valencia
Instituto de Biología Molecular y Celular de
Plantas
Avenida de los Naranjos s/n
46022 Valencia
Spain

Benoît Masquida
CNRS – UPR 9002
Institut de Biologie Moléculaire et Cellulaire
15 rue René Descartes
67084 Strasbourg
France

Jason A. Mears
Georgia Institute of Technology
Department of Biology
310 Ferst Dr.
Atlanta, GA 30332
USA

Jan Medenbach
Justus-Liebig-Universität Giessen

Institut für Biochemie
Heinrich-Buff-Ring 58
35392 Giessen
Germany

Agnès Méreau
UMR 7567 CNRS-UHP
Université Herni Poincaré Nancy 1
Boulevard des Aiguillettes, BP 239
54506 Vandoeuvre-Lès-Nancy
France

Mario Mörl
Max-Planck-Institute for Evolutionary
Anthropology
Deutscher Platz 6
04103 Leipzig
Germany

Hervé Moine
CNRS – UPR 9002
Institut de Biologie Moléculaire et Cellulaire
15 rue René Descartes
67084 Strasbourg
France

Annie Mougin
UMR 7567 CNRS-UHP
Université Herni Poincaré Nancy 1
Boulevard des Aiguillettes, BP 239
54506 Vandoeuvre-Lès-Nancy
France

Wolfgang Nellen
Universität Kassel
Abteilung Genetik
Heinrich-Plett-Straße 40
34132 Kassel
Germany

Karla M. Neugebauer
Max-Planck-Institute of Molecular Cell Biology
and Genetics
Pfotenhauerstrasse 108
01307 Dresden
Germany

Tatyana Novoyatleva
University of Erlangen
Institute of Biochemistry
Fahrstrasse 17
91054 Erlangen
Germany

Filipp Oesterhelt
Heinrich-Heine-Universität Düsseldorf
Institute for Molecular Physical Chemistry
Universitätsstr. 1
40225 Düsseldorf
Germany

Zsofia Palfi
Justus-Liebig-Universität Giessen
Institut für Biochemie
Heinrich-Buff-Ring 58
35392 Giessen
Germany

Tao Pan
University of Chicago
Department of Biochemistry and Molecular
Biology
920 E 58th St
Chicago, IL 60637
USA

Maria Parisova
Biochemie
Universität Bayreuth
Universitätsstraße 30
95447 Bayreuth
Germany

Tina Persson
Lund University
Department of Chemistry
P.O. Box 124
221 00 Lund
Sweden

Maria Possedko
CNRS – UPR 9002
Institut de Biologie Moléculaire et Cellulaire
15 rue René Descartes
67084 Strasbourg
France

Werner G. Purschke
NOXXON Pharma AG
Max-Dohrn-Strasse 8–10
10589 Berlin
Germany

Anna Marie Pyle
Yale University
Department of Molecular Biophysics and
Biochemistry
266 Whitney Ave

New Haven, CT 06520
USA

Yan-Guo Ren
Genomics Institute of the Novartis Research Foundation
10675 John Jay Hopkins Dr.
San Diego, CA 92121
USA

Detlev Riesner
Heinrich-Heine-Universität Düsseldorf
Institut für Physikalische Biologie
Universitätsstr. 1
40225 Düsseldorf
Germany

Pascale Romby
CNRS – UPR 9002
Institut de Biologie Moléculaire et Cellulaire
15 rue René Descartes
67084 Strasbourg
France

Sylvia Rösch
Universität Ulm
Molekulare Botanik
Albert-Einstein-Allee 11
89069 Ulm
Germany

Steffen Schiffer
Universität Ulm
Molekulare Botanik
Albert-Einstein-Allee 11
89069 Ulm
Germany

Astrid Schön
Universität Leipzig
Institut für Biochemie
Brüderstr. 34
04103 Leipzig
Germany

Silke Schreiner
Justus-Liebig-Universität Giessen
Institut für Biochemie
Heinrich-Buff-Ring 58
35392 Giessen
Germany

Renée Schroeder
Department of Microbiology and Genetics
University of Vienna
Dr. Bohrgasse 9/4
1030 Vienna
Austria

Enno Schweinberger
Heinrich-Heine-Universität Düsseldorf
Institute for Molecular Physical Chemistry
Universitätsstr. 1
40225 Düsseldorf
Germany

Felicia Scott
University of Michigan
Department of Biological Chemistry
Ann Arbor, MI 48109-0606
USA

Claus Seidel
Heinrich-Heine-Universität Düsseldorf
Institute for Molecular Physical Chemistry
Universitätsstr. 1
40225 Düsseldorf
Germany

Snorri Th. Sigurdsson
University of Iceland
Science Institute
Dunhaga 3
107 Reykjavik
Iceland

Tobin R. Sosnick
University of Chicago
Department of Biochemistry and Molecular Biology
920 E 58th St
Chicago, IL 60637
USA

Bettina Späth
Universität Ulm
Molekulare Botanik
Albert-Einstein-Allee 11
89069 Ulm
Germany

Brian S. Sproat
RNA-TEC NV
Minderbroedersstraat 17–19
3000 Leuven
Belgium

List of Contributors

Scott M. Stagg
Georgia Institute of Technology
Department of Biology
310 Ferst Dr.
Atlanta, GA 30332
USA

Stefan Stamm
University of Erlangen
Institute of Biochemistry
Fahrstrasse 17
91054 Erlangen
Germany

Gerhard Steger
Heinrich-Heine-Universität Düsseldorf
Institut für Physikalische Biologie
Universitätsstr. 1
40225 Düsseldorf
Germany

Martin Tabler
Foundation for Research and Technology Hellas
Institute of Molecular Biology and Biotechnology
P.O.Box 1527
71110 Heraclion (Crete)
Greece

Yesheng Tang
University of Erlangen
Institute of Biochemistry
Fahrstrasse 17
91054 Erlangen
Germany

Alphonse Thanaraj
European Bioinformatics Institute
Wellcome Trust Genome Campus
Hinxton
Cambridge, CB10 1SD
United Kingdom

Janne J. Turunen
University of Helsinki
Institute of Biotechnology
Viikinkaari 9
000014 University of Helsinki
Finland

Henning Urlaub
Max-Planck-Institute for Biophysical Chemistry
Department of Cellular Biochemistry
Am Fassberg 11
37077 Göttingen
Germany

Andrea C. Vaiana
CNRS – UPR 9002
Institut de Biologie Moléculaire et Cellulaire
15 rue René Descartes
67084 Strasbourg
France

Anders Virtanen
Uppsala University
Department of Cell and Molecular Biology
BMC Box 596
751 24 Uppsala
Sweden

Jörg Vogel
Max Planck Institute for Infection Biology
Schumannstr. 21/22
10117 Berlin
Germany

Hans-Peter Vornlocher
Alnylam Europe AG
Fritz-Hornschuch-Str. 9
95326 Kulmbach
Germany

C. Stefan Vörtler
Biochemie
Universität Bayreuth
Universitätsstraße 30
95447 Bayreuth
Germany

E. Gerhart H. Wagner
Uppsala University Biomedical Center
Department of Cell and Molecular Biology
Box 596
751 24 Uppsala
Sweden

Christina Waldsich
Yale University
Department of Molecular Biophysics and Biochemistry
266 Whitney Ave
New Haven, CT 06520
USA

Nils G. Walter
University of Michigan
Department of Chemistry
Ann Arbor, MI 48109
USA

Andreas Werner
CNRS – UPR 9002
Institut de Biologie Moléculaire et Cellulaire
15 rue René Descartes
67084 Strasbourg
France

Eric Westhof
CNRS – UPR 9002
Institut de Biologie Moléculaire et Cellulaire
15 rue René Descartes
67084 Strasbourg
France

Jeremy Widmann
University of Colorado
Department of Molecular, Cellular and Developmental Biology
Boulder, CO 80309-0347
USA

Cindy L. Will
Max-Planck-Institute for Biophysical Chemistry
Department of Cellular Biochemistry
Am Fassberg 11
37077 Göttingen
Germany

Dagmar K. Willkomm
Philipps-Universität Marburg
Institut für Pharmazeutische Chemie
Marbacher Weg 6
35037 Marburg
Germany

Flore Winter
CNRS – UPR 9002
Institut de Biologie Moléculaire et Cellulaire
15 rue René Descartes
67084 Strasbourg
France

Michael Yarus
University of Colorado
Department of Molecular, Cellular and Developmental Biology
Boulder, CO 80309-0347
USA

Nathan H. Zahler
University of Michigan
Department of Chemistry
930 N. University
Ann Arbor, MI 48109-1055
USA

Part I
RNA Synthesis

I.1
Enzymatic RNA Synthesis, Ligation and Modification

1
Enzymatic RNA Synthesis using Bacteriophage T7 RNA Polymerase

Heike Gruegelsiepe, Astrid Schön, Leif A. Kirsebom and Roland K. Hartmann

1.1
Introduction

Bacteriophage T7 RNA polymerase (T7 RNAP) was first cloned and overexpressed from bacteriophage T7-infected *Escherichia coli* cells in 1984 [1]. In contrast to multi-subunit DNA-dependent RNA polymerases from eukaryotes and prokaryotes, T7 RNAP consists of a single subunit of about 100 kDa [2]. The subdomains adopt a hand-like shape with the palm, thumb and fingers around a central cleft where the active site containing the functionally essential amino acid residues is located, creating a binding cavity for magnesium ions and ribonucleotide substrates. For RNA synthesis, the unwound template strand is positioned such that the template base −1 is anchored in a hydrophobic pocket in direct vicinity to the active site [3].

T7 RNAP is highly specific for its own promoters and exhibits no affinity even to closely related phage T3 promoters, although the 23-bp consensus sequences are very similar (Fig. 1.1A). During the initiation process, the polymerase goes through several elongation attempts, generating short abortive oligoribonucleotides. Only when the nascent RNA transcript exceeds 9–12 nt do initiation complexes convert to stable elongation complexes. Transcription proceeds with an average rate of 200–260 nt/s until the elongation complex encounters a termination signal or falls off the template end during *in vitro* run-off transcription [4, 5]. The error frequency in transcripts of wild-type (wt) T7 RNAP is about 6×10^{-5} [6].

In the following sections, we will describe protocols that have been used routinely for T7 transcription. Further, a robust and simple protocol for the partial purification of T7 RNAP is included, which yields an enzyme preparation that fully satisfies all *in vitro* transcription demands. The transcription protocols given suffice for most purposes. However, in special cases, such as the synthesis of milligram quantities, modified RNAs or very A,U-rich RNAs, it may be worthwhile to further optimize transcription conditions. We also draw the reader's attention to the paper by Milligan and Uhlenbeck [7], which briefly discusses many fundamental aspects of T7 transcription and is still handed out to every new member of our groups.

Handbook of RNA Biochemistry. Edited by R. K. Hartmann, A. Bindereif, A. Schön, E. Westhof
Copyright © 2005 WILEY-VCH Verlag GmbH & Co. KGaA, Weinheim
ISBN: 3-527-30826-1

A

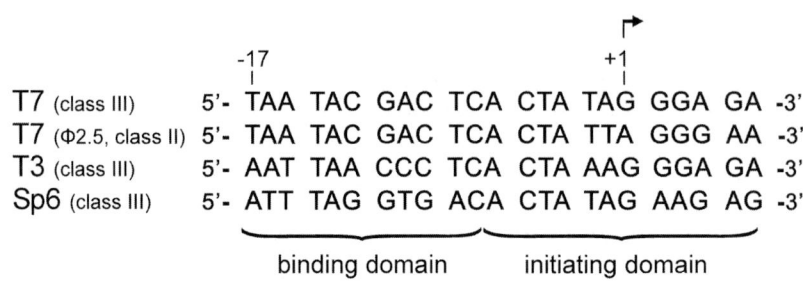

	-17					+1		
T7 (class III)	5'- TAA	TAC	GAC	TCA	CTA	TAG	GGA	GA -3'
T7 (Φ2.5, class II)	5'- TAA	TAC	GAC	TCA	CTA	TTA	GGG	AA -3'
T3 (class III)	5'- AAT	TAA	CCC	TCA	CTA	AAG	GGA	GA -3'
Sp6 (class III)	5'- ATT	TAG	GTG	ACA	CTA	TAG	AAG	AG -3'

binding domain   initiating domain

B

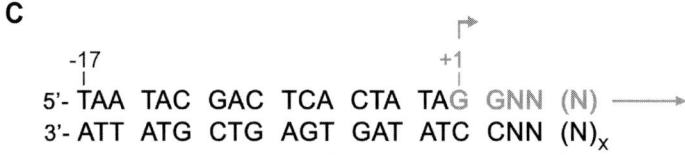

+1   +6
pppGGGAGA

smaller effects

Nucleotide +1	Relative yield	Nucleotide +2	Relative yield
C	0.1	C	0.5
A	0.2	A	0.5
U	n.d.	U	n.d.
G	1.0	G	1.0

C

```
       -17                               +1
        |                                 |
5'- TAA TAC GAC TCA CTA TAG GNN (N) ────▶
3'- ATT ATG CTG AGT GAT ATC CNN (N)ₓ
```

Fig. 1.1. (A) Consensus sequences of class III promoters of bacteriophages T7, T3 and SP6, and sequence of the T7 φ2.5 class II promoter [5, 34–36]. Phage polymerase initiating domains also include the first 5–6 nt of the transcribed template strand. The transcription start (position +1) is indicated by the arrow. The phage T7 genome encodes a total of 17 promoters, including five class III promoters and one replication promoter (φOR), which are all completely conserved in the region from nt −17 to +6. In addition, there are 10 T7 class II promoters plus one more replication promoter (φOL); among these 11 promoters, which display some sequence variation within the −17 to +6 region, only the φ2.5 and φOL promoter initiate transcription with an A instead of G residue [35] (B) Effect of sequence variations in the +1 to +6 region of the T7 class III promoter on transcription efficiency (adapted from Milligan and Uhlenbeck [7]); n.d.: not determined. (C) T7 class III promoter region with the recommended G identities at positions +1 and +2 of the RNA transcript shown in grey.

1.2
Description of Method – T7 Transcription *in vitro*

T7 RNAP can be used *in vitro* to produce milligram amounts of RNA polymers ranging from less than 100 to 30 000 nt [7, 8]. Since the commonly used T7 class III promoter, usually referred to as the T7 promoter, is also strictly conserved in the transcribed region of nt +1 to +6, sequence variations especially at nt +1 and +2 influence transcription yields significantly (Fig. 1B and C [7]).

1.2.1
Templates

Templates can be generated in three different ways: by insertion into a plasmid [double-stranded (ds) DNA], by polymerase chain reaction (PCR) (dsDNA) or by annealing a T7 promoter DNA oligonucleotide to a single-stranded template DNA oligonucleotide.

Strategy (a) Insertion into a Plasmid
We prefer to work with plasmid dsDNA templates, because once the correct sequence of a plasmid clone has been confirmed, the DNA can be conveniently amplified by *in vivo* plasmid replication exploiting the high fidelity of bacterial DNA polymerases. The RNA expression cassette (either with or without the T7 promoter sequence) is usually obtained by PCR and cloned into a bacterial plasmid. Since PCR amplification is error-prone, plasmid inserts ought to be sequenced. When the T7 RNAP promoter region from -17 to -1 is not encoded in the PCR fragment, one can use commercially available T7 transcription vectors (e.g. pGEM®3Z and derivatives from Promega or the pPCR-Script series from Stratagene) containing the T7 promoter and a multiple cloning site for insertion of the RNA expression cassette. If there are no sequence constraints at the transcript 5′ end, we routinely design templates encoding 5′-GGA at positions +1 to +3 of the RNA transcript, which usually results in high transcription yields. Whenever possible, at least the nucleotide preferences at positions +1 and +2 should be taken into account (Fig. 1.1B and C). The plasmid-encoded RNA expression cassette ought to be followed by a single restriction site (avoid restriction enzymes yielding 3′ overhangs [7]) for producing run-off transcripts. Templates with 5′ overhangs have successfully been used in several laboratories. Among those have been templates linearized with restriction enzymes that cleave several residues away from their binding/recognition site, such as *Fok*I. The advantage of using this type of restriction enzymes is independence from the sequence at the cleavage site. This permits the design of RNA transcript 3′ ends of complete identity to natural counterparts. Individual steps of template preparation are (1) ligation of (PCR) insert into plasmid, (2) cloning in *E. coli*, purification and sequencing of plasmid, (3) linearization of plasmid DNA for run-off transcription, (4) phenol/chloroform extraction and ethanol precipitation of template DNA before (5) use in T7 transcription assays.

Strategy (b) Direct use of Templates Generated by PCR
Direct use of PCR fragments as templates is faster than insertion into a plasmid and preferred if only minor amounts of RNA are required.

Strategy (c) Annealing of a T7 Promoter DNA Oligonucleotide to a Single-stranded Template
This strategy is the fastest and we have used it to synthesize small amounts of an RNA 31mer for 5′-end-labeling purposes (see Protocol 6).

1.2.2
Special Demands on the RNA Product

1.2.2.1 Homogeneous 5′ and 3′ Ends, Small RNAs, Functional Groups at the 5′ End

While T7 RNAP usually initiates transcription at a defined position, it tends to append one or occasionally even a few more non-templated nucleotides to the product 3′ terminus [7, 9]. Also, 5′ end heterogeneity may become a problem when the template encodes a transcript with more than three consecutive guanosines at the 5′ end [10], as well as in the case of unusual 5′-terminal sequences, such as 5′-CACUGU, 5′-CAGAGA or 5′-GAAAAA [11]. Yet, 5′ end heterogeneity seems to be a problem associated with T7 class III promoters (Fig. 1.1A). Almost complete 5′ end homogeneity of T7 transcripts has been achieved with templates directing transcription from the more rarely used T7 ϕ2.5 class II promoter (Fig. 1.1A), at which T7 RNAP initiates synthesis with an A instead of a G residue. Transcription yields from this promoter were reported to equal those of the commonly used T7 class III promoter [12].

However, for the production of RNAs with 100% 5′ and 3′ end homogeneity, several methods are available (see Chapters 2 and 3). For example, hammerhead or Hepatitis delta virus (HDV) ribozymes can be tethered to the RNA of interest on one or both sides (see Chapter 2). The ribozyme(s) will release the RNA product by self-cleavage during transcription. Such a *cis*-acting ribozyme placed upstream releases the RNA of interest with a 5′-OH terminus directly accessible to 5′-end-labeling (see Chapter 9) and simultaneously eliminates the problem of 5′ end heterogeneity as well as constraints on the identity of the 5′-terminal nucleotide of the RNA of interest (Chapter 2). The same strategy may also be considered for synthesis of large amounts of smaller RNAs. Chemical synthesis and purification of 10 mg of, for example, an RNA 15mer by a commercial supplier can be quite expensive. In such a case, a cheaper alternative would be to transcribe the 15mer sandwiched between two *cis*-cleaving ribozymes, resulting in post-transcriptional release of the 15mer with uniform 5′ and 3′ ends. Purification of the 15mer (and separation from the released ribozyme fragments) can then be achieved either by denaturing polyacrylamide gel electrophoresis (PAGE), UV shadowing or staining and gel elution (see Chapters 3 and 9), or by preparative HPLC if available (see Chapters 7 and 27). If T7 RNAP is self-prepared according to the protocol described in this article, synthesis of 10 mg of a 15mer will become quite affordable.

Normally, transcription by T7 RNAP is initiated with GTP, resulting in 5′-triphosphate ends. If, however, 5′-OH ends or 5′-monophosphate termini are preferred, T7 RNAP can be prompted to initiate transcripts with guanosine or 5′-ApG (to generate 5′-OH ends for direct end-labeling with ^{32}P) or 5′-GMP (to generate 5′-monophosphates), when these components are added to reaction mixtures in excess over GTP [13]. RNA transcripts with 5′-GMP ends are preferred when the RNA is used for ligation with other RNA molecules.

Tab. 1.1. Modified substrates for T7 transcription.

NTP	wt T7 RNAP	Reference
NTPαS (Sp)	+	14
NTPαS (Rp)	−	14
5-Br-UTP	+	7
5-F-UTP	+	7
5-Hexamethyleneamino-UTP	+	7
6-Aza-UTP	+	7
4-Thio-UTP	+	7
Pseudo-UTP	+	7
8-Br-ATP	+	7
7-Me-GTP	−	7
ITP (with initiator)[1]	+	15
2′-dNTP	+/−	7
2′-dNTPαS	+/−	16, 17
2′-O-Me-NTP or -NTPαS	+/−	16
GTPγS	+	18

+/−: low incorporation efficiency.
[1] Inosine triphosphate (ITP) cannot be used to start transcription, but can substitute for GTP during elongation if a primer, such as 5′-ApG or 5′-GMP, is present as initiator of transcription.

1.2.2.2 Modified Substrates

There are a number of modified nucleoside-5′-triphosphates known to be substrates for T7 RNAP. Table 1.1 has been adopted from Milligan and Uhlenbeck [7] and expanded by addition of more recent information.

Due to discrimination of rNTPs and dNTPs by wt T7 RNAP, the polymerase incorporates rNTPs 70- to 80-fold more efficiently than dNTPs in the presence of Mg^{2+} as the metal ion cofactor. However, a T7 RNAP mutant (Y639F) carrying a tyrosine to phenylalanine exchange at position 639 [19] was shown to have only about 4-fold higher preference for rNTPs than dNTPs [19, 20] and thus permits more efficient incorporation of substrates lacking the ribose 2′-hydroxyl, such as 2′-deoxy-2′-fluoro or 2′-deoxy-2′-amino nucleotides [20]. Incorporation of substrate analogs with 2′-ribose modifications can also be stimulated to some extent in reactions catalyzed by wt T7 RNAP upon addition of Mn^{2+} [16]. Likewise, dNTPαS analogs were partially incorporated into RNase P RNA in a sequence-specific manner under mixed metal ion conditions (Mg^{2+}/Mn^{2+} [21]). Despite these achievements, the Y639F mutant T7 RNAP is nowadays the enzyme of choice for the incorporation of all nucleotides with 2′-ribose modifications (available from Epicentre, WI, USA). For detailed protocols, the reader is referred to [20, 22].

Further modifications can be introduced into transcripts by tailored initiator (oligo)nucleotides. Di- to hexanucleotides with a 3′-terminal guanine base, including di- to tetranucleotides with internal or terminal 2′-deoxy- or 2′-O-methylated residues, were tested as initiators of transcription by wt T7 RNAP [23]. 5′-Terminal incorporation varied between 20% (hexamer) and 80–95% in the case of 5′-ApG or a 5′-biotinylated ApG (e.g. custom-synthesized by IBA, Göttingen, Germany). Also, transcription by T7 RNAP, in this case from the T7 ϕ2.5 class II promoter, was initiated with coenzymes containing an adenosine moiety, such as CoA (3′-dephospho-coenzyme A), NAD or FAD. Reduced NADH and oxidized FAD are highly fluorescent, which opens up the perspective to employ coenzyme-linked RNAs for the study of RNA–RNA or RNA–protein interactions by fluorescence techniques [24].

1.3
Transcription Protocols

1.3.1
Transcription with Unmodified Nucleotides

The protocols given below have been applied to template DNAs directing transcription from the T7 class III promoter. Transcription yields can differ substantially, depending on the individual DNA template and the origin of T7 RNAP. In Protocols 1–6 (Hartmann lab), T7 RNAP from MBI Fermentas has been used. These protocols may be suboptimal with T7 RNAP from other sources. This has been accounted for by including protocols from the Kirsebom (Protocols 7 and 8) and Schön (Protocol 9) labs. Nevertheless, it is advisable to put some effort into the optimization of the transcription protocol if large amounts of RNA are to be produced or if the transcript represents a "standard RNA" in the laboratory, used over longer periods, which may require repeated synthesis. In addition to commercially available T7 RNAP, protocols for partial purification of T7 RNAP from bacterial overexpression strains are available ([1, 25, 26] and Section 1.5).

When a new DNA template is used for the first time, one approach is to perform small test transcriptions on a 50-µl scale according to different basic protocols (e.g. Protocols 1–3). Reaction mixtures should be prepared at room temperature, since DNA may precipitate in the presence of spermidine at low temperatures. In the case of plasmid DNA, template amounts of 40–80 µg/ml (final assay concentration) are used as a rule of thumb, whereas a PCR template of 400 bp is adjusted to about 5 µg/ml final assay concentration. In the Hartmann lab, we usually incubate transcription mixes for 4–6 h at 37 °C, although overnight incubations have been used as well. A variation is to add another aliquot (e.g. 2 U/µl) of T7 RNAP after 2 h at 37 °C, followed by a further 2-h incubation period at 37 °C. In transcriptions according to Protocol 3, a white precipitate will appear because of pyrophosphate accumulation. This is avoided in Protocols 1 and 2 where pyrophosphate is hydrolyzed due to the presence of pyrophosphatase. Extension of incubation periods beyond 4–6 h did not prove advantageous in our hands, and may be associated with some product degradation since T7 RNAP has a DNase and RNase function, which is normally inhibited by NTP substrates added in excess to *in vitro* transcription assays. However, after extended incubation periods, the NTP concentration may drop under a critical limit, thereby favoring RNA degradation [27]. Protocol 4 represents an inexpensive strategy to incorporate a 5′-terminal guanosine. Since guanosine has a low solubility, a 30 mM solution is prepared and kept at 75 °C; the reaction mixture – except for guanosine and T7 RNAP – is prepared at room temperature and prewarmed to 37 °C before addition of guanosine.

We like to note here that transcription from the T7 ϕ2.5 class II promoter is initiated with A instead of G (Fig. 1.1A), opening up the perspective to incorporate an adenosine at the 5′ end. Adenosine would fulfill the same purpose as the aforementioned guanosine used in the case of the class III promoter, but is advantageous because of its better water solubility [12, 24].

As an alternative to starting class III-promoter-directed transcripts with guanosine, the dinucleotide 5′-ApG (see above) may be employed. The 5′-terminal incorporation of the dinucleotide leads to a −1 adenosine extension of the transcript (Protocol 5). ApG, which is more convenient to use than guanosine, is also available from Sigma, but about 10 000-fold more expensive than guanosine. Transcripts can further be initiated with 5′-GMP if present in excess over 5′-GTP (15 versus 3.75 mM). Although the majority of RNA products should possess a 5′-terminal monophosphate, a dephosphorylation/phosphorylation strategy (see Chapter 3) may be preferred to obtain RNA products with 100% 5′-monophosphates.

Protocol 6 is a quick protocol for the synthesis of small amounts of shorter RNAs for 5′-end-labeling purposes. The promoter and template DNA oligonucleotides are simply added to the reaction mixture which is then preincubated for 1 h at 37 °C before starting transcription by addition of T7 RNAP.

Protocol 7 (from the Kirsebom lab) has the characteristics of high T7 RNAP concentrations and the presence of RNase inhibitor, suitable for large-scale transcriptions. Protocol 8 is used in the Kirsebom lab for the production of internally labeled RNA.

Protocol 9 (from the Schön lab) has been used for standard transcriptions as well as for the synthesis of extremely A,U-rich RNAs, with the ratio of individual NTPs adapted to their proportion in the final transcript.

Protocol 1: Hartmann lab

	Final concentration	1000 μl
HEPES pH 7.5, 1 M	80 mM	80 μl
DTT 100 mM	5 mM	50 μl
MgCl$_2$ 3 M	22 mM	7.3 μl
Spermidine 100 mM	1 mM	10 μl
BSA[1] 20 mg/ml	0.12 mg/ml	6 μl
rNTP mix (25 mM each)	3.75 mM (each)	150 μl
Template (linearized plasmid 3.2 kb) 1 μg/μl	40 μg/ml	40 μl
Pyrophosphatase[2] 200 U/ml	1 U/ml	5 μl
T7 RNAP 200 U/μl	1000–2000 U/ml	5–10 μl
RNase-free water		to 1000 μl

For small-scale transcriptions (50 µl final volume), reaction mixes are incubated for 4–6 h at 37 °C. For preparative transcription according to this protocol and Protocols 2–5, usually 1-ml reaction mixtures are prepared and then incubated in 200-µl aliquots (for better thermal equilibration) for 2 h at 37 °C; then a second aliquot of T7 RNAP is added (400 U/200 µl reaction mix), followed by another 2 h of incubation at 37 °C. Efficient transcription reactions in the 1-ml scale result in a product yield of about 3 nmol.

[1] BSA (Sigma, minimum purity 98% based on electrophoretic analysis, pH 7).
[2] Pyrophosphatase from yeast (Roche, EC 3.6.1.1, 200 U/mg, <0.01% ATPase and phosphatases each).

Protocol 2: Hartmann lab

	Final concentration	1000 µl
HEPES pH 7.5, 1 M	80 mM	80 µl
DTT 100 mM	15 mM	150 µl
MgCl$_2$ 3 M	33 mM	11 µl
Spermidine 100 mM	1 mM	10 µl
rNTP mix (25 mM each)	3.75 mM (each)	150 µl
Template (linearized plasmid 3.2 kb) 1 µg/µl	80 µg/ml	80 µl
Pyrophosphatase 200 U/ml	2 U/ml	10 µl
T7 RNAP 200 U/µl	2000–3000 U/ml	10–15 µl
RNase-free water		to 1000 µl

For incubation, see Protocol 1.

Protocol 3: Hartmann lab

	Final concentration	1000 µl
5 × transcription buffer (MBI)[1]	1 × buffer	200 µl
MgCl$_2$ 3 M	40 mM	13.3 µl
rNTP mix (25 mM each)	3 mM (each)	120 µl
Template (linearized plasmid 3.2 kb) 1 µg/µl	80 µg/ml	80 µl
T7 RNAP 200 U/µl	2000–3000 U/ml	10–15 µl
RNase-free water		to 1000 µl

[1] 5 × transcription buffer (MBI Fermentas): 200 mM Tris–HCl (pH 7.9 at 25 °C), 30 mM MgCl$_2$, 50 mM DTT, 50 mM NaCl and 10 mM spermidine. For incubation, see Protocol 1.

Protocol 4: Hartmann lab

	Final concentration	1000 µl
HEPES pH 7.5, 1 M	80 mM	80 µl
DTT 100 mM	5 mM	50 µl
MgCl$_2$ 3 M	22 mM	7.3 µl

Spermidine 100 mM	1 mM	10 µl
BSA 20 mg/ml	0.12 mg/ml	6 µl
rNTP mix (25 mM each)	3.75 mM (each)	150 µl
Template (linearized plasmid 3.2 kb) 1 µg/µl	40 µg/ml	40 µl
Pyrophosphatase 200 U/ml	5 U/ml	25 µl
RNase-free water		321.7 µl
• Prewarm mixture to 37 °C, then add:		
guanosine (30 mM, kept at 75 °C)	9 mM	300 µl
T7 RNAP 200 U/µl	2000 U/ml	10 µl

For incubation, see Protocol 1.

Protocol 5: Initiation with 5'-GMP or 5'-ApG (Hartmann lab)

	Final concentration	1000 µl
HEPES pH 7.5, 1 M	80 mM	80 µl
DTT 100 mM	5 mM	50 µl
MgCl$_2$ 3 M	22 mM	7.3 µl
Spermidine 100 mM	1 mM	10 µl
BSA 20 mg/ml	0.12 mg/ml	6 µl
rNTP mix (25 mM each)	3.75 mM (each)[1]	150 µl
5'-GMP 100 mM (initiator)[1]	15 mM	150 µl
Template (linearized plasmid 3.2 kb) 1 µg/µl	40 µg/ml	40 µl
Pyrophosphatase 200 U/ml	5 U/ml	25 µl
T7 RNAP 200 U/µl	2000 U/ml	10 µl
RNase-free water		to 1000 µl

[1] When 5'-GMP is replaced with the dinucleotide 5'-ApG for transcription initiation, adjust 5'-ApG to 7.5 mM and rNTPs to 2.5 mM each (final concentrations). For incubation, see Protocol 1.

Protocol 6: Hartmann lab

	Final concentration	500 µl
HEPES pH 8.0, 1 M	160 mM	80 µl
DTT 100 mM	15 mM	75 µl
MgCl$_2$ 3 M	33 mM	5.5 µl
Spermidine 100 mM	1 mM	5 µl
rNTP mix (25 mM each)	3.75 mM (each)[3]	75 µl
Promoter DNA oligonucleotide 3.3 µg/µl[1]	132 µg/ml	20 µl
Template DNA oligonucleotide 2.1 µg/µl[2]	84 µg/ml	20 µl
BSA 20 mg/ml	0.12 mg/ml	3 µl
Pyrophosphatase 200 U/ml	2 U/ml	5 µl

RNase-free water		51.5 µl

- Prewarm mixture to 37 °C

Then add guanosine (30 mM, kept at 75 °C)	9 mM	150 µl

- Mix and preincubate for 1 h at 37 °C

Then add T7 RNAP 200 U/µl	4000 U/ml	10 µl

Incubate at 37 °C for 4 h.

[1] Promoter DNA oligonucleotide: 5'-TAA TAC GAC TCA CTA TAG.

[2] In this example, the template DNA oligonucleotide had the sequence 5'-GGT CAT AGG TAT TCC CCC TCT CTC CAT TCC TAT AGT GAG TCG TAT TAA, resulting in an RNA product with the sequence 5'-GGA AUG GAG AGA GGG GGA AUA CCU AUG ACC.

[3] To increase the percentage of transcripts initiated with guanosine, the ratio of guanosine to rNTPs may be increased, e.g. by reducing the rNTP concentration to 1.5 mM each.

10 × transcription buffer (TRX), Kirsebom lab, for transcription Protocols 7 and 8

	Final concentration	1000 µl
Tris–HCl pH 7.5, 1 M	200 mM	200 µl
Tris–HCl pH 7.9, 1 M	200 mM	200 µl
MgCl$_2$ 3 M	240 mM	80 µl
Spermidine 100 mM	20 mM	200 µl
RNase-free water		320 µl

Protocol 7: Kirsebom lab, non-radioactive transcription, volume sufficient for 4 reactions; however, note that the mix is prepared for 4.5 reactions

	Final concentration[1]	432 µl
10 × TRX	1×	45 µl
DTT 0.5 M	10 mM	9 µl
0.2% Triton X-100	0.01%	22.5 µl
ATP (100 mM)	2 mM	9 µl
GTP (100 mM)	2 mM	9 µl
CTP (100 mM)	2 mM	9 µl
UTP (100 mM)	2 mM	9 µl
RNase inhibitor 24 U/µl	32 U/ml	0.6 µl
T7 RNAP 200 U/µl	10 000 U/ml	22.5 µl
RNase-free water		296.4 µl

- To 96 µl of this mix add:

template (linearized plasmid ≈ 3.2 kb) 1 µg/µl	40 µg/ml	4 µl

Incubate at 37 °C for ≤10 h.

[1,2,3] Final concentrations after addition of template.

Protocol 8: Kirsebom lab; internal radioactive labeling mix, volume sufficient for 9 reactions; however, note that the mix is prepared for 10 reactions

	Final concentration[1]	230 µl
10 × TRX	1×	25 µl
DTT 0.5 M	10 mM	5 µl
0.2% Triton X-100	0.01%	12.5 µl
ATP (100 mM)	2 mM	5 µl
GTP (100 mM)	2 mM	5 µl
CTP (100 mM)	2 mM	5 µl
UTP 1 mM	0.2 mM	50 µl
[α-^{32}P]UTP 800 Ci/mmol (20 mCi/ml)	10 Ci/mmol	25 µl
RNase inhibitor 24 U/µl	31.7 U/ml	0.33 µl
T7 RNAP 200 U/µl	10 000 U/ml	12.5 µl
RNase-free water		84.7 µl

- To 23 µl of this mix add:

template (linearized plasmid ≈ 3.2 kb) 1 µg/µl	80 µg/ml	2 µl

Incubate at 37 °C for ≤10 h.
[1] Final concentrations after addition of template.

To account for a severely biased nucleotide composition of the template, such as in RNase P RNAs from the *Cyanophora paradoxa* cyanelle [28] or from a plant-pathogenic phytoplasma [29], the relative concentrations of rNTPs are adjusted accordingly. For phytoplasma RNase P RNA (around 73% A + U), the composition of the nucleotide mix was calculated as follows:

	Calculated mol% of each nucleotide in transcript	Concentration of each rNTP in nucleotide mix (mM)	Final concentration of each rNTP in reaction mix (mM)
rATP	41.08	33	3.3
rCTP	11.06	9	0.9
rGTP	16.03	12.5	1.25
rUTP	31.83	25.5	2.55
Total	100	80	8.0

The following sample protocol is routinely used for preparation of large amounts of RNA and can be easily adjusted to the transcription of templates with a biased nucleotide composition. In such cases, the "standard" rNTP mix (20 mM each rNTP) is replaced by the template-specific rNTP mix with adjusted nucleotide concentrations.

Protocol 9: Preparative transcription, nucleotide composition from the example above (Schön lab)

	Final concentration	250 μl
10 × transcription buffer[1]	1 × buffer	25 μl
rNTP mix (here: 33 mM ATP/9 mM CTP/12.5 mM GTP/25.5 mM UTP)	0.1 × rNTP mix	25 μl
Template (linearized plasmid 3.2 kb) 0.1 μg/μl	50 μg/ml	125 μl
T7 RNAP (own preparation, 5–10 μg/μl total protein; see Section 1.5)	40–200 μg/ml[2]	2–5 μl[2]
RNase-free water		to 250 μl

[1] 10 × transcription buffer: 400 mM Tris–HCl (pH 7.9 at 25 °C), 120 mM $MgCl_2$, 50 mM DTT, 50 mM NaCl and 10 mM spermidine.

[2] Depending on the specific activity of the individual T7 RNAP preparation; see Section 1.5.2.2.

For phytoplasma RNase P RNA, a 250-μl reaction performed under these conditions results in up to 150 μg RNA after gel purification. Note that due to the high concentrations of rNTPs and Mg^{2+} in the reaction, insoluble precipitates may form if the complete mix is kept on ice. It is thus advisable to start with water before adding the other components, and to prewarm the mix to 37 °C before the addition of template and polymerase. We let the reactions proceed for at least 2 h (preferably overnight) at 37 °C and quench the excess Mg^{2+} by addition of Na_2EDTA to a final concentration of 25 mM before phenol extraction and EtOH precipitation (see Section 1.3.3, Step 1).

After transcription, check product yield and quality (5–10-μl aliquot plus equal volume gel loading buffer) by PAGE in the presence of 8 M urea and stain the gel with ethidium bromide. Load at least one reference RNA on the gel to identify the genuine product, since sometimes a complex mixture of bands is observed. A low number of bands in addition to the product band points to good transcription performance, and high yields of transcription correlate with the observation that the RNA product appears as a prominent band, while the DNA template is faintly visible. However, with some templates one has to be satisfied with product amounts exceeding that of the template by a factor of only 5. Aberrant transcripts of similar size and abundance as the desired product, sometimes even appearing as a smear, can make identification and gel purification of the RNA product of interest impossible. In view of such potential problems, the best transcription protocol will be the one generating the highest amount of specific product at the lowest cost of incorrect products. If RNA yields are not satisfactory, vary concentrations of template, $MgCl_2$, T7 RNAP or DTT for further optimization.

1.3.2
Transcription with 2′-Fluoro-modified Pyrimidine Nucleotides

To produce nuclease-resistant 2′-fluoro-modified RNAs with the Y639F mutant T7 RNAP, we replaced rUTP and rCTP with the corresponding 2′-fluoro-analogs (IBA,

Göttingen, Germany; or Epicentre, WI, USA). Protocol 10 has been employed in an *in vitro* selection study using a 117-bp double-stranded PCR template including an internal segment of 60 randomized positions. Transcription assays were incubated for 3–4 h at 37 °C.

Protocol 10: Transcription of 2′-fluoro-modified RNA (Hartmann lab)

		150 µl
10 × transcription buffer[1]	1 × buffer	15 µl
DTT 100 mM	5 mM	7.5 µl
2′-F-CTP, 2′-F-UTP (10 mM each)	1.25 mM (each)	18.75 µl
rATP, rGTP (10 mM each)	1.25 mM (each)	18.75 µl
[α-^{32}P]ATP 800 Ci/mmol (10 mCi/ml)	0.16 Ci/mmol	3 µl
PCR template	1–3.33 nmol/ml	0.15–0.5 nmol
Y639F mutant T7 RNAP (3.7 µg/µl)	93.73 µg/ml	3.8 µl
RNase-free water		to 150 µl

[1] 10 × transcription buffer: 400 mM Tris–HCl (pH 8.0), 200 mM $MgCl_2$, 10 mM spermidine, 0.1% Triton X-100.

1.3.3
Purification

Some purification steps are optional, and depend on transcription quality and the demands on product purity. Often, a purification procedure only including Steps 5–8 is sufficient. Another protocol for gel purification of RNA is described in Chapter 3.

(1) Insoluble pyrophosphate complexes. In transcription reactions without pyrophosphatase or when pyrophosphatase activity is low, a white pyrophosphate precipitate may form. In such cases, remove the precipitate before Step 2 by centrifugation at 14 000 g for about 5 min directly after transcription. Carefully remove the clear supernatant and transfer to a new Eppendorf tube for further sample processing. Also, Na_2EDTA (500 mM, pH 7.5) may be added immediately after transcription to give a final concentration of 50–100 mM. By chelating Mg^{2+}, the formation of insoluble precipitates is substantially reduced.
(2) DNase I digestion to remove template DNA. Add 10 U DNase I (RNase-free, Roche) per 200 µl and incubate for 20 min at 37 °C.
(3) Phenol and chloroform extractions to remove the enzyme(s). For extraction with phenol (Tris-saturated, stabilized with hydroxy-quinoline, pH 7.7; Biomol), mix the sample with 0.5–1.0 volumes phenol, vortex for 30 s, centrifuge 1–5 min (until phases have cleared) at 12 000 g, withdraw the aqueous upper phase and mix it with 0.5–1.0 volumes chloroform, vortex for 30 s, centrifuge 3 min at 12 000 g and transfer the aqueous upper phase to a new tube, avoiding to withdraw any chloroform.

(4) Removal of salt for better gel resolution. For example, use NAP 10 columns (Amersham Biosciences, now part of GE Healthcare), column material: Sephadex G-25.

(5) Ethanol precipitation. Ethanol precipitation is performed to remove residual chloroform and salts, and to concentrate the RNA. Mix sample with 2.5 volumes ethanol, 0.1 volumes 3 M NaOAc (pH 4.7) and 1 µl glycogen (20 µg/µl); leave for 10–20 min at −70 °C or at least 2 h at −20 °C. Centrifuge for 30–45 min at 4 °C and 16 000 g. Wash the pellet with 70% ethanol and centrifuge again for 10 min. After ethanol precipitation and air-drying of the pellet, redissolve it in a small volume of RNase-free water and add an equal volume of gel loading buffer [0.33 × TBE (see Step 6), 2.7 M urea, 67% formamide, 0.01% (w/v) each bromophenol blue (BPB) and xylene cyanol blue (XCB)].

(6) Preparative denaturing PAGE. A gel well, 6- to 7-cm wide and 1-mm thick, is appropriate for loading the product RNA from an efficient 1-ml transcription reaction. The pocket size is of some importance, as an overloaded gel may cause separation problems; on the other hand, if the pocket is too large, RNA bands may be barely visible and elution efficiency may decrease. After electrophoresis, the desired RNA band is visualized at 254 nm by UV-shadowing and marked for gel excision (for details, see Chapter 3); gel running buffer: 1 × TBE (89 mM Tris base, 89 mM boric acid, 2 mM EDTA).

(7) Elution of RNA product.
  (a) *Diffusion elution.* Cover the excised gel pieces with elution buffer and shake overnight at 4 °C; in the case of efficient transcriptions, a second gel elution step in fresh elution buffer may substantially increase the yield. Different elution buffers can be used: buffer A: 1 mM EDTA, 200 mM Tris–HCl (pH 7)/buffer B: 1 M NaOAc (pH 4.7)/buffer C (successfully used for the elution of phosphorothioate-modified RNAs): 1 M NH_4OAc (pH 7). Usually, buffer A is used; buffer B was found to be advantageous in cases where elution efficiency with buffer A was low. After elution, RNA is concentrated by ethanol precipitation.
  (b) *Alternatively: electro-elution.* Excised gel pieces containing the RNA are placed in a BIOTRAP chamber (Schleicher & Schuell, Dassel, Germany; in USA and Canada: ELUTRAP) following the manufacturer's protocol. The RNA is eluted in 0.5 × TBE buffer (see Step 6). The final volume of RNA solution after elution is approximately 600 µl, depending on the extent of evaporation during the elution process. The elution is permitted to proceed overnight at 150 V/20 mA or for 4–6 h at 200–300 V/30 mA. During the elution process there is evaporation resulting in condensation on the lid of the BIOTRAP chamber. This lid has to be closed when the BIOTRAP is running. To minimize evaporation/condensation, the BIOTRAP should not be run at higher voltage than 150 V overnight. After elution, the RNA is extracted once with phenol and twice with chloroform/isoamylalcohol, followed by ethanol precipitation.

(8) Quantification (UV spectroscopy, see Chapter 4 and Appendix) and quality check (denaturing PAGE).

1.4
Troubleshooting

1.4.1
Low or No Product Yield

- If product yields are low with a protocol that had already been successfully used for the same template, repeat transcription assay once without any alteration on 50-μl scale; if unsuccessful as well, test different enzyme batches or enzymes from alternative suppliers. Differences between enzyme preparations can be considerable.
- Be sure that all components (except enzymes) have been warmed up to ambient temperature before preparation of reaction mixtures at ambient temperature.
- Check that thawed stock solutions, particularly concentrated transcription buffers, do not contain precipitated ingredients. For nucleotide solutions, limit freeze–thawing cycles, store in aliquots at −20 °C, and adjust stock solutions (in H_2O) to pH 7.0 or consider to buffer with 10–40 mM Tris–HCl adjusted to the pH used in transcription reactions; use lithium salts if available and be aware that diluted working solutions may degrade rapidly.
- Prepare new template DNA; take particular care to effectively remove salts as well as traces of phenol and chloroform.
- For templates with a highly biased nucleotide composition (e.g. coding for RNAs with extremely high $A + U$ content [28, 29]), adjust the composition of the rNTP solution according to the nucleotide ratio of the RNA. However, do not alter the total nucleotide concentration of the reaction mix.

1.4.2
Side-products and RNA Quality

- Usually we get the least artifact products, in addition to the correct RNA product, with Protocol 3, but redissolving the RNA after ethanol precipitation may become a severe problem due to pyrophosphate precipitates. To alleviate such solubility problems, see Section 1.3.3, Step 1.
- Gel entry problems or smear on gel: perform Steps 1–5 of Section 1.3.3 before proceeding to gel purification (Section 1.3.3, Step 6).

1.5
Rapid Preparation of T7 RNA Polymerase

This protocol is based on the publications of Grodberg and Dunn [30] and Zawadzki and Gross [25], and provides a fast and efficient procedure for the preparation of a highly stable T7 RNAP which is sufficiently pure for most purposes. The chromatography is described for FPLC, but any standard low-pressure equipment will give satisfactory results if the procedure is adapted accordingly.

1.5.1
Required Material

E. coli BL21 pAR1219 (obtained from F. W. Studier, Biology Department, Brookhaven National Laboratory, Upton, NY 11973, USA).

Medium
LB (Luria-Bertani) medium [31] supplemented with 50 µg/ml Ampicillin.

Buffers and solutions
100 mM IPTG
TEN buffer (50 mM Tris–HCl, pH 8.1; 2 mM Na$_2$EDTA; 20 mM NaCl)
Phenylmethylsulfonyl fluoride (PMSF), 20 mg/ml in isopropanol
Leupeptin, 5 mg/ml
Egg white lysozyme, 1.5 mg/ml in TEN (freshly prepared)
0.8% sodium deoxycholate solution
2 M ammonium sulfate (enzyme grade)
Polymin P: 10% solution, adjusted to pH 8 with HCl
Saturated ammonium sulfate solution (4.1 M; adjust pH to 7 with some drops of concentrated Tris base, keep at 4 °C where a precipitate will form)
Buffer C [20 mM sodium phosphate pH 7.7; 1 mM Na$_2$EDTA; 1 mM DTT; 5% glycerol (w/v)]
4 × Laemmli gel loading buffer: 100 mM Tris–HCl, pH 6.8, 8% (w/v) SDS, 30% glycerol (w/v), 8% (v/v) β-mercaptoethanol, 0.04% (w/v) bromophenol blue; adjust pH before addition of bromophenol blue
Buffer C-10, C-100: buffer C supplemented with 10 or 100 mM NaCl, respectively.

Electrophoresis and chromatography
Laemmli-type SDS gel for protein separation under denaturing conditions (10% PAA)
Merck EMD Fractogel SO$_3^-$, equilibrated in buffer C-100 in a 2 × 10 cm column

1.5.2
Procedure

1.5.2.1 Cell Growth, Induction and Test for Expression of T7 RNAP

(1) Inoculate 25 ml LB supplemented with 50 µg/ml Ampicillin with a colony from a fresh plate culture and grow overnight at 37 °C.
(2) Four 2-l flasks with 500 ml of the same medium are inoculated 1:100 from this culture; grow at 37 °C under vigorous shaking.
(3) When the cultures have reached an OD$_{600}$ of about 0.6 (which should take not longer than 3 h), transfer 1 ml to an Eppendorf tube, centrifuge for 5 min at 5000 r.p.m. in a desktop centrifuge and keep the sediment as a control.
(4) Then induce the remaining culture for expression by addition of IPTG to a

final concentration of 0.5 mM. At 1.5, 2 and 3 h after induction, take 1-ml samples as in Step 3.
(5) Harvest the flask cultures by centrifugation (10 min, 5000 g), wash once with TEN buffer, shock-freeze in liquid N_2 or dry ice and keep at −80 °C until needed.
(6) Analyze the 1-ml samples from Steps 3 and 4 for expression of T7 RNAP as follows: resuspend the cell sediment in 100 µl of 1 × concentrated Laemmli gel loading buffer (note that this buffer is usually prepared as a more concentrated stock solution; see Section 1.5.1) and denature for 2 min at 95 °C. Then load 10–25 µl of each sample onto an SDS–10% polyacrylamide gel with appropriate size markers. If expression is sufficient (a strong band of about 100 kDa should appear 2–3 h after induction), proceed with enzyme purification.

1.5.2.2 Purification of T7 RNAP

Generally, all steps are performed on ice or at 4 °C and all buffers are supplemented with the protease inhibitor PMSF (20 µg/ml, if not stated otherwise). From each purification step, a small sample should be retained for the determination of protein concentration and purity. Protein concentrations are most conveniently determined by dye binding [32] or by direct measurement of extinction at 280 nm.

(1) Resuspend cells in 24 ml of TEN buffer supplemented with 50 µl of PMSF and 20 µl of leupeptin stock solutions (see Section 1.5.1).
(2) Add 6 ml lysozyme solution; after 20 min of incubation, add 2.5 ml of 0.8% sodium deoxycholate solution and incubate for another 20 min.
(3) Shear the DNA in this viscous lysate by sonication (4 × 15 s with an immersible probe; 2–5 min for sonication in a water bath).
(4) Add 5 ml 2 M ammonium sulfate and adjust the total volume to 50 ml with TEN buffer.
(5) Remove DNA by slow addition of 5 ml Polymin P and stirring for 20 min.
(6) After centrifugation (15 min, 39 000 g), keep the supernatant and determine its volume.
(7) Precipitate the enzyme from the supernatant by slow addition of 0.82 volumes of saturated ammonium sulfate and stirring for another 15 min.
(8) After centrifugation for 15 min at 12 000 g, resuspend the sediment in 15 ml buffer C-100, and dialyze for at least 8 h against 2 × 1 l of the same buffer. Dialysis should be extensive in order to completely remove the ammonium sulfate contained in the sediment; otherwise, T7 RNAP will not bind to the cation exchange column.
(9) Remove insoluble material by centrifugation as in Step 8, and apply the supernatant to the EMD-SO_3^- column at a flow rate of about 20 ml/h. Wash the column with 10 volumes of buffer C-100 or until protein is no longer detectable in the flow-through. Then apply a 250-ml gradient from 100 to 500 mM NaCl in buffer C and collect 5-ml fractions. T7 RNAP elutes between 300 and 400 mM NaCl from the EMD-SO_3^- column, as visible by the high

protein content in these fractions. Finally, wash the column with 1 M NaCl in buffer C; the resin can be reused after equilibration in buffer C-100.
(10) 10–20 µl of each fraction and the flow-through, and 2–5 µl of the applied sample, are then analyzed by SDS–PAGE as described in Section 1.5.2.1, Step 6.
(11) The fractions containing T7 RNAP are pooled and dialyzed for at least 8 h against 2 × 2 l buffer C-10.
(12) The resulting precipitate, enriched in T7 RNAP, is collected by centrifugation as in Step 8, resuspended in 1–2 ml of buffer C-100 and adjusted to 50% glycerol (w/v) for storage at $-20\,°C$.

Specific activity of T7 RNAP may be determined by incorporation of ^{3}H- or ^{32}P-labeled nucleotides into acid-precipitable material and can reach 400 000 U/mg (1 U is defined as the incorporation of 1 nmol AMP into acid precipitable material in 1 h at 37 °C [9]). For most purposes, it is sufficient to titrate the amount of enzyme preparation needed to give good transcription yields without too many side products. To test a typical preparation of T7 RNAP (5–10 µg/µl total protein content), we vary the amount of polymerase (usually between 0.1 and 1 µl) in a series of analytical transcriptions, for example, in a variation of Protocol 9 scaled down to 25 µl, containing 2 mM of each rNTP and a standard template DNA (around 50% G + C). Transcription products are then separated by gel electrophoresis and the efficiency of transcription is evaluated after toluidine blue staining of the gel (see Chapter 9). Alternatively, for easier detection of abortive transcripts and degradation products, a radioactive tracer (e.g. 10 µCi of $[\alpha\text{-}^{32}P]GTP$) can be added to the transcription reaction and the products visualized on a PhosphoImager or by autoradiography (see Protocol 8).

1.5.3
Notes and Troubleshooting

(1) If protein gel electrophoresis of crude cell samples (Section 1.5.2.1, Step 6) yields badly smeared bands, try to shear the DNA by sonication as described for the enzyme purification. Alternatively, samples can be squeezed with a syringe through a thin (0.7 mm, or 22 gauge) needle.
(2) If expression of T7 RNAP is not sufficient, vary ampicillin or IPTG concentration, change growth times before and after induction, or switch to other growth media, such as TB or 2 × YT [31], or M9 Minimal Medium supplemented with trace elements [33].
(3) The Fractogel EMD-SO$_3^-$ column can be substituted by any strong cation exchanger of the SO$_3^-$ type. However, with most other column matrices, T7 RNAP elutes much earlier (around 200 mM NaCl).
(4) If T7 RNAP does not precipitate after the dialysis step with buffer C-10 (Section 1.5.2.2, Step 11), repeat dialysis with fresh buffer C (without NaCl).
(5) If a substantial portion of the precipitate from Step 12 above (Section 1.5.2.2) cannot be re-dissolved in buffer C-100, one may gradually increase the NaCl concentration up to 500 mM.

Acknowledgement

We thank Dagmar K. Willkomm for critical reading of the manuscript.

References

1. P. Davanloo, A. H. Rosenberg, J. J. Dunn, F. W. Studier, *Proc. Natl. Acad. Sci. USA* **1984**, *81*, 2035–2039.
2. R. H. Ebright, *J. Mol. Biol.* **2000**, *304*, 687–698.
3. G. M. T. Cheetham, T. A. Steitz, *Curr. Opin. Struct. Biol.* **2000**, *10*, 117–123.
4. G. A. Diaz, M. Rong, W. T. McAllister, R. K. Durbin, *Biochemistry* **1996**, *35*, 10837–10843.
5. S. N. Kochetkov, E. E. Rusakova, V. L. Tunitskaya, *FEBS Lett.* **1998**, *440*, 264–267.
6. S. Brakmann, S. Grzeszik, *ChemBioChem* **2001**, *2*, 212–219.
7. J. F. Milligan, O. C. Uhlenbeck, *Methods Enzymol.* **1989**, *180*, 51–63.
8. I. D. Pokrovskaya, V. V. Gurevich, *Anal. Biochem.* **1994**, *220*, 420–423.
9. J. F. Milligan, D. R. Groebe, G. W. Witherell, O. C. Uhlenbeck, *Nucleic Acids Res.* **1987**, *15*, 8783–8798.
10. J. A. Pleiss, M. L. Derrick, O. C. Uhlenbeck, *RNA* **1998**, *4*, 1313–1317.
11. M. Helm, H. Brulé, R. Giegé, C. Florentz, *RNA* **1999**, *5*, 618–621.
12. T. M. Coleman, G. Wang, F. Huang, *Nucleic Acids Res.* **2004**, *32*, e14.
13. J. R. Sampson, O. C. Uhlenbeck, *Proc. Natl. Acad. Sci. USA* **1988**, *85*, 1033–1037.
14. A. D. Griffiths, B. V. Potter, I. C. Eperon, *Nucleic Acids Res.* **1987**, *15*, 4145–4162.
15. V. D. Axelrod, F. D. Kramer, *Biochemistry* **1985**, *24*, 5716–5723.
16. F. Conrad, A. Hanne, R. K. Gaur, G. Krupp, *Nucleic Acids Res.* **1995**, *23*, 1845–1853.
17. R. K. Gaur, G. Krupp, *FEBS Lett.* **1993**, *315*, 56–60.
18. N. Logsdon, C. G. L. Lee, J. W. Harper, *Anal. Biochem.* **1992**, *205*, 36–41.
19. R. Sousa, R. Padilla, *EMBO J.* **1995**, *14*, 4609–4621.
20. Y. Huang, F. Eckstein, R. Padilla, R. Sousa, *Biochemistry* **1997**, *36*, 8231–8242.
21. W. D. Hardt, V. A. Erdmann, R. K. Hartmann, *RNA* **1996**, *2*, 1189–1198.
22. S. P. Ryder, S. A. Strobel, *Methods* **1999**, *18*, 38–50.
23. C. Pitulle, R. G. Kleineidam, B. Sproat, G. Krupp, *Gene* **1992**, *112*, 101–105.
24. F. Huang, *Nucleic Acids Res.* **2003**, *31*, e8.
25. V. Zawadzki, H. J. Gross, *Nucleic Acids Res.* **1991**, *19*, 1948.
26. T. Ellinger, R. Ehricht, *BioTechniques* **1998**, *24*, 718–720.
27. S. S. Sastry, B. M. Ross, *J. Biol. Chem.* **1997**, *272*, 8644–8652.
28. M. Baum, A. Cordier, A. Schön, *J. Mol. Biol.* **1996**, *257*, 43–52.
29. M. Wagner, C. Fingerhut, H. J. Gross, A. Schön, *Nucleic Acids Res.* **2001**, *29*, 2661–2665.
30. J. Grodberg, J. J. Dunn, *J. Bacteriol.* **1988**, *170*, 1245–1253.
31. J. Sambrook, E. F. Fritsch, T. Maniatis, *Molecular Cloning: A Laboratory Manual*, Cold Spring Harbor Laboratory Press, Cold Spring Harbor, NY, **1989**.
32. M. M. Bradford, *Anal. Biochem.* **1976**, *72*, 248–254.
33. D. E. Mossakawska, R. A. G. Smith, in: *Protein NMR Techniques*, D. G. Reid (ed.), Humana Press, Totowa, NJ, **1997**.
34. F. Huang, C. W. Bugg, M. Yarus, *Biochemistry* **2000**, *39*, 15548–15555.
35. J. J. Dunn, F. W. Studier, *J. Mol. Biol.* **1983**, *166*, 477–535.
36. S. S. Lee, C. Kang, *J. Biol. Chem.* **1993**, *268*, 19299–19304.

2
Production of RNAs with Homogeneous 5′ and 3′ Ends

Mario Mörl, Esther Lizano, Dagmar K. Willkomm and Roland K. Hartmann

2.1
Introduction

Synthesis of RNA molecules by *in vitro* transcription, primarily utilizing T7 RNA polymerase (T7 RNAP) for reasons of price and efficiency, is a widely used method in biochemistry and molecular biology. For most purposes, it is sufficient to purify transcripts by electrophoretic separation on denaturing PAA (polyacrylamide) gels. Thereby, template DNA, enzyme, nucleotides as well as incomplete transcripts (resulting from premature transcription termination) can be removed efficiently. However, the seemingly uniform population of full-length transcripts is in most cases heterogeneous because T7 RNAP tends to add one or occasionally even a few additional, non-encoded nucleotides to the 3′ end [1, 2]. Also, non-templated nucleotide incorporation at the 5′ end can affect up to 30% of the transcribed RNA molecules when the template encodes a T7 transcript with more than three consecutive guanosines at the 5′ end [3, 4]. Such micro-heterogeneous full-length transcripts can rarely be separated efficiently by denaturing PAGE. Thus, if applications require complete homogeneity at the transcript 5′ and 3′ ends, e.g. in the context of *in vitro* aminoacylation of tRNAs, ligation of RNA fragments (see Chapters 3 and 4), RNA crystallization or NMR studies, refined approaches are needed.

Apart from product end heterogeneity, another problem associated with *in vitro* transcription stems from the fact that the sequence requirements of the core promoter regions recognized by phage RNA polymerases also put some constraints on the 5′-terminal nucleotide identities of the transcript. For example, transcription by T7 RNAP from the predominantly used (class III) T7 promoter only proceeds with reasonable efficiency if the transcript is initiated with a G residue (for more details, see Chapter 1). As a consequence, synthesis yields will be low for transcripts with a different nucleotide identity at the +1 position.

An approach toward eliminating 5′ and 3′ end heterogeneity as well as sequence constraints at the 5′ end is the use of catalytic RNA entities acting in *cis* or in *trans*, which precisely release the RNA of interest from a primary transcript with extra

5′- and 3′-flanking sequences. These strategies permit to produce RNA without any sequence restrictions and with essentially complete 5′ and 3′ end homogeneity.

2.2
Description of Approach

2.2.1
Cis-cleaving Autocatalytic Ribozyme Cassettes

2.2.1.1 The 5′ Cassette
An elegant and efficient way to produce transcripts with uniform 5′ ends is to use a construct consisting of a self-cleaving hammerhead sequence immediately upstream of the RNA sequence of interest. The hammerhead ribozyme is a small structural element, originally identified in pathogenic plant viroids and virusoids, which catalyzes phosphodiester bond hydrolysis at a single defined position. With a length of about 50 nt, the hammerhead sequence can easily be inserted as a DNA cassette at the 5′ end of the DNA sequence to be transcribed (Fig. 2.1). In the primary transcript (Fig. 2.1, middle), the 5′ part of the RNA of interest is integrated into the secondary structure of the hammerhead ribozyme. Since this ribozyme sequence is able to fold into the catalytically active form immediately after synthesis, cleavage occurs already during transcription – provided that Mg^{2+} ions are present at sufficiently high concentrations (5 mM or above), as is the case under standard transcription conditions. In addition to 5′ end uniformity, another advantage of a 5′ hammerhead is that optimal transcription start sequences can be used upstream of the ribozyme cassette in favor of high transcription yields.

Within a 5′-flanking hammerhead cassette, the sequence of helix P1 is dictated by the 5′-terminal nucleotides of the RNA of interest (Fig. 2.1, middle). Since the only requirement for P1 is helicity, there are basically no sequence constraints for this structural element. Thus, any efficient minimal hammerhead variant matching the consensus sequence shown in Fig. 2.1 can be utilized [5–8]. Two 5′-hammerhead constructs which in our hands have proved efficient in such strategies are shown in Fig. 2.2(A and B).

2.2.1.2 The 3′ Cassette
For the production of RNAs with homogeneous 3′ ends, again hammerhead cassettes (Fig. 2.2C) can be tethered to the RNA of interest. A disadvantage of these structures are some sequence constraints imposed on the product RNA 3′ end, such as the requirement for a 3′-terminal NUH (N = any nucleotide; H = C, U or A), preferentially GUC, sequence motif in hammerhead constructs (Fig. 2.2C [5]).

The autocatalytic domain of the Hepatitis delta virus (HDV ribozyme), however, has no such sequence requirements upstream of its cleavage site and can therefore be employed for any transcript sequence of interest (Fig. 2.1 and 2.2A). This ribo-

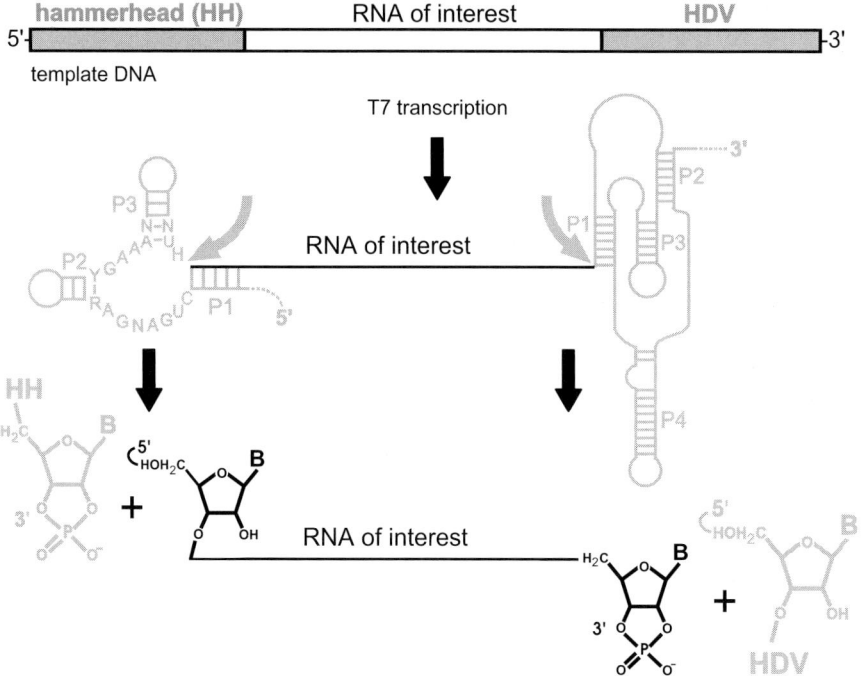

Fig. 2.1. The 5′ and 3′ cassettes for the creation of homogeneous RNA ends. The DNA molecule to be transcribed (linearized plasmid or PCR product) includes extra sequence cassettes upstream and downstream of the sequence encoding the RNA of interest. These cassettes (indicated in grey at the top) are transcribed into self-cleaving ribozyme structures (hammerhead at the 5′ end, HDV ribozyme at the 3′ end) flanking the RNA of interest. Conserved nucleotides in the hammerhead core structure are indicated; Y = pyrimidine; R = purine; NUH (N = any nucleotide; H = C, U or A), preferentially GUC. Both cassettes are designed to permit insertion of virtually any sequence between the ribozymes, rendering this strategy applicable to essentially any RNA of interest. The resulting cleavage product carries a 5′-hydroxyl group, which can be converted to a (radioactively labeled) 5′-phosphate. At the 3′ terminus, a 2′,3′-cyclic phosphate group is generated which interferes with some RNA functions. Protocols for the removal of the 3′-terminal phosphate group are described in Section 2.3.2.

zyme folds into four helical domains that form a pseudoknotted structure, with the cleavage site located immediately upstream of helix P1 at the 5′ end of the ribozyme [9, 10]. Using an optimized version of the HDV ribozyme, homogeneous RNA 3′ ends can easily be generated [11]. The HDV ribozyme is already sufficiently active at Mg^{2+} concentrations of about 1 mM, usually resulting in efficient self-cleavage during transcription. A frequently observed problem is that sequences of the RNA of interest interfere with proper folding of the HDV ribozyme, leading to reduced cleavage efficiency. In our hands, this problem has been solved by incubating the primary transcript repeatedly (10 cycles) for 3 min at 60 °C and 3 min at 25 °C. This procedure allows the ribozyme cassette of the transcript to adopt,

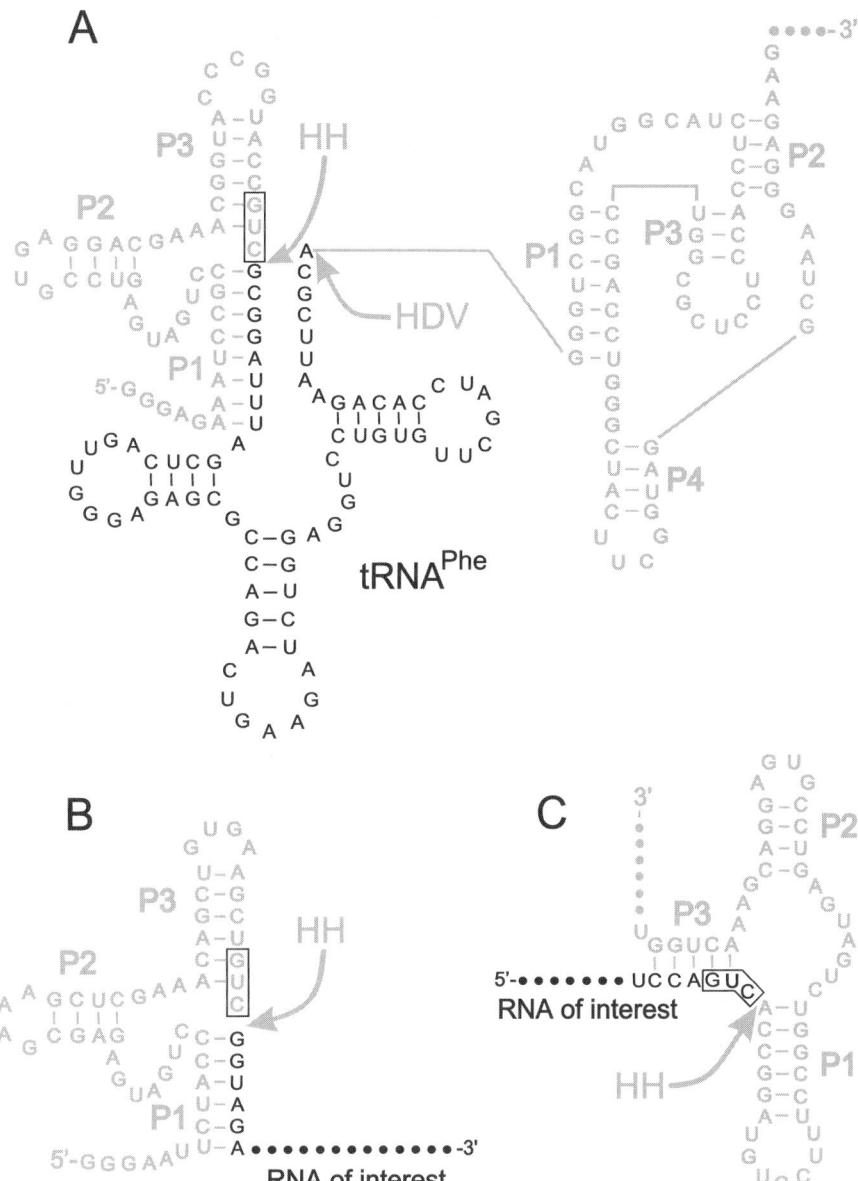

Fig. 2.2. Examples of terminal ribozyme cassettes in primary transcripts. The RNA of interest is shown in black, the ribozyme modules are depicted in grey. Ribozyme self-cleavage sites are indicated by arrows. The preferred GUC triad of hammerhead ribozymes is boxed. The constructs shown have displayed efficient and precise autocatalytic cleavage in the authors' laboratories. (A) Sequence of a primary transcript consisting of a 5'-hammerhead, a central tRNA^{Phe} as the RNA of interest and a 3'-terminal HDV ribozyme. (B and C) Examples of a 5'- and 3'-terminal hammerhead, respectively.

at least transiently, its active structure and has in our hands led to quantitative cleavage.

2.2.1.3 Purification of Released RNA Product and Conversion of End Groups

After co-transcriptional self-cleavage of the terminal ribozyme cassette(s), it is advisable to purify the released RNA of interest by denaturing PAGE and subsequent gel elution in, for example, 500 mM ammonium acetate, pH 5.7, 0.1 mM EDTA and 0.1% SDS at 4 °C overnight, followed by ethanol precipitation ([12]; for detailed protocols, see Chapters 1 and 3). If, for example, a tRNA product is released from the primary transcript, the ribozyme cassettes will have similar lengths of about 50–70 nt, complicating purification of the tRNA. A practical solution to the problem is to extend the 5′ and/or 3′ termini of the primary transcripts, thereby increasing the size of the released terminal fragments containing the ribozyme core structures.

Another important point to be considered is the nature of the 5′ and 3′ ends generated by these ribozymes. Both hammerhead and HDV ribozyme cassettes catalyze transesterification reactions induced by a nucleophilic attack of the 2′-OH group of the neighboring ribose, resulting in 5′-hydroxyl and 2′,3′-cyclic phosphate termini instead of 5′-phosphates and 3′-hydroxyls. A terminal 5′-OH group is for most purposes neutral to RNA function and may even be advantageous, since it is directly accessible to 5′-end-labeling with [γ-$^{32/33}$P]ATP by T4 polynucleotide kinase (T4 PNK). The 3′-terminal cyclophosphate, however, interferes with a variety of downstream experiments, such as aminoacylation of tRNAs, ligation of RNA molecules or 3′-end-labeling using [5′-^{32}P]pCp. To remove the 2′,3′-cyclic phosphate, the phosphatase activity of T4 PNK can be employed with excellent efficiency (see protocols below and [13, 14]).

2.2.2
Trans-cleaving Ribozymes for the Generation of Homogeneous 3′ Ends

As an alternative to the *cis*-cleaving ribozyme cassettes, homogeneous 3′ ends can be obtained by *trans*-cleavage, e.g. using a *trans*-acting hammerhead ribozyme [15], the *Neurospora* Varkud satellite (VS) ribozyme [15] or the catalytic RNA subunit of a bacterial RNase P [16] as detailed below. A *trans*-acting VS ribozyme has the advantage that the RNA of interest requires only a short stem–loop 3′-extension of 24 nt which serves as the ribozyme substrate recognized by tertiary interactions [15]. Like the HDV ribozyme, the VS ribozyme has no specific sequence requirements upstream of the cleavage site and generates 2′,3′-cyclic phosphate ends [17]. Similar to the VS ribozyme strategy, it is possible to utilize the HDV ribozyme in a *trans*-cleavage reaction. Here, the transcripts are extended by as few as seven nucleotides complementary to the 3′ part of helix P1 of the ribozyme. A correspondingly 5′-truncated HDV RNA can base-pair with this extension, thereby restoring the P1 helix and a functional HDV ribozyme structure. Cleavage of the target RNA then occurs seven nucleotides upstream from its 3′ end [18].

In vitro, the roughly 380-nt long bacterial RNase P RNA, with and without the

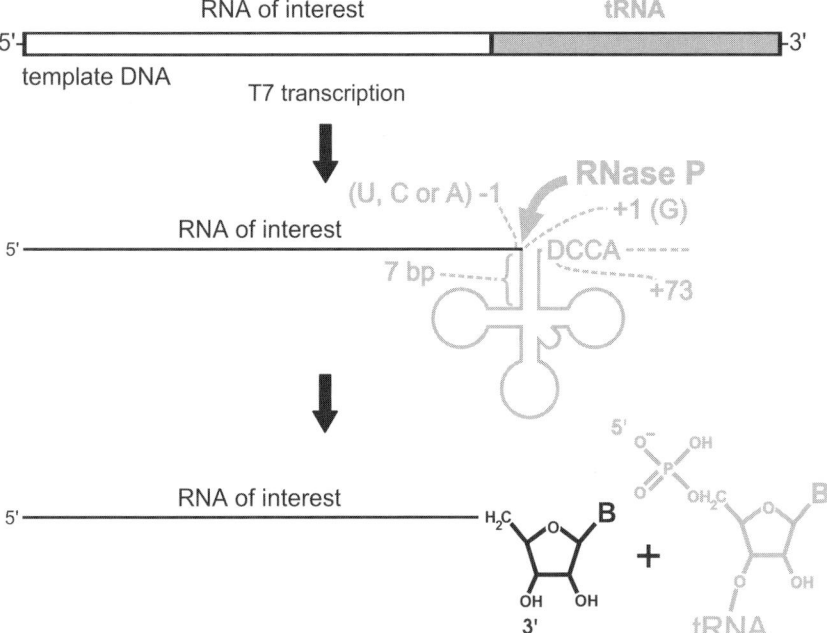

Fig. 2.3. RNase P cleavage to generate homogeneous 3' ends. An alternative strategy to generate homogeneous 3' ends is the use of bacterial RNase P (RNA). The DNA template for transcription is designed in a way that a tRNA gene (in grey) is positioned immediately downstream of the sequence encoding the RNA molecule of interest. By incubating the primary transcript, derived from run-off transcription, with RNase P (RNA), the tRNA molecule is cleaved off at its 5' end and the upstream portion of the primary transcript is released. Aspects of importance: avoid a G residue at position −1; avoid base-pairing between nt −1 and discriminator (+73); the discriminator should be D (= G, A or U) and not C when using *E. coli* RNase P RNA [21]; the first acceptor stem base pair should be G_{+1}–C_{+72}; choose a class I tRNA with 7-bp acceptor stem and short variable arm; encode CCA at the tRNA 3' terminus; a 3' extension of up to 6 nt beyond CCA does not interfere with RNase P cleavage, but a second CCA trinucleotide should be avoided within this extension.

RNase P protein, catalyzes accurate and efficient removal of the 5'-flanking sequences of tRNA precursor transcripts. To exploit this system for the production of RNAs with homogeneous 3' ends, the RNA of interest is fused to a downstream tRNA sequence. As a result, the construct mimics a tRNA precursor molecule, whose 5'-leader sequence represents the RNA of interest (Fig. 2.3). Precise endonucleolytic cleavage by the RNase P ribozyme then releases the RNA product with uniform 3'-OH termini (Fig. 2.3) which, in contrast to the 2',3'-cyclic phosphates generated by hammerhead and HDV ribozymes, are suitable for many downstream applications.

Since the substrate recognition elements for RNase P RNA are located predominantly in the tRNA structure, there are little upstream sequence requirements to be fulfilled. One restriction is to avoid a G residue at position −1 relative to the tRNA, which otherwise may cause some aberrant cleavage between positions -1 and -2. Also, for reasons of cleavage fidelity, use of a bacterial class I tRNA moiety with a 7-bp acceptor stem and short variable arm is recommended (such as *Bacillus subtilis* tRNA^{Asp} [16, 19] or *Thermus thermophilus* tRNA^{Gly} [20]). Further, any base-pairing potential between nt -1 and the discriminator base ($+73$) should be avoided, the discriminator should be D ($=$ G, A or U) and not C when using *Escherichia coli* RNase P RNA [21], and the first acceptor stem base pair should be G$_{+1}$–C$_{+72}$. However, note that some class I tRNA transcripts, despite meeting all the aforementioned requirements, may elicit some miscleavage, depending on their sequence context [22]. We have successfully used RNase P RNAs from *E. coli* and *T. thermophilus*, which are available from the authors as T7 expression plasmids. Yet, since achieving 100% 3′ end homogeneity – and not 99% – by use of this approach may require some optimization, we suggest considering the use of an HDV-based strategy as a quicker alternative if such an extent of homogeneity is essential for downstream procedures.

In applications where a short 3′ appendage to the RNA of interest is preferable (e.g. when incorporating costly isotopically labeled nucleotides for NMR studies), the tethered RNase P substrate can be reduced to a hairpin structure of less than half the tRNA size. One such substrate is pATSerU$_{-1}$G$_{+73}$ and its variants pATSerU$_{-1}$U$_{+73}$, pATSerC$_{-1}$A$_{+73}$ and pATSerA$_{-1}$U$_{+73}$, which release an RNA of interest with a 3′-terminal U, C or A residue, respectively [23]. These substrates showed >99% cleavage at the canonical RNase P cleavage site under conditions of 50 mM Tris–HCl, pH 7.2, 5% (w/v) PEG 6000, 100 mM NH$_4$Cl and 40 mM MgCl$_2$ ([23]; Leif A. Kirsebom, personal communication).

The RNase P RNA-based approach could have advantages over a *cis*-cleaving HDV cassette in cases where interactions between the RNA of interest and the HDV cassette prevent the ribozyme from adopting an active conformation. Such folding interference is less likely for a tRNA cassette because tRNAs are among the most stable autonomous RNA folding units. If folding of the downstream tRNA is impeded by the RNA of interest at 37 °C, thermostable RNase P RNA from *T. thermophilus* may be used for cleavage at elevated temperatures, assuming that folding interference is abolished under such conditions. This ribozyme, when acting on a transcript with, for example, a GC-rich tRNA^{Gly} cassette from the same organism [24], will cleave off the tRNA moiety with high precision at temperatures of up to 75 °C [25].

The *trans*-cleaving ribozyme approach may be somewhat laborious if the ribozyme is prepared independently of the transcript of interest and if *in vitro* transcription is followed by a second, independent incubation step for the cleavage reaction. However, the procedure can be simplified by transcribing both RNAs simultaneously from two different templates added to the same reaction mix [15]. Irrespective of whether the ribozyme is co-transcribed or added post-transcriptionally, the transcription mixture containing the RNase P substrate transcript has to be adjusted to RNase P cleavage assay conditions, since RNase P RNA requires elevated mono- and divalent metal ion concentrations for efficient processing (see Section 2.3.3). Alternatively, substrate transcripts may be concentrated by ethanol precipitation before RNase P RNA processing. As for the *cis*-ribozyme cassette systems, purification of the desired cleavage product by denaturing PAGE is the method of choice.

2.2.3
Further Strategies Toward Homogeneous Ends

In addition to those described above, other approaches for generating homogeneous ends are available as well. Two methods, an RNase H-based strategy and another making use of T7 transcription templates with two consecutive 2′-O-methyl nucleotides at the 5′ end of the template strand, are detailed in Chapter 3. The RNase H-based strategy can be employed to generate homogeneous 5′ and 3′ ends, while the 2′-O-methyl approach is suited for the production of homogeneous 3′ ends only.

Recently, almost complete 5′ end homogeneity of T7 transcripts was demonstrated with templates directing transcription from the less frequently used T7 class II promoter, at which T7 RNAP initiates synthesis with an A instead of a G residue. Transcription yields from this promoter were reported to equal those of the commonly used T7 class III promoter (see Chapter 1 and [26]).

An elegant variation of *trans*-cleavage concepts, usually relying on ribozymes, involves a 10–23 DNA enzyme [26–28]. This type of DNAzyme has a 15-nt core DNA sequence flanked by two arms that form 8- to 10-bp long hybrid helices with the substrate RNA. Cleavage occurs within a 5′-RY motif (R = A, G; Y = C, U; for details, see [29]) at the junction of the two hybrid helices, resulting in 5′-OH and 2′,3′-cyclic phosphate termini. Sequence-tailored versions of the 10–23 DNAzyme, representing simple DNA oligonucleotides (around 30 nt), can be easily obtained from commercial suppliers and provide an inexpensive and effortless, although less explored alternative to ribozymes in the production of RNAs with homogeneous ends.

2.3
Critical Experimental Steps, Changeable Parameters, Troubleshooting

2.3.1
Construction of *Cis*-cleaving 5′ and 3′ Cassettes

A critical step in the construction of hammerhead and/or HDV ribozyme cassettes is to establish an efficient overlap extension PCR (Fig. 2.4). Usually, the 5′ hammerhead cassette is created by two overlapping oligonucleotides that cover the complete hammerhead domain (Fig. 2.4A). In addition, the upstream primer can carry the sequence for the T7 RNAP promoter at the 5′ end (and/or a terminal restriction enzyme cleavage site if plasmid cloning of the PCR fragment is intended). A minor disadvantage is that such an oligonucleotide will be extended by at least 17 nt, which is associated with lower yields of chemical synthesis. Therefore, a

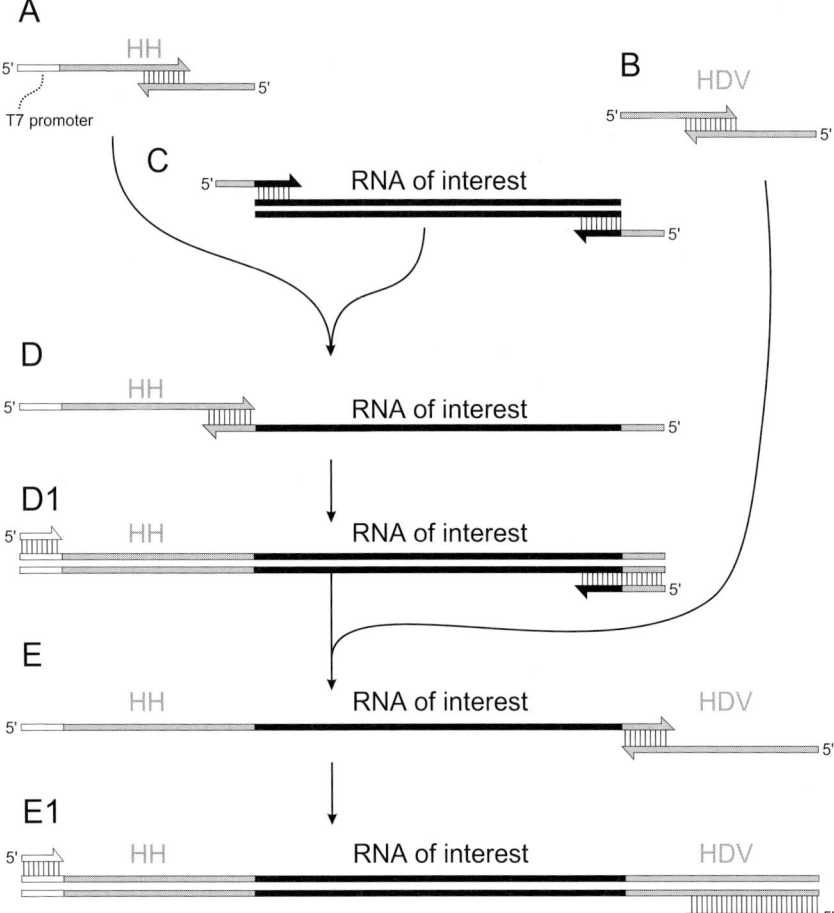

Fig. 2.4. Construction of transcription templates carrying autocatalytic ribozyme cassettes at the 5' and 3' termini. (A and B) By PCR extension of overlapping primer pairs, the initial cassettes (HH, hammerhead; HDV, HDV ribozyme) are created. (C) Using primers with 5' extensions overlapping the ribozyme cassettes, the sequence of interest is amplified in a third PCR reaction. (D) Subsequently, PCR products from reactions A and C are combined (without addition of primers), leading to the fusion of the hammerhead sequence to the sequence encoding the RNA of interest. (D1) The resulting overlap extension product is further amplified using the indicated primers. (E) Eventually, the same strategy is applied to append the HDV sequence. (E1) The final product carrying both cassettes at the corresponding ends is further amplified with terminal primers.

reasonable alternative is to use an upstream primer without the T7 promoter sequence and to insert the PCR product into a cloning vector that encodes the T7 promoter immediately upstream of the cloning site. Otherwise, the T7 promoter can be introduced at a later PCR step.

In a second PCR reaction, the 3'-cassette representing the HDV ribozyme domain is synthesized, again by the use of overlapping primers (Fig. 2.4B). In a third step, the sequence encoding the RNA of interest is amplified (Fig. 2.4C). Here, the upstream primer includes a region overlapping the hammerhead sequence, such that the resulting product can be used for an overlap extension in combination with the hammerhead PCR product (Fig. 2.4D). Likewise, the downstream primer introduces an extension corresponding to the 5′ part of the HDV ribozyme cassette. In a final overlap extension (Fig. 2.4E), the amplified product from Fig. 2.4(D1) is combined with the HDV PCR product in order to generate the full-length construct. Before cloning this product into an appropriate plasmid, it is further amplified using terminal primers (Fig. 2.4E1). After cloning, inserted sequences need to be verified, since the numerous PCR steps involved may lead to aberrant products or sequence deviations.

A complete protocol representing an example of the PCR strategy depicted in Fig. 2.4 is detailed below. The resulting final primary transcript with upstream hammerhead and downstream HDV ribozyme cassettes is illustrated in Fig. 2.2(A). PCR reactions outlined below have been successfully performed with *Taq* DNA polymerase. However, one may consider to use a thermostable DNA polymerase with 3′–5′ proofreading activity, such as *Pfu* polymerase, for the generation of all PCR products that are subsequently used in overlap extension reactions. The reason is that *Taq* polymerase (which has no proofreading activity) tends to add a single non-templated A residue to the 3′ end of PCR products. While this activity is exploited in some cloning strategies (TA-cloning kits), it potentially interferes with overlap extensions: the additional 3′-terminal A does not base pair with the complementary strand. As a consequence, the fraction of strands carrying this extra A residue may not be extended, thus decreasing the overall yield of extension product [30].

PCR protocols
PCR reactions were performed with 2.5 U *Taq* DNA polymerase per 50 µl standard reaction volume in 10 mM Tris–HCl, pH 8.3, 1.5 mM MgCl$_2$, 50 mM KCl. For overlap extensions, the complementary stretches of primer pairs are underlined. The T7 promoter is given in italics, a terminal *Bgl*II site in lowercase letters. Reactions A–E1 correspond to the steps shown in Fig. 2.4. Note that when including terminal restriction enzyme recognition sites in the PCR product for cloning purposes, a few flanking nucleotides beyond the restriction sites have to be added for efficient restriction enzyme cleavage after PCR amplification (for details, see New England Biolabs catalogue "Reference Appendix").

(A) Overlap extension of regions P2, P3 and the 5′ part of P1 of the hammerhead cassette (see Fig. 2.1), including an upstream T7 promoter and a *Bgl*II site

Primer 1 *sense*: 5′-GGa gat ctA *ATA CGA CTC ACT ATA* GGG AGA AAT CCG CCT GAT GAG-3′

Primer 2 *antisense*: 5′-GAC GGT ACC GGG TAC CGT TTC GTC CTC ACG GAC TCA TCA GGC GGA-3′

PCR profile: 20 cycles: 1 min 94 °C/1 min 40 °C/30 s 72 °C

Resulting sequence: 5′-GGa gat ctA ATA CGA CTC ACT ATA GGG AGA AAT CCG CCT GAT GAG TCC GTG AGG ACG AAA CGG TAC CCG GTA CCG TC-3′; 77 bp

(B) Overlap extension of the HDV ribozyme cassette
Primer 3 *sense*: 5′-GGG TCG GCA TGG CAT CTC CAC <u>CTC CTC GCG GTC CGA CCT GGG CTA</u>-3′
Primer 4 *antisense*: 5′-CTT CTC CCT TAG CCT ACC GAA <u>GTA GCC CAG GTC GGA CCG CGA GGA</u>-3′
PCR profile: 20 cycles: 1 min 94 °C/1 min 60 °C/30 s 72 °C
Resulting sequence: 5′-GGG TCG GCA TGG CAT CTC CAC CTC CTC GCG GTC CGA CCT GGG CTA CTT CGG TAG GCT AAG GGA GAA G-3′; 67 bp

(C) Amplification of the template encoding the RNA of interest (here: yeast tRNA^{Phe}, Fig. 2.2A) using primers overlapping with hammerhead and HDV sequence
Primer 5 *sense* (5′ extension into the hammerhead sequence underlined): 5′-<u>GTA CCC GGT ACC GTC</u> GCG GAT TTA GCT CAG-3′
Primer 6 *antisense* (5′ extension into the HDV ribozyme sequence underlined): 5′-<u>TGG AGA TGC CAT GCC GAC CCT</u> GCG AAT TCT GTG G-3′
PCR profile: 2 min 94 °C
30 cycles: 1 min 94 °C/1 min 42 °C/30 s 72 °C
Resulting sequence (regions overlapping with hammerhead and HDV ribozyme sequences are underlined): 5′-<u>GTA CCC GGT ACC GTC</u> GCG GAT TTA GCT CAG TTG GGA GAG CGC CAG ACT GAA GAT CTG GAG GTC CTG TGT TCG ATC CAC AGA ATT CGC A<u>GG GTC GGC ATG GCA TCT CCA</u>-3′; 108 bp

(D) Overlap extension of products from A and C
PCR profile: 4 min 94 °C
10 cycles: 1 min 94 °C/2 min 40 °C/45 s 72 °C

(D1) Addition of primers, product amplification
Primer 7 *sense*: 5′-GGa gat ctA ATA CGA CTC ACT ATA GGG-3′
Primer 6 *antisense*: 5′-TGG AGA TGC CAT GCC GAC CCT GCG AAT TCT GTG G-3′
PCR profile: 30 cycles; 1 min 94 °C/2 min 55 °C/45 s 72 °C
Resulting sequence (hammerhead region and overlap with HDV ribozyme cassette underlined): 5′-GGa gat ct A ATA CGA CTC ACT ATA <u>GGG AGA AAT CCG CCT GAT GAG TCC GTG AGG ACG AAA CGG TAC CCG GTA CCG TCG</u> CGG ATT TAG CTC AGT TGG GAG AGC GCC AGA CTG AAG ATC TGG AGG TCC TGT GTT CGA TCC ACA GAA TTC GCA <u>GGG TCG GCA TGG CAT CTC CA</u>-3′; 170 bp

(E) Overlap extension using the product obtained in D1 (carrying a 3′ extension into the HDV-coding sequence) and the PCR product for the HDV ribozyme cassette (B)
PCR profile: 4 min 94 °C
10 cycles: 1 min 94 °C/2 min 60 °C/45 s 72 °C

(E1) Addition of primers, product amplification
Primer 7 *sense*: 5'-GGa gat ctA ATA CGA CTC ACT ATA GGG-3'
Primer 4 *antisense*: 5'-CTT CTC CCT TAG CCT ACC GAA GTA GCC CAG GTC GGA CCG CGA GGA-3'
PCR profile: 30 cycles; 1 min 94 °C/2 min 60 °C/45 s 72 °C
Resulting sequence (hammerhead and HDV ribozyme regions underlined): 5'-GGa gat ctA ATA CGA CTC ACT ATA GGG AGA AAT CCG CCT GAT GAG TCC GTG AGG ACG AAA CGG TAC CCG GTA CCG TCG CGG ATT TAG CTC AGT TGG GAG AGC GCC AGA CTG AAG ATC TGG AGG TCC TGT GTT CGA TCC ACA GAA TTC GCA GGG TCG GCA TGG CAT CTC CAC CTC CTC GCG GTC CGA CCT GGG CTA CTT CGG TAG GCT AAG GGA GAA G-3'; 217 bp

2.3.2
Dephosphorylation Protocols

As discussed above, the activities of both hammerhead and HDV ribozymes lead to the release of RNA molecules that carry 5'-OH and 2',3'-cyclic phosphate groups at their termini. While the 5' ends can be phosphorylated by standard T4 PNK procedures, several efficient and robust protocols can be used in order to remove the 2',3'-cyclic phosphate group [11]. However, the efficiency of these protocols may vary with the RNA substrate to be dephosphorylated. Hence, it is recommended to test different dephosphorylation procedures if one method does not give satisfying results. An easy test for the removal of the terminal 2',3'-cyclic phosphate group is to analyze aliquots of the RNA before and after treatment with T4 PNK by denaturing PAGE: The removal of the phosphate group leads to a reduced net charge of the transcript, which, for small RNAs (less than 100 nt), can be monitored by a lower electrophoretic mobility of the RNA in comparison to the untreated RNA (see Chapter 6).

Dephosphorylation protocol 1
Up to 50 pmol RNA are incubated with 6 U T4 PNK (New England Biolabs) in a final volume of 50 µl for 6 h at 37 °C in the following buffer:
100 mM Tris–HCl, pH 6.5
100 mM magnesium acetate
5 mM β-mercaptoethanol

Dephosphorylation protocol 2
100 pmol RNA are incubated with 1 U T4 PNK (New England Biolabs) in a final volume of 50 µl for 6 h at 37 °C in a buffer containing:
100 mM imidazole–HCl, pH 6.0
10 mM $MgCl_2$
0.1 mM ATP
10 mM β-mercaptoethanol
20 µg/ml BSA (RNase–free)

Dephosphorylation protocol 3
Up to 300 pmol RNA are incubated with 10 U T4 PNK (New England Biolabs) in a final volume of 20 µl for 6 h at 37 °C in the following buffer:
100 mM morpholinoethanesulfonate/NaOH, pH 5.5
300 mM NaCl
10 mM $MgCl_2$
10 mM β-mercaptoethanol

2.3.3
Protocols for RNase P Cleavage

Preincubation of *E. coli* RNase P RNA
The preincubation step described below is not essential when using *E. coli* RNase P RNA, but it can increase the proportion of ribozyme molecules competent for substrate binding [31]. The procedure is the following:

20 pmol of *E. coli* RNase P RNA (produced by *in vitro* transcription) incubated in a volume of 15 µl for 1 h at 37 °C in RNase P cleavage buffer:
50 mM Tris–HCl, pH 7.5
0.1 mM EDTA
100 mM ammonium acetate
100 mM magnesium acetate
5% PEG 6000

Cleavage reaction
10 pmol of an RNA–tRNA primary transcript of the type shown in Fig. 2.3 (in 15 µl RNase P cleavage buffer) is added to the preincubation mixture and incubated for 1.5 h at 37 °C. It is recommended to adjust the ratio of ribozyme:substrate concentration to 2:1 as a compromise between cleavage efficiency and saving of ribozyme material. When using RNase P RNA from *T. thermophilus*, cleavage is usually performed in the same buffer as used for *E. coli* RNase P RNA (see above). However, a typical incubation temperature is 55 °C [31]. Preincubation of *T. thermophilus* RNase P RNA (20 min at 55 °C) in RNase P cleavage buffer is essential for ribozyme activation if the cleavage reaction is to be performed at 37 °C, but can be omitted for cleavage assays at 55–75 °C.

2.3.4
Potential Problems

Although the described systems do not have any known restrictions concerning the sequence of the RNA of interest, one should keep in mind that some primary structures might interfere with the correct folding of the ribozyme cassettes or the linked tRNA molecule. Such misfolding may result in a low reaction efficiency or even in no cleavage at all. As a first approach, a temperature cycling procedure, such as the one described in Section 2.2.1.2, should be attempted. If unsuccessful, it is advisable to change the linked tRNA sequence (in the case of an RNase P substrate) or to switch from *cis*-cleaving ribozyme cassettes to the *trans*-cleaving RNase P strategy or *vice versa*. To avoid such failures from the beginning, we recommend readers scrutinize the structure of the primary transcript by Mfold [32] before experimental work.

References

1 D. E. DRAPER, S. A. WHITE, J. M. KEAN, *Methods Enzymol.* **1988**, *164*, 221–237.
2 J. F. MILLIGAN, O. C. UHLENBECK, *Methods Enzymol.* **1989**, *180*, 51–62.
3 J. A. PLEISS, M. L. DERRICK, O. C. UHLENBECK, *RNA* **1998**, *4*, 1313–1317.
4 M. HELM, H. BRULÉ, R. GIEGÉ, C. FLORENTZ, *RNA* **1999**, *5*, 618–621.
5 K. R. BIRIKH, P. A. HEATON, F. ECKSTEIN, *Eur. J. Biochem.* **1997**, *245*, 1–16.
6 B. CLOUET-D'ORVAL, O. C. UHLENBECK, *Biochemistry* **1997**, *36*, 9087–9092.
7 T. K. STAGE-ZIMMERMANN, O. C. UHLENBECK, *RNA* **1998**, *4*, 875–889.
8 T. PERSSON, R. K. HARTMANN, F. ECKSTEIN, *ChemBioChem* **2002**, *3*, 1066–1071.
9 M. D. BEEN, A. T. PERROTTA, S. P. ROSENSTEIN, *Biochemistry* **1992**, *31*, 11843–11852.
10 B. M. CHOWRIRA, P. A. PAVCO, J. A. MCSWIGGEN, *J. Biol. Chem.* **1994**, *269*, 25856–25864.
11 H. SCHÜRER, K. LANG, J. SCHUSTER, M. MÖRL, *Nucleic Acids Res.* **2002**, *30*, e56.
12 D. A. PEATTIE, *Proc. Natl. Acad. Sci. USA* **1979**, *76*, 1760–1764.
13 V. CAMERON, O. C. UHLENBECK, *Biochemistry* **1977**, *16*, 5120–5126.
14 L. F. POVIRK, R. J. STEIGHNER, *Biotechniques* **1990**, *9*, 562.
15 A. R. FERRÉ-D'AMARÉ, J. A. DOUDNA, *Nucleic Acids Res.* **1996**, *24*, 977–978.
16 W. A. ZIEHLER, D. R. ENGELKE, *Biotechniques* **1996**, *20*, 622–624.
17 R. A. COLLINS, *Biochem. Soc. Trans.* **2002**, *30*, 1122–1126.
18 A. WICHLACZ, M. LEGIEWICZ, J. CIESIOLKA, *Nucleic Acids Res.* **2004**, *32*, E39.
19 B. K. OH, D. N. FRANK, N. R. PACE, *Biochemistry* **1998**, *37*, 7277–7283.
20 S. BUSCH, L. A. KIRSEBOM, H. NOTBOHM, R. K. HARTMANN, *J. Mol. Biol.* **2000**, *299*, 941–951.
21 L. A. KIRSEBOM, *Biochem. Soc. Trans.* **2002**, *30*, 1153–1158.
22 A. LORIA, T. PAN, *Biochemistry* **1998**, *37*, 10126–10133.
23 M. BRÄNNVALL, B. M. PETTERSSON, L. A. KIRSEBOM, *J. Mol. Biol.* **2003**, *325*, 697–709.
24 W. D. HARDT, J. SCHLEGL, V. A. ERDMANN, R. K. HARTMANN, *J. Mol. Biol.* **1995**, *247*, 161–172.
25 R. K. HARTMANN, V. ERDMANN, *Nucleic Acids Res.* **1991**, *19*, 5957–5964.
26 T. M. COLEMAN, G. WANG, F. HUANG, *Nucleic Acids Res.* **2004**, *32*, e14.
27 S. W. SANTORO, G. F. JOYCE, *Proc. Natl. Acad. Sci. USA* **1997**, *94*, 4262–4266.
28 S. W. SANTORO, G. F. JOYCE, *Biochemistry* **1998**, *37*, 13330–13342.
29 M. J. CAIRNS, A. KING, L. Q. SUN, *Nucleic Acids Res.* **2003**, *31*, 2883–2889.
30 N. A. SHEVCHUK, A. V. BRYKSIN, Y. A. NUSINOVICH, F. C. CABELLO, M. SUTHERLAND, S. LADISCH, *Nucleic Acids Res.* **2004**, *32*, E19.
31 W. D. HARDT, J. SCHLEGL, V. A. ERDMANN, R. K. HARTMANN, *Nucleic Acids Res.* **1993**, *21*, 3521–3527.
32 M. ZUKER, *Nucleic Acids Res.* **2003**, *31*, 3406–3415.

3
RNA Ligation using T4 DNA Ligase

Mikko J. Frilander and Janne J. Turunen

3.1
Introduction

Efficient RNA ligation methods for generating site-specifically modified long RNA molecules using the T4 DNA ligase were initially described some 10 years ago [1]. More recently, this method has been widely used to provide chimeric RNAs to study various RNA–RNA and RNA–protein interactions in diverse biochemical reconstitution systems as well as in live cells (e.g. in *Xenopus* oocytes). The modifications introduced include simple insertion of a single radioactive group at a specific location of an RNA molecule and more complicated alterations in which different nucleotide analogs, crosslinking groups or RNA backbone modifiers have been inserted into long RNA molecules [2–10].

Currently the efficiency of chemical RNA synthesis is such that high-quality RNA molecules up to 80-nt long can be obtained from commercial sources. Since the yield of chemical synthesis is significantly higher than that of RNA ligation, the latter should be carried out only when chemical synthesis cannot be used to obtain the desired molecules. Such cases include (1) the introduction of radioactive groups or modified nucleotides in the middle of a long RNA molecule, (2) the need for capped RNAs, or (3) requirement for modified RNA molecules that are longer than what can be synthesized chemically. As the repertoire of various phosphoramidite monomers used in chemical RNA synthesis is constantly growing, RNA oligonucleotides containing modified nucleotides are frequently utilized in the construction of a long chimeric RNA molecule. In such cases, the RNA oligonucleotides are ligated to other RNA molecules, which are often produced by *in vitro* transcription reactions. This combination provides a way to modify functional groups in virtually any position in a given RNA molecule.

Despite the seemingly simple overall reaction, i.e. the joining of two or three RNA molecules together with the aid of T4 DNA ligase, the execution of a ligation experiment can be a laborious task if large amounts of the ligated products are needed. Here we will review the current methods for RNA ligation. As several excellent reviews on this subject have already been published [11–13], we will here, in addition to concentrating on the RNA ligation itself, describe special methods to

Handbook of RNA Biochemistry. Edited by R. K. Hartmann, A. Bindereif, A. Schön, E. Westhof
Copyright © 2005 WILEY-VCH Verlag GmbH & Co. KGaA, Weinheim
ISBN: 3-527-30826-1

generate high-quality *in vitro* transcripts with homogeneous 5′ or 3′ termini, which are required for large-scale production of chimeric RNAs. Other widely used strategies for the generation of homogeneous 5′ and 3′ ends are detailed in Chapter 2.

3.2
Overview of the RNA Ligation Method using the T4 DNA Ligase

Although T4 DNA ligase is ordinarily used to ligate DNA molecules, it can also catalyze the formation of a phosphodiester bond between two RNA molecules or between RNA and DNA molecules, as its original name "T4 polynucleotide ligase" indicates [14]. The principle of the RNA ligation by T4 DNA ligase is depicted in Fig. 3.1. Typical applications are the so-called two-way (Fig. 3.1A) and three-way

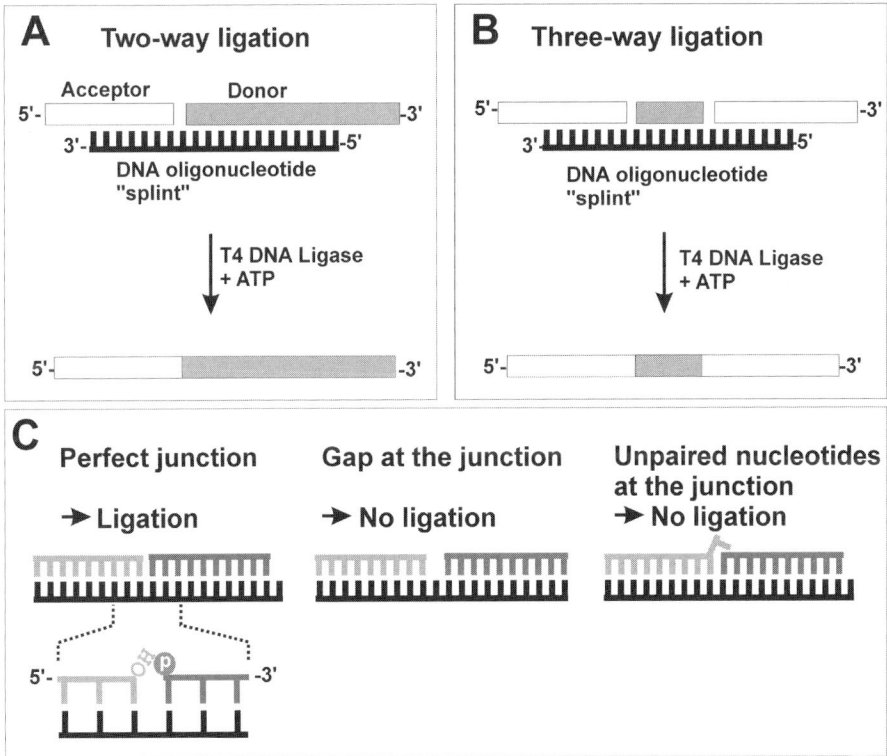

Fig. 3.1. The principle of the RNA ligation with T4 DNA ligase. DNA splint oligonucleotide (black) hybridizes with two (A: two-way ligation) or three (B: three-way ligation) RNA molecules (grey and white) and forms a double-helical structure. (C) Ligation requires 3′-OH and 5′-monophosphate on acceptor and donor molecules, respectively. A gap in the double-stranded helix structure or unpaired nucleotides at the junction (such as $n+1$ products resulting from T7 transcription) will inhibit the ligation.

(Fig. 3.1B) ligations, in which either two or three RNA pieces, respectively, are joined together with the aid of T4 DNA ligase. Normally one of the pieces contains the desired modification(s), while the other(s) are used to reconstitute the full-length RNA molecule under study. The RNA pieces to be ligated are aligned and held together with a complementary bridging DNA oligonucleotide, also known as "DNA splint" or "cDNA template". The ligase catalyses phosphodiester bond formation between the 5'-phosphate of the donor (3' substrate RNA) and the 3'-hydroxyl of the acceptor (5' substrate RNA). Therefore, RNAs containing, for example, a 3'-phosphate or a 5'-triphosphate are not ligated. Furthermore, the RNA/DNA double helix formed by the two RNA pieces and the DNA splint has to be consecutive without any bulges or gaps, especially at the point of the junction of the two RNA molecules.

Although the RNA molecules can also be joined with T4 RNA ligase (see Chapter 4), the use of T4 DNA ligase has several advantages in the construction of long RNA molecules. The main advantage is that, as a consequence of the strict requirement for uninterrupted/unbulged double-helical structure by the T4 DNA ligase, the ligation reaction takes place only for such RNA molecules that have correct termini at the point of the junction (Fig. 3.1C). Furthermore, the use of T4 DNA ligase does not lead to an unwanted formation of circular RNA structures that can be a problem when using T4 RNA ligase with RNA molecules containing unprotected 3' or 5' ends. Finally, as compared to the T4 RNA ligase, the T4 DNA ligase has a lower K_m for polynucleotides, resulting in higher ligation efficiency at lower RNA concentrations, and it does not display any sequence specificity on either donor or acceptor molecules.

A simplified flowchart for a three-way ligation experiment is presented in Fig. 3.2 to illustrate the typical steps needed for the preparation of individual RNA pieces for the RNA ligation. In this example, the final product has been capped with GpppG and the modified nucleotides are located in the middle of the molecule. The modified nucleotides introduced into the RNA molecule are incorporated in the central piece, which is chemically synthesized. The 5' and 3' pieces are produced by *in vitro* transcription using T7 RNA polymerase. In this scheme the preparation of the 5' piece is relatively simple, while the 3' piece requires further processing to generate a 5' monophosphate and/or correct 5'-terminal sequence (dephosphorylation followed by phosphorylation to generate 5'-monophosphate termini or, alternatively, site-specific cleavage with RNase H). As the quality of the chemically synthesized pieces is often relatively high and does not necessarily require extensive processing, we will concentrate in the following sections on the production of high-quality transcripts by T7 RNA polymerase.

3.3
Large-scale Transcription and Purification of RNAs

When preparing RNA for the ligation it is necessary to start with a relatively large initial amount of RNA to account for the unavoidable losses at the various stages of

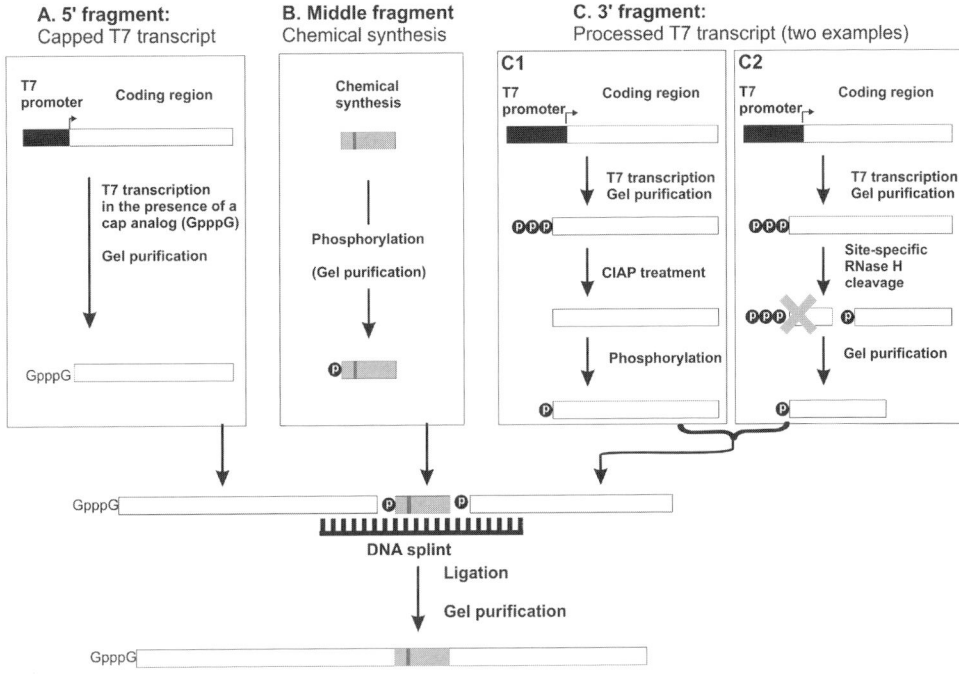

Fig. 3.2. An example of steps needed for the production of RNA pieces for a three-way ligation with T4 DNA ligase. (A) 5′ Fragment: capped RNA is produced by transcription with T7 RNA polymerase. (B) Middle fragment: RNA oligonucleotide containing a modified nucleotide (black stripe) is produced by chemical synthesis. (C) 3′ Fragment: two alternative examples are presented. (C1) The 3′ fragments is produced by T7 transcription, followed by dephosphorylation (to remove the 5′ triphosphate) and phosphorylation (to add a single phosphate to the 5′ terminus). (C2) The 3′ fragment is initially produced as a longer precursor, which is subsequently cleaved and gel purified.

the procedure (substrate manipulations, RNA ligation itself and the subsequent purification steps). A practical rule of thumb used in our laboratory is to start with an at least 10-fold excess of each individual RNA piece compared to the amount of the ligated product needed in the final experiments. As described in Chapter 1, transcription by T7 phage polymerase can be used to generate large quantities of RNAs from defined DNA templates containing phage-specific promoters. The templates can be linearized plasmids or PCR products or even annealed oligonucleotides [15]. In ligation experiments, the use of PCR products is preferred because they provide an easy way to specify nucleotides at the point of junction of the two RNAs to be ligated. Here we describe conditions for generating large amounts of RNA (several nanomoles) from a single transcription reaction (Protocol 1). This reaction is suitable for RNA molecules that fit the promoter consensus of the phage polymerase, where at least the first transcribed nucleotide or, if possible, both the first and the second nucleotides should be G residues. If other than a G residue is required as the initial nucleotide, the transcript can be initially produced as a longer precursor which is subsequently cleaved at a specific site with a ribozyme (Chapter 2) or RNase H (see below). Alternatively, the transcription can be primed with a suitable dinucleotide ([1, 16, 17] and Chapter 1).

We have successfully used the buffer and reaction conditions indicated in Protocol 1 with T7 RNA polymerase, with yields up to 4–5 nmol of gel-purified RNA from a 200-μl reaction. The reaction conditions described in Protocol 1 are not recommended for the production of capped RNAs as the high concentrations of divalent cations and spermidine tend to precipitate the cap analogs. Instead, modified conditions described in Protocol 2 should be used for the transcription of capped RNAs.

Following the transcription reaction, the DNA template is degraded with DNase to ensure that the contaminating template DNA (which will have almost the same mobility in the gel as the transcript itself) will not interfere with the further steps of the ligation procedure. Subsequently, the full-length transcript RNA will be purified from prematurely terminated products by denaturing gel electrophoresis. Following electrophoresis, the bands are visualized using the "UV shadow" technique, excised, and eluted by passive diffusion (see Protocol 3).

3.4
Generating Homogeneous Acceptor 3′ Ends for Ligation

PCR-based template generation followed by transcription with phage polymerases is a simple and efficient method for generating large amounts of RNA fragments for ligation purposes. However, a problem with the phage polymerases is that they can add non-templated nucleotides to the 3′ end of the synthesized RNA molecule. In the worst cases more than 50% of the synthesized RNA molecules can contain these so-called $n+1$ and $n+2$ nucleotides [18, 19]. As the T4 DNA ligase requires an absolute match between the DNA splint and the two RNA molecules to be ligated, the non-template addition of extra nucleotides can significantly reduce the efficiency of the RNA ligation. This is not necessarily a severe problem in small-scale or initial screening experiments, in which limited amounts of the ligated products are often sufficient. However, if large amounts of the ligated products are needed, the inefficient ligation may become a major limitation. In such cases methods producing RNAs with specific 3′ terminus can often lead to a several-fold increase in the quantity of the ligated products.

A simple method for reducing non-templated nucleotide addition has been described recently [20]. In this method the very 5′ end of the downstream PCR primer used in the synthesis of the template DNA for T7 transcription is modified: instead of standard deoxyribonucleotides, it contains two 2′-O-methyl RNA residues (Fig. 3.3A). During the PCR reaction the primers are incorporated into the synthesized DNA fragments which subsequently serve as templates in transcription by T7 RNA polymerase. During transcription the modified nucleotides at the 5′ end of the template strand will lead to a significant reduction of the

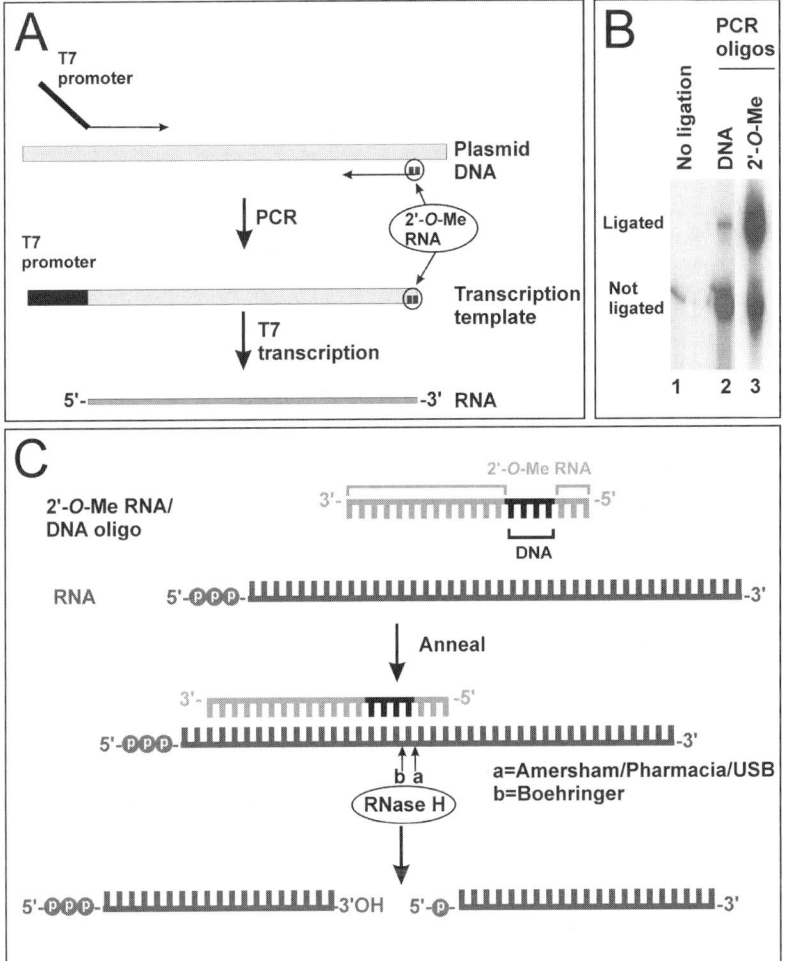

Fig. 3.3. (A) Production of T7 transcripts with homogeneous 3′ ends. The template for T7 transcription is produced by PCR. The upstream primer contains a promoter for T7 RNA polymerase, while the downstream primer contains two 2′-O-methyl RNA residues at the 5′ end of the oligonucletotide. Both primers are incorporated into the PCR product and during transcription with T7 RNA polymerase, the 2′-O-methyl RNA residues prevent the addition of non-template nucleotides to the 3′ end of the RNA molecules. (B) Comparison of the ligation efficiencies when a standard DNA oligonucleotide and a hybrid DNA/2′-O-Methyl RNA oligonucleotide (depicted in panel A) were used as the downstream primer in the PCR reaction to produce a template for T7 transcription. Lane 1: control lane containing a ^{32}P-labeled donor molecule; lane 2: ligation with an RNA fragment derived from transcription using an all-DNA template; lane 3: ligation with an RNA fragment derived from transcription using a template with two 5′ terminal 2′-O-methyl modifications. In the ligation reactions (lanes 2 and 3) the donor RNA fragments were radioactively labeled, while the acceptor RNAs were unlabeled. (C) The principle of site-specific RNase H cleavage. The cleavage sites of Amersham/Pharmacia/USB (a) or Boehringer (b) RNase H [24] are indicated.

non-templated addition of extra nucleotides to the 3′ end of transcripts [20]. We have used this method successfully, and have been able to raise the ligation efficiency from 20 to 70% with no other alterations in the procedure (Fig. 3.3B). The drawback of the hybrid DNA/2′-O-methyl RNA oligonucleotides is that they are more expensive and are readily available only from few commercial sources. We have purchased our oligonucleotides either from Dharmacon Research (www.dharmacon.com) or from Keck oligonucleotide synthesis facility at Yale University (info.med.yale.edu/wmkeck/oligos.htm).

Another way of creating homogenous 3′ ends is to synthesize the RNA as a longer precursor and cut it into the desired length using site-directed cleavage with RNase H or with a specific ribozyme. The ribozyme cleavage has been described in Chapter 2, while the site-specific RNase H cleavage, which can be used to trim both the 3′ and 5′ end of the RNA molecule, will be described in detail in the next paragraph.

3.5
Site-directed Cleavage with RNase H

At times it may not be possible to transcribe the desired RNA directly, e.g. when the first nucleotide is not a guanosine. One possible solution for this is the site-specific cleavage of a longer precursor RNA molecule using RNase H and chimeric 2′-O-methyl RNA/DNA oligonucleotides. This method may also be used to solve problems with 5′ or 3′ end heterogeneity.

RNase H recognizes and binds nucleic acids that are duplexes of DNA and RNA and cleaves the backbone of the RNA strand leaving a 5′-phosphate and a 3′-hydroxyl [21]. The site of cleavage may be specified when using oligonucleotides containing a short DNA stretch (3 or 4 nt) which is flanked by 2′-O-methyl-RNA sequences [22, 23]. We have successfully used hybrid 20mer oligonucleotides which contain three 2′-O-methyl RNA residues at the 5′ end, followed by four DNA residues and thirteen 2′-O-methyl residues (see Fig. 3.3C). An important observation to be noted [24] is that the exact position of cleavage is, for unknown reasons, dependent on the commercial source of the RNase H. The enzymes supplied by Pharmacia, Sigma and Takarashuzo were reported to cleave the phosphodiester bond which is located 3′ to the ribonucleotide base-paired with the 5′-most deoxyribonucleotide of the oligonucleotide, whereas the enzyme from Boehringer Mannheim cleaved the bond located one nucleotide upstream (5′ direction) in the RNA molecule (Fig. 3.3C). After the report was made, however, there have been mergers of the aforementioned companies, and it is unclear which enzyme sources and purification protocols are now used by the merged companies. In our studies we have used RNase H supplied by Amersham and found that it functions as the one supplied earlier by Pharmacia. If other enzyme sources are used it is advisable to map the exact cleavage site by primer extension analysis.

Protocol 4 describes a general cleavage strategy using RNase H and hybrid DNA/

2′-O-methyl RNA oligonucleotides. In the first step the RNA and the hybrid oligonucleotides are allowed to anneal, after which the appropriate buffers and enzyme are added. For efficient annealing, the RNA and oligonucleotides should be initially denatured completely by heating them to 95 °C and then, by lowering the temperature slowly, allowed to anneal. Addition of a monovalent salt, such as KCl, further enhances the annealing, but the concentration in the final cleavage reaction should not exceed 50 mM. Small amounts of EDTA are included to chelate any traces of divalent cations, e.g. Mg^{2+}, to reduce chemical degradation of the RNAs at high temperatures. The amounts of RNA and oligonucleotides can be adjusted to suit the particular experiment, but the reaction should always contain close to equimolar amounts of the oligonucleotide relative to the RNA to be cleaved. Large excess of the hybrid oligonucleotide may lead to aberrant cleavage at additional sites. We typically use 10–15% excess of the oligonucleotide to ensure efficient annealing. If the 5′ end of the fragment is to be dephosphorylated for a subsequent labeling with a radioactive phosphate, the final yield of the cleaved RNA product may be increased by carrying out the dephosphorylation prior to the gel purification.

3.6
Dephosphorylation and Phosphorylation of RNAs

The 5′-triphosphate resulting from the transcription reaction (see Fig. 3.2) must be converted to 5′-monophosphate if the RNA is to be used as a donor RNA in the ligation reactions. This is achieved by first dephosphorylating the RNA and then phosphorylating the resulting 5′-hydroxyl with unlabeled or radioactively labeled phosphate. The dephosphorylation catalyzed by the calf intestinal alkaline phosphatase (CIAP) is carried out at 50 °C (see Protocol 5). The elevated temperature is used to reduce the effect of RNA secondary structure on dephosphorylation. Following the dephosphorylation it is necessary to completely remove the CIAP, as it could seriously inhibit the further downstream steps. As very large amounts of CIAP are used, we typically remove it by proteinase K digestion. It is also possible to use other phosphatases [such as shrimp alkaline phosphatase (SAP)] which are easier o inactivate compared to CIAP. However, our experience with SAP is that at least with some substrates it does not work as efficiently as CIAP.

As mentioned previously, the donor RNA must have a 5′-phosphate for successful ligation. Dephosphorylated transcripts or chemically synthesized oligonucleotides can be phosphorylated by T4 polynucleotide kinase (PNK) in the presence of ATP (note that RNA oligonucleotides can also be phosphorylated during the chemical synthesis). T4 PNK can also be used in site-specific labeling to insert a single radioactive phosphorus at the junction between the acceptor and donor. Protocol 6 describes phosphorylation with $[\gamma$-$^{32}P]$ATP resulting in a specific activity of approximately 0.5×10^6 c.p.m./pmol. The protocol can also be used for non-radioactive phosphorylation to create the 5′ phosphate required for ligation, in which case 500 μM unlabeled ATP should be used.

3.7
RNA Ligation

Following successful production of the individual RNA pieces they will finally be joined by RNA ligation. The ligation protocol, much like the RNase H cleavage discussed previously, is divided into two parts. First, the RNA fragments are aligned together with a bridging oligonucleotide, also known as the "DNA splint" or "cDNA template". Second, the reaction mix is added and the RNA ends at the junction are joined by the T4 DNA ligase in the presence of ATP. The major consideration when planning both steps is to know the molar concentrations of each individual RNA piece and the DNA splint oligonucleotide. Furthermore, the ligation volume should be kept as small as possible, as the ligation proceeds more efficiently when the reactants are concentrated. The final volume of the reaction should therefore be kept at approximately 10–20 μl. Finally, to help the detection of the ligated products in purification gels and their subsequent quantification one should always include trace amounts of radioactively labeled RNAs in the ligation reaction. This should be done even when the aim is to produce unlabeled chimeric RNAs.

Let us consider first the annealing of the RNA fragments and the splint. Splints spanning the sequence for 20 nt on both sides of the junction align the substrates efficiently, although splints down to about 20 nt (10 on each side) can be used [11]. For efficient ligation, the three polynucleotides should be present in equal molar amounts. If one of the RNA fragments is scarce, the splint and the other fragment may be added in excess to drive the reaction. Adding donor RNA in excess can also be used to decrease the negative effect of the 3′ end heterogeneity of the acceptor RNA. However, it is important that the concentration of at least one of the RNA fragments is greater than that of the DNA splint. If large amounts of the splint are used, the individual RNA fragments may hybridize to different splint molecules and thus be sequestered from their ligation partners.

Some monovalent salt and EDTA may be added to the annealing reaction, for the same reasons as previously described for annealing prior to RNase H cleavage (see Section 3.5). Performing the annealing in a thermal cycler with a heated lid has also the added advantage that little condensation of water occurs on the lid of the tube, which is an important consideration when small volumes, which are susceptible to drying, are used.

The buffer used for the ligation reaction may be made by the researcher, as the one described in Protocol 7, or a commercial one supplied with the enzyme can be used. Macromolecular crowding agents, such as polyvinyl alcohol (PVA), polyethylene glycol (PEG) or polyvinyl pyrrolidine (PVP) should be added to increase the effective concentration of the reactants. Some thought should be given to the amount of ligase used, as it has been reported that the T4 DNA ligase does not turn over efficiently on RNA-containing duplexes [11]. Therefore, a stoichiometric amount of T4 DNA ligase should be used, with 1 Weiss unit corresponding approximately to 1 pmol.

Finally, the incubation time and temperature should be considered. Incubating the reaction at 30 °C for 4 h is a widely used approach. In our research we have also performed the incubation overnight at room temperature (around 20–25 °C) and this seems to result in somewhat higher yields, probably due to both increased incubation time and the reduced temperature which can stabilize the double-stranded structures at the junction.

3.8
Troubleshooting

A typical problem with ligations is a low yield of the final ligated product. With two-way ligation the efficiency should be at least 20%, but even nearly stoichiometric ligations are possible with high-quality RNA. With three-way ligations the efficiencies can sometimes be less than 10%, but one should be able to increase the efficiency to approximately 30% relatively easily. If the ligation yield is very low, the first thing to do is to determine which one of the RNA or DNA fragments is responsible for the low efficiency. This can be done easily by setting up small-scale test ligations which contain only about 1 pmol of each fragment. Short DNA oligonucleotides can also be used as acceptors or donors during troubleshooting instead of the actual RNA fragments.

If the problem with the ligation can be pinpointed to the acceptor RNA the most obvious question would be the quality of the 3′ end of the RNA molecule: is it homogeneous or does it contain non-templated nucleotides? Non-templated nucleotide additions can often be avoided using the techniques described in Sections 3.4 and 3.5. Another possible problem with the acceptor is the presence of a stable RNA secondary structure that could prevent the hybridization of the DNA splint oligonucleotide. The best way to resolve this is to destabilize the structure with site-specific mutations. The mutations near the junction are often easy to incorporate by means of the PCR oligonucleotides that are used in the generation of templates for T7 transcription. If the strategy does not allow mutations, an alternative is to use an additional "disrupter" oligonucleotide which binds adjacent to the splint oligonucleotide and prevents the formation of stable RNA secondary structures.

A typical problem with the donor RNA, in addition to the stable RNA secondary structure described above, is inefficient dephosphorylation (or phosphorylation) at the 5′ end of the RNA. This can be caused for example by a stable secondary structure at the 5′ end of the molecule. A larger amount of phosphatase (or kinase) may be used to overcome this problem or, alternatively, an analogous disrupter oligo-nucleotide strategy could be designed.

Apart from the problems with ligations, one can also experience difficulties at stages that are further downstream of the actual RNA ligation reaction, but which are the results of the ligation procedure. One, at least theoretical, possibility is the cleavage of the ligated RNA during incubation with cellular extract due to endogenous RNase H activity and residual amounts of the DNA splint oligonucleotide in the ligated RNA sample. We have not observed any RNase H activity resulting from a contaminating DNA splint in any of our studies. However, if this is of concern, it can be avoided by treating the ligation reactions with DNase before purification.

3.9
Protocols

Protocol 1: Transcription
5 × T7 transcription buffer: 600 mM HEPES–KOH, pH 7.5; 120 mM MgCl$_2$; 100 mM DTT; 5 mM spermidine.

T7 transcription 200 μl		Final concentration
40 μl	5 × T7 transcription buffer	1×
40 μl	25 mM each rNTP	5 mM
5 μl	40 U/μl RNase inhibitor (Promega)	1 U/μl
2–10 μl	100 U/μl T7 RNA polymerase	1–5 U/μl
40 μl	50–250 ng/μl PCR-product	10–50 ng/μl
65–73 μl	RNase-free water	

(1) Combine reaction components, add the DNA last to avoid precipitation by high concentrations of spermidine present in the transcription buffer.
(2) Incubate at 37 °C for 1.5 h, afterwards add more enzyme (0.5–1 × the original amount) and incubate for an additional 1.5 h. Typically a white pyrophosphate precipitate will start to accumulate at the later stages of the transcription reaction.
(3) Add 1 U of RNase-free DNase, such as RQ1 DNase (Promega) for each microgram of template DNA and continue the incubation for an additional 15–30 min.
(4) Extract RNA once by phenol:chloroform:isoamyl alcohol (25:24:1).
(5) Precipitate RNA by adding 20 μl 3 M NaOAc (pH 5.2) or NaCl and 2.5 volumes of ethanol.
(6) Dissolve the pellet in 5 μl water and gel-purify the RNA by denaturing polyacrylamide gel electrophoresis as described in Protocol 3.

Protocol 2: Transcription of capped RNAs
5 × transcription buffer for capped RNA transcription (Promega): 200 mM Tris-HCl, pH 7.9; 30 mM MgCl$_2$; 10 mM spermidine; 50 mM NaCl.

T7 transcription 200 μl		Final concentration
40 μl	5 × transcription buffer	1×
2 μl	100 mM ATP	1 mM
2 μl	100 mM CTP	1 mM
2 μl	100 mM UTP	1 mM
1 μl	100 mM GTP	0.5 mM
40 μl	10 mM G(5′)ppp(5′)G	2 mM[1]

40 µl	50–250 ng/µl PCR product	10–50 ng/µl
5 µl	40 U/µl RNase inhibitor (Promega)	1 U/µl
2–10 µl	100 U/µl T7 RNA polymerase	1–5 U/µl
58–66 µl	RNase-free water	

[1] Cap analogs should be at least in 4-fold molar excess relative to GTP to ensure efficient initiation with the cap analog. Similarly, 5'-hydroxyl- or 5'-monophosphate-containing RNAs can be produced by priming the transcription reaction with guanosine or GMP, respectively, under reaction conditions comparable to those used with the cap analog (for details, see Chapter 1).

Carry out the reaction as described in Protocol 1 and gel-purify according to Protocol 3.

Protocol 3: Purification of RNA by denaturing gel electrophoresis
Gel loading buffer: 0.01% bromophenol blue; 0.006% xylene cyanol in 7.5 M urea/ 1 × TBE.
RNA elution buffer: 50 mM Tris–HCl, pH 7.5; 10 mM EDTA; 0.1% SDS; 0.3 M NaCl.

Before electrophoresis, the reaction mixture should be extracted once with an equal volume of phenol:chloroform:isoamyl alcohol (25:24:1) followed by ethanol precipitation, as proteins in the sample may cause smearing of the bands and retain some of the RNA in the wells. With large amounts (several nanomoles) of RNA, the pellet should be dissolved in a small volume of water (5–10 µl) before applying the gel loading buffer in order to minimize loss.

Prepare the polyacrylamide gel (19:1 acrylamide:N,N'-methylene bisacrylamide) in 7.5 M urea/1 × TBE. The percentage of acrylamide should be adjusted to the size of RNA fragment to be purified. The thickness of the gel should be adjusted according to the amount of RNA to be loaded. As the transcription reactions usually contain several nanomoles of RNA, a relatively thick gel (approximately 1 mm) should be used to avoid smearing of the bands. Additionally, RNA samples should be distributed between several wells, even though this may reduce the RNA yields after the elution. In later steps, when dealing with smaller amounts of RNA, a thinner gel (0.3–0.5 mm) may be used to increase the elution yields.

(1) Pre-run the denaturing gel at least 20–30 min at 60 W.
(2) Dissolve the RNA pellet in a small amount of water (5–10 µl) and 1 volume of gel loading buffer. The final urea concentration should be at least 3.5 M but loading buffers containing up to 7.5 M urea can be used.
(3) Heat the RNA samples at 95 °C for 3–5 min. Immediately put the samples on ice. Centrifuge the samples if there is water condensation on the lid. This is especially important if a high concentration urea loading buffer is used, as otherwise the loading buffer (and the sample) can crystallize in the pipette tip during loading.
(4) Load the samples and run the gel at approximately 60 W for at least 30–60 min, or longer if necessary.

(5) Separate the gel plates. Place the gel between two Saran wrap (or alike) sheets.
(6) Visualize the RNA bands using UV shadowing. The gel (between two sheets of Saran wrap) is placed on a fluorescent TLC plate or intensifying screen (a sheet of white paper will do as well) and illuminated briefly with UV light (254 nm). RNA bands are visualized as dark bands on a fluorescent background. Use a pen or marker to indicate the location of each band. The exposure time should be minimized to avoid damage to RNA by the UV light.
(7) Cut out the bands and add approximately 5 volumes of elution buffer to the excised bands. Carry out the elution overnight at room temperature using a tube rotator, "rocking table" or a similar device to provide for gentle shaking during elution.
(8) Collect the supernatant. For increased yield, replace the buffer and continue elution for a further 4–6 h. Extract the eluates once with phenol:chloroform: isoamyl alcohol and once with chloroform:isoamyl alcohol. This helps to remove any impurities present in the gel, as well as any remaining gel fragments. The RNA is concentrated by ethanol precipitation. Note that there is no need to add salt as the elution buffer already contains a sufficient amount of salt for the precipitation. Wash the pellet at least 3 times with 70% ethanol to remove any traces of SDS and dissolve it in RNase-free water. After purification the concentration of the transcript should be measured by UV spectrometry (see Appendix).
(9) If purifying radiolabeled RNAs, the bands can be visualized by autoradiography instead of UV shadowing. In this case the gel should be left on one of the glass plates and covered with Saran wrap. Pieces of fluorescent tape (such as Rad-Tape from Diversified Biotech) serving as alignment marks are attached to the wrapped gel and illuminated briefly with light. In the darkroom, place an X-ray film on top of the gel. Expose the film (depending on the activity of the RNA to be purified, this can be anything from 10 s to several hours). Use the markings from the fluorescent tape to align the gel and the film, mark the bands, cut them out and elute as described in Step 7.

Protocol 4: Site-directed cleavage with RNase H
5 × RNase H buffer (Amersham): 100 mM Tris–HCl, pH 7.5; 100 mM KCl; 50 mM $MgCl_2$; 0.5 mM EDTA; 0.5 mM DTT.

Annealing reaction 13 µl		Final concentration
5 µl	400 µM RNA (2 nmol)	154 µM
5.5 µl	400 µM 2′-O-methyl RNA/DNA oligo (2.2 nmol)	169 µM
1.3 µl	1 M KCl	100 mM
1.2 µl	1 mM EDTA	92 µM

In a thermal cycler, run the following program: 95 °C 5 min, 85 °C 10 s, decrease the temperature with a slope of −0.1 °C/s until the temperature of 35 °C is reached. Alternatively, the annealing may be carried out in a heating block. In this case place the samples first in a hot heating block (95 °C) for 3–5 min, then remove the block from the heating unit and allow to cool slowly to room temperature. In each case, the tubes should be checked for any condensed water on the lid after the annealing and centrifuged if necessary.

Cleavage reaction 50 µl		Final concentration
13 µl	annealing mix	
10 µl	5 × RNase H buffer	1×
1.5 µl	40 U/µl RNase inhibitor (Promega)	~1 U/µl
0.5 µl	100 mM DTT	1 mM
15 µl	5 U/µl RNase H (Amersham)	1.5 U/µl
10 µl	RNase-free water	

(1) Incubate at 37 °C for 3–4 h, extract once with phenol:choloroform:isoamyl alcohol and ethanol-precipitate. If dephosphorylation is to be performed before electrophoresis, carry out chloroform:isoamyl alcohol extraction after the phenol extraction to remove any traces of the phenol.
(2) Separate the cleavage products in a denaturing polyacrylamide gel and visualize the bands by UV shadowing. Use approximately 200 pmol of uncut RNA as a control to distinguish the full-length RNA. Elute and purify as described in Protocol 3.

Protocol 5: Dephosphorylation
10 × CIAP buffer (Finnzymes): 100 mM Tris–HCl, pH 7.9; 100 mM MgCl$_2$; 10 mM DTT; 500 mM NaCl.

Dephosphorylation reaction 100 µl		Final concentration
10 µl	10 × CIAP buffer	1×
10 µl	200 µM RNA (2 nmol)	20 µM
2.5 µl	40 U/µl RNase inhibitor (Promega)	1 U/µl
10 µl	10 U/µl CIAP (Finnzymes)	1 U/µl
68 µl	RNase-free water	

(1) Incubate at 50 °C for 1 h.
(2) Add at least 1 volume of an appropriate 1 × buffer for proteinase K (the RNA elution buffer described earlier also works very well) and 100 µg of proteinase K to the reaction and continue incubation at 50 °C for an additional hour.
(3) Extract RNA with phenol:chloroform:isoamyl alcohol followed by extraction with chloroform: isoamyl alcohol, and ethanol precipitation. If the RNA fragment has been cleaved with RNase H prior to the dephosphorylation and has not been purified by gel electrophoresis, this should be performed at this stage.

Protocol 6: Site-specific labeling with radioactive phosphorus at the donor 5' end
10 × T4 polynucleotide kinase buffer: 700 mM Tris–HCl, pH 7.5; 100 mM MgCl$_2$; 50 mM DTT.

Phosphorylation reaction 40 µl		Final concentration
4 µl	10 × T4 polynucleotide kinase buffer	1×
5 µl	20 µM RNA (100 pmol)	2.5 µM
6 µl	200 µM ATP[1]	30 µM
1 µl	40 U/µl RNase inhibitor (Promega)	1 U/µl
20 µl	[γ-^{32}P]ATP (10 µCi/µl, 6000 Ci/mmol)[1]	1.7 µM
1 µl	10 U/µl T4 polynucleotide kinase	0.25 U/µl
3 µl	RNase-free water	

[1] The ratio of radioactively labeled versus cold ATP can be adjusted depending on the particular experiments. However, the total concentration of ATP in the reaction should be kept at least 2× higher than the concentration of the RNA. If large amounts of highly radioactive RNA are needed, more crude, but concentrated [γ-^{32}P]ATP can be used [such as NEG 035C (New England Nuclear) or PB15068 (Amersham) – both having activities of about 150 µCi/µl]. If RNA is not to be radioactively labeled, the ^{32}P-labeled ATP can be replaced with 500 µM unlabeled ATP.

(1) Incubate at 37 °C for 30–60 min. Subsequently raise the concentration of cold ATP to 45 µM and continue incubation for 15 min.
(2) Extract once with phenol:chloroform:isoamyl alcohol and once with chloroform, and ethanol-precipitate.
(3) Dissolve the pellet in a few microliters of water. Alternatively, the dry pellet can be directly used in the ligation if dividing the RNA donor into smaller aliquots is not needed.

Protocol 7: Ligation
10 × ligation buffer: 500 mM Tris–HCl, pH 7.5; 100 mM MgCl$_2$; 200 mM DTT; 10 mM ATP.

Annealing mix for a two-way ligation, 5 µl[1]		Final concentration
1.2 µl	50 µM donor RNA (50 pmol)	10 µM[2]
1.4 µl	50 µM acceptor RNA (70 pmol)	14 µM[2]
1.2 µl	50 µM splint DNA oligo (60 pmol)	12 µM
0.5 µl	1 mM EDTA	0.1 mM
0.5 µl	500 mM KCl	50 mM

[1] Even if the aim is to produce unlabeled product it is a good idea to include trace amounts of a radioactively labeled piece to the ligation reaction corresponding to one of the RNA fragments. This provides an easy way to quantify the total yield of the ligation reaction. Simply compare the amount of radioactivity included in the reaction with the amount of radioactivity after the gel purification (by measuring small samples with liquid scintillation counting) and multiply this percentage with the total amount of this particular RNA fragment in the reaction.
[2] We have successfully used RNA concentrations ranging from 100 nM to 20 µM for each individual RNA segment in the ligation reactions.

(1) In a thermal cycler, run the following program: 95 °C 5 min, 85 °C 10 s, a slope of −0.1 °C/s until the temperature of 35 °C is reached.
(2) Check for condensation at the lid and centrifuge the tube(s) if necessary. Combine the ligation reaction as follows:

Ligation reaction 10 μl		Final concentration
5 μl	annealing reaction	
1 μl	10 × ligation buffer	1×
1.5 μl	13% polyvinyl alcohol (PVA)	~2%
0.5 μl	40 U/μl RNase inhibitor	2 U/μl
2 μl	30 Weiss U/μl T4 DNA ligase (Fermentas)	6 U/μl

(3) Incubate at room temperature overnight or at 30 °C for at least 4 h.
(4) Extract once with phenol:chloroform:isoamyl alcohol and ethanol-precipitate. Separate the ligated products from the unligated donor and acceptor in a denaturing polyacrylamide gel. Use either radioactively labeled donor, acceptor, or a full-length transcript as a control. Due to the small amount of RNA present, the bands must be visualized using autoradiography. Elute and purify the excised samples as described in Protocol 3. As the amount of the RNA can be relatively small after the ligation, care must be taken to maximize the yields of the gel-purified product. At this stage we have used siliconized microcentrifuge tubes in the elution step and carried out the RNA precipitation in the presence of carrier (10 μg of glycogen).

Acknowledgments

We thank Heli Pessa, Minni Laurila and Xiaojuan Meng for critical reading of the manuscript. This work was supported by Academy of Finland (grant 50527 to M. J. F.).

References

1. M. J. MOORE, P. A. SHARP, *Science* 1992, *256*, 992–997.
2. A. M. MACMILLAN, C. C. QUERY, C. R. ALLERSON, S. CHEN, G. L. VERDINE, P. A. SHARP, *Genes Dev.* 1994, *8*, 3008–3020.
3. C. C. QUERY, M. J. MOORE, P. A. SHARP, *Genes Dev.* 1994, *8*, 587–597.
4. O. GOZANI, R. FELD, R. REED, *Genes Dev.* 1996, *10*, 233–243.
5. Z. PASMAN, M. A. GARCIA-BLANCO, *Nucleic Acids Res.* 1996, *24*, 1638–1645.
6. H. LE HIR, E. IZAURRALDE, L. E. MAQUAT, M. J. MOORE, *EMBO J.* 2000, *19*, 6860–6869.
7. H. LE HIR, M. J. MOORE, L. E. MAQUAT, *Genes Dev.* 2000, *14*, 1098–1108.
8. P. A. MARONEY, C. M. ROMFO, T. W. NILSEN, *Mol. Cell* 2000, *6*, 317–328.
9. M. J. FRILANDER, J. A. STEITZ, *Mol. Cell* 2001, *7*, 217–226.
10. T. S. MCCONNELL, J. A. STEITZ, *EMBO J.* 2001, *20*, 3577–3586.
11. M. J. MOORE, C. C. QUERY, in: *RNA: Protein Interactions. A Practical*

Approach, C. W. J. SMITH (ed.), Oxford University Press, Oxford, **1998**, pp. 75–108.
12 R. REED, M. D. CHIARA, *Methods Enzymol.* **1999**, *18*, 3–12.
13 Y.-T. YU, *Methods Enzymol.* **1999**, *18*, 13–21.
14 K. KLEPPE, J. H. VAN DE SANDE, H. G. KHORANA, *Proc. Natl. Acad. Sci. USA* **1970**, *67*, 68–73.
15 J. F. MILLIGAN, O. C. UHLENBECK, *Methods Enzymol.* **1989**, *180*, 51–62.
16 J. R. WYATT, E. J. SONTHEIMER, J. A. STEITZ, *Genes Dev.* **1992**, *6*, 2542–2553.
17 E. J. SONTHEIMER, J. A. STEITZ, *Science* **1993**, *262*, 1989–1996.
18 J. F. MILLIGAN, D. R. GROEBE, G. W. WITHERELL, O. C. UHLENBECK, *Nucleic Acids Res.* **1987**, *15*, 8783–8798.
19 G. KRUPP, *Gene* **1988**, *72*, 75–89.
20 C. KAO, M. ZHENG, S. RÜDISSER, *RNA* **1999**, *5*, 1268–1272.
21 I. BERKOWER, J. LEIS, J. HURWITZ, *J. Biol. Chem.* **1973**, *248*, 5914–5921.
22 H. INOUE, Y. HAYASE, S. IWAI, E. OHTSUKA, *FEBS Lett.* **1987**, *215*, 327–330.
23 J. LAPHAM, D. M. CROTHERS, *RNA* **1996**, *2*, 289–296.
24 J. LAPHAM, Y.-T. YU, M.-D. SHU, J. A. STEITZ, D. M. CROTHERS, *RNA* **1997**, *3*, 950–951.

/ # 4
T4 RNA Ligase

Tina Persson, Dagmar K. Willkomm and Roland K. Hartmann

4.1
Introduction

The growing interest in RNA–RNA and RNA–protein interactions has led to an increased demand for the production of RNA molecules with chain lengths usually in the range of 75–500 nt. In particular, chemogenetics (exchange of single functional groups) has become more and more popular in studies of RNA function. The most efficient RNA synthesis method used today is *in vitro* run-off transcription by T7 RNA polymerase (see Chapter 1). Although this technique produces large amounts of RNA at relatively low cost with reasonable effort, synthesis of RNA by *in vitro* transcription suffers from essentially two drawbacks: (1) potential 5′ and/or 3′ end heterogeneity of the product, and (2) the inability to introduce internal modifications at specific sites, apart from the limited scope of modifications that can be introduced by the polymerase. End heterogeneity can be overcome by several approaches described in detail in Chapters 2 and 3, but for the site-specific incorporation of nucleotide modifications, chemical RNA synthesis is in most instances inevitable [1].

Development of new commercially available phosphoramidites of natural and unnatural nucleosides, improvement of 2′-OH protecting groups, and the use of more efficient activators are innovations that have paved the way for chemical synthesis of RNA molecules in better and more reproducible yields [2–5]; see also Chapters 7 and 8). Despite these advances, substantial amounts of longer RNA molecules are still very difficult and expensive to synthesize. Current techniques permit efficient routine chemical synthesis of RNA molecules of up to 50 nt. As a consequence, longer RNA molecules with site-specific modifications are usually prepared by chemical synthesis of an RNA oligonucleotide (less than 50 nt) carrying the modification(s), which is then ligated to one or two other RNA molecules preferably produced by *in vitro* transcription. For such applications, the "DNA splint" ligation technique employing T4 DNA ligase (described in Chapter 3) is widely used. We will focus here on an alternative method based on T4 RNA ligase.

T4 RNA ligase (EC 6.5.1.3; also named Rnl1 or RnlA RNA ligase 1), a 347-aa

Handbook of RNA Biochemistry. Edited by R. K. Hartmann, A. Bindereif, A. Schön, E. Westhof
Copyright © 2005 WILEY-VCH Verlag GmbH & Co. KGaA, Weinheim
ISBN: 3-527-30826-1

polypeptide encoded by gene 63 of bacteriophage T4, belongs to a family of oligonucleotide end-joining enzymes involved in RNA repair, splicing and editing pathways [6–8]. The enzyme, introduced into molecular biology laboratories about 30 years ago [9–11], catalyzes formation of phosphodiester bonds between 5′-phosphate and 3′-hydroxyl ends of preferentially single-stranded RNA (ssRNA) and, less efficiently, ssDNA. The oligo(ribo)nucleotide carrying the terminal 3′-hydroxyl group is termed acceptor substrate, and the one providing the terminal 5′-monophosphate is described as the donor substrate (Fig. 4.1A). In intramolecular circularization reactions, both end groups (3′-OH, 5′-phosphate) are located on the same oligo(ribo)nucleotide molecule. The biological role of T4 RNA ligase seems to have its seeds in the intricate antagonisms of T4 phage and *Escherichia coli* host strains: T4 infection was shown to induce activation of a nuclease that cleaves bacterial tRNA^{Lys} in the anticodon loop. T4 RNA ligase in concert with T4 polynucleotide kinase catalyze the repair of the damaged tRNA^{Lys} [12, 13].

4.2
Mechanism and Substrate Specificity

4.2.1
Reaction Mechanism

The reaction catalyzed by T4 RNA ligase consists of three distinct and reversible steps (Fig. 4.1A [14]). In the first step, the ligase reacts with ATP to form an adenylated enzyme intermediate. In the second step, a donor substrate with a 5′ terminal

Fig. 4.1. (A) Mechanism of the ligation reaction catalyzed by T4 RNA ligase. In the ligation reaction, an acceptor substrate with a 3′-hydroxyl group reacts with a donor substrate carrying a 5′-phosphate group, resulting in a 3′–5′-phosphodiester bond. A reaction cycle consists of three distinct and reversible steps [14]: initially, the ligase reacts with ATP to form an adenylated enzyme intermediate (at lysine 99; [44, 45]), with concomitant release of pyrophosphate (1.). Then a donor substrate with a 5′-terminal monophosphate is bound by the enzyme and converted to the adenylated donor A(5′)pp(5′)Np(Np)$_n$, an intermediate in which the terminal adenosine moiety is attached via a 5′,5′ diphosphate bridge to the donor RNA (2.). In a final transesterification step, the phosphodiester bond connecting the two 5′,5′-linked phosphates is cleaved and a phosphodiester bond is formed between the donor and acceptor substrate, with concomitant release of AMP (3.). (B) End groups of donor and acceptor substrates that prevent formation of alternative products (intramolecular donor or acceptor cyclization, or formation of donor or acceptor tandems) in intermolecular ligation reactions of two oligo(ribo)nucleotides. In the case of the acceptor substrate, a 5′-OH terminus excludes that this substrate can act as a donor. A 3′-OH terminus is mandatory for acceptor function, whereas blockage of this end group in the case of donor substrates precludes that they can function as an acceptor; 3′ end blockage is achieved by introducing a terminal 2′,3′-cyclic phosphate (a), 3′-phosphate (b), a dideoxy residue (c; [25]), a 3′-inverted deoxythymidine (d; modification available from Dharmacon) or a periodate-oxidized 3′-terminal ribose (e; for details, see Chapter 6).

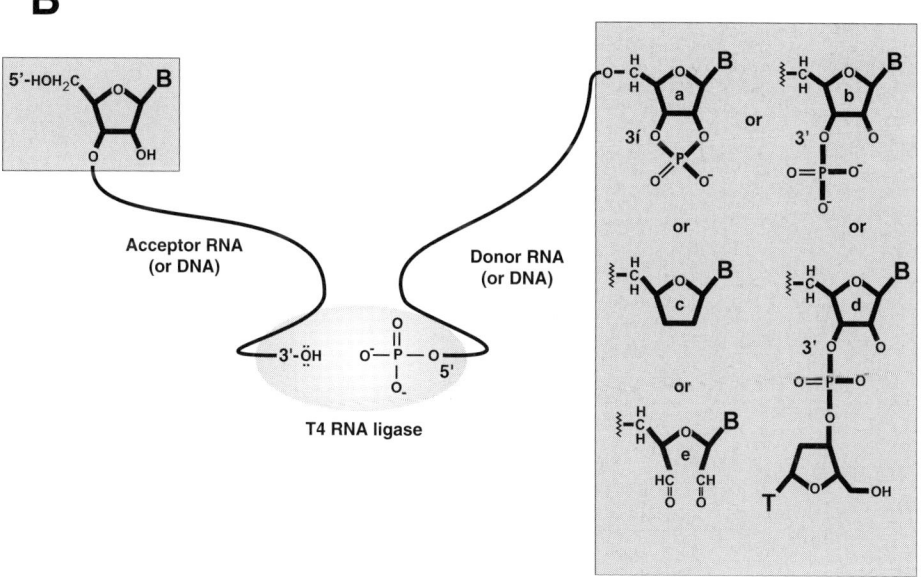

monophosphate is bound by the enzyme and the enzyme-linked 5'-AMP moiety is now transferred to the 5'-phosphate of the donor substrate, yielding the adenylated donor product in which the terminal adenosine moiety is attached via a 5',5'-diphosphate bridge [A(5')pp(5')Np(Np)$_n$]. In a final transesterification step, the phosphodiester bond connecting the two 5',5'-linked phosphates is cleaved and a phosphodiester bond is formed between the donor and acceptor substrate, with concomitant release of AMP.

4.2.2
Early Studies

Toward understanding the substrate specificity of T4 RNA ligase, it is instructive to briefly review results of early studies. The enzyme was first described by its ability to circularize 5'-^{32}P-labeled tRNA and polyhomoribonucleotides, such as poly(A), in a reaction requiring ATP and Mg^{2+} [11]. Relative to poly(A), the reaction occurred about 4-fold less efficient with poly(I), around 100-fold less efficient with poly(C) and poly(U), and at least 800-fold less efficient with poly(dA). Circularization of poly(A) molecules with an average chain length between 34 and 40 nt was about twice as efficient as for those 70–100 nt in length and no intermolecular ligation products were observed in this early study [11]. Subsequently, the shortest circularizing polyadenylate was shown to be (pA)$_8$, the optimal chain length for this reaction being 10–30 [15]. Cyclization was generally found to be the preferred reaction by orders of magnitude over intermolecular joining [16, 17], with four exceptions representing conditions that favor intermolecular ligation: (1) when the donor is too short to cyclize, (2) when a DNA donor is combined with a 5'-dephosphorylated RNA acceptor, (3) when the acceptor carries 5'- and 3'-hydroxyls and the donor a 5'-phosphate and a blocked 3'-terminus (Fig. 4.1B [18]), and (4) when the donor 5'-phosphate of one nucleic acid molecule is juxtaposed to the 3'-hydroxyl of a second acceptor nucleic acid molecule by base-pairing interactions, resulting in a *quasi*-intramolecular reaction (see below).

For ligation of DNA, studies with 5'-^{32}P-labeled oligodeoxythymidylates of various length ([5'-^{32}P]dT$_n$) revealed that ligase-catalyzed cyclization requires a minimal chain length of 6 dT residues and the best efficiency was obtained with chains of 20 [17]. Since DNA is a less efficient acceptor than RNA, adenylated A(5')pp(5')dT$_n$ intermediate accumulated to some extent – an observation not (or less) observed with RNA substrates. Interestingly, addition of the ribotrinucleotide ApApA, beyond serving as acceptor substrate, also stimulated cyclization of [5'-^{32}P]dT$_n$, suggesting that acceptors not only function as substrates for ligation, but also as cofactors for adenylation of the donor. ApA instead of ApApA neither stimulated donor adenylation nor was it joined [16, 17]. Likewise, ApA, IpI or UpU were found to be inactive as acceptors in the overall reaction with pAp as donor [18].

Analysis of minimal donor substrates of the pNp type (nucleoside-3',5'-bisphosphate) in the adenylation partial reaction revealed the highest efficiency of A(5')pp(5')Np formation for pCp, whereas the reaction with pUp and pAp was 3-fold and with pGp 10-fold less efficient [14]. A similar hierarchy was seen in the overall reaction [18, 19]. On the acceptor side, ApApA was a much better substrate than UpUpU. Ligation of the "poor" pGp donor to the "good" ApApA acceptor could be stimulated to proceed to almost completion by increasing the pGp concentration from 1 to 10 mM [14]. Furthermore, this study led to the conclusion that the enzyme exhibits more specificity in the donor adenylation reaction than in the subsequent joining of adenylated donor to acceptor. Formation of the UpUpUpGp product was much more efficient in the reaction of UpUpU with the adenylated donor A(5′)pp(5′)Gp than in the overall reaction starting from UpUpU, pGp and the ATP cofactor. Thus, pre-adenylation of the donor substrate may serve as a strategy to improve product yields when dealing with "poor" substrates [14]. However, this principle may not be generalized since results of two other studies investigating the synthesis of UpUpUpAp from UpUpU and pAp suggested rate limitation at the level of transfer of adenylated donor A(5′)pp(5′)Ap to acceptor [18, 19].

4.2.3
Substrate Specificity and Reaction Conditions

Substrate specificity of T4 RNA ligase can be summarized as follows. The minimal donor substrate is a nucleoside-3′,5′-bisphosphate, with efficiency decreasing in the order pCp > pUp ≈ pAp > pGp. Isocytidine-3′,5′-bisphosphate was reactive as well, indicating that also modified bases are tolerated [20]. Except for p(dCp), the deoxyribonucleoside-3′,5′-bisphosphates were found to be poorer donor substrates than the corresponding pNp ribonucleosides in the overall reaction [18]. 5′-AMP (pA) is not a donor and also nucleoside-2′,5′-bisphosphates are neither donor substrates nor effective inhibitors. Thus, in the donor substrate the enzyme specifically recognizes the 5′-terminal phosphate and ribonucleoside plus the next 3′-linked phosphate; the chain length of the donor exerts only marginal effects on reaction extent [18].

The smallest reactive acceptors are trinucleoside diphosphates (NpNpN) with a 3′-terminal hydroxyl [16]. Recently, also dinucleoside polyphosphates with at least four bridging phosphates, such as Gp_4G, were identified as acceptor substrates [21]. The 3′-terminal ribose moiety is important for acceptor recognition, and a 3′-terminal adenosine is preferred over cytidine and guanosine, showing intermediate reactivity, while a uridine residue is a relatively poor substrate (A > C ≥ G > U [18, 19]). However, intermolecular ligation yields are not simply dependent on the identity of the 3′-terminal base. For example, lower yields were observed with trimeric NpNpN acceptors containing a U residue at any of the three positions [18]. In another study, comparing trimeric acceptors equal in base composition, two consecutive purines enhanced ligation yields with pCp as donor (e.g. GpApU > UpApG > ApUpG [19]).

Regarding DNA oligonucleotides as substrates for T4 RNA ligase, combined appreciation of several studies [17, 22, 23] revealed the following rules of thumb: DNAs are less efficient substrates than RNAs, but discrimination against DNA occurs mainly at the acceptor substrate level.

Despite these substrate preferences, several strategies have been successfully used to optimize reaction conditions for poorer substrates. Reductions in Mg^{2+} concentration or addition of dimethylsulfoxide (DMSO) improved product (UpUpUpAp) yields for a UpUpU acceptor and a pAp donor [18]. Stimulation by DMSO, however, was not observed in ligation of UpUpU with pCp [19], suggesting that the effect of such additives may not be generalized. Also, joining of UpUpU to pUpUpUpCp at high enzyme concentration (350 U/ml) was stimulated from 6 to 60% product yield when ligation mixtures were incubated for 18 h at 15 °C instead of 1 h at 37 °C [19]. Ligation efficiencies with DNA were increased to some extent by elevated enzyme concentration, (partial) replacement of Mg^{2+} with Mn^{2+}, reduction of incubation temperature to around 17 °C, variation of donor:substrate:ATP ratio as well as their individual concentrations or low ATP concentration plus an ATP regeneration system [22, 23].

In summary, the substrate specificity of T4 RNA ligase is rather broad, permitting to ligate essentially any RNA or DNA sequence, mostly with satisfactory efficiency. Best reaction yields are commonly obtained at pH 7.2–7.8, 10–20 mM Mg^{2+}, 10–20% DMSO and often at temperatures as low as 5 °C, with substantial activity even exerted at 0 °C [24]. Several of the abovementioned parameters may be varied to optimize ligation yields. Furthermore, additives beyond DMSO, such as PEG 8000 or hexamine cobalt chloride, have been shown to improve product formation [25]. Another aspect is to prevent formation of unwanted byproducts. Thus, for intermolecular joining of two oligo(ribo)nucleotides, one should generally bear the following aspects in mind, illustrated in Fig. 4.1(B): the acceptor substrate ought to carry a 5′-hydroxyl terminus to prevent acceptor cyclization or joining of two acceptor substrates; likewise, the donor substrate should be blocked at its 3′-terminus by a 2′,3′-cyclic phosphate, a 3′-phosphate, a dideoxy residue [25], a 3′-inverted deoxythymidine or a periodate-oxidized terminal ribose (see Chapter 6) to avoid donor cyclization or formation of donor tandems.

4.3
Applications of T4 RNA Ligase

4.3.1
End-labeling

A common application of T4 RNA ligase is 3′-end-labeling with [5′-^{32}P]pCp. A ribocytidine dinucleotide bearing a 5′-phosphate and a 3′-terminal non-radioactive label, such as a fluorescein group (pCpC$_{3'\text{-fluorescein}}$), was also shown to be efficiently attached to the 3′ end of RNA substrates by the enzyme [26]. In a related application, a 5′-phosphorylated pCpC dinucleotide with a 3′-terminal polyethylene glycol linker including an internal photocleavage site and a terminal primary aliphatic amino group for coupling purposes was attached to RNA acceptor 3′ ends [27]. Kinoshita et al. [28] have made use of the fact that the ligase attaches an

AMP residue via a 5′,5′-pyrophosphate linkage to the 5′-phosphate end of the donor RNA or DNA, representing the intermediate donor activation step on the reaction pathway (Fig. 4.1A). In the absence of an acceptor nucleic acid, this intermediate was shown to be stable, and the authors demonstrated that fluorescent 2-aminopurine riboside triphosphate or 3′-amino ATP (for subsequent biotinylation) could replace the normal ATP in this 5′-end-labeling reaction catalyzed by T4 RNA ligase.

4.3.2
Circularization

The enzyme has further been exploited for intramolecular circularization of linear RNA to produce an authentic infectious circular RNA (371 nt) of the citrus exocortis viroid strain A (CEV-A [29]). One site of ligation was located within a dinucleotide internal loop of the viroid's rod-like structure ([29, 30] and Mfold secondary structure prediction of CEV-A), juxtaposing the reacting end groups (Fig. 4.2A). T4 RNA ligase was further employed to produce circular versions of hammerhead ribozyme strands as small as 15 nt, which exhibited increased activity, a reduced requirement for divalent metal ions, as well as increased resistance against nucleolytic degradation [31]. Interestingly, efficient T4 RNA ligase-catalyzed circularization of such hammerhead ribozyme oligoribonucleotides was achieved after their internal 7–8 nt had been annealed to a complementary DNA oligonucleotide, either linear or presented within a DNA hairpin loop (Fig. 4.2B and C). This setup favored circularization over formation of linear dimers, and the short central RNA–DNA duplex constrained the overall flexibility of the RNA oligonucleotide, while simultaneously juxtaposing the single-stranded 5′ and 3′ ends to be ligated.

4.3.3
Intermolecular Ligation of Polynucleotides

Ligation of RNA oligonucleotides to the 5′-end of mRNAs or other RNAs is used in so-called 5′-RACE (rapid amplification of cDNA ends) strategies to map RNA 5′ ends [32, 33]. Likewise, ligation of DNA or RNA oligonucleotides to RNA 3′ ends is the initial step before reverse transcription and PCR in approaches to map RNA 3′ ends (3′-RACE) or to determine the length of poly(A) tails and to identify polyadenylation sites [32, 33].

Nishigaki et al. [34] utilized T4 RNA ligase to tie two DNA single strands together (the 3′-terminal nucleotide of the acceptor oligonucleotide was a riboC in these experiments to increase ligation efficiency). To bring the reacting ends in proximity to each other, they equipped the two oligonucleotides with 12-nt long complementary sequences at one terminus, such that they formed "Y"-like hybrid structures, with the blunt-ended helix representing the stem and the two single-stranded arms presenting the donor and acceptor groups at their tips (Fig. 4.2D). This setup converted the intermolecular into a *quasi*-intramolecular reaction, and

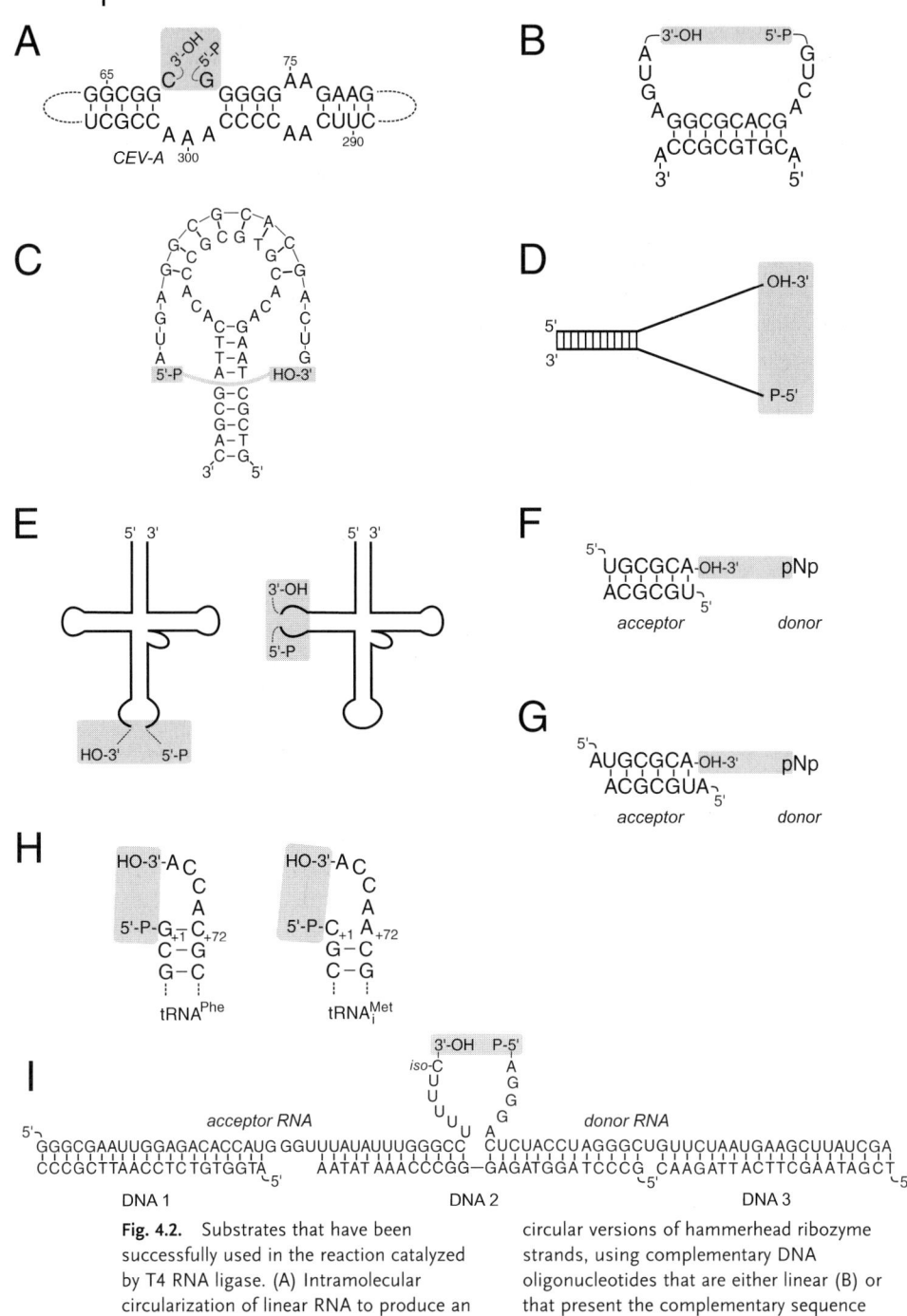

Fig. 4.2. Substrates that have been successfully used in the reaction catalyzed by T4 RNA ligase. (A) Intramolecular circularization of linear RNA to produce an authentic infectious circular RNA (371 nt) of CEV-A [29]. (B and C) Setups to produce circular versions of hammerhead ribozyme strands, using complementary DNA oligonucleotides that are either linear (B) or that present the complementary sequence within a hairpin loop (C) [31]. (D) "Y"-shape design for intermolecular ligation [34]. (E)

the blunt-ended stem of the "Y" ensured that only the unpaired end of each oligonucleotide reacted. Even dangling "Y" arms of up to ca. 50 nt each still gave ligation yields of around 20%.

Tessier et al. [25] have optimized the T4 RNA ligase reaction for the joining of pure DNA oligonucleotides, in this case a 25mer and a 23mer. Ligation yields of more than 50% were achieved by including PEG 8000 and hexamine cobalt chloride in the reaction. Oligonucleotide joining was favored over accumulation of the adenylated donor intermediate by restricting the ATP cofactor concentration to 20 µM. In this setup, the acceptor carried 5'- and 3'-OH end groups, whereas the donor oligonucleotide carried the 5'-phosphoryl group, but its 3' end was blocked via a single dideoxy analog (Fig. 4.1B), added by terminal transferase, to avoid ligation of two donor oligonucleotides.

4.4
T4 RNA Ligation of Large RNA Molecules

For the ligation of larger RNAs, several aspects should be kept in mind.

1. Proximity of Ends: To increase the probability for an enzyme molecule to simultaneously bind both reacting end groups, single-stranded RNA acceptor and donor ought to be brought in close proximity to each other. Generally, this is accomplished by letting the ends protrude from a helical region, resulting in a hairpin loop as the reaction product. Such a design converts intermolecular into *quasi-intramolecular* reactions. The importance of the structural context of the ligation site is easily illustrated for tRNA molecules. The secondary structure of tRNA is built from three stem–loop structures and one stem, which together form what is known as a cloverleaf structure (Fig. 4.3). Based on the enzyme's preference for single-stranded RNA termini, the three loops are expected to be favorable ligation areas in a tRNA molecule. Indeed, corresponding ligation strategies have been established for the anticodon loop (Fig. 4.2E and 4.3A [35–37]) and the D loop (Fig. 4.2E and 4.3B [38]), while so far T loop ligation in the context of a full-length tRNA structure has not been described. The size of the product hairpin loop will

---

Documented strategies used for the ligation of broken tRNA structures. (F and G) Double-stranded acceptor substrates [39] with blunt ends (F) or with a single nucleotide 5' overhang (G). (H) Donor termini in the context of tRNA structures. The 5'-terminal phosphate of tRNA^{Phe} is an inefficient donor; in contrast, the 5'-terminal phosphate of tRNA$_i^{Met}$ has excellent donor substrate quality due to the mispairing between C$_{+1}$ and A$_{+72}$ [24]. (I) Adaptation of the "DNA splint oligonucleotide" principle to the T4 RNA ligase reaction [20]. DNA oligonucleotides DNA 1 and DNA 3, used to prevent formation of unwanted ligation byproducts, are likely to be dispensable when the acceptor RNA carries a 5'-OH terminus and the donor RNA is blocked at its 3' end, as specified in Fig. 4.1(B). IsoC: *iso*-cytidine used instead of a natural nucleoside as the 3'-terminal nucleotide of the acceptor RNA in this particular experiment.

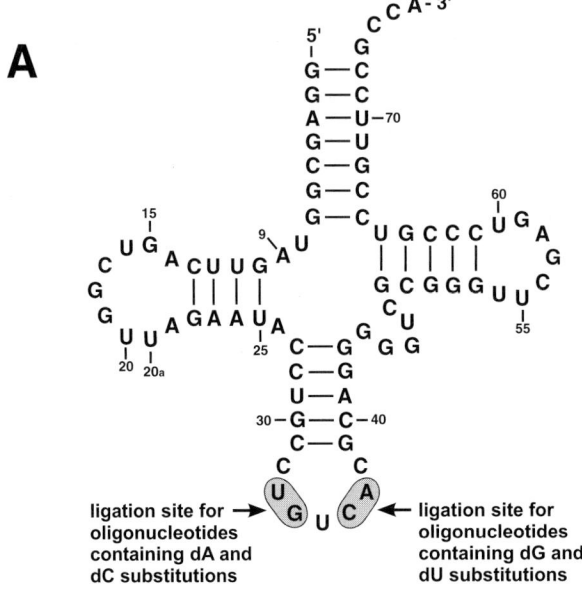

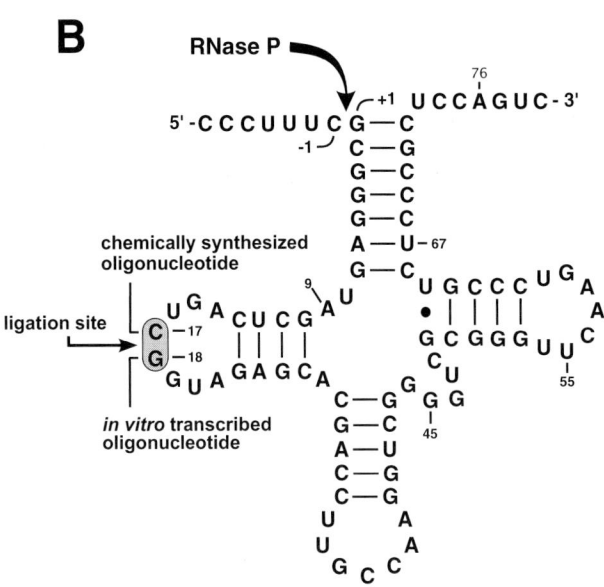

Fig. 4.3. (A) Schematic illustration of the T4 RNA ligation strategy for *E. coli* tRNA[Asp]. The two alternative ligation sites (highlighted in grey) were placed in the anticodon loop, between U33 and G34 for oligonucleotides containing 2′-deoxyA or 2′-deoxyC modifications, and between C36 and A37 for those containing 2′-deoxyG or 2′-deoxyU modifications. (B) Schematic representation of the T4 RNA ligation strategy for *Thermus thermophilus* ptRNA[Gly]. The arrow between positions −1 and +1 marks the canonical RNase P cleavage site; the ligation site between C17 and G18 is highlighted in grey.

influence ligation efficiency: increased loop size means more flexibility of the dangling single strands and increased average distance of the reactive groups before ligation. This is expected to reduce the efficiency of product formation. On the other hand, ligation within a tetraloop may occur less efficiently because the acceptor and donor ends may already be conformationally restricted before end joining.

2. Acceptor Substrates: Earlier studies with short acceptor oligoribonucleotide duplexes (6 bp) and pNp donors have indicated that acceptor substrates can be double stranded or even blunt ended (Fig. 4.2F and G) [39]. Moreover, duplexes with single nucleotide 5′ overhangs reacted most efficiently with the pNp donor that can base-pair with the duplex overhang (Fig. 4.2G). For example, with single cytidine 5′ overhangs on each end of the acceptor duplex, best yields were obtained for the pGp donor [39]. Regarding blunt-ended acceptor duplexes, it should, however, be mentioned that a dA_8 DNA acceptor oligonucleotide was joined less efficiently to a donor substrate in the presence of a complementary dT_8 DNA oligonucleotide [23], suggesting that a blunt-ended duplex, which is expected to form from the two oligonucleotides, is a less efficient type of acceptor substrate.

3. Donor Substrates: Donor termini that are part of a helix are inefficient substrates [24]. For example, the 5′-terminal phosphate of G_{+1} of yeast $tRNA^{Phe}$ was a relatively inefficient donor because G_{+1} is base-paired to C_{+72}, forming the terminal acceptor stem base pair (Fig. 4.2H). In comparison, the 5′-terminal monophosphate at C_{+1} of E. coli initiator $tRNA^{Met}$, carrying a single mismatch at the acceptor stem terminus (C_{+1} and A_{+72}, Fig. 4.2H), was a much better donor in the T4 RNA ligase reaction. Here, intramolecular cyclization between the 5′-terminal phosphate and the 3′-hydroxyl at the 3′ end (A_{+76}) was the favored reaction, already occurring with high efficiency at very low enzyme concentrations [24].

4. Accessibility of Ligation Sites: In the case of large RNAs, the helix structure that clamps the preferably single-stranded acceptor and donor substrates should be positioned at the surface of the RNA complex to ensure enzyme access. In the case of RNAs for which the ligation site is embedded in higher order structures, experimenters usually switch to the "DNA splint" ligation technique using T4 DNA ligase, since it involves disruption of RNA structure (see Chapter 3 and [40]). However, it may sometimes have escaped attention that the "splint principle" has also been successfully adapted to the T4 RNA ligase reaction by annealing donor and acceptor RNAs to a bridging DNA oligonucleotide for juxtaposition of reacting end groups [20]. Yet, in this setup the bridging oligonucleotide design excluded 5–6 nt of each, acceptor 3′ end and donor 5′ end, from the RNA–DNA hybrid, creating a broken bulge loop structure in the RNA strand, while the DNA strand was entirely engaged in base pairing (Fig. 4.2I).

5. RNA End Homogeneity: A problem connected with *in vitro* transcription is 3′ and/or 5′ heterogeneity of RNA products, which may reduce ligation efficiency

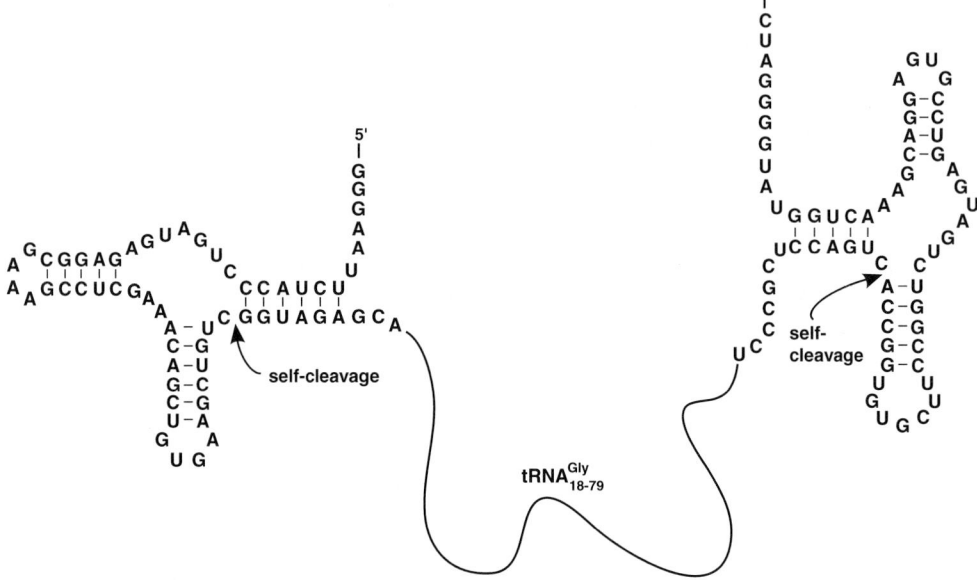

Fig. 4.4. A primary transcript with self-cleaving hammerhead ribozyme structures at its 5' and 3' termini. During transcription, the flanking regions are removed by ribozyme self-cleavage, releasing the internal RNA molecule of interest, in this case the donor substrate for the ligation illustrated in Fig. 4.3(B) (nt 18–79 of tRNA^{Gly}).

and compromise product homogeneity (for details, see Chapter 1). This problem can be solved either by sandwiching the RNA of interest between terminal *cis*-cleaving ribozymes (Fig. 4.4 and Chapter 2), or by use of alternative approaches described in Chapter 3. When involving *cis*-cleaving ribozymes, however, it is important to keep in mind that the cleavage reaction produces a 2',3'-cyclic phosphate at the 3' end and a hydroxyl group at the 5'-terminus. A 5'-hydroxyl is optimal for acceptor substrates, but requires phosphorylation when present on the donor substrate. Conversely, a 2',3'-cyclic phosphate nicely blocks the donor 3' end, but has to be removed when present on an acceptor substrate. Protocols to remove 2',3'-cyclic phosphates are described in Chapters 2 and 6.

4.5
Application Examples and Protocols

4.5.1
Production of Full-length tRNAs

In our hands, the T4 RNA ligation procedure was successfully used for the production of ca. 80-nt long tRNA derivatives with site-specific modifications. In the first case, a 77-nt long *E. coli* tRNA^{Asp} was prepared by enzymatic ligation of two chemically synthesized oligonucleotides, each between 34 and 43 nt in length. The ligation sites were placed in the anticodon loop, the most explored region of tRNA for ligation by T4 RNA ligase. In this study [37] analyzing the effect of 2′-deoxy modifications on aminoacylation, tRNA variants were prepared that either contained single-site 2′-deoxy modifications or had all, for example, A residues in the 5′ or 3′ half or even in the entire tRNA replaced with the 2′-deoxyA analog. The ligation position for tRNA halves containing 2′-deoxyA- and 2′-deoxyC modifications at every A and C position, respectively, was placed between anticodon nucleotides U33 and G34, while nucleotides C36 and A37 were selected as ligation site when the substrate halves contained 2′-deoxyG- and 2′-deoxyU modifications (Fig. 4.3A). Ligation yields were between 30 and 50%. These findings illustrate that satisfactory product yields can be obtained despite seemingly unfavorable identities (see Section 4.2) of the acceptor 3′ terminus (U33) and the donor 5′ terminus (G34).

In the second application, a bacterial precursor tRNA^{Gly} (ptRNA^{Gly} from *Thermus thermophilus*) was prepared from a 24-nt acceptor substrate obtained by chemical synthesis (representing the 5′ portion of the ptRNA) and a 62-nt donor substrate representing the 3′-proximal portion of the ptRNA and generated by T7 RNA transcription (Fig. 4.3B). The 24-nt acceptor oligonucleotide carried 5′- and 3′-terminal hydroxyl groups, as routinely present in chemically synthesized RNAs. The 62-nt donor RNA was released from a primary transcript with terminal *cis*-hammerheads (Fig. 4.4), generating the aforementioned 5′-hydroxyl and 2′,3′-cyclic phosphate end groups. Before ligation, the 5′ end was phosphorylated (Protocol 2; see also Chapter 3) using T4 polynucleotide kinase. Here, the purpose was to study the effect of ribose modifications at nt -1 of ptRNA^{Gly} on catalysis by *E. coli* RNase P RNA [38]. The ligation site was placed in the D loop, between positions C17 and G18 (Fig. 4.3B), to minimize the length of the chemically synthesized RNA oligonucleotides carrying the single-site modification and thus to reduce the costs of chemical synthesis. The ligation yield was about 50% (see Protocol 3), again despite unfavorable identity of the donor 5′-terminus (G18).

4.5.2
Specific Protocols

In addition to the specific protocols given below, some routine buffers and procedures used in these protocols are detailed in Section 4.5.3, *General methods*, referred to by the abbreviation "GM" in the following.

Protocol 1: *In vitro* transcription
(1) *In vitro* transcription reactions (1 ml) using T7 RNA polymerase (T7 RNAP) include the following components:

		Final concentration
80 μl	1 M HEPES–KOH, pH 7.5	80 mM
11 μl	2 M MgCl$_2$	22 mM

6 µl	20 mg/ml BSA[1]	120 µg/ml
10 µl	100 mM spermidine	1 mM
50 µl	100 mM DTT	5 mM
37.5 µl	100 mM each NTP (pH 7)	3.75 mM of each NTP
5 µl	1 U/µl pyrophosphatase[2]	5 U/ml
30 µl	template 1 µg/µl (linearized plasmid 3.2 kb)	30 µg/ml
760.5 µl	RNase-free water	

[1] BSA (Sigma, minimum purity 98% based on electrophoretic analysis, pH 7).
[2] Pyrophosphatase from yeast (Roche, EC 3.6.1.1, 200 U/mg, <0.01% each ATPase and phosphatases).

The total volume of this reaction mix without enzyme is 990 µl. Before start of transcription, the reaction mix is divided into 5 portions of 198 µl, to each of which 2 µl T7 RNAP (MBI Fermentas, stock solution 100 U/µl; final assay concentration 1 U/µl) is added, followed by incubation at 37 °C overnight. Shorter incubation periods (2–5 hours) have been used as well and may even be preferable (Chapters 1 and 3).

In transcription reactions lacking pyrophosphatase or when pyrophosphate activity is insufficient, a white pyrophosphate precipitate may be observed. In such cases, it is advisable to remove the precipitate before Step 2 by centrifugation at 14 000 g for about 5 min directly after incubation of transcription mixtures has been stopped. The clear supernatant is then carefully removed and transferred to a new Eppendorf tube for further sample processing. Another way to reduce the amount of precipitate is to add Na_2EDTA immediately after transcription to give a final concentration of 50–100 mM (use a 500 mM stock solution at pH 7.5). By chelating Mg^{2+}, the formation of insoluble precipitates is substantially reduced.

(2) Before isolation of transcription product, the DNA template is degraded by addition of 1 µl DNase I (RNase-free, Roche; stock solution 10 U/µl; final concentration 50 U/ml) to 200 µl reaction mix and incubation at 37 °C for 20 min.

(3) Extract the RNA solution (200 µl) once with a phenol/chloroform mixture (GM). It may further be advisable to remove excess NTPs and/or salt components before EtOH precipitation (Step 4) by use of NAP 10 columns (Pharmacia Biotech) or equivalent matrices. Removal of excess NTPs is in general necessary only when using ^{32}P-radiolabeled nucleotides to avoid background radioactivity in purification gels. Salt components may impair band separation by denaturing polyacrylamide gel electrophoresis (PAGE), particularly when optimal gel resolution is essential, such as in separations of full-length RNA product from $n + 1$ and $n - 1$ species.

(4) Precipitate the RNA with ethanol (GM).

(5) Dissolve the RNA pellet in 20–40 µl RNase-free water and purify the RNA of interest by denaturing PAGE according to Protocol 4.

(6) Finally, determine the RNA concentration (GM).

Protocol 2: 5′-Phosphorylation of donor oligonucleotide

Many manufacturers provide convenient protocols for 5′-phosphorylation. We prefer to use a T4 polynucleotide kinase (T4 PNK) protocol from New England Biolabs. T4 PNK requires a free hydroxyl group at the 5′ terminus.

The protocol for the phosphorylation reaction is based on a total volume of 50 μl.

Phosphorylation buffer (10×)	Final concentration
500 mM Tris–HCl, pH 7.6	50 mM
100 mM MgCl$_2$	10 mM
50 mM DTT	5 mM
5 mM ATP	0.5 mM

(1) Combine 3 nmol of donor RNA with 5 μl 10 × phosphorylation buffer, 1 μl RNase Inhibitor (MBI Fermentas; stock solution 25 U/μl; final concentration 0.5 U/μl), 2.5 μl ATP (100 mM) and RNase-free water to a final volume of 47.5 μl; mix gently.

(2) Add 2.5 μl T4 PNK (MBI Fermentas; stock solution 10 U/μl; final concentration 0.5 U/μl) and incubate the mixture at 37 °C for 30 min.

(3) To increase 5′-phosphorylation efficiency, add another 2.5 μl of ATP (100 mM) and also another 2.5 μl T4 PNK (stock solution 10 U/μl), and incubate the mixture for another 20 min at 37 °C. Either directly add 55 μl 2 × gel loading buffer (GM) for PAGE purification (Protocol 4) or proceed to Step 4.

(4) Precipitate the RNA with ethanol (GM).

(5) Dissolve the RNA pellet in about 20 μl RNase-free water and purify the RNA by denaturing PAGE according to Protocol 4.

(6) Determine the RNA concentration (GM).

Protocol 3: Enzymatic ligation

The protocol described below was used to generate full-length *E. coli* tRNA^{Asp} and *T. thermophilus* ptRNA^{Gly}. The tRNA^{Asp} was generated by ligation of pairs of chemically synthesized oligoribonucleotides, 34–43 nt in length (Fig. 4.3A); ptRNA^{Gly} was generated by ligation of its 3′ portion (nt 18–79), transcribed *in vitro*, to a chemically synthesized 24meric oligoribonucleotide contributing the ptRNA^{Gly} 5′ portion (Fig. 4.3B).

(1) If the donor 3′-oligonucleotide does not contain a 5′-phosphate, 5′-phosphorylate according to Protocol 2 prior to ligation. It is recommended to check the purity of the oligoribonucleotides to be used in the ligation reaction on an analytical denaturing polyacrylamide (PAA) gel. If necessary, also gel-purify the freshly phosphorylated donor oligoribonucleotide as described in Protocol 4.

(2) Ligation reactions are performed in a total volume of 200 μl (reached in step 5). As the first step, combine 3 nmol of 5′-phosphorylated donor oligoribonucleotide and 4.5 nmol of the corresponding acceptor oligoribonucleotide with:

		Final concentration in 200 μl
20 μl	1 M HEPES–KOH, pH 7.5	100 mM
20 μl	100 mM DTT	10 mM
x μl	RNase-free water to a volume of 156 μl	

(3) Denature for 3 min at 90 °C, followed by incubation for 10 min at 65 °C and slow cooling (45 min) to ambient temperature in a metal block removed from the heating apparatus to anneal the oligoribonucleotides.
(4) After the cooling step, add:

		Final concentration in 200 µl
12 µl	250 mM MgCl$_2$	15 mM
1 µl	100 mM ATP	0.5 mM
20 µl	100% DMSO	10% (v/v)
3.75 µl	25–40 U/µl RNase inhibitor (MBI Fermentas or Promega)	0.5–0.75 U/µl

Mix gently.

(5) Add 7 µl of T4 RNA ligase (stock solution 20 000 U/ml; final concentration 0.7 U/µl, New England Biolabs), resulting in the final volume of 200 µl. Incubate at 16 °C for about 12–15 h.
(6) To analyze ligation efficiency, withdraw 1.5 µl from the ligation reaction and mix with 10 µl loading buffer; load onto a denaturing 8–12% PAA gel, stain the gel with ethidium bromide and visualize RNA bands by exposure to UV light.
(7) Ethanol precipitate the bulk of the ligation reaction (GM).
(8) Dissolve the resulting RNA pellet in 15–30 µl RNase-free water and purify the RNA by denaturing PAGE according to Protocol 4.
(9) Calculate the RNA concentration (GM).

Protocol 4: Preparative purification of RNA by denaturing PAGE
The appropriate gel concentration depends, as usually, on the size of the RNA; a 12% PAA/8 M urea sequencing gel (1-mm thick) was used for purification of ligated tRNA^{Asp} (77 nt), a corresponding 10% gel for purification of ligated ptRNA^{Gly} (86 nt). A somewhat lower urea concentration, such as 7.5 M (Chapter 3), may be used, but we recommend concentrations of at least 7 M. To facilitate localizing the RNA product on the gel, it is advisable to increase the concentration of the product band by pooling three separate ligation reactions (3 × 15–30 µl), each based on 3 nmol of input donor RNA (Step 8 above).

(1) Mix the 45–90 µl of pooled ligation product with 45–90 µl 2 × gel loading buffer and pipette into a 2-cm broad gel pocket of the PAA/8 M urea gel. As an example, the above-mentioned 86-nt ligation product was run on a roughly 40-cm long, 1-mm thick 10% PAA/8 M urea gel until the xylene cyanol marker had migrated 22–25 cm from the top.
(2) Detect the ligated product by UV shadowing, excise from the gel and elute from crushed gel slices in elution buffer overnight at 4 °C. A detailed description of this procedure is given in Chapter 3, Protocol 3.
(3) Concentrate the eluted RNA by ethanol precipitation and redissolve in 20–40 µl RNase-free water (GM).
(4) Determine the concentration by UV spectroscopy (GM).

4.5.3
General Methods (GM)

Several of the methods described here are detailed in the handbook *Molecular Cloning: A Laboratory Manual* [41]. Also, for some of the methods given below, slightly variant protocols exist and might be used just as well.

Preparation of RNase-free water
A major cause of RNA degradation is due to ribonuclease contamination in the water used for the preparation of buffers and solutions. It is therefore recommended to use double-distilled water, which may be further treated with DEPC (add 1/1000 volume DEPC and stir vigorously for 2 h followed by autoclaving for DEPC decomposition). For longer storage, the water may be additionally filtered through a 0.6-µm filter and kept in 1 ml aliquots in Eppendorf tubes, preferably at $-20\ °C$. It should be noted that DEPC remnants may interfere with enzymatic reactions. Thus, fresh double-distilled water, frozen at $-20\ °C$ for storage, may be preferred; store water at ambient temperature only when freshly autoclaved, but freeze in aliquots once the bottles have been opened.

Gel running buffer (5 × TBE)
54 g of Tris base (446 mM), 27.5 g of boric acid (445 mM) and 20 ml of 0.5 M EDTA (pH 8.0)/l; no pH adjustment required; store at room temperature.

2 × gel loading buffer for denaturing PAGE
Weigh 48 g urea, 50 mg bromophenol blue, 50 mg xylene cyanol and 3.72 g EDTA; add 20 ml 5 × TBE (see above), adjust the volume to 80 ml with double-distilled water and finally to 100 ml after complete dissolving [41]. Store the solution at room temperature.

Staining buffer (ethidium bromide solution)
PAA/8 M urea gels are stained in 1 × TBE containing 0.5–1.0 µg/ml ethidium bromide (EtBr). An EtBr stock solution at a concentration of 10 mg/ml in water is preferred. The solution should be stored in dark bottles (e.g. bottles covered with aluminum foil) at room temperature or 4 °C.

Phenol/chloroform extraction
Add 200 µl of phenol/chloroform (ratio 5:1, pH 4.7, AMRESCO), cooled to 4 °C, to 200 µl of aqueous RNA solution. Vortex the resulting mixture vigorously before centrifugation at 14 000 g for about 2 min. Transfer the aqueous phase carefully to a new Eppendorf tube and extract twice with 100 µl chloroform in the same manner. Remove the RNA-containing aqueous phase (upper layer) and recover the RNA by ethanol precipitation.

Ethanol precipitation of RNA

- *Method A*: RNA precipitation by method A applies to RNA in buffer solution or in RNase-free double-distilled water. Add 100 μl NaOAc (3 M, pH 4.7), 1 μl glycogen (20 mg/ml) and 900 μl EtOH to 100 μl of RNA solution and mix vigorously. Store the sample at −20 °C for 2–3 h or at −80 °C for 30 min. Centrifuge the cooled sample at 14 000 g for about 10–30 min at around 10 °C. Carefully remove the supernatant and wash the RNA pellet with ice-cold 70 or 80% EtOH, followed by a short centrifugation step (e.g. 5 min, as above). Prior to dissolving the pellet in RNase-free water, air-dry the pellet for 10–15 min.
- *Method B*: This method of RNA precipitation is used for RNA recovered by gel elution in buffer B (see below). Add 1 μl glycogen (20 mg/ml) and 1 ml EtOH to 450 μl of the RNA solution and proceed as in Method A.

Elution of RNA from PAA/8 M urea gel slices

We routinely use two different elution buffers (A and B, see below). There is no general rule for the choice between these buffers and no systematic differences in yield have been noticed. However, if the RNA shows some degradation on the analytical gel after elution, it is recommended to switch to buffer B that has a lower pH. Both buffers should be stored at 4 or −20 °C.

Elution buffer A: 200 mM Tris–HCl (pH 7).
Elution buffer B: 1 M NaOAc (pH 4.7).

Calculation of RNA concentration

Two alternative formulas can be used for the calculation of RNA concentrations based on UV spectrometry measurement (for more details, see Appendix):

(a) $c = A_{260} \times$ dilution factor of cuvette solution$/(\alpha_{m260} \times$ number of nucleotides $\times l)$, where A_{260} is the absorbance at 260 nm, α_{m260} is the average molar absorption coefficient of the 4 nucleotides at 260 nm (for single-stranded DNA and RNA, an average α_{m260} of $8\,500$ M^{-1} cm^{-1} is appropriate) and l is the path length of cuvette (normally 1 cm).

(b) 1 A_{260} unit (absorbance of 1 measured at 260 nm in a 1-cm cuvette) corresponds to approximately 40 μg/ml single-stranded RNA. Total amount of RNA (μg) = $40 \times A_{260}$ units $\times$ dilution factor of cuvette solution $\times$ total volume of RNA stock solution in milliliters.

4.5.4
Chemicals and Enzymes

4.5.4.1 Chemical Synthesis and Purification of Oligoribonucleotides

Oligo(ribo)nucleotides were purchased from IBA (Göttingen, Germany) or self-synthesized using standard phosphoramidite chemistry on an Applied Biosystems 394A DNA synthesizer. The coupling time was 16.6 min for ribonucleoside and 30 s for 2′-deoxyribonucleoside building blocks. Oligoribonucleotides were removed from the solid support and purified according to the protocol outlined below [42]. An alternative work-up procedure for RNAs longer than 50 nt has been described [2].

Purification of oligoribonucleotides

(1) Transfer the polymer-bound oligoribonucleotide from the column to a 4-ml vial and suspend in 3:1 (v/v) NH_3:EtOH at 55 °C for 16 h. Before opening the vial, cool the mixture on ice for about 15 min; then reduce the volume to around 1 ml by use of a Speed Vac system. Remove the clear aqueous supernatant carefully and transfer to a new 2-ml Eppendorf tube; wash the solid support with 100 µl water. Combine the aqueous phases and concentrate to dryness in a Speed Vac; add 500 µl EtOH and again concentrate the oligoribonucleotide to dryness to remove any water left from the preceding steps. Resuspend the base-deprotected oligoribonucleotide in 500 µl 1 M tetrabutylammonium fluoride (TBAF) in tetrahydrofurane (THF) and let the mixture react under gentle shaking at room temperature for at least 20 h. Then add 500 µl 2 M NaOAc (pH 6), and concentrate to a volume of about 0.5–0.6 ml (30–60 min in a Speed Vac). Extract the resulting mixture with 2–3 × 800 µl ethyl acetate (EtOAc) and centrifuge again in the Speed Vac for about 5–15 min until a clear solution is obtained. Do not concentrate the mixture to dryness in this step. Add 1.6 ml EtOH and store the mixture at −20 °C for 2 h or overnight. Centrifuge at 14 000 g for 15 min and remove the supernatant carefully. Air-dry the oligoribonucleotide pellet for 15 min and dissolve in 600 µl RNase-free water.

(2) Purify the oligoribonucleotide on a PAA/8 M urea sequencing gel (2-mm thick); a 12–15% PAA/8 M urea gel should be used for purification of 30- to 40-nt long oligoribonucleotides, a 20% gel for shorter ones. Mix the 600 µl product sample with 600 µl loading buffer and apply to the gel.

(3) Detect the oligoribonucleotide by UV shadowing, excise from the gel and elute from gel pieces by using a Biotrap system (Schleicher & Schuell Biotrap elution chambers and membranes BT1 and BT2 [42]; see also Chapter 1). Alternatively, diffusion elution from crushed gel slices may be employed (see Section 4.5.3). However, in our hands the Biotrap technique was more reliable and efficient when eluting oligoribonucleotides from a preparative gel (gel thickness of 2 mm).

(4) After gel elution, further purify the oligoribonucleotide by using Sep-Pak cartridges according to the manufacturer's instructions (Waters Sep-Pak cartridges, Millipore). In this step, any salt components left from the previous steps are removed. Then concentrate the oligoribonucleotide to dryness using a Speed Vac system.

(5) Dissolve the resulting pellet in 400–600 µl RNase-free water and determine the concentration as described in Section 4.5.3.

4.5.4.2 Chemicals

Ribonucleoside phosphoramidites for chemical nucleic acid synthesis were purchased from PerSeptive Biosystems (Hamburg, Germany); NTPs and glycogen were obtained from Roche; 48% polyacrylamide/bisacrylamide and the phenol/chloroform mix (5:1, pH 4.7) were purchased from AMRESCO; diethylpyrocarbonate (DEPC) was obtained from FLUKA; 1 M TBAF in THF was purchased from Aldrich.

4.5.4.3 Enzymes

T4 polynucleotide kinase and T4 RNA ligase were purchased from New England Biolabs. RNase Inhibitor and T7 RNA polymerase were purchased from MBI Fermentas. DNase I and pyrophosphatase were obtained from Roche.

4.6
Troubleshooting

Low yields of the ligation reaction may have the following reasons:

(1) Check for unfavorable secondary structure formation of the RNA fragments, particularly at the ligation joint. For this purpose, software such as OLIGO version 4.0 (National Bioscience) or Mfold [43] can be employed.
(2) Heterogeneous 3′ ends of RNA transcripts: for RNA fragments of up to about 40 nt it is usually sufficient to purify the RNA by preparative PAGE in the presence of 8 M urea prior to the ligation reaction. However, for RNA transcripts longer than around 50 nt, RNAs slightly differing in length from the main product are hard to get rid of by preparative gel purification. Methods to eliminate the problem of 5′- and 3′ end heterogeneities are described in Chapters 2 and 3.
(3) RNA degradation – this is usually due to RNase contamination in water or solutions. Prepare and store RNase-free water as described in Section 4.5.3. Check individual solutions for RNase activity; prepare fresh buffers and solutions.

Acknowledgments

The authors thank Erik Tullberg, Ulf Berg and Anders Barfod for critical reading of the manuscript. This work was supported by Crafoordska stiftelsen, FLÄK (Forskarskolan i Läkemedelsvetenskap), Carl Tryggers Stiftelse, Kungliga Fysiografiska Sällskapet i Lund, Schybers stiftelse and the Deutsche Forschungsgemeinschaft.

References

1 S. VERMA, F. ECKSTEIN, *Annu. Rev. Biochem.* **1998**, *67*, 99–134.
2 T. PERSSON, U. KUTZKE, S. BUSCH, R. HELD, R. K. HARTMANN, *Bioorg. Med. Chem.* **2001**, *9*, 51–56.
3 C. VARGEESE, J. CARTER, J. YEGGE, S. KRIVJANSKY, A. SETTLE, E. KROPP, K. PETERSON, W. PIEKEN, *Nucleic Acids Res.* **1998**, *26*, 1046–1050.
4 R. I. HOGREFE, A. P. MCCAFFREY, L.

U. Borozdina, E. S. McCampbell, M. M. Vaghefi, *Nucleic Acids Res.* **1993**, *21*, 4739–4741.

5 F. Wincott, A. DiRenzo, C. Shaffer, S. Grimm, D. Tracz, C. Workman, D. Sweedler, C. Gonzales, S. Scaringe, N. Usman, *Nucleic Acids Res.* **1995**, *23*, 2677–2684.

6 E. A. Arn, J. Abelson, RNA ligases: function, mechanism, and sequence conservation, in: *RNA Structure and Function*, R. W. Simons, M. Grunberg-Manago (eds), Cold Spring Harbor Laboratory Press, Cold Spring Harbor, NY, **1998**, pp. 695–726.

7 A. Schnaufer, A. K. Panigrahi, B. Panicucci, R. P. Igo, Jr, E. Wirtz, R. Salavati, K. Stuart, *Science* **2001**, *291*, 2159–2162.

8 J. Abelson, C. R. Trotta, H. Li, *J. Biol. Chem.* **1998**, *273*, 12685–12688.

9 G. Kaufmann, U. Z. Littauer, *Proc. Natl. Acad. Sci. USA* **1974**, *71*, 3741–3745.

10 G. C. Walker, O. C. Uhlenbeck, E. Bedows, R. I. Gumport, *Proc. Natl. Acad. Sci. USA* **1975**, *72*, 122–126.

11 R. Silber, V. G. Malathi, J. Hurwitz, *Proc. Natl. Acad. Sci. USA* **1972**, *69*, 3009–3013.

12 M. Amitsur, R. Levitz, G. Kaufmann, *EMBO J.* **1987**, *6*, 2499–2503.

13 C. Tyndall, J. Meister, T. A. Bickle, *J. Mol. Biol.* **1994**, *237*, 266–274.

14 L. W. McLaughlin, N. Piel, E. Graeser, *Biochemistry* **1985**, *24*, 267–273.

15 G. Kaufmann, T. Klein, U. Z. Littauer, *FEBS Lett.* **1974**, *46*, 271–275.

16 G. Kaufmann, N. R. Kallenbach, *Nature* **1975**, *254*, 452–454.

17 A. Sugino, T. J. Snopek, N. R. Cozzarelli, *J. Biol. Chem.* **1977**, *252*, 1732–1738.

18 T. E. England, O. C. Uhlenbeck, *Biochemistry* **1978**, *17*, 2069–2076.

19 E. Romaniuk, L. W. McLaughlin, T. Neilson, P. J. Romaniuk, *Eur. J. Biochem.* **1982**, *125*, 639–643.

20 J. D. Bain, C. Switzer, *Nucleic Acids Res.* **1992**, *20*, 4372.

21 E. A. Atencia, M. Montes, M. A. G. Sillero, A. Sillero, *Eur. J. Biochem.* **2000**, *267*, 1707–1714.

22 D. M. Hinton, J. A. Baez, R. I. Gumport, *Biochemistry* **1978**, *17*, 5091–5097.

23 M. I. Moseman McCoy, R. I. Gumport, *Biochemistry* **1980**, *19*, 635–642.

24 A. G. Bruce, O. C. Uhlenbeck, *Nucleic Acids Res.* **1978**, *5*, 3665–3677.

25 D. C. Tessier, R. Brousseau, T. Vernet, *Anal. Biochem.* **1986**, *158*, 171–178.

26 G. L. Igloi, *Anal. Biochem.* **1996**, *233*, 124–129.

27 F. Hausch, A. Jäschke, *Bioconjugate Chem.* **1997**, *8*, 885–890.

28 Y. Kinoshita, K. Nishigaki, Y. Husimi, *Nucleic Acids Res.* **1997**, *25*, 3747–3748.

29 J. E. Rigden, M. A. Rezaian, *Virology* **1992**, *186*, 201–206.

30 D. Skoric, M. Conerly, J. A. Szychowski, J. S. Semancik, *Virology* **2001**, *280*, 115–123.

31 L. Wang, D. E. Ruffner, *Nucleic Acids Res.* **1998**, *26*, 2502–2504.

32 L. Argaman, R. Hershberg, J. Vogel, G. Bejerano, E. G. H. Wagner, H. Margalit, S. Altuvia, *Curr. Biol.* **2001**, *11*, 941–950.

33 X. Liu, M. A. Gorovsky, *Nucleic Acids Res.* **1993**, *21*, 4954–4960.

34 K. Nishigaki, K. Taguchi, Y. Kinoshita, T. Aita, Y. Husimi, *Mol. Diversity* **1998**, *4*, 187–190.

35 T. Ohtsuki, G. Kawai, K. Watanabe, *J. Biochem.* **1998**, *124*, 28–34.

36 L. D. Sherlin, T. L. Bullock, T. A. Nissan, J. J. Perona, F. J. Lariviere, O. C. Uhlenbeck, S. A. Scaringe, *RNA* **2001**, *7*, 1671–1678.

37 C. S. Vörtler, O. Fedorova, T. Persson, U. Kutzke, F. Eckstein, *RNA* **1998**, *4*, 1444–1454.

38 T. Persson, S. Cuzic, R. K. Hartmann, *J. Biol. Chem.* **2003**, *278*, 43394–43401.

39 N. Sugimoto, A. Matsumura, K. Hasegawa, M. Sasaki, *Bull. Chem. Soc. Jpn.* **1991**, *64*, 2978–2982.

40 M. J. Moore, P. A. Sharp, *Science* **1992**, *256*, 992–997.

41 J. Sambrook, D. W. Russel, *Molecular Cloning: A Laboratory Manual*, Cold Spring Habor Laboratory Press, Cold Spring Harbor, NY, **2001**.
42 T. Tuschl, F. Eckstein, *Proc. Natl. Acad. Sci. USA* **1993**, *90*, 6991–6994.
43 M. Zuker, *Nucleic Acids Res.* **2003**, *31*, 3406–3415.
44 S. Heaphy, M. Singh, M. J. Gait, *Biochemistry* **1987**, *26*, 1688–1696.
45 L. K. Wang, C. K. Ho, Y. Pei, S. Shuman, *J. Biol. Chem.* **2003**, *278*, 29454–29462.

5
Co- and Post-Transcriptional Incorporation of Specific Modifications Including Photoreactive Groups into RNA Molecules

Nathan H. Zahler and Michael E. Harris

5.1
Introduction

A great deal of modern RNA biochemistry and molecular biology involves the incorporation of modified nucleotides into RNA molecules. Nucleotides bearing base and backbone modifications can be introduced randomly or at specific locations, and the variety of modifications available can facilitate both mechanistic and structural studies of RNA. The goal of this chapter is to outline some common techniques by which modified nucleotides and photo-crosslinking agents can be introduced into an RNA molecule and to provide examples of experiments in which these techniques have proven useful. Techniques discussed in detail include 5′-end modification by transcription priming, generation of nucleotide monophosphates and monophosphorothioates for use in 5′-end modification, derivatization of a 5′-phosphorothioate modification with a photo-crosslinking agent, and post-transcriptional 3′ end modification.

5.1.1
Applications of RNA Modifications

The availability of a wide range of modified nucleotides provides several key advantages for the study of RNA. Modified nucleotides can be used to alter individual functional groups, or in some cases individual atoms, and as such allow for more straightforward interpretation of experimental results than is possible with base mutation alone. Available analogs can be used to alter RNA base and backbone functional groups, and even remove nucleotide bases altogether [1]. Additionally, modifications with useful chemistry, such as photo-crosslinking agents, can be introduced. Combined with techniques to either randomly modify a population of RNAs or position modifications in a site-specific manner, the use of chemical modifications is fundamental to the study of RNA.

Site-specific incorporation of modified nucleotides is useful when a single nucleotide or functional group is the focus of study. Site-specific placement of modified nucleotides is therefore an essential first step in a variety of further tech-

Handbook of RNA Biochemistry. Edited by R. K. Hartmann, A. Bindereif, A. Schön, E. Westhof
Copyright © 2005 WILEY-VCH Verlag GmbH & Co. KGaA, Weinheim
ISBN: 3-527-30826-1

niques. For example, site-specific fluorescent markers can be introduced into RNA molecules for kinetic and folding studies (see Chapter 28 [2–4]). Site-specific placement of phosphorothioate modifications is also necessary for many thiophillic metal ion rescue experiments (see Chapter 19). 5′-End modification with m^7Gp^3G or other cap analog is an essential step in the use of *in vitro* transcribed RNAs in translation experiments [5–8]. In addition, affinity tags can also be placed at specific positions in an RNA structure to facilitate isolation of ribonucleoprotein complexes (see Part IV.1).

Site-specific modification of RNA can also be used to position photo-crosslinking agents for studies of RNA structure. A number of modifications are available for RNA which can form crosslinks to both RNA and proteins [9–12]. Several commonly used photo-crosslinking agents such as 4-thiouridine and 6-thioguanosine can be internally incorporated into an RNA molecule (see Chapter 22, e.g. [13, 14]). In addition, the 5′ and 3′ ends of RNA molecules can be post-transcriptionally modified with arylazide containing photo-crosslinking agents (see below and [11, 15, 16]). Together with the use of circular permutation to move the 5′ end of complex RNA molecules to different positions in the RNA structure, these modifications have been used to provide distance constraints for the structures of complex RNAs such as bacterial ribonuclease P RNA and 16S ribosomal RNA [17–19].

Another powerful aspect of site-specific modification is the ability to systematically vary nucleotide functional groups. For example, this approach has been used to examine the effects of phosphorothioate modification on MS2 coat protein binding to RNA and to investigate the contributions of 2′-hydroxyl groups to group I intron substrate binding [20–22]. Similarly, if a nucleotide base is known to contribute to structure or function, an array of modified nucleotides can be used to examine the contributions of individual base functional groups. Such studies have been undertaken to investigate the role of the conserved GU base pair immediately adjacent to the group I intron cleavage site and to assess the similarities in branch point adenosine recognition by the spliceosome and the group II intron [23–25]. Furthermore, this approach can be extended to the level of individual functional groups for 2′-hydroxyls and other functional groups which can be replaced with a number of chemically distinct modifications (e.g. [26, 27]).

Unlike site-specific modification, which is useful when a location or functional group is known to be of interest, random incorporation of modified nucleotides can be used to survey for important nucleotides and functional groups. The primary example of this approach is nucleotide analog interference mapping (NAIM), which has been used successfully in a number of systems to identify functionally important base and backbone functional groups (see Chapters 17 and 18 [28–31]). Photo-crosslinking agents can also be positioned randomly throughout an RNA. 4-Thiouridine and 6-thioguanosine can be randomly incorporated through transcription and have been used in the study of RNA structure and RNA–protein interactions in such diverse systems as the HIV-I Rev protein, the ribosome, RNA polymerase and the group II self-splicing intron [10, 32]. Finally, an RNA can be completely substituted with modifications. This approach can be used to introduce modifications that protect therapeutic RNAs from ribonucleases or increase their cellular uptake (for review, see [33]).

5.1.2
Techniques for Incorporation of Modified Nucleotides

The choice between techniques for incorporation of modifications is dependent primarily on the location of the desired modification and thus on the experiment in which the modified RNA will be used. Site-specific incorporation of modified nucleotides can be accomplished by a number of methods. Internal modifications are often introduced by chemical synthesis of all or part of the desired RNA. If the RNA in question is short enough, the simplest solution for internal site-specific incorporation is often chemical synthesis of the entire RNA. If not, chemical synthesis can be combined with RNA ligation (see Chapters 3 and 4). Ligation, however, has limitations which can often lead to low yields and circular permutation combined with either 5'- or 3'-end modification can be a viable alternative, especially for highly structured RNAs [11].

RNAs with site-specific 5'-end modifications can be prepared in a straightforward manner by transcription priming, an example protocol for which is given below. During transcription with T7 RNA polymerase, nucleotides lacking a 5'-triphosphate cannot be incorporated into an elongating RNA chain, but can be used to initiate a transcript ([34]; see also Chapter 1). Therefore modified nucleosides or nucleotide monophosphates included in an *in vitro* transcription reaction will be incorporated only at the 5' ends of transcripts. However, transcription priming can produce a mixed population of modified and unmodified RNAs. To maximize analog incorporation, it is essential to include a large excess of modified nucleotide over the corresponding unmodified nucleotide triphosphate. While this has the effect of increasing the fraction of the population with the desired 5' modification, it also tends to lower transcription efficiency. It is therefore generally necessary to empirically determine the appropriate analog concentration to balance these factors.

Random incorporation of modified nucleotides into RNA molecules can also be performed co-transcriptionally (e.g. [28–32]). In this case, modified nucleotide triphosphates are included in the reaction mixture and are added randomly to the elongating RNA chain. As with transcription priming, the modified nucleotide competes with the corresponding unmodified nucleotide for incorporation. Unlike transcription priming, however, the goal of random incorporation is generally not to produce a completely modified RNA. For experiments such as NAIM, which call for interference or selection, it is preferable to have at most one modification per transcript.

It is also important to note that the range of modified nucleotides that can be incorporated during transcription, both internally and at the 5' terminus, is limited by the specificity of T7 RNA polymerase. Wild-type T7 RNA polymerase is unable to efficiently incorporate some modified nucleotides, including those with 2' modifications that alter hydrogen-bonding ability or introduce large substituents. However, some of these restrictions can be relaxed through the use of mutant polymerases which allow for the use of 2' modifications and possibly minor groove modifications [35, 36].

Site-specific 3'-terminal modifications are generally introduced post-transcriptionally. T4 RNA ligase can be used to add a modified nucleotide to the 3' end of a RNA (see Chapter 4). In addition, post-transcriptional chemical modification can be used to attach affinity selection or photo-crosslinking agents to the 3' terminus using the method of Oh and Pace [16], an example protocol for which is given below. In this method, the unique 2',3'-cis-diol of the 3' terminus is oxidized to form a dialdehyde, which is then reacted with an alkyldiamine under reducing conditions to yield a unique primary aliphatic amine (Fig. 5.1; see also Chapter 6). This primary amine can then be further derivatized with N-hydroxysuccinimidyl esters to introduce photo-crosslinking agents or other useful modifications. The modifica-

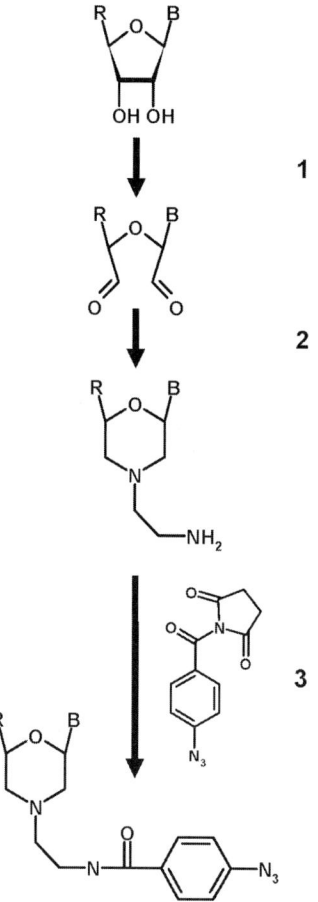

Fig. 5.1. 3'-End attachment. Chemical scheme for 3'-end modification by the method of Oh and Pace showing the three reactions steps. (1) Oxidation of the 2',3'-cis-diol of the 3'-terminal ribose to a dialdehyde. (2) Reduction of the dialdehyde in the presence of ethylenediamine to yield a primary amine with a two-carbon linker. (3) Reaction with the N-hydroxysuccinimidyl ester derivative of an azidophenacyl crosslinking agent to yield the final, 3'-crosslinking construct.

tions that can be introduced to the 3' end in this manner are only limited by the availability of reagents which will react with primary amines.

5.2
Description

5.2.1
5'-End Modification by Transcription Priming

As described above, 5'-end modification can be accomplished co-transcriptionally by transcription priming [34]. The following protocol describes the procedure for incorporation of a 5'-phosphorothioate modification with guanosine 5'-monophosphorothioate (GMPS). GMPS modified RNA can be further derivatized and an example of modification with an arylazide photo-crosslinking agent is described below. Other nucleotide monophosphates or nucleosides can be substituted in this protocol if a different 5' modification is required. GMPS to GTP ratios of 10:1 to 40:1 result in a 70–90% yield of 5'-modified transcripts [11]. The following protocol was designed to maximize transcription efficiency of a 76-nt bacterial tRNA and utilizes a ratio of 4.8:1. If modified nucleotides other than GMPS are used, the optimal ratio for efficient incorporation and maximal transcription efficiency will have to be determined empirically.

This protocol predominantly requires widely available reagents. Nucleotide triphosphates, as well as T7 RNA polymerase and its associated buffer are available from a number of sources including Ambion. Yeast pyrophosphatase is available from Sigma-Aldrich. GMPS is not currently commercially available, but can be generated by chemical phosphorylation of guanosine (see below).

Begin by mixing the following:

10 × transcription buffer	10 μl
0.2 M DTT	3 μl
1 M MgCl$_2$	2 μl
100 mM ATP	4 μl
100 mM CTP	4 μl
100 mM UTP	4 μl
100 mM GTP	1.25 μl
30 mM GMPS	20 μl
Linearized DNA template	5 μg
40 U/μl RNase inhibitor	1 μl
5 U/μl yeast pyrophosphatase	5 μl
40 U/μl T7 RNA polymerase	5 μl
Water	to 100 μl total reaction volume

Incubate the reaction mixture overnight at 37 °C, recover products by ethanol precipitation and gel purify on an appropriate percentage denaturing polyacrylamide gel.

5.2.2
Chemical Phosphorylation of Nucleosides to Generate 5′-Monophosphate or 5′-Monophosphorothioate Derivatives

Generation of 5′-phosphorylated nucleosides can be achieved in a straightforward manner and in high yield by chemical phosphorylation of nucleosides with phosphoryl chloride and its derivatives. The example protocol given below describes the generation of GMPS by reacting unphosphorylated guanosine with thiophosphoryl chloride. The reaction of phosphoryl chloride with nucleosides in triethylamine occurs almost exclusively with the 5′-hydroxyl, making protection of the 2′- and 3′-hydroxyl unnecessary [37]. In addition, workup to generate nucleotide triphosphates from the resulting monophosphates has been well described [38]. Because many useful analogs are only available as nucleosides, this general procedure can provide opportunities for probing RNA structure and function that might not otherwise be available.

This protocol begins with the generation of a saturated solution and slurry of the nucleoside in triethylamine. Phosphoryl chloride or one of its derivatives is then added to the slurry. A nucleophilic reaction results in the 5′ attachment of phosphoryl chloride to the nucleoside. As the nucleoside becomes phosphorylated its solubility increases and the pH of the solution is lowered, resulting in more of the nucleoside becoming soluble until a clear solution is achieved. When this reaction is complete, the nucleotide 5′-phosphoryl chloride is hydrolyzed to the nucleotide monophosphate by simple addition of an excess of water. The nucleotide is then purified by ion exchange chromatography [39, 40].

A wide variety of nucleoside analogs and phosphoryl chloride derivatives are commercially available and can be substituted for thiophosphoryl chloride in this protocol. Methyl-, phenyl- and ethyl-dichlorophosphite, thiophosphoryl chloride, and 4-nitrophenyl phosphodichloridate are available from Sigma-Aldrich, and can be used to yield useful analogs. For example, in addition to GMPS, we have used this procedure to generate other 5′-phosphorylated guanosine derivatives including 6-thioguanosine monophosphate and guanosine 5′-p-nitrophenylphosphate [39, 40]. All other reagents used in the following procedure are commonly available from a variety of sources.

To synthesize GMPS, begin by mixing 2 mmol of guanosine and 5 ml triethylamine in a small round-bottom flask. Use a heating mantle to gently warm the flask to 50 °C and stir at that temperature for 10 min. Next, cool the guanosine solution on wet ice for at least 10 min. With the resulting guanosine slurry on ice, add 0.6 ml (5.8 mmol) of thiophosphoryl chloride. Continue stirring on ice in a cold room at 4 °C overnight (or at least 7 h) until the guanosine slurry has become a clear solution. The reaction mixture is next mixed with 500 ml of water to hydrolyze the resultant to guanosine 5′-thiophosphoryl chloride to GMPS. Because the hydrolysis reaction is exothermic, the reaction mixture should be added to the water in small aliquots and the solution should be stirred on ice between additions. Prior to chromatographic purification, adjust the pH of the solution to 7.5.

For purification of GMPS and other analogs, we have used a 2.5 × 18-cm col-

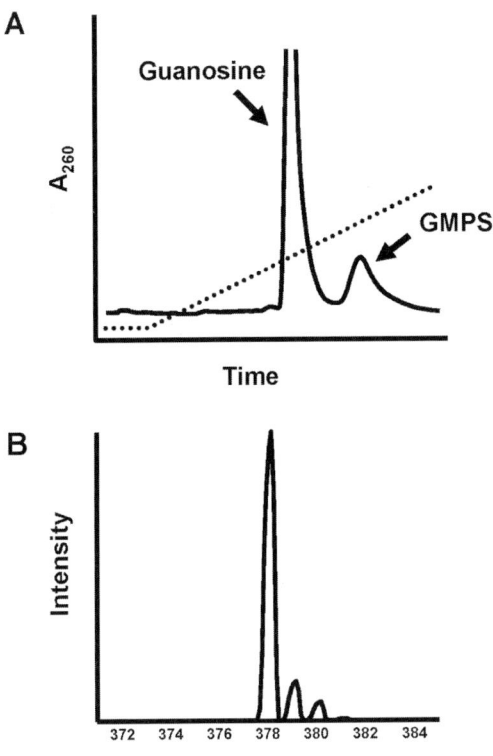

Fig. 5.2. GMPS synthesis. (A) Absorbance trace showing the purification of GMPS from guanosine by liquid chromatography. The dotted line indicates the fraction of buffer B (0.5 M ammonium bicarbonate, pH 7.0) in water. (B) Mass spectrographic analysis of GMPS from the indicated peak in (A). Data were acquired with a Thermoquest TSQ quadrapole mass spectrometer equipped with an electrospray ion source.

umn containing Supelco TSK-gel Toyopearl DEAE-650M resin prepared according to the manufacturers instructions. Load the sample in 50-ml aliquots at a flow rate of 1 ml/min and wash the column with three column volumes of water. Elute the nucleoside and phosphorylated compound with a linear gradient of 0–100% B in 60 ml (A = water; B = 0.5 M ammonium bicarbonate, pH 7.0). A sample chromatogram with good separation between the nucleoside, which elutes first, and the thiophosphorylated nucleotide is shown in Fig. 5.2(A). The column can be regenerated by washing with several column volumes of 100% B followed by water. Pool peak fractions and recover products by rotary evaporation. Next, resuspend the residue in 200 ml of 10% ethanol and dry again. Repeat this step at least 4 times. Finally, resuspend the GMPS product in 5 ml of water.

The identity of the individual peaks can be confirmed by running the appropriate fractions on polyethyleneimine TLC plates developed in 1 M lithium chloride relative to standards. In addition, samples can be analyzed using mass spectrome-

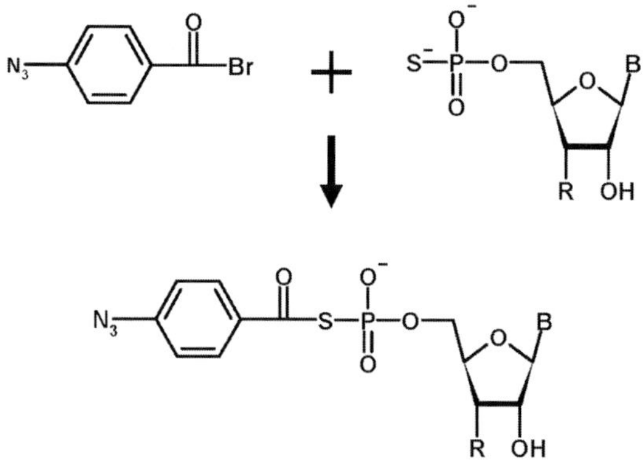

Fig. 5.3. Azidophenacyl bromide attachment to a 5′-phosphorothioate. Reaction scheme for modification of a 5′-monophosphorothioate modification with azidophenacyl bromide. Reaction with an acid bromide generates a linkage through the sulfur atom of the phosphorothioate.

try. A sample of mass chromatogram of final pooled GMPS fractions is shown in Fig. 5.2(B). A clear peak at 378 m.u. indicates that GMPS was the sole component of the second peak shown in Fig. 5.2(A). Peaks at 379 and 380 m.u. are due to the natural abundance of carbon and oxygen isotopes. The final product concentration can be determined by UV absorbance ($\varepsilon_{260} = 11.7 \times 10^3$ M^{-1} cm^{-1} at pH 7 [41]).

5.2.3
Attachment of an Arylazide Photo-crosslinking Agent to a 5′-Terminal Phosphorothioate

RNAs which are modified with a 5′-terminal phosphorothioate as described above can be further derivatized with reagents containing an acid bromide [11, 15]. The following example describes the attachment of an azidophenacyl photo-crosslinking agent to a 5′-phosphorothioate modified RNA (see Fig. 5.3). This crosslinking agent has been used to investigate the structures of various RNAs and has an effective crosslinking radius of approximately 10 Å.

Sodium bicarbonate, Tris(hydroxymethyl)aminomethane (Tris), ethylenediamine-tetraacetic acid (EDTA), sodium dodecyl sulfate (SDS) and methanol are commonly available from a number of suppliers. Azidophenacyl bromide (APBr) is available from Sigma-Aldrich. The methanol solution of APBr used for this reaction should be freshly prepared. In addition, reducing agents such as DTT may reduce the azide moiety of this crosslinker to an amine, and should be avoided. Finally, as this protocol involves the use of photosensitive reagents, care should be taken to avoid exposure to light.

Begin by resuspending 10–40 μg of GMPS primed RNA in 10 μl of water. Add the following:

Water	24 µl
100 mM sodium bicarbonate, pH 9.0	20 µl
0.05% SDS	22 µl
3 mg/ml APBr in methanol	40 µl

Incubate the reaction mixture in the dark at room temperature 1 h. Add 100 µl of 10 mM Tris, 1 mM EDTA, pH 8.0. Next, extract the reaction mixture with an equal volume of 1:1 phenol:chloroform to remove any unreacted APBr. Finally, extract the reaction with an equal volume of chloroform and recover the RNA products by ethanol precipitation.

5.2.4
3'-Addition of an Arylazide Photo-crosslinking Agent

Photo-crosslinking agents and other modification can be post-transcriptionally added to the 3' end of RNA molecules by chemical modification [11, 42]. As described above, this procedure takes advantage of the unique *cis*-diol of the 3'-terminal ribose. The procedure described below is a three-step process. In the first step, the 3'-*cis*-diol is oxidized to a dialdehyde (Fig. 5.1) [43]. In the second step, a primary amine is introduced at the 3' end. The example below uses ethylenediamine in this step of the reaction [44, 45]. If a longer carbon chain is desired, other compounds such as 1,6-diaminohexane can be used [42]. In the third and final step, an N-hydroxysuccinimidyl containing reagent is reacted with the modified RNA. The example given here utilizes N-hydroxysuccinimidyl 4-azidobenzoate for the addition of the arylazide crosslinking agent. Other N-hydroxysuccinimidyl containing reagents, such as a number of available fluorescent labels, should also be usable in this protocol.

As with the other protocols presented in this section, most of the reagents called for are commonly available. N-hydroxysuccinimidyl 4-azidobenzoate is available from Sigma (http://www.sigmaaldrich.com). Like azidophenacyl bromide used in the protocols above, care should be taken to avoid exposure of the photoagent to light and to reducing agents such as DTT. In addition, appropriate safety precautions should be taken when handling ethylenediamine.

Begin by mixing the following:

1 M sodium acetate, pH 5.4	10 µl
30 mM NaIO$_4$	10 µl
RNA	5–10 µg
Water	to 100 µl

Incubate the reaction mixture for 1 h at room temperature in the dark. Next, recover the RNA by ethanol precipitation. Resuspend the precipitate in 72 µl and add the following:

200 mM imidazole, pH 8.0	10 µl
50 mM NaCNBH$_3$	10 µl
100 mM EDTA	1 µl
15 mM ethylenediamine	6.7 µl

Incubate this solution at 37 °C for 1 h. Add 10 μl of 50 mM NaBH$_4$ and continue incubating at 37 °C for an additional 10 min. Once again, recover reaction products by ethanol precipitation. Resuspend the precipitate in 50 μl of 100 mM HEPES, pH 9.0 and add 50 μl of 20 mM N-hydroxysuccinimidyl 4-azidobenzoate. Allow this mixture to react at room temperature for 1 h in the dark. Finally, ethanol precipitate to recover the 3′-modified RNA product.

5.3
Troubleshooting

The 5′-end modification protocol described in this section is dependent on T7 RNA polymerase to incorporate modified nucleotides. As such, a major concern is likely to be the efficiency of incorporation. As described above, the level of incorporation can be adjusted to suit the experiment at hand by altering the ratio of modified nucleotide to its unmodified counterpart in the transcription reaction. However, increasing concentrations of modified nucleotide, especially if the modification does not lend itself to efficient incorporation is likely to also reduce overall yield. Mutant versions of T7 RNA polymerase may also be useful in increasing the incorporation of modifications, especially those with 2′ or minor groove constituents [35, 36].

References

1 M. Takeshita, C. N. Chang, F. Johnson, S. Will, A. P. Grollman, *J. Biol. Chem.* **1987**, *262*, 10171–10179.
2 N. G. Walter, *Methods* **2001**, *25*, 19–30.
3 X. Zhuang, L. E. Bartley, H. P. Babcock, R. Russell, T. Ha, D. Herschlag, S. Chu, *Science* **2000**, *288*, 2048–2051.
4 K. K. Singh, T. Rucker, A. Hanne, R. Parwaresch, G. Krupp, *Biotechniques* **2000**, *29*, 344–348, 350–341.
5 R. Contreras, H. Cheroutre, W. Degrave, W. Fiers, *Nucleic Acids Res.* **1982**, *10*, 6353–6362.
6 M. M. Konarska, R. A. Padgett, P. A. Sharp, *Cell* **1984**, *38*, 731–736.
7 J. K. Yisraeli, D. A. Melton, *Methods Enzymol.* **1989**, *180*, 42–50.
8 J. Jemielity, T. Fowler, J. Zuberek, J. Stepinski, M. Lewdorowicz, A. Niedzwiecka, R. Stolarski, E. Darzynkiewicz, R. E. Rhoads, *RNA* **2003**, *9*, 1108–1122.
9 S. H. Hixson, S. S. Hixson, *Biochemistry* **1975**, *14*, 4251–4254.
10 A. Favre, C. Saintome, J. L. Fourrey, P. Clivio, P. Laugaa, *J. Photochem. Photobiol. B* **1998**, *42*, 109–124.
11 M. E. Harris, E. L. Christian, *Methods* **1999**, *18*, 51–59.
12 M. M. Hanna, S. Dissinger, B. D. Williams, J. E. Colston, *Biochemistry* **1989**, *28*, 5814–5820.
13 J. R. Wyatt, E. J. Sontheimer, J. A. Steitz, *Genes Dev.* **1992**, *6*, 2542–2553.
14 E. L. Christian, M. E. Harris, *Biochemistry* **1999**, *38*, 12629–12638.
15 A. B. Burgin, N. R. Pace, *EMBO J.* **1990**, *9*, 4111–4118.
16 B. K. Oh, N. R. Pace, *Nucleic Acids Res.* **1994**, *22*, 4087–4094.
17 M. E. Harris, A. V. Kazantsev, J. L. Chen, N. R. Pace, *RNA* **1997**, *3*, 561–576.
18 M. E. Harris, N. R. Pace, *Mol. Biol. Rep.* **1995**, *22*, 115–123.
19 A. Montpetit, C. Payant, J. M.

Nolan, L. Brakier-Gingras, *RNA* **1998**, *4*, 1455–1466.
20 D. Dertinger, L. S. Behlen, O. C. Uhlenbeck, *Biochemistry* **2000**, *39*, 55–63.
21 A. M. Pyle, T. R. Cech, *Nature* **1991**, *350*, 628–631.
22 S. A. Strobel, T. R. Cech, *Biochemistry* **1993**, *32*, 13593–13604.
23 S. A. Strobel, T. R. Cech, *Science* **1995**, *267*, 675–679.
24 S. A. Strobel, T. R. Cech, *Biochemistry* **1996**, *35*, 1201–1211.
25 R. K. Gaur, L. W. McLaughlin, M. R. Green, *RNA* **1997**, *3*, 861–869.
26 D. J. Earnshaw, M. L. Hamm, J. A. Piccirilli, A. Karpeisky, L. Beigelman, B. S. Ross, M. Manoharan, M. J. Gait, *Biochemistry* **2000**, *39*, 6410–6421.
27 D. Herschlag, F. Eckstein, T. R. Cech, *Biochemistry* **1993**, *32*, 8312–8321.
28 R. K. Gaur, G. Krupp, *Nucleic Acids Res.* **1993**, *21*, 21–26.
29 S. A. Strobel, *Curr. Opin. Struct. Biol.* **1999**, *9*, 346–352.
30 E. L. Christian, N. H. Zahler, N. M. Kaye, M. E. Harris, *Methods* **2002**, *28*, 307–322.
31 Y. L. Chiu, T. M. Rana, *RNA* **2003**, *9*, 1034–1048.
32 S. Dokudovskaya, O. Dontsova, O. Shpanchenko, A. Bogdanov, R. Brimacombe, *RNA* **1996**, *2*, 146–152.
33 J. Kurreck, *Eur. J. Biochem.* **2003**, *270*, 1628–1644.
34 J. R. Sampson, O. C. Uhlenbeck, *Proc. Natl. Acad. Sci. USA* **1988**, *85*, 1033–1037.
35 R. Padilla, R. Sousa, *Nucleic Acids Res.* **2002**, *30*, e138.
36 R. Padilla, R. Sousa, *Nucleic Acids Res.* **1999**, *27*, 1561–1563.
37 M. Yoshikawa, T. Kato, T. Takenishi, *Tetrahedron Lett.* **1967**, *50*, 5065–5068.
38 A. Arabshahi, P. A. Frey, *Biochem. Biophys. Res. Commun.* **1994**, *204*, 150–155.
39 E. L. Christian, D. S. McPheeters, M. E. Harris, *Biochemistry* **1998**, *37*, 17618–17628.
40 A. G. Cassano, V. E. Anderson, M. E. Harris, *J. Am. Chem. Soc.* **2002**, *124*, 10964–10965.
41 R. M. C. Dawson, D. C. Elliott, W. G. Elliott, K. M. Jones, *Data for Biochemical Research*, Clarendon Press, Oxford, **1986**.
42 B. K. Oh, D. N. Frank, N. R. Pace, *Biochemistry* **1998**, *37*, 7277–7283.
43 S. B. Easterbrook-Smith, J. C. Wallace, D. B. Keech, *Eur. J. Biochem.* **1976**, *62*, 125–130.
44 U. C. Krieg, P. Walter, A. E. Johnson, *Proc. Natl. Acad. Sci. USA* **1986**, *83*, 8604–8608.
45 R. Rayford, D. D. Anthony, Jr, R. E. O'Neill, Jr, W. C. Merrick, *J. Biol. Chem.* **1985**, *260*, 15708–15713.

6
3′-Terminal Attachment of Fluorescent Dyes and Biotin

Dagmar K. Willkomm and Roland K. Hartmann

6.1
Introduction

A large number of experimental approaches in RNA biochemistry are based on some sort of label or tag within RNA molecules. Several of the methods at hand for RNA labeling are widely used, such as incorporation of modified nucleotides during solid-phase synthesis (Chapters 7 and 8) or transcription (Chapter 1), as well as post-transcriptional random chemical labeling by commercially available kits (e.g. Biotin Chem-Link from Roche or ULYSIS Nucleic Acid Labeling Kits from Molecular Probes). In contrast, the selective chemical attachment of a label to periodate-oxidized RNA 3′ ends described here is less common, despite some advantages over alternative labeling techniques. Primarily, the procedure is efficient, inexpensive and suitable for large-scale preparations. Further, it is not restricted to newly synthesized RNA and can therefore also be applied to commercially available RNAs or molecules isolated from cells and tissue. Moreover, the tag is incorporated at the RNA end, where it is less likely to impede proper folding than elsewhere in the molecule. Finally, as a single labeling group is added to each RNA molecule's 3′ end, this technique yields a rather uniform RNA population in terms of structure and RNA:label ratio.

Both fluorescent and biotin RNA tags are versatile tools in a broad array of techniques. One of the traditional domains of 3′ fluorescence of labeling RNA is fluorescence resonance energy transfer (see also Chapter 28) and related methods which have initially been used to probe RNA positioning within the ribosome [1, 2] and to analyze RNA–protein interactions within the signal recognition particle [3, 4]; further applications range from RNA–RNA binding measurements [5] to the use of fluorescent RNAs as hybridization probes [6, 7] and for microinjection experiments [8, 9].

Biotin as a tag is generally used because of its exceptionally tight binding to streptavidin, a 60-kDa bacterial protein, and avidin, the related protein from egg white. With an estimated dissociation constant of about 10^{-14} M, the complex formation between biotin and streptavidin is essentially irreversible under a wide variety of conditions [10 and references therein, 11]. More recently, modified avidins which allow reversible binding have also become available [6]. Accordingly, biotin

---

Handbook of RNA Biochemistry. Edited by R. K. Hartmann, A. Bindereif, A. Schön, E. Westhof
Copyright © 2005 WILEY-VCH Verlag GmbH & Co. KGaA, Weinheim
ISBN: 3-527-30826-1

tags have their uses in immobilization of RNA to solid supports as a prerequisite for capture assays and affinity chromatography [12, 13] or surface plasmon resonance measurements [14, 15]. Finally, among many further applications (reviewed in [16]), biotin-labeled RNAs are also employed as probes, with a subsequent detection mostly based on streptavidin-alkaline phosphatase conjugates [17].

6.2
Description of Method

The underlying reaction mechanism of 3′-label chemical attachment is based on the selective periodate-mediated oxidation of the RNA 3′-terminal ribose cis-diol (Fig. 6.1A; [18] and references therein). Oxidation results in a dialdehyde which is highly susceptible to nucleophilic attack and will therefore readily react with nucleophilic amino components such as hydrazine derivatives. As a final step, the reaction product can be stabilized by borohydride reduction.

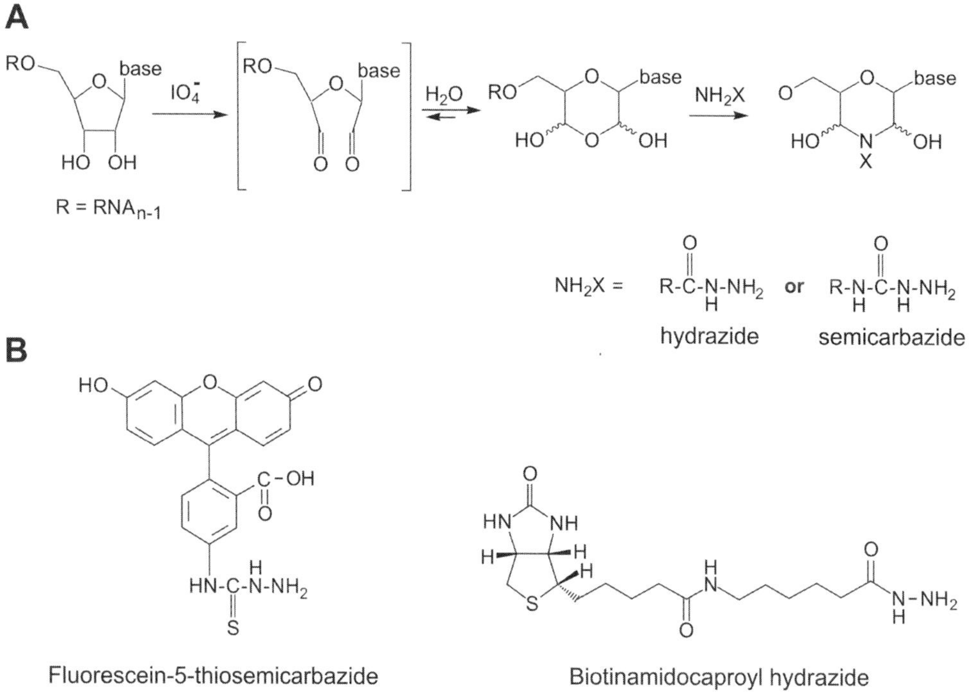

Fig. 6.1. (A) Reaction mechanism for modifications at the 3′-terminal ribose of RNAs. Oxidation of the ribose cis-diol with periodate results in a reactive dialdehyde which is then attacked by the hydrazide amino group (adapted from [18]). (B) Examples of coupling reagents: fluorescein-5-thiosemicarbazide, the hydrazide most frequently used as an RNA fluorescence tag, and biotinamidocaproyl hydrazide, to be used for biotin labeling.

With regard to the requirements of the labeling chemistry, a variety of hydrazine derivatives have been developed and are commercially available. Biotin can be purchased as biotin hydrazide, a direct hydrazide conjugate, from a number of sources. In two other biotin derivatives on the market, biotinamidocaproyl hydrazide (= biotin-X-hydrazide = biotin-6-aminohexanoic hydrazide, Fig. 6.1(B); from Sigma-Aldrich or Calbiochem) and biotin-XX-hydrazide (Calbiochem, Molecular Probes), the carboxylic acid hydrazide moiety serving as nucleophilic reactant is separated from the biotin moiety by a 7- or 14-atom spacer, respectively, to minimize steric interference between RNA and tag. Biotin-X hydrazide is the more commonly used reagent due to better water solubility.

Regarding fluorescent dyes, the fluorophore is in most cases directly conjugated to the semicarbazide or hydrazide moiety, as for example in fluorescein-5-thiosemicarbazide (Fig. 6.1B) and Alexa Fluor™ hydrazides (both from Molecular Probes), as well as eosin-5-thiosemicarbazide (Sigma-Aldrich). However, reagents with a short extra spacer between the fluorophore and the reactive group, advantageous to some applications [19], are also available. For a compilation of the diverse fluorophore structures and fluorescence properties, see [6] and Molecular Probes' extensive online list at http://www.interchim.fr/bio/molprobes/cd/docs/tables/0301.htm.

6.3
Protocols

6.3.1
3′ Labeling

6.3.1.1 Biotin Attachment [12]

(1) Incubate up to about 3–4 nmol RNA in a total volume of 100 μl 40 mM KIO_4 for 1 h at room temperature in the dark.
(2) Stop the reaction with 100 μl of 50% ethylene glycol, then add 1/10th volume of 3 M NaOAc (pH 5.2) and 2.5 volumes of 96% ethanol to precipitate the RNA. After centrifugation, wash the pellet with 70% ethanol and dry.
(3) Dissolve the dried pellet in 100 μl of 10 mM biotinamidocaproyl hydrazide and incubate for 2 h at 37 °C.
(4) Add 100 μl of 0.2 M $NaBH_4$ and 200 μl of 1 M Tris–HCl, pH 8.2. Incubate for 30 min on ice in the dark.
(5) Purify the RNA (see Section 6.3.3) to remove salt and coupling reagent.

Special care has to be taken with some of the solutions:

- KIO_4 needs to be prepared as a 50 mM aqueous solution adjusted to pH 7.0 with 10 N NaOH. The KIO_4 will dissolve only upon NaOH addition; at pH 7.0,

50 mM is at the limit of solubility. A more alkaline pH, although helpful for dissolving the KIO_4, reduces the yield of the overall reaction.
- Biotinamidocaproyl hydrazide has low solubility in water; to prepare the required 10 mM solution, incubate for 2 min at 95 °C.
- 0.2 M $NaBH_4$: the solution sets free hydrogen. Therefore, prepare small portions (Eppendorf tube scale is fine).

6.3.1.2 Fluorescence Labeling [5]

While similar in chemistry, a slightly different protocol has been used in our lab for fluorescence labeling. In order to minimize photobleaching of the fluorescent dye, all reactions need to be carried out in the dark.

(1) Incubate 20 nmol of RNA in a volume of 400 μl 2.5 mM $NaIO_4$, 100 mM NaOAc (pH 5.0) for 50 min on ice.
(2) Precipitate with ethanol, wash the pellet with 70% ethanol and dry.
(3) For the coupling reaction, dissolve the RNA in 400 μl of 100 mM NaOAc (pH 5.0) and 1 mM fluorescein-5-thiosemicarbazide (200 mM stock solution in dimethyl formamide) and incubate on ice overnight.
(4) Ethanol-precipitate and redissolve in 50 μl double-distilled water.
(5) Purify (see Section 6.3.3).

6.3.2
Preparatory Procedures: Dephosphorylation of RNA Produced with 3' Hammerheads

Transcription of RNA with 3' *cis*-hammerheads is particularly attractive because it gives defined homogeneous 3' ends as opposed to the heterogeneous 3' ends which result from run-off transcription (see Chapter 1). 3' Labeling of these RNAs released by hammerhead self-cleavage, however, poses a problem because their 3' end is masked by a 2',3'-cyclic phosphate. Prior to the labeling reaction, these RNAs therefore require 3' dephosphorylation, which can be done in a kinase reaction at low concentration of ATP, thus making use of the T4 polynucleotide kinase phosphatase activity [20]. The method can also be applied to other RNA substrates with a 3'-phosphate.

(1) Incubate 300 pmol RNA with 2',3'-cyclic phosphate ends for 6 h at 37 °C in a total volume of 100 μl containing:
   0.1 mM ATP
   100 mM imidazole–HCl (pH 6.0)
   10 mM $MgCl_2$
   10 mM β-mercaptoethanol
   2 μg BSA
   20 U T4 polynucleotide kinase
(2) Extract first with an equal volume of phenol/chloroform (1:1) and then with chloroform.
(3) Desalt by gel chromatography (see below) and precipitate with ethanol.

6.3.3
RNA Downstream Purifications

6.3.3.1 Gel Chromatography

Among the diverse options, Sephadex G-50 gel exclusion chromatography with self-made spin columns [21] has in our hands proved a cheap and efficient means to remove unincorporated label, particularly biotin, with almost no loss of RNA material.

(1) Prepare Sephadex G-50 slurry: add an equal volume of double-distilled water to the Sephadex, let swell for several hours with occasional gentle shaking. Wash twice with double-distilled water: let sediment (or spin briefly at low speed), exchange the water, resuspend and repeat once more. Perform a third wash with $0.1 \times$ TE (1 mM Tris, 0.1 mM EDTA) pH 7.4, let sediment and adjust the final volume of the supernatant liquid phase to 30% of the total volume (Sephadex plus supernatant). Resuspend before use.
(2) Remove the barrel from a 2-ml syringe, plug the syringe with a tiny amount of siliconized glass wool and fill with 1.5 ml Sephadex G-50 slurry.
(3) Place in a 15-ml disposable tube and centrifuge in a swinging-bucket rotor for 2 min at 550 g. Discard flowthrough.
(4) Place a decapped Eppendorf tube at the bottom of the 15-ml tube. Apply 200 µl of sample to the column and spin exactly as before. The flowthrough collected in the Eppendorf contains the purified RNA.

Depending on the centrifuge, the centrifugation step might need optimization. The aim is that the volume of flowthrough finally collected will be identical to the sample volume applied to the column. Alternatively, ready-made columns are available from a number of suppliers (e.g. Amersham Biosciences).

6.3.3.2 Purification on Denaturing Polyacrylamide Gels

With chromatographic purifications, there is the risk of low amounts of coupling reagents not being removed. For many applications it might therefore prove useful to check the purification by running a sample on a denaturing polyacrylamide gel, where unincorporated fluorescent dye can be seen to migrate below the bromophenol blue band, or to do a gel purification in the first place.

The method of gel purification, described extensively elsewhere in this handbook (e.g. Chapter 3), is convenient and efficient for RNAs of 400 nt or less. In addition to complete removal of coupling reagent, for shorter RNAs, such as tRNAs, it also allows elimination of unmodified RNA from the RNA pool because the attached dye slows down RNA migration (Fig. 6.2). For larger RNAs, the difference in electrophoretic mobility caused by the 3' label might well be too small to allow discrimination of modified and unmodified molecules.

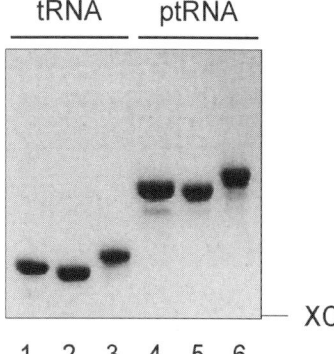

Fig. 6.2. Denaturing 9% polyacrylamide gel showing a tRNA^{Gly} (79 nt, lanes 1–3) and its 93-nt precursor (lanes 4–6) – both with homogeneous 3′ ends due to release from a primary transcript carrying a 3′-terminal *cis*-hammerhead – after hammerhead self-removal and gel purification (lanes 2 and 5), after subsequent 3′ dephosphorylation (lanes 1 and 4) and finally after 3′ labeling with fluorescein-5-thiosemicarbazide (lanes 3 and 6). Both the dephosphorylated molecules (lacking the negative charge of one phosphate group) and the tagged molecules (of increased molecular weight) run slightly slower than the unmodified RNAs. XC: position of xylene cyanol blue, at 20 cm from the slot.

6.3.4
Quality Control

Particularly when setting up the procedure in one's lab, it is advisable to run controls for labeling efficiency. For fluorescence labeling, the expected efficiency of the reaction is over 90%, and for homogeneous populations of small RNAs this can be monitored by denaturing polyacrylamide gel electrophoresis (Fig. 6.2). Biotin attachment should also proceed almost quantitatively. Biotinylation efficiency can be analyzed by gel shift assays with saturating amounts of streptavidin (commercially available for example from Sigma-Aldrich) on native agarose gels (see Fig. 6.3). As opposed to fluorescence labeling analysis, the shift caused by streptavidin is fairly large, so that the labeling of slightly heterogeneous RNA populations and larger RNAs can also be monitored.

6.4
Troubleshooting

When labeling and/or downstream application efficiencies are low, check the following aspects.

6.4.1
Problems Caused Prior to the Labeling Reaction

Quality of the RNA 3′ Ends
While the RNA 2′- and 3′-hydroxyls are a prerequisite for the reaction chemistry, a number of cleavage activities generate RNA 3′-phosphate or 2′,3′-cyclic phosphate

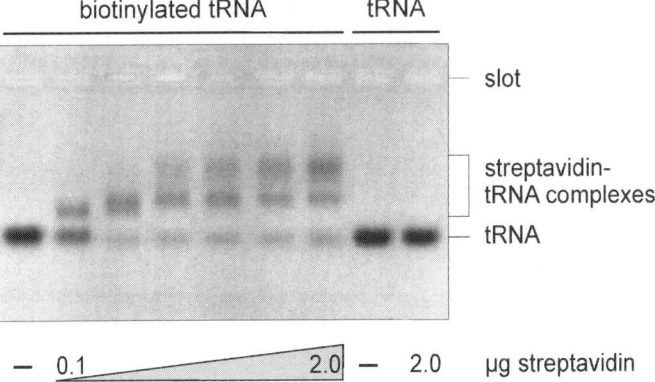

Fig. 6.3. Streptavidin retardation experiment to monitor the efficiency of biotinylation of yeast total tRNA (Roche). 140 ng of biotinylated tRNA were incubated with up to 2 μg of streptavidin for 15 min in 10 mM Tris, pH 7.4, 2.5 mM MgCl$_2$ and 100 mM NaCl. Samples were run on a 0.8% ethidium bromide-stained agarose gel. Different streptavidin–tRNA complexes were resolved on the gel, which may be related to the fact that streptavidin forms tetramers. At saturating amounts of streptavidin, over 80% of the tRNA showed reduced gel mobility. Non-biotinylated tRNA incubated with 2 μg streptavidin is shown as a control on the right.

ends, including several RNases (e.g. RNase A and RNase T1 [22]), metal ions [23] and small ribozymes (reviewed in [24]). The respective RNAs will need to be dephosphorylated prior to the labeling reaction (see Section 6.3.2). If the labeling efficiency is low even after a dephosphorylation step, the extent of dephosphorylation can be checked by denaturing PAGE (Fig. 6.2). However, PAGE resolution limits this kind of analysis to RNAs smaller than 100 nt.

Purity of the RNA to be Labeled
Nucleotides also react with the labeling reagents. When preparing the RNA to be labeled by transcription, thorough purification by either Sephadex columns or, preferably, polyacrylamide gel electrophoresis is therefore recommended. Impurities such as proteins or salts might impede the labeling reaction as well.

When the labeling efficiency for the RNA of interest is low, even though 2′,3′-hydroxyl ends should be present and the RNA has been thoroughly purified and desalted prior to labeling, a most likely cause is poor RNA quality due to degradation. In this case, an entirely new preparation of RNA should be used.

6.4.2
Problems with the Labeling Reaction Itself

pH of Reagents
A crucial and limiting aspect to the overall yield of the labeling reaction is the lability of the dialdehyde reaction intermediate. Since it is destabilized at basic pH, the

reaction conditions need to be kept non-alkaline. In particular, for the periodate oxidation according to the biotin-labeling protocol, neutral pH of the 50 mM KIO$_4$ solution is essential. At higher pH the reaction will be markedly less efficient.

Stability of Reagents
In general, it is advisable to store all labeling reagents at −20 °C. Nevertheless, especially the sodium borohydride solution will suffer from repeated freeze–thaw cycles and should be stored in aliquots. We have further prepared the periodate solutions freshly after three freeze–thaw cycles, and have kept the borohydride and periodate solutions for no longer than a few months.

6.4.3
Post-labeling Problems

Removal of Labeling Reagents
For most downstream applications, special care has to be taken to efficiently remove excess labeling reagents. Ethanol precipitation, even repeatedly, will generally not suffice. Substantial quantities of (anionic) contaminant fluorescent dyes will be visible on polyacrylamide gels, utilizing an appropriate excitation light source. Such gels need to be run very shortly in order to prevent the low molecular weight dye from exiting the gel and to minimize lateral diffusion. Unincorporated biotin derivatives can be detected by competition with the biotinylated RNA for binding to streptavidin, resulting in a considerable increase in the amount of streptavidin needed for saturation in a control shift assay.

Loss of RNA Material during Downstream Purification
The risk of losing RNA during purification is not significantly altered compared to unmodified RNA. A very important exception is phenol extraction of biotinylated material: since biotin increases hydrophobicity, the biotinylated RNA may be retained to some extent at the water/phenol interphase. Phenol extraction of biotinylated RNA should therefore be avoided.

Stability of Labeled RNA
Thiosemicarbazide adducts tend to degrade above pH 8 and at elevated temperatures [25], and thus should strictly be kept cold and dark, as well as below pH 8.

References

1 O. W. ODOM, D. J. ROBBINS, J. LYNCH, D. DOTTAVIO-MARTIN, G. KRAMER, B. HARDESTY, Biochemistry **1980**, 19, 5947–5954.
2 M. STOFFLER-MEILICKE, G. STOFFLER, O. W. ODOM, A. ZINN, G. KRAMER, B. HARDESTY, Proc. Natl. Acad. Sci. USA **1981**, 78, 5538–5542.
3 F. JANIAK, P. WALTER, A. E. JOHNSON, Biochemistry **1992**, 31, 5830–5840.
4 G. LENTZEN, B. DOBBERSTEIN, W.

WINTERMEYER, *FEBS Lett.* **1994**, *348*, 233–238.
5 S. BUSCH, L. A. KIRSEBOM, H. NOTBOHM, R. K. HARTMANN, *J. Mol. Biol.* **2000**, *299*, 941–951.
6 R. P. HAUGLAND, *Handbook of Fluorescent Probes and Research Products*, 9th edn, Molecular Probes, Eugene, OR, **2002**. This handbook, available from Molecular Probes or online (www.probes.com/handbook) provides comprehensive up-to-date information on labeling and detection reagents and techniques currently used.
7 D. EGGER, R. BOLTEN, C. RAHNER, K. BIENZ, *Histochem. Cell Biol.* **1999**, *111*, 319–324.
8 C. KRUSE, D. K. WILLKOMM, A. GRÜNWELLER, T. VOLLBRANDT, S. SOMMER, S. BUSCH, T. PFEIFFER, J. BRINKMANN, R. K. HARTMANN, P. K. MÜLLER, *Biochem. J.* **2000**, *346*, 107–115.
9 T. PEDERSON, *Nucleic Acids Res.* **2001**, *29*, 1013–1016.
10 M. L. JONES, G. P. KURZBAN, *Biochemistry* **1995**, *34*, 11750–11756.
11 M. WILCHEK, E. A. BAYER (eds), *Avidin–Biotin Technology (Methods Enzymol 184)*, Academic Press, San Diego, CA, **1990**.
12 U. VON AHSEN, H. F. NOLLER, *Science* **1995**, *267*, 234–237.
13 J.-L. JESTIN, E. DÈME, A. JACQUIER, *EMBO J.* **1997**, *16*, 2945–2954.
14 M. BUCKLE, R. M. WILLIAMS, M. NEGRONI, H. BUC, *Proc. Natl. Acad. Sci. USA* **1996**, *93*, 889–894.
15 M. HENDRIX, E. S. PRIESTLEY, G. F. JOYCE, C.-H. WONG, *J. Am. Chem. Soc.* **1997**, *119*, 3641–3648.
16 M. WILCHEK, E. A. BAYER, *Biomol. Eng.* **1999**, *16*, 1–4.
17 F. M. AUSUBEL, R. BRENT, R. E. KINGSTON, D. D. MOORE, J. G. SEIDMANN, J. A. SMITH, K. STRUHL, *Current Protocols in Molecular Biology*, Wiley, New York, **1994**.
18 F. HANSSKE, F. CRAMER, *Methods Enzymol.* **1979**, *59*, 172–181.
19 D. KLOSTERMEIER, D. P. MILLAR, *Methods* **2001**, *23*, 240–254.
20 V. CAMERON, O. C. UHLENBECK, *Biochemistry* **1977**, *16*, 5120–5126.
21 J. SAMBROOK, E. F. FRITSCH, T. MANIATIS, *Molecular Cloning: A Laboratory Manual*. Cold Spring Harbor Laboratory, Cold Spring Harbor Press, NY, **1989**.
22 J. N. DAVIDSON, *The Biochemistry of the Nucleic Acids*, 7th edn. Academic Press, New York, **1972**.
23 T. PAN, D. M. LONG, O. C. UHLENBECK, Divalent metal ions in RNA folding and catalysis, in: *The RNA World*, R. F. GESTELAND, J. F. ATKINS (eds), Cold Spring Harbor Laboratory Press, Cold Spring Harbor Press, NY, **1983**.
24 Y. TAKAGI, M. WARASHINA, W. J. STEC, K. YOSHINARI, K. TAIRA, *Nucleic Acids Res.* **2001**, *29*, 1815–1834.
25 R. DULBECCO, J. D. SMITH, *Biochim. Biophys. Acta* **1960**, *39*, 358–361.

I.2
Chemical RNA Synthesis

7
Chemical RNA Synthesis, Purification and Analysis

Brian S. Sproat

7.1
Introduction

The interest in chemically synthesized RNA took a dramatic leap forward with the discovery and application of the small interfering RNA (siRNA) technology [1, 2], a technique which has revolutionized functional genomics and target validation during the past 2–3 years and equals antisense technology in its applicability. However, the chemical synthesis of RNA until recently lagged a long way behind the well-established DNA synthesis technology. A few pioneers in the field have contributed to the three solid-phase RNA synthesis chemistry variants that are now used by commercial suppliers of RNA, i.e. the 2′-O-TBDMS method [3], the TOM method [4], which is a variant of the TBDMS method, and the 2′-ACE method [5, 6]. The abbreviations refer to the ether protecting groups used for the ribose 2′-hydroxyl group: TBDMS is tert-butyldimethylsilyl, TOM is triisopropylsilyloxymethyl and ACE is bis(acetoxyethoxy)methyl. The TOM and ACE variants are quite recent. In the past, problems with RNA synthesis were largely caused by poor-quality RNA phosphoramidites (the building blocks for solid phase synthesis), inappropriate protecting groups taken from DNA synthesis, poor activating agents and suboptimal deprotection protocols. This combination of largely unavoidable obstacles combined with the intrinsic chemical and biological instability of RNA led in most cases to failed syntheses. However, the boom in usage of synthetic RNA both for siRNA and other applications such as ribozymes and aptamers has had a very positive effect in that the speciality reagent suppliers have improved the quality of the building blocks leading to healthy competition and affordable prices. Moreover, the use of optimized protecting groups, coupling agents and deprotection protocols has revolutionized chemical RNA synthesis.

Since most commercially available synthesizers are not compatible with the highly specialized 2′-ACE chemistry, the methods described here have been restricted to the standard TBDMS chemistry, but also apply to the closely related TOM chemistry. The synthesis method described here is of course one of many variant methods, but all methods are in the end a variation of the basic methods described here. Since synthesis will be performed in the solid-phase using well-

Handbook of RNA Biochemistry. Edited by R. K. Hartmann, A. Bindereif, A. Schön, E. Westhof
Copyright © 2005 WILEY-VCH Verlag GmbH & Co. KGaA, Weinheim
ISBN: 3-527-30826-1

established phosphoramidite chemistry [7, 8] it can be performed manually or on any of the commercially available instruments. Synthesis starts from the 3′ terminus starting with the 3′-terminal nucleoside anchored most commonly via a succinyl linkage to an insoluble matrix, generally aminopropyl or long-chain alkylamine functionalized controlled pore glass, or polystyrene, contained in an appropriate reaction vessel. The nucleobases of the phosphoramidites and functionalized supports are preferably protected with N-phenoxyacetyl (pac) [9] or N-tert-butylphenoxyacetyl (tac) [10] groups enabling very mild deprotection of the RNA at the end of the synthesis, however the use of N^6-benzoyl A, N^4-acetyl C and N^2-isobutyryl G phosphoramidites leads to similar results regarding yield and purity of the RNA. The structure of the tac-protected cytidine building block is illustrated in Fig. 7.1. The advantage of the solid-phase method is that reagents are introduced into the vessel for removing protecting groups and enabling chain extension of the RNA 1 nt at a time and excess reagents are simply flushed away with a suitable solvent, in this case acetonitrile. The cyclical process is repeated until the desired length of RNA is obtained. Since no intermediate purification steps are possible, purification is done at the end of the assembly when most of the protecting groups have been removed. In practice, all reactions proceed close

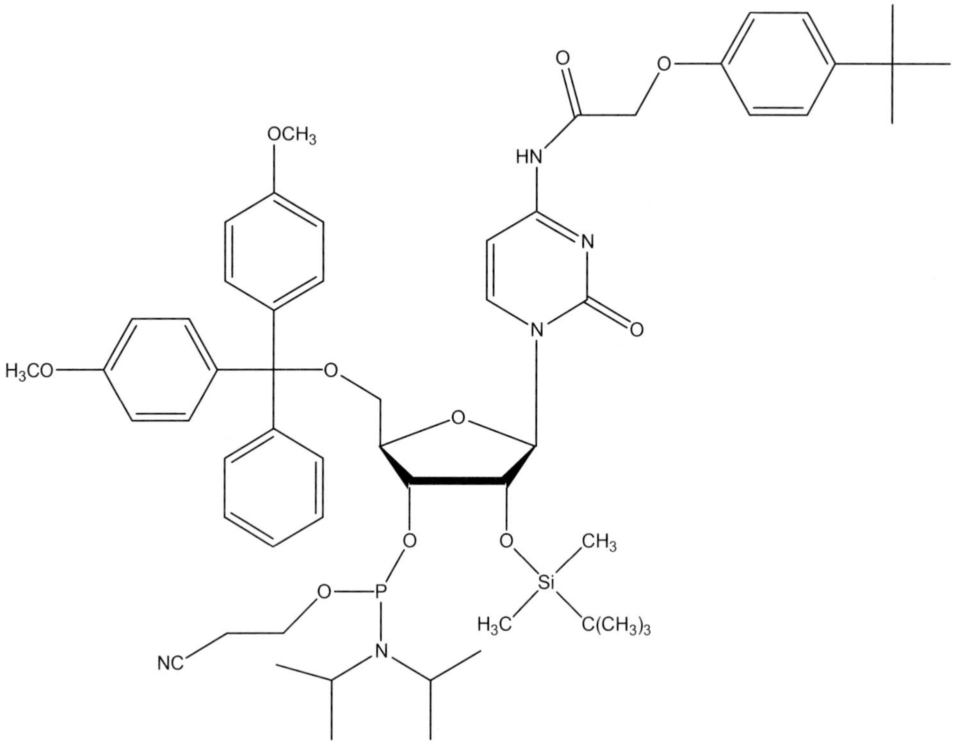

Fig. 7.1. Structure of a standard cytidine phosphoramidite building block, carrying 4-t-butylphenoxyacetyl protection of the exocyclic amino group.

to 100% yield and the chain extension reaction has a yield in the range of 98.5–99%, thus enabling in most cases a straightforward purification of the crude product.

Upon completion of the synthesis the fully protected support-bound RNA is deprotected in a stepwise fashion. In the first step the linkage to the solid-phase and the nucleobase and phosphate protecting groups are cleaved. In the second step the TBDMS groups are cleaved using triethylamine *tris*(hydrofluoride) [11, 12]. When RNAs longer than about 25 nt are synthesized it is best to leave the 5′-terminal dimethoxytrityl group attached as it is lipophilic and can be used as a purification aid. The crude RNA is then purified by anion-exchange and/or reversed-phase HPLC according to the length of the RNA and the purity required. For applications such as NMR spectroscopy and X-ray crystallography, purities of greater than 98% are desirable.

7.2 Description

7.2.1 The Solid-phase Synthesis of RNA

This section is devoted to the synthesis of the fully protected RNA in the solid-phase. The various steps involved in each cycle of the synthesis are illustrated in Fig. 7.2. Each cycle comprises a detritylation step that unmasks the 5′-hydroxyl group for chain extension, a coupling step in which the desired nucleotide as a phosphoramidite building block activated with 5-(benzylthio)-1*H*-tetrazole [13] is added, a capping step that acylates any unreacted 5′-hydroxyl group, an oxidation step that converts the phosphite triester to a phosphate triester, a second capping step that removes any occluded iodine and of course in between washing steps with acetonitrile to remove excess reagents. 5-(Benzylthio)-1*H*-tetrazole (BTT) for activation of the sterically hindered 2′-*O*-TBDMS-protected phosphoramidites is strongly preferred over conventional 1*H*-tetrazole with regard to both speed and coupling efficiency [13]; however, 4,5-dicyanoimidazole (DCI) and 5-(ethylthio)-1*H*-tetrazole can also be used with similar efficiency. Moreover, syntheses can be performed manually or machine-assisted using the following reagents and equipment:

(1) RNA monomers: 5′-*O*-Dimethoxytrityl-*N*(pac or tac)-2′-*O*-TBDMS-3′-*O*-(β-cyanoethylphosphoramidites) of adenosine (A), uridine (U), cytidine (C) and guanosine (G). These compounds are available for instance from Pierce (Milwaukee, USA) or Proligo (Hamburg, Germany) and should be stored dry under argon at −20 °C. Alternative suppliers of fast deprotecting RNA phosphoramidites are Transgenomic, Promega JBL, Glen Research and ChemGenes.
(2) Solid-phase supports, either CPG (Proligo, Pierce, ChemGenes, Glen Research and Transgenomic) or polystyrene (available from Amersham Biosciences, now part of GE Healthcare) functionalized with A, U, C and G.

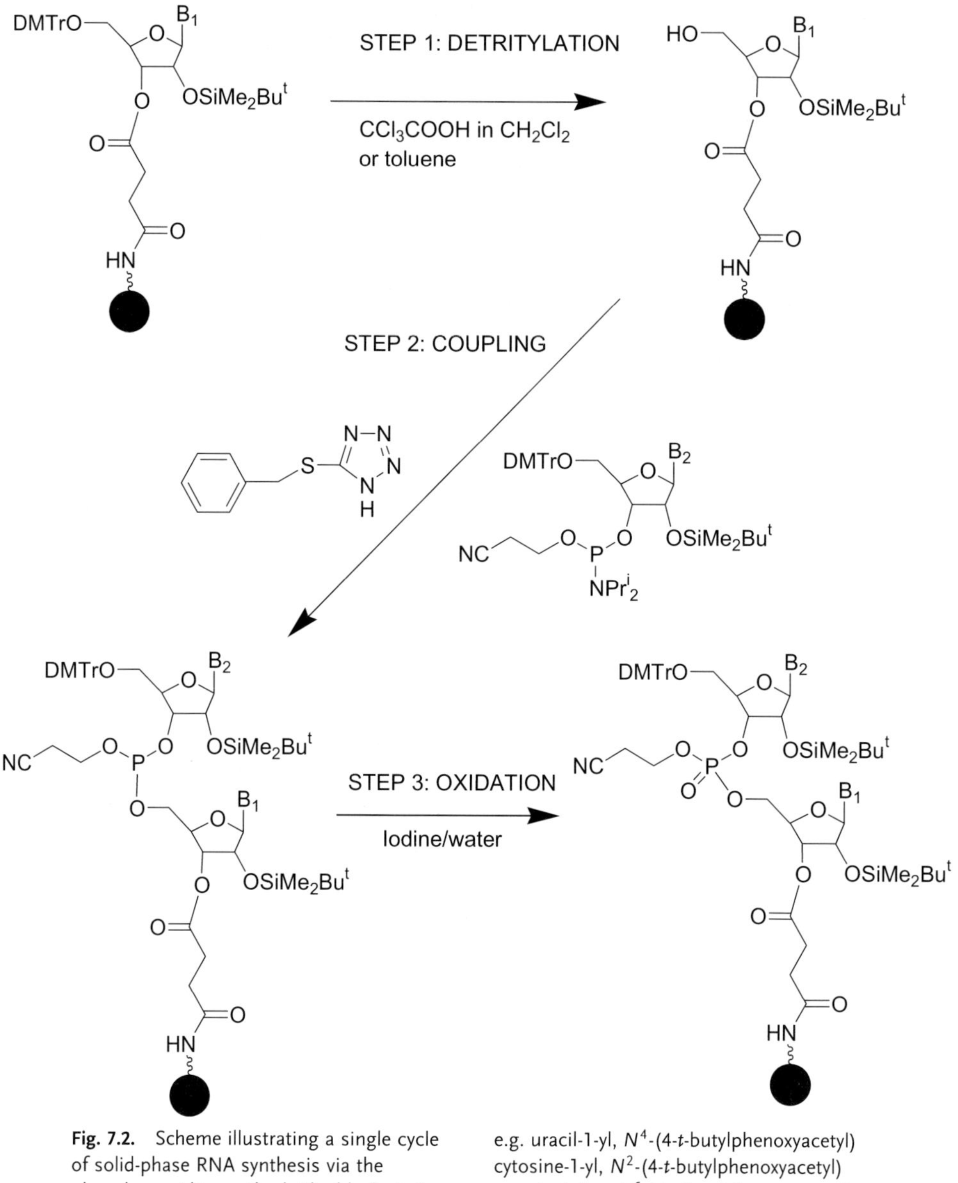

Fig. 7.2. Scheme illustrating a single cycle of solid-phase RNA synthesis via the phosphoramidite method. The black circle represents the controlled pore glass support. B_1 and B_2 represent protected nucleobases, e.g. uracil-1-yl, N^4-(4-t-butylphenoxyacetyl) cytosine-1-yl, N^2-(4-t-butylphenoxyacetyl) guanin-9-yl or N^6-(4-t-butylphenoxyacetyl) adenin-9-yl.

(3) 5-Benzylthio-1H-tetrazole (BTT), the activating agent which is available with a very low residual water content from emp Biotech (Berlin, Germany) or CMS (Oxford, UK).
(4) Capping solutions A (fast deprotection since it is based on 4-*tert*-butylphenoxy-

acetic anhydride) and B from Proligo. Capping solutions are also available from Merck, Riedel-de-Haen, Biosolve and Malinckrodt-Baker, for example.

(5) Oxidation solution containing iodine from Proligo, Merck, Riedel-de-Haen, Biosolve or Malinckrodt-Baker, for example. Since the iodine concentration is often only 17 mM, the concentration should be adjusted to 50 mM when performing large-scale syntheses by adding the correct amount of high-purity iodine.

(6) Deblock solution comprising 3% trichloroacetic acid in dichloromethane from Proligo or another supplier. For large-scale syntheses an alternative deblock solution containing up to 6% dichloroacetic acid in toluene is usually used, in particular on the Äkta OligoPilot 10.

(7) DNA synthesis grade acetonitrile containing less than 30 p.p.m. water (Malinckrodt-Baker, Merck, Riedel-de-Haen and Biosolve, for example).

(8) Assorted 1000 series gas-tight syringes with volumes of 0.5, 1 and 2.5 ml, which can be purchased from the Hamilton Company (Reno, NV, USA).

(9) DNA/RNA synthesizer (Applied Biosystems, Amersham Biosciences or other manufacturer) or, for manual synthesis, a glass reaction vessel fitted with a B14 ground glass joint at the top and a fine porosity glass frit and a tap at the bottom [14]. A set of suitable vessels can be made by any laboratory glass blower.

7.2.1.1 Manual RNA Synthesis

(1) Weigh out the requisite amounts of the tac- or pac-protected monomers required in small vials that can be closed with a septum and dry them overnight *in vacuo* over separate containers of phosphorus pentaoxide and potassium hydroxide pellets to remove traces of water. Suitable vials for this purpose are those amber glass bottles, which are used by suppliers of DNA and RNA phosphoramidites. To perform syntheses in the range of 1–3 µmol it is recommended to use 8–10 equivalents of monomer per coupling relative to the amount of support used. For synthesis scales above 5 µmol the monomer excess can be reduced to 5-fold.

(2) Carefully release the vacuum with dry argon and seal the bottles with tight fitting rubber septa.

(3) Using a gas tight syringe dissolve each of the monomers in the requisite volume of dry acetonitrile to give a 0.1 M solution and seal with Parafilm. It is not recommended to store the monomer solutions for more than 2–3 days at room temperature.

(4) Prepare an adequate volume of a 0.3 M solution of BTT in very dry acetonitrile in a tightly closed bottle under argon.

(5) Prepare 100 ml of capping mixture comprising 1 volume of fast deprotection capping solution A and 1.1 volumes of capping solution B in a tightly stoppered flask. Fresh capping mixture should be made each day.

(6) Weigh out the requisite amount of CPG carrying the desired 3′-terminal ribonucleoside into the glass reaction vessel. For a 1-µmol scale synthesis the ves-

sel should have a volume of about 5 ml, whereas for 10-μmol scale a volume of 20 ml is more appropriate to allow good washing.

(7) Using a Pasteur pipette add 3% TCA in dichloromethane (deblock solution) to the support and let it percolate through. Immediately a deep orange color is produced, characteristic of the released dimethoxytrityl cation. Continue to add acid until the effluent is colorless.

(8) Now drain the support using a slight pressure of dry nitrogen or better argon.

(9) Wash the support 8–10 times in a batchwise fashion with acetonitrile using a Teflon wash bottle, removing the supernatant each time with argon pressure.

(10) Just prior to the coupling step, wash the CPG once with very dry acetonitrile containing less than 30 p.p.m. water, flush away with argon pressure, close the tap and stopper the vessel.

(11) Using two gas tight syringes add the desired monomer as a 0.1 M solution in acetonitrile and an equal volume of 0.3 M BTT solution in acetonitrile in a second gas tight syringe to the CPG, stopper the vessel and agitate several times during a period of 6 min.

(12) Whilst the coupling step is in progress clean both syringes thoroughly with acetonitrile and store them in a desiccator.

(13) Flush away the coupling mixture, wash the CPG once with acetonitrile and flush away with argon pressure.

(14) Add a few milliliters of capping mixture to the reaction vessel, stopper and agitate for 1 min and then drain.

(15) Wash the CPG once with acetonitrile and flush away with argon pressure.

(16) Add a few milliliters of oxidation mixture and allow it to slowly percolate through the CPG during 2 min. This step oxidizes the phosphite triester to a phosphate triester.

(17) Drain the CPG and wash once with acetonitrile and drain with argon pressure.

(18) Once again add a few milliliters of capping mixture, agitate for 30 s and then drain using argon pressure.

(19) Now wash the CPG thoroughly with acetonitrile 6 times, draining each time in between using argon pressure.

(20) Repeat Steps 7–19 as many times as necessary until the desired sequence has been reached.

(21) If the RNA is longer than about 25 nt the final trityl group should be left on as a lipophilic purification handle. For RNAs less than 25 nt in length remove the final trityl group as in Step 7 and wash the CPG very thoroughly with acetonitrile.

(22) Finally dry the CPG using a stream of argon.

7.2.1.2 Automated RNA Synthesis

In order to perform automated RNA synthesis follow the instructions for the particular synthesizer plus the programme for the RNA synthesis scale you intend to use. Now the CPG or polystyrene support is placed inside a small plastic cartridge. All the reagents that you will need, including prepacked columns, are commercially available in the correct bottles to fit the various instruments on the market. Activated molecular sieves or trap bags can be added to ensure that reagents stay dry during the synthesis.

7.2.2
Deprotection

In the first deprotection step the succinate linkage connecting the 3' terminus of the protected RNA to the solid support is cleaved, the β-cyanoethyl phosphate protecting groups are removed by β-elimination and in addition the exocyclic nucleobase protecting groups are cleaved. In our hands this step is best performed with a 1:1 mixture of concentrated aqueous ammonia and 8 M ethanolic methylamine, which prevents premature loss of the TBDMS groups, that would otherwise lead to degradation of the RNA under basic conditions. *Warning, methylamine is not compatible with N^4-benzoyl-protected C.* In the second deprotection step the TBDMS groups are removed using triethylamine *tris*(hydrofluoride) in an appropriate solvent. At this point you will need the following reagents:

(1) High-purity concentrated aqueous ammonium hydroxide. This solution is highly irritating to the eyes and respiratory system and must only be used in a well-ventilated fume cupboard.
(2) Anhydrous 8 M methylamine in ethanol. This compound is also highly irritating to the eyes and respiratory system and must only be used in a well-ventilated fume cupboard.
(3) Anhydrous dimethylsulfoxide (e.g. Fluka, Biotech. Grade).
(4) Triethylamine *tris*(hydrofluoride) (e.g. Aldrich). This compound is hazardous and toxic, and should only be handled wearing full protection and used only in a well-ventilated fume cupboard.
(5) Anhydrous triethylamine.
(6) *N*-Methylpyrrolidone, peptide synthesis grade.
(7) Prop-2-yl trimethylsilyl ether prepared according to Jones [15].
(8) Diethyl ether.

7.2.2.1 Deprotection of Base Labile Protecting Groups

(1) Transfer the support obtained from Section 7.2.1 to a small screw top vial or Duran bottle equipped with a tight fitting screw top.
(2) Add a mixture of concentrated aqueous ammonia and 8 M ethanolic methylamine, 1:1 by volume. A volume of 2 ml is adequate for a 0.2–1-μmol scale synthesis. Otherwise use 2 ml/μmol of support.
(3) Close the vial or bottle tightly and seal further with Parafilm.
(4) Place the vial or bottle in a preheated oven at 65 °C for 20 min for small vials, but 40 min for larger bottles, which take longer to equilibrate thermally.

(5) Allow the vial/bottle to cool completely to room temperature before opening in a fume cupboard.
(6) Carefully remove the supernatant and wash the support several times with a few milliliters of ethanol/sterile water (1:1 by volume).
(7) Combine the supernatant and washings in a Falcon tube and dry in a Speed Vac or, for bigger volumes, evaporate to dryness on a rotary evaporator. Do not use water bath temperatures above 30 °C for trityl-on material.
(8) Dry the residue once by evaporation of absolute ethanol.

7.2.2.2 Desilylation of Trityl-off RNA

(1) Add a 1:1 mixture of dry DMSO and triethylamine *tris*(hydrofluoride) [16], using 600 μl/μmol, to the trityl-off residue in the Falcon tube obtained in section 7.2.2.1 above and sonicate briefly. If you have dried down the oligoribonucleotide in a glass flask, dissolve it in the required volume of dry DMSO using gentle warming of the flask using a hair dryer, transfer the solution to a Falcon tube and add an equal volume of the fluoride reagent.
(2) Close the tube, seal with Parafilm and place it in a preheated oven at 65 °C for 2.5 h.
(3) Cool the tube to room temperature.
(4) Add 2 volumes of isopropyl trimethylsilyl ether [15] to destroy the excess fluoride reagent, close the tube and shake vigorously at intervals during 10 min. At this point a white precipitate appears.
(5) Open the tube carefully and add 5 volumes of diethyl ether, close and agitate vigorously.
(6) Collect the precipitate by centrifugation at 4000 r.p.m. at 4 °C for 5 min.
(7) Remove the supernatant by careful decantation.
(8) Resuspend the pellet in diethyl ether, close the tube, agitate and again collect the precipitate by centrifugation.
(9) Repeat Steps 7 and 8 twice more.
(10) Finally dry the RNA pellet carefully *in vacuo*.

7.2.2.3 Desilylation of Trityl-on RNA

(1) Prepare a solution of N-methylpyrrolidone/triethylamine/triethylamine *tris*(hydrofluoride) (6:3:4 by volume) [17] immediately before use and add to the trityl-on residue obtained in section 7.2.2.1 above using 600 μl/μmol. For material that has been dried down in a glass flask, dissolve the residue in the minimum volume of dry DMSO, transfer the solution to a Falcon tube and add the freshly prepared desilylation solution.
(2) Perform Steps 2–8 as described in Section 7.2.2.2 above.
(3) Finally dry the RNA pellet very briefly using a stream of argon gas, then dissolve it immediately in sterile 0.1 M aqueous ammonium bicarbonate and purify immediately by reversed phase HPLC as the DMTr group has a limited half-life under these conditions.

7.2.3
Purification

This entire section is devoted to the anion-exchange HPLC purification of fully deprotected RNA using gradients of sodium perchlorate [17] or lithium perchlorate [18] as chaotropes, the reversed-phase HPLC purification of trityl-on RNA, detritylation and desalting. In this section you will need the following items:

(1) A biocompatible HPLC system (Amersham Biosciences or other).
(2) A set of anion-exchange HPLC columns, e.g. MonoQ 5/5, Source 15Q 16/10 and/or a FineLINE 35 pilot column packed with Source 15Q (Amersham Biosciences) or Dionex DNAPac PA-100 columns (Dionex, Sunnyvale, CA, USA).
(3) Sodium perchlorate. *Please note that this salt is toxic and corrosive.*
(4) Disodium EDTA.
(5) 1 M sterile Tris–HCl buffer, pH 7.4.
(6) A Hi-Prep 26/10 desalting column (Amersham Biosciences).
(7) Reversed-phase HPLC columns, for instance Hamilton PRP-1, 7 × 305 mm, XTerra^{TM} RP$_8$, 4.6 × 250 mm (Waters), XTerra^{TM} RP$_8$, 19 × 300 mm and/or a FineLINE 35 pilot column packed with Source 15RPC (Amersham Biosciences).
(8) HPLC grade acetonitrile.
(9) High-purity ammonium bicarbonate.
(10) Glacial acetic acid.

7.2.3.1 Anion-exchange HPLC Purification

The purity of oligoribonucleotides less than 25 nt in length, obtained by anion-exchange HPLC as the only purification step is perfectly adequate for most biological applications. It generally results in a purity in the range of 95–98%. Longer RNAs must be purified in the trityl-on mode, see Section 7.2.3.2 below. It is recommended to use a gradient of sodium perchlorate in sterile water/acetonitrile (9:1 v/v) containing 50 mM Tris–HCl buffer pH 7.6 and 50 µM EDTA for anion-exchange HPLC. The reason for adding EDTA is to complex traces of heavy metals that could otherwise lead to cleavage and degradation of the RNA. Recommended columns are the Source 15Q 16/10 columns for purification of 1-µmol scale syntheses with a flow rate of 5 ml/min. For syntheses in the 10–100-µmol scale, purification is best achieved using a FineLINE Pilot 35 column packed with Source 15Q and eluted at 20 ml/min. The low salt or A buffer preferably contains 10 mM sodium perchlorate and the high salt or B buffer contains 600 mM sodium perchlorate. It has been found that a gradient from 10–60% B during 40 min gives good resolution. When not in use the columns should be stored in 20% ethanol in sterile water to prevent microbial growth. For long-term storage of columns it is advisable to add 0.2% sodium azide as an antimicrobial. Prior to using a column that has been stored wash it with several column volumes of sterile water. The column is then equilibrated by washing it with several column volumes of buffer B followed by several column volumes of 90% buffer A plus 10% buffer B before injecting the

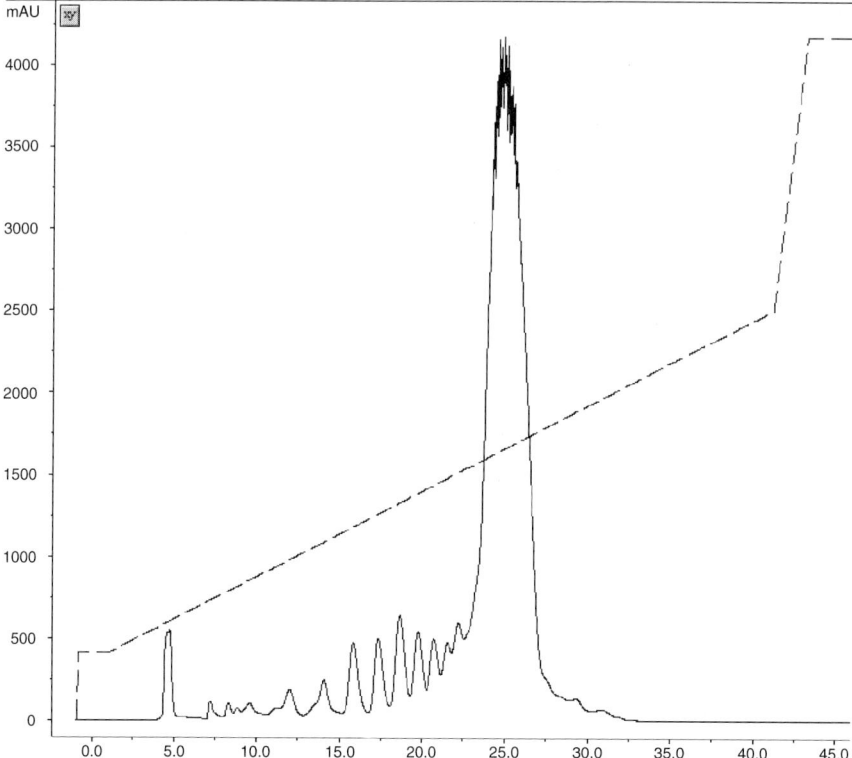

Fig. 7.3. Preparative anion-exchange HPLC trace of a 21mer oligoribonucleotide synthesized manually on 20-µmol scale and purified on Source 15Q packed in a FineLINE Pilot 35 column. The column was eluted with a linear gradient from 10 to 60% B during 40 min at a flow rate of 20 ml/min. Buffer A was 10 mM sodium perchlorate, 50 µM EDTA and 50 mM Tris–HCl, pH 7.6 in sterile water/acetonitrile (9:1 v/v), and buffer B was the same as buffer A except that the sodium perchlorate concentration was 600 mM. Absorbance was monitored at 280 nm. The x-axis is in minutes.

sample of fully deprotected RNA as obtained in Section 7.2.2.2 dissolved in buffer A and running the salt gradient. The desired product peak is the late eluting major component. This material is then desalted as described in Section 7.2.3.4 below. A typical trace of an anion-exchange HPLC purification is shown in Fig. 7.3. In this example the oligomer is a 21mer synthesized manually on a 20-µmol scale and purified on Source 15Q packed in a FineLINE Pilot 35 column. As can be seen the failure peaks are very small compared to the product peak which elutes at about 25 min.

7.2.3.2 Reversed-phase HPLC Purification of Trityl-on RNA

The highly lipophilic dimethoxytrityl group profoundly retards the full-length trityl-on RNA when purification is performed on a reversed phase HPLC column.

Although a better separation of failure peaks from the desired trityl-on product peak is obtained using aqueous triethylammonium acetate/acetonitrile buffers, for ease of salt removal and minimal damage to the RNA the use of ammonium bicarbonate instead of triethylammonium acetate is strongly preferred. However, make up the ammonium bicarbonate buffer fresh, otherwise store it cold, since it has a limited stability at room temperature in contrast to triethylammonium acetate.

Columns recommended for trityl-on purification are the Hamilton PRP-1, 7×305 mm for purifications on the scale of a few micromoles or a 21.5×250 mm column for 10–20-μmol scale purifications. As an alternative for larger-scale purifications, a FineLINE Pilot 35 column packed with Source 15RPC can be used. The buffers required are 0.1 M ammonium bicarbonate prepared in sterile water (buffer A) and 0.1 M aqueous ammonium bicarbonate/acetonitrile (1:1 by volume), which is buffer B. A useful gradient to use is 0–90% B during 40 min. The failure peaks elute early and are well separated from the desired trityl-on product peak which elutes last. The pure product fraction should be collected in a polypropylene Falcon tube and dried down in a Speed Vac. Residual ammonium bicarbonate is then removed by lyophilization of the product, which is then ready for detritylation. A typical trace of a trityl-on RNA purification by reversed phase HPLC is shown in Fig. 7.4. The example shows a trityl-on 34mer oligoribonucleotide synthesized by machine on a 20-μmol scale and purified on a FineLINE Pilot 35 column packed with Source 15RPC. As can be seen the desired product peak elutes at 20–25 min well separated from the trityl-off failure sequences which elute between 7 and 12 min.

7.2.3.3 Detritylation of Trityl-on RNA

(1) Dissolve the HPLC-purified trityl-on RNA, obtained in Section 7.2.3.2 above, in 3% sterile aqueous acetic acid (200 μl/μmol) and keep for 45 min at room temperature. The pH should be about 3.5.
(2) Neutralize the solution by careful addition of solid ammonium bicarbonate until evolution of carbon dioxide ceases. The pH will now be about 7.8.
(3) Dry the sample in a Speed Vac.
(4) Repurify the product by anion-exchange HPLC, as described in Section 7.2.3.1 above, which in addition converts the RNA from the ammonium form into the biologically useful sodium form.
(5) Desalt according to Section 7.2.3.4 below.

As an alternative to Steps 4 and 5 the salt exchange can be performed in a reliable and high yielding fashion by dissolving the residue from Step 3 in sterile 0.3 M aqueous sodium acetate (400 μl/μmol synthesis scale) and adding 2.5 volumes of absolute ethanol. After mixing and storage at −70 °C for 20 min the precipitated RNA is recovered by centrifugation, washed once with absolute ethanol and then dried carefully *in vacuo*. To ensure a complete exchange of cation from ammonium to sodium the precipitation procedure should be repeated once more. This protocol

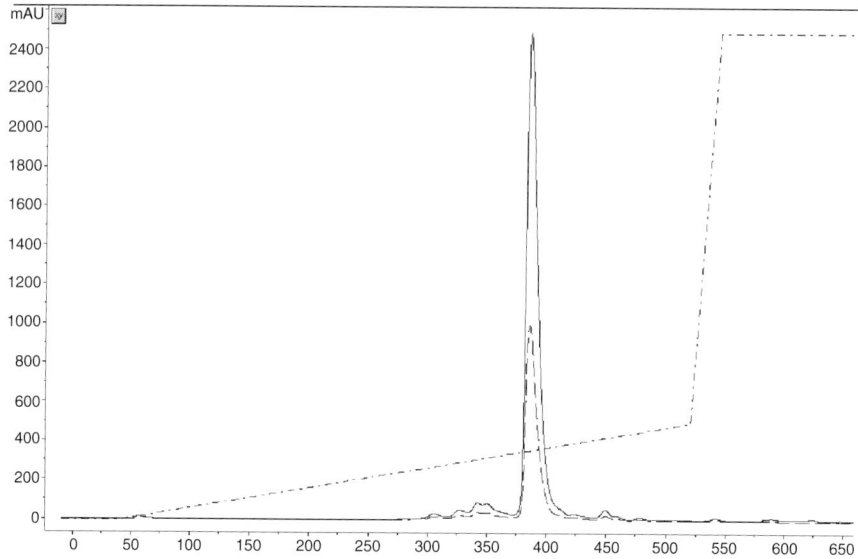

Fig. 7.5. Preparative reversed phase HPLC trace of a 27mer oligoribonucleotide on a 7-μm XTerra™ RP$_8$, 19 × 300 mm column. The compound was initially purified trityl-on by anion-exchange HPLC. The RP$_8$ column was eluted with a linear gradient from 0 to 40% B during 40 min at a flow rate of 12 ml/min. Buffer A was 0.1 M sterile aqueous ammonium bicarbonate and B was acetonitrile. The solid line traces absorption at 260 nm, the dashed line that at 280 nm. The x-axis is in milliliters.

be heeded during the solid-phase synthesis of RNA. The coupling step in the solid-phase synthesis is very sensitive to traces of water and it is essential to use very dry acetonitrile for monomer dissolution. It is also imperative to allow monomers to reach room temperature before opening and weighing out material, otherwise condensation of atmospheric water will occur leading to eventual degradation. The bottle contents should also be put back under argon before sealing and storing again at −20 °C. Addition of activated molecular sieves or trap bags to the monomer and activator solutions ensures that they stay dry during the synthesis. It is also critical to wash away the acid from the detritylation step with copious acetonitrile washes, otherwise residual acid will cause serious problems with the coupling step.

Incomplete oxidation will cause serious problems with the overall synthesis yield and quality as any residual phosphite triester is cleaved at the internucleotide linkage by the acid used in the detritylation step. As mentioned the standard oxidation mixture from several commercial suppliers is only 17 mM in iodine, i.e. 17 μmol/ml, so use enough solution for larger-scale syntheses to ensure that there is an ample excess of reagent and/or increase the iodine concentration to 50 mM. For a 20-μmol scale synthesis use 5 ml of the oxidation mixture. At the end of the solid-phase assembly the CPG or polystyrene bearing trityl-off protected RNA can be stored cold and dry ready for deprotection at an appropriate time; however, CPG

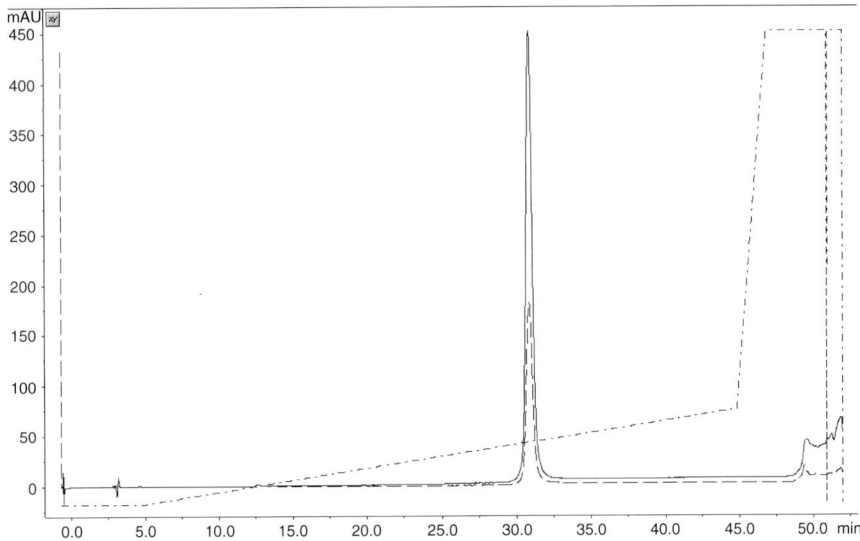

Fig. 7.6. Analytical reversed-phase HPLC trace of double HPLC-purified 27mer oligoribonucleotide run on a 5-µm XTerra™ RP$_8$, 4.6 × 250 mm column. The column was eluted with a linear gradient from 0–40% B during 40 min at a flow rate of 1 ml/min. Buffer A was 0.1 M sterile aqueous ammonium bicarbonate and B was acetonitrile. The solid line traces absorption at 260 nm, the dashed line that at 280 nm. The x-axis is in minutes.

bearing trityl-on RNA must be deprotected and purified immediately upon completion of the synthesis, otherwise there will be partial or complete loss of the trityl group during storage. This is particularly bad for RNAs that terminate at the 5′ end with one or more G residues.

For safety reasons handle all chemicals in a well-ventilated fume cupboard and wear suitable resistant disposable gloves, particularly when handling toxic materials such as triethylamine *tris*(hydrofluoride). In addition, make sure that you have read handling protocols for all chemicals – in particular those with which you are not familiar. Concerning the deprotection step with ammonia/methylamine, avoid too great an air space in the vial or bottle, otherwise most of the ammonia and methylamine will end up in the vapor phase. In the worst case this could lead to an incomplete deprotection. Take care when handling trityl-on RNA, do not over dry or let it get too hot, and purify immediately to avoid partial detritylation that will result in an unnecessary loss of product. This problem seems to be particularly serious with sequences that have one or more Gs at the 5′ end.

To avoid inadvertent degradation of unprotected RNA by RNases use RNase-free salts, sterilize all buffers by autoclaving and sterilize all glassware in a 180 °C oven. In addition, wear disposable gloves and as much as possible use sterile plasticware such as Eppendorf and Falcon tubes. Concerning anion-exchange HPLC purification, oligoribonucleotides that contain four or more consecutive Gs are notoriously difficult to purify since they form tetraplexes and higher aggregates in solution. Such RNAs are best purified using a lithium perchlorate gradient since these structures do not form if lithium ions are present instead of sodium or potassium ions. Of course prior to use in biological experiments the lithium ions must be replaced by sodium ions since lithium ions are toxic in many biological systems. Denaturants such as formamide can also be added to the salt gradient to reduce problems caused by strong secondary structures.

In large-scale purifications, to avoid product shoot through due to the ionic strength of the applied sample solution being too high, it is advisable to apply the crude RNA sample to the FineLINE Pilot 35 column dissolved in a volume of 10–50 ml of 50 mM Tris–HCl buffer, pH 7.4, using a 10- or 50-ml superloop. As an alternative desalt the sample prior to purification.

It is important to note that RNA samples as sodium salts are not suitable for mass spectrometry. Mass spectrometry samples of RNA are best prepared as ammonium salts. This can be done in several ways. One way is to do anion-exchange HPLC using ammonium sulfate for elution, followed by desalting on a small NAP cartridge. In this case store all buffers in plastic bottles and collect the product in an Eppendorf tube. A second way is to take a small aliquot of the RNA in its sodium form and exchange the sodium ions with ammonium ions by using ammonium form Dowex 50 cation exchange resin. A third way is to purify a small sample of RNA by reversed-phase HPLC using the aqueous ammonium bicarbonate/acetonitrile system followed by lyophilization to remove the residual salt. Once in the ammonium form the RNA should not be in contact with glass surfaces, otherwise sodium and potassium ions will be picked up that will severely degrade the quality of the mass spectra. This latter point is of great importance when trying to analyze very long RNAs, e.g. in the 40–70mer range.

Following the protocols and troubleshooting hints given above, the reader should be in a position to synthesize and purify RNA with success.

References

1 G. Ramaswamy, F. J. Slack, Chem. Biol. 2002, 9, 1053–1055.
2 J. Couzin, Science 2002, 298, 2296–2297.
3 N. Usman, K. K. Ogilvie, M.-Y. Jiang, R. J. Cedergren, J. Am. Chem. Soc. 1987, 109, 7845–7854.
4 S. Pitsch, Helv. Chim. Acta 1997, 80, 2286–2314.
5 S. A. Scaringe, F. E. Wincott, M. H. Caruthers, J. Am. Chem. Soc. 1998, 120, 11820–11821.
6 S. A. Scaringe, Methods Enzymol. 2000, 317, 3–18.
7 M. D. Matteucci, M. H. Caruthers, J. Am. Chem. Soc. 1981, 103, 3185–3191.
8 N. D. Sinha, J. Biernat, J. McManus, H. Köster, Nucleic Acids Res. 1984, 12, 4539–4557.
9 C. Chaix, D. Molko, R. Téoule, Tetrahedron Lett. 1989, 30, 71–74.
10 N. D. Sinha, P. Davis, N. Usman, J. Pérez, R. Hodge, J. Kremsky, R. Casale, Biochimie 1993, 75, 13–23.
11 D. Gasparutto, T. Livache, H. Bazin, A.-M. Duplaa, A. Guy, A. Khorlin, D. Molko, A. Roget, R. Téoule, Nucleic Acids Res. 1992, 20, 5159–5166.

12 E. Westman, R. Strömberg, *Nucleic Acids Res.* **1994**, *22*, 2430–2431.
13 R. Welz, S. Müller, *Tetrahedron Lett.* **2002**, *43*, 795–797.
14 B. S. Sproat, M. J. Gait, *Oligonucleotide Synthesis: A Practical Approach*, IRL Press, Oxford, **1994**, p. 92.
15 Q. Song, R. A. Jones, *Tetrahedron Lett.* **1999**, *40*, 4653–4654.
16 R. Vinayak, A. Andrus, A. Hampel, *Biomedical Peptides, Proteins & Nucleic Acids* **1995**, *1*, 227–230.
17 F. Wincott, A. DiRenzo, C. Shaffer, S. Grimm, D. Tracz, C. Workman, D. Sweedler, C. Gonzalez, S. Scaringe, N. Usman, *Nucleic Acids Res.* **1995**, *23*, 2677–2684.
18 B. Sproat, F. Colonna, B. Mullah, D. Tsou, A. Andrus, A. Hampel, R. Vinayak, *Nucleosides & Nucleotides* **1995**, *14*, 255–273.
19 U. Pieles, W. Zürcher, M. Schär, H. E. Moser, *Nucleic Acids Res.* **1993**, *21*, 3191–3196.

8
Modified RNAs as Tools in RNA Biochemistry

Thomas E. Edwards and Snorri Th. Sigurdsson

8.1
Introduction

RNA displays a vast variety of functions in that it carries genetic information, regulates gene expression, catalyzes reactions and participates in all facets of protein expression [1]. In addition to the four basic nucleosides (adenosine, guanosine, cytidine and uridine), many RNA molecules contain modified nucleosides essential for function. The fact that these modifications are essential for function in some RNAs, but entirely absent in others, indicates a significant layer of complexity in the hierarchy of RNA structure. With progress in the chemical synthesis of RNA over the last 15 years, modified nucleosides can now be readily incorporated at specific positions in RNA. These advances in solid-phase synthesis have promoted a cornucopia of experiments examining the influence of single-functional-group modification on the biological function of RNA.

Modified nucleosides have also been site-specifically incorporated into RNA as reporter groups for biochemical and biophysical structure–function analysis. There is a large diversity in these approaches. For example, fluorescent probes have been used to report internal changes during RNA folding [2] as well as to measure interhelical distances for determining the global structure of RNA [3–5]. Disulfide crosslinks have been used to restrict RNA helical elements to validate structural models based on other techniques [6]. These are but a few examples of site-specific incorporation of RNA structure–function probes.

The major goals of this chapter are to review the various types of modifications that can be incorporated site-specifically into RNA by chemical synthesis, and to provide a general method for the incorporation of reporter groups into RNA for biochemical and biophysical analysis. This includes comparison of the two central strategies for the incorporation of modified nucleosides into RNA, i.e. the phosphoramidite strategy and post-synthetic labeling. The phosphoramidite strategy utilizes chemical synthesis of a modified nucleoside phosphoramidite in conjunction with solid-phase synthesis, whereas post-synthetic labeling utilizes incorporation by the phosphoramidite method of a convertible nucleoside containing a reactive group, which is selectively modified after oligonucleotide synthesis with a labeling reagent. The advantages and disadvantages of each modification strategy will be described. Finally, a general and efficient modification strategy will be presented that utilizes post-synthetic labeling of 2′-amino groups with a wide range of reporter groups through a number of different coupling chemistries.

8.1.1
Modification Strategy: The Phosphoramidite Method

While enzymatic synthesis can be used to prepare uniformly labeled RNA, modified nucleosides can be incorporated site-specifically into RNA by solid-phase chemical synthesis using modified nucleoside phosphoramidites [7]. The main advantage of this method is that it allows for the incorporation of a desired modification or reporter group at a specific position in the RNA. While this is a highly effective and powerful method, it has several disadvantages. In most cases the synthesis of a modified phosphoramidite requires a lengthy and costly synthetic route. Furthermore, the reporter group must be stable to the conditions used in solid-phase oligonucleotide synthesis (e.g. incubation with acid, base and oxidizing solutions) as well as the deprotection conditions. However, phosphoramidites of many desirable modified nucleosides are commercially available, providing rapid, cost-effective access to a variety of modified RNAs.

There are four basic categories of RNA modifications that can be incorporated into RNA via the phosphoramidite method: end (5′ and 3′), base, phosphate and sugar modifications. Figure 8.1 shows selected examples of such modifications, and Table 8.1 lists several of the RNA modifications that are commercially available as phosphoramidites and/or modified RNAs. Of those not commercially available (many of which are reviewed in [8–11]), other notable examples of RNA modification by the phosphoramidite method include the base modifications 2-deoxyribonolactone [12] and 5-ketone pyrimidine derivatives [13]. Internucleotide linkages include boranophosphates [14] and phosphoroselenoates [15]. Sugar modifications

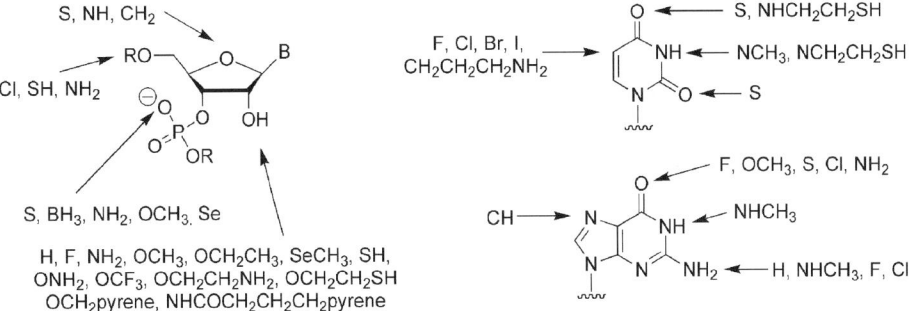

Fig. 8.1. Selected examples of modifications that can be incorporated at the sugar (left), phosphodiester backbone (left) and base (right) using the phosphoramidite method.

Tab. 8.1. Commercially available modifications that can be incorporated into RNA by the phosphoramidite method

Site	Modification	Commercial Source[1]	Reference[2]
End-labeling			reviewed in 74–76
5′ end	fluorescent dyes	CG, Dh, GR	3–5
	amino groups	CG, Dh, GR	77, 78
	biotin	CG, Dh, GR	
	photo-cleavable biotin	Dh, GR	
	5′-thiol	Dh, GR	
	acridine	GR	
3′ end	fluorescent dyes	CG, Dh, GR	
	amino groups	CG, Dh, GR	
	inverted abasic	Dh	
	puromycin	CG, Dh	79
	dideoxy G,C	Dh	
	biotin	CG, GR	
	acridine	GR	
	psoralen	CG, GR	
	cholesterol	CG, GR	
	DNP	CG, GR	
Internucleotide			
	S (non-bridging)	CG, Dh, GR	29–31
	3, 9, 18 atom spacers	CG, Dh, GR	
Sugar			reviewed in 9, 10
1′	abasic	Dh, GR	
2′	NH_2 U,C	CG	37
	F U,C	CG, Dh	
	OCH_3	CG, Dh, GR	
	SCH_3 U	GR	80
	$OCH_2CH_2CH_2NH_2$	CG	65
	$NHCOCH_2CH_2CH_2pyr$ U	Dh	36
	LNA	GR	81
Purine			reviewed in 9, 10
	N^6,N^6-dimethyl A	Dh	
	inosine	CG, Dh, GR	
	purine ribonucleoside	CG, Dh	
	ribavirin	Dh	
	7-deaza A,G	CG	
	2-aminopurine	CG, Dh, GR	2
	2,6-diaminopurine	Dh	
	8-bromo A	CG	
Pyrimidine			reviewed in 9, 10
	N^3-methyl U,rT	CG	
	N^3-thiobenzoyl ethyl U	CG	
	4-triazoylyl U	CG	
	N^4-ethyl C	CG	
	pyridine-2-one	CG	
	pyrrolo-C	GR	

Tab. 8.1. (continued)

Site	Modification	Commercial Source[1]	Reference[2]
	2,2′-anhydro U	CG	
	5-methyl U,C	Dh	
	4-thio uridine	CG, Dh, GR	71
	5-fluoro U	CG, Dh, GR	
	5-bromo U,C	CG, Dh, GR	
	5-iodo U	CG, Dh, GR	
	pseudouridine	CG, Dh, GR	
	5-CHCHCH$_2$NH$_2$ U	Dh	

[1] For commercial sources: CG, ChemGenes; Dh, Dharmacon; GR, Glen Research. Please note that chemical suppliers are subject to change and this list is a representative example at time of publication. Several other companies exist which sell modified RNA and modified RNA phosphoramidites.
[2] References are select examples and may be reviews, applications or synthetic procedures.

include 1′-deutero [16], 2′-modifications (*O*-(2-thioethyl) [17] and *O*-(2-aminoethyl) [18]), 5′-modifications (tallo or C-methyl [19, 20], chloro [21], amino [21]) and perdeuterated ribose [22]. Fluorescent labels have also been synthetically attached to the 2′-position via an ether linkage [23], a carbamoyl linker [24], an arabino carbamoyl linker [25] and an amido linkage [26].

The phosphoramidite method has been particularly useful in the investigation of ribozyme cleavage mechanisms. For example, incorporation of a 5′-C-methyl-modified nucleoside near the cleavage site of the hammerhead ribozyme resulted in a kinetically trapped intermediate in a crystal and provided information about a conformational change along the reaction pathway prior to transition state formation [20]. In another example, a crystal structure of the hairpin ribozyme containing a 5′-chloro group at the cleavage site provided structural information for comparison with the non-cleaved state (all RNA) and a vanadyl transition state mimic, providing valuable information about the entire mechanistic pathway [27].

8.1.2
Modification Strategy: Post-synthetic Labeling

Post-synthetic modification of convertible nucleosides enables the site-specific incorporation of a wide variety of reporter groups into RNA. The main advantage of this approach is that once the RNA has been prepared, it enables the rapid and efficient production of a wide variety of modified RNAs. Another advantage is that sensitive reporter groups, which would otherwise be unstable to the conditions of solid-phase oligonucleotide synthesis, can be incorporated into RNA. Possible disadvantages of this strategy are that in some cases additional purification steps are necessary and that it may be necessary to synthesize the convertible nucleoside phosphoramidite if the desired one is not commercially available.

Post-synthetic modification strategies have been developed for attachment of re-

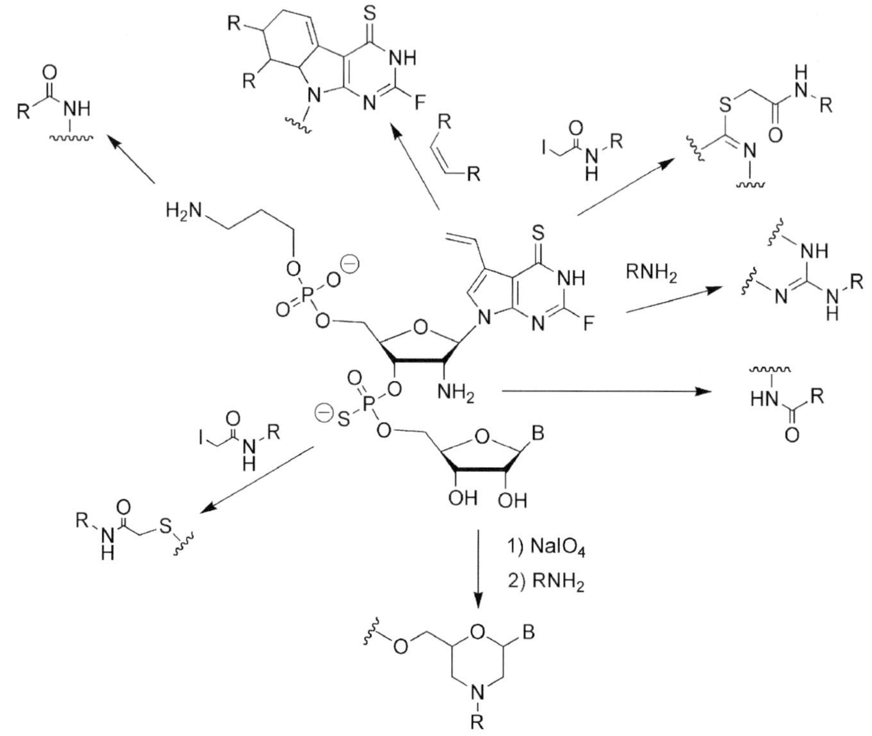

Fig. 8.2. Selected examples of RNA post-synthetic labeling.

porter groups to the 5′ and 3′ ends and at internal sites on the base, phosphate and sugar (Fig. 8.2 and Table 8.2). The main focus of this chapter is the general strategy of post-synthetic labeling of 2′-amino containing RNA and this approach will be described in detail in the next section. In addition to the modifications shown in Fig. 8.2 and Table 8.2, various groups can be attached to the 5-position of pyrimidines via on-column Pd-catalyzed coupling reactions [28]. A variety of molecules have also been attached to the phosphodiester backbone through phosphorothioate [29–31] or phosphoramidate linkages [32]. However, these labeling strategies are problematic for RNA due to the inherent instability of these linkages in the presence of 2′-hydroxyl groups; consequently, this problem is overcome by incorporation of a 2′-deoxy or 2′-methoxy group at the nucleotide 5′ of the modification.

8.2
Description of Methods

8.2.1
Post-synthetic Modification: The 2′-Amino Approach

Post-synthetic labeling of the 2′-amino group (Fig. 8.3a and b) has emerged as an effective approach for the site-specific incorporation of reporter groups into RNA.

Tab. 8.2. Select examples of modifications for post-synthetic RNA derivatization

Modification	Molecular handle	Commercially available?[1]	Labeling reactants	Reference[2]
1	3-amino modifiers	CG, Dh, GR	activated esters	77, 78
2[3]	diene	No	dienophile	82
3	sulfur-containing bases	CG, Dh, GR	iodoacetamides, disulfides	69–71
4	convertible F or ClΦ nucleosides	No	amines	83
5	2'-amino	CG, Dh	isothiocyanates, NHS esters, isocyanates	37
6	2',3'-diols	All (RNA)	NaIO$_4$, amines	84
7	non-bridging phosphorothioates	CG, Dh, GR	iodoacetamides	29–31

[1] For commercial sources: CG, ChemGenes; Dh, Dharmacon; GR, Glen Research. Please note that chemical suppliers are subject to change and this list is a representative example at time of publication. Several other companies exist which sell modified RNA and modified RNA phosphoramidites.
[2] References are select examples and may be reviews, applications or synthetic procedures.
[3] This modification strategy has only been applied to DNA thus far, but is of select interest.

Several notable examples include the incorporation of disulfide crosslinking reagents for the evaluation of RNA helical orientation [6, 33, 34], the incorporation of fluorescent probes for the study of RNA folding and ligand binding [35, 36] and the incorporation of EPR active probes [37] for the study of RNA internal dynamics [38–40] and for distance measurements [41]. RNAs containing 2'-amino groups at specific pyrimidine nucleotides (Dharmacon) and 2'-amino-modified pyrimidine phosphoramidites (ChemGenes) are now commercially available. Because the 2'-amino group is an aliphatic amine, it is more reactive (i.e. nucleophilic) than the aromatic amines or hydroxyl groups native to RNA, making this method of post-synthetic labeling highly selective. The major advantage of the 2'-amino group over other amino-based modifiers (e.g. 5'- and 3'-amino modifiers, 5-alkylamino-modified pyrimidines) is that it offers a minimal linker length. Several chemical conjugation approaches exist, including reaction with succinimidyl esters (often referred to as NHS esters) to produce amide-modified RNA [33], reaction with aromatic isothiocyanates to form thiourea-linked RNA [6, 42] and reaction with aliphatic isocyanates to prepare urea-tethered RNA [42, 43]. These three methods will be described in detail below, and examples employing these methods to address biochemical and biophysical questions will be provided. Of notable importance for this modification strategy is an alternative 2' protection approach that has been developed based on a photo-cleavable protecting group in place of the standard 2'-trifluoroacetyl group; after removal of the protecting group, the 2'-amino group may be derivatized on-column, providing many advantages over solution-based post-synthetic modification of deprotected oligonucleotides [44]. Other

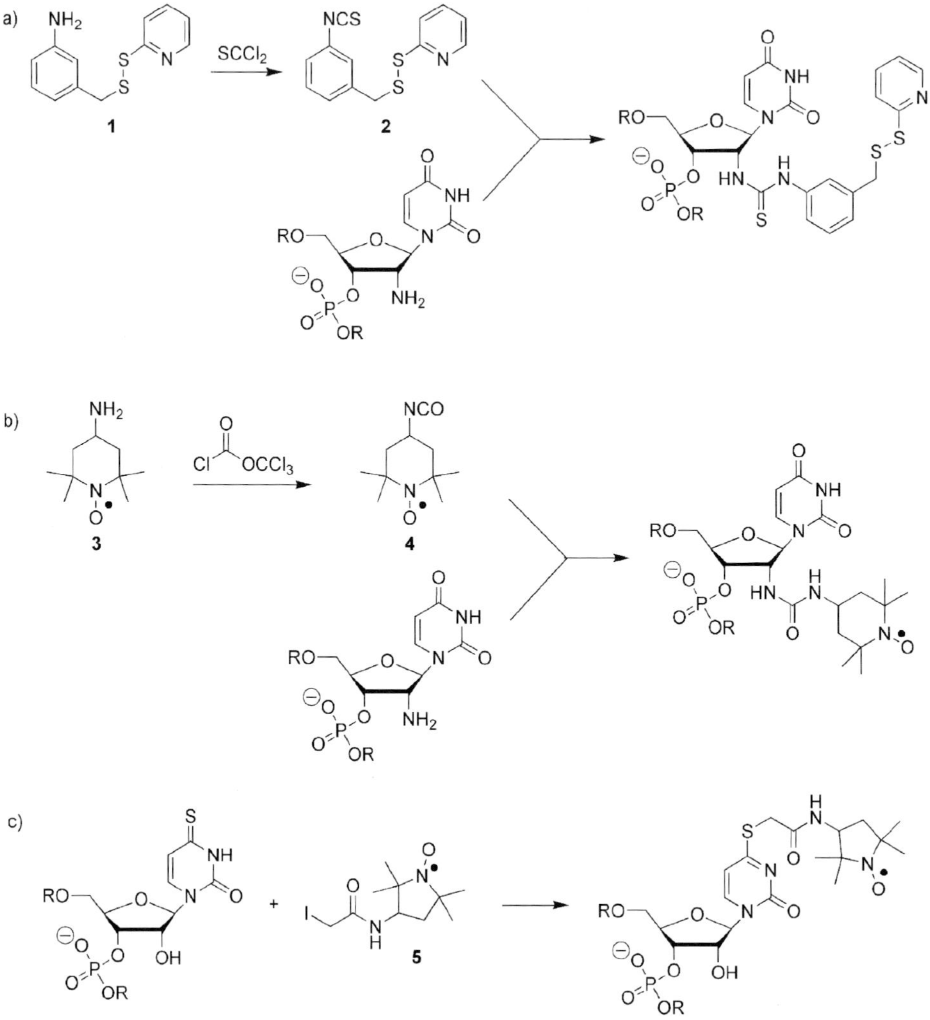

Fig. 8.3. Preparation of an isothiocyanate crosslinking reagent **2** (a) and an EPR spin-labeling reagent 4-isocyanato TEMPO **4** (b) from the corresponding amines using thiophosgene and diphosgene, respectively, and their subsequent incorporation into 2′-amino-modified RNA. (c) Post-synthetic modification of 4-thiouridine by alkylation with spin-label **5**.

approaches for the attachment of reporter groups to the 2′-position of oligonucleotides that will not be addressed in depth include chelation of metal ions such as ruthenium to 2′-amino-modified oligonucleotides [45, 46], incorporation of fluorescamine at a 2′-amino group using a Michael addition and rearrangement reaction [47], reaction of amines with a 2′-O-(acetaldehyde) group [48] and reduction of thiol-containing compounds with 2′-O-(2-thioethyl) to form disulfide-linked modified RNAs [17]. The attachment of sterically hindered compounds to the 2′-amino group may be difficult and can be overcome by use of the 2′-O-(2-aminoethyl) modification [18].

8.2.1.1 Reaction of 2′-Amino Groups with Succinimidyl Esters

The reaction of 2′-amino groups with succinimidyl esters to produce amido-linked modified RNAs has been used to incorporate disulfide crosslinks to evaluate RNA conformational dynamics [33, 49], to convert the hammerhead ribozyme from a ribonuclease to a ligase ribozyme [50], to incorporate photocrosslinking reagents to evaluate RNA tertiary structure [51], to identify base-pair mismatches [52, 53], and to incorporate fluorescent pyrene labels to study RNA folding and ligand binding [36, 54–56]. In conjunction with the isocyanate method described below, this method has been used to probe steric interference in the hammerhead ribozyme [57]. Catalysis of this chemistry by the phosphodiester on the 3′-position adjacent to the 2′-amino-containing nucleoside and/or the 3′-oxygen has been reported [58]. One advantage of this method is that many succinimidyl esters are commercially available (e.g. Molecular Probes and ChemGenes have a wide variety of amine-reactive succinimidyl ester dyes available). The major drawback of this method is that this chemistry often suffers from low labeling efficiency, e.g. the pyrene succinimidyl esters typically coupled with only 20–26% yield after purification [36]. However, in some cases it is possible to overcome this low coupling efficiency by the use of the corresponding carboxylic acid with an activating agent, such as a carbodiimide, which may result in nearly quantitative coupling [18, 44, 49]. Another disadvantage of this modification approach is that 2′-amido modifications destabilize RNA when incorporated at internal positions ($\Delta T_m \sim -5$ to 12 °C per modification) [26, 59, 60]. However, some 2′-amido-linked modifications located at end positions increase RNA stability [26, 61], which is likely a result of these particular modifications that contain large aromatic groups (e.g. fluorescent labels), which may stack onto the end of the helix. Nevertheless, this destabilizing effect of 2′-amido groups at internal sites should be kept in mind when incorporating reporter groups into RNA.

8.2.1.2 Reaction of 2′-Amino Groups with Aromatic Isothiocyanates

The reaction of 2′-amino groups with aromatic isothiocyanates (Fig. 8.3a) has been used to incorporate fluorescent probes [26, 35], disulfide crosslinks [6, 34] and photocrosslinking agents [62] into RNA. The main advantage of this method is the highly efficient chemistry, which has resulted in reported conversion yields in excess of 90% in all cases. In addition, fewer equivalents of the isothiocyanate labeling reagent are required than for the succinimidyl ester chemistry. The main drawback is that although some isothiocyanates are commercially available, most must be prepared from the corresponding amine and thiophosgene; however, this synthetic conversion is relatively straightforward [42, 63]. There is limited available UV thermal stability data for 2′-thioureido modifications and all of the data involves incorporation of large fluorescent probes (fluorescein and rhodamine). Following a similar pattern to that observed for the 2′-amido modifications, these modifications are rather destabilizing at internal positions, but have a stabilizing effect at end positions [26, 35]. Isothiocyanates have also been used to selectively incorporate metal ion chelators at 5-amino-derived pyrimidines [64] and at 2′-(O-propylamino)-derived nucleotides [65].

8.2.1.3 Reaction of 2′-Amino Groups with Aliphatic Isocyanates

The reaction of 2′-amino groups with aliphatic isocyanates (Fig. 8.3b) is a versatile platform for the incorporation of biochemical and biophysical reporter groups into RNA. This method has been used to incorporate disulfide crosslinks [34, 43, 66]; an activated disulfide that can be used to conjugate a wide variety of groups such as cholesterol [67], glutathione and bimane [43]; nitrophenol [44], pyrene [18] and nitroxide spin-labels [37]. Like the isothiocyanate coupling, this chemistry is highly efficient, typically displaying quantitative yields for unhindered isocyanates [43, 44]. Unlike the succinimidyl ester coupling, there is no leaving group for the isocyanate (and isothiocyanate) coupling chemistry, and therefore good yields have been reported for structurally hindered isocyanates (e.g. 90% yields are routinely observed for the secondary isocyanate, 4-isocyanato-TEMPO [37, 38]). Due to the high selectivity and efficiency of this reaction, the crude, deprotected RNA can be labeled directly, allowing for only one purification and therefore high yields. Another advantage is the relatively fast coupling time (15–60 min). In addition, 2′-ureido modifications are not as destabilizing as 2′-amido modifications [43, 66], e.g. incorporation of the EPR spin probe TEMPO through a 2′-urea linkage at internal base-pairing nucleotides resulted in a minor decrease in stability as measured by a small decrease in melting temperature of 1–3 °C [37]. The main drawback is that usually the isocyanate labeling reagent must be prepared from the corresponding amine; however, this chemistry is straightforward and pure isocyanates are obtained in high yields after purification by extraction [42, 43, 68].

8.3
Experimental Protocols

The general experimental protocols for two representative examples of RNA labeling by the 2′-amino approach will be detailed: incorporation of a crosslinking reagent (Fig. 8.3a) for validation of existing structural models [6, 42] and incorporation of an EPR spin-probe (Fig. 8.3b) for biophysical analysis of structure [41] and dynamics of RNA molecular recognition [37–40]. We have also included an example of base-labeling using 4-thiouridine for RNAs that cannot be modified in the 2′-position due to loss of function (Fig. 8.3c).

8.3.1
Synthesis of Aromatic Isothiocyanates and Aliphatic Isocyanates

The 2′-amino post-synthetic labeling approach often requires the synthesis of the desired labeling reagent from the corresponding amine, which may be commercially available. The isothiocyanate crosslinking reagent **2** was prepared from the corresponding amine **1** and thiophosgene (Fig. 8.3a) [6, 42].

(1) Add a solution of amine **1** (for synthesis, see [6]; 8.20 g, 33 mmol) in chloroform (250 ml) drop-wise to a solution of thiophosgene (4.17 g, 36.3 mmol) in chloroform (50 ml) over 10 min at room temperature.
(2) Stir for 1 h at room temperature.
(3) Dilute the mixture with methylene chloride (330 ml).
(4) Wash with NaOH (1 M aq, 165 ml).
(5) Extract the aqueous phase with additional methylene chloride (40 ml).
(6) Combine the organic phases.
(7) Dry the combined organic phases (Na_2SO_4) and filter off the salt.
(8) Remove the solvent *in vacuo*.
(9) Purify the crude product by flash column chromatography (CH_2Cl_2).
(10) This procedure produces an oil (in our hands 8.80 g, 92% yield).

The isocyanate spin-labeling reagent, 4-isocyanato TEMPO **4**, was prepared from 4-amino TEMPO **3** (Acros and Sigma-Aldrich) and diphosgene (Fig. 8.3b) [37].

(1) Pre-cool a solution of **3** (198 mg, 1.15 mmol) in anhydrous CH_2Cl_2 (1.5 ml) in a rock salt/ice water bath at −8 °C.
(2) Separately, pre-cool in the same bath a solution of trichloromethyl chloroformate (diphosgene, 25 µl, 0.21 mmol) in CH_2Cl_2 (1.5 ml).
(3) Rapidly (around 8 s), add the solution of amine under a positive pressure of nitrogen to the stirred solution of trichloromethyl chloroformate at −8 °C.
(4) Remove the cooling bath and allow the reaction to stir for 2 min.
(5) Dilute the crude reaction mixture to 20 ml with CH_2Cl_2.
(6) Wash the organic layer successively with NH_4Cl (1 M aq, 4 × 20 ml) and NaOH (1 M aq, 20 ml).
(7) Dry the organic layer with Na_2SO_4 and filter off the salt.
(8) Remove the solvent *in vacuo*.
(9) This protocol typically yields a peach-colored solid (66 mg, 29% based on starting amine or 87% maximum theoretical yield).
(10) Store the isocyanate desiccated at −20 °C in CH_2Cl_2 (0.5 mg/50 µl). Small quantities of isocyanates hydrolyze slowly when stored concentrated at −20 °C (around 30% after 4 weeks) and rapidly when stored in dimethyl formamide (DMF) at −20 °C [43]. However, isocyanates can be stored in CH_2Cl_2 solutions as described above for several months after preparation.

The syntheses of the isothiocyanate and isocyanate can be readily performed on a scale ranging from 25 mg to several grams. Preparation of isothiocyanates and isocyanates produces acid (HCl), which combines with the starting amine to produce an unreactive ammonium salt. This is particularly problematic for isocyanates where it is only possible to convert one-third of the amine to the corresponding isocyanate using this protocol. Alternatively, the non-nucleophilic base Proton

Sponge® (1,8-bis((dimethyl)amino)naphthylene, 2.5 equivalents; Sigma-Aldrich) can be used, which is especially advantageous if the starting amine is expensive or only available in small quantities. If the light-sensitive Proton Sponge® is used, the reaction should be performed in the dark. After the reaction, Proton Sponge® can be removed by extraction using the protocol described above.

Like most chemicals commonly used in any chemistry laboratory, thiophosgene and diphosgene are harmful, but since they are liquids and used in small quantities, they are relatively simple to handle. However, these reactions should be carried out in a well-ventilated area. Likewise, the isothiocyanates and isocyanates are toxic chemicals, but they are simple to use and require only standard laboratory safety equipment (e.g. nitrile gloves).

8.3.2
Post-synthetic Labeling of 2'-Amino-modified RNA

RNA modified with 2'-amino groups can be purchased from several companies. The standard 2'-trifluoroacetyl protecting group is readily cleaved under standard RNA deprotection conditions and thus no additional deprotection step is necessary. Reaction of 2'-amino-modified RNA with isothiocyanates or isocyanates is typically done under conditions where the RNA is denatured, in aqueous DMF and/or formamide. The organic solvents also act as co-solvents for dissolving the isothiocyanates or isocyanates. It is pertinent that highly pure amine-free anhydrous DMF be used in these reactions, due to the high reactivity of succinimidyl esters, isothiocyanates and isocyanates toward amines. Furthermore, we recommend ethanol precipitation of 2'-amino-modified RNAs, effectively converting ammonium salts of RNA from chemical synthesis into sodium salts, prior to reaction with these amine-reactive compounds as a precaution against unwanted side reactions.

Labeling of 2'-amino-modified RNA with aromatic isothiocyanates

(1) Dissolve the 2'-amino-containing RNA in 5 µl of 50 mM borate buffer, pH 8.6 (RNA concentration around 2 mM).
(2) Add **2** (100 mM in DMF, 5 µl).
(3) Incubate at 37 °C for 28 h (final concentrations: 1 mM 2'-amino RNA, 50 mM isothiocyanate **2**; 50% aqueous DMF, v/v). This reaction proceeds more slowly at room temperature.

Labeling reactions with the aliphatic isocyanates were carried out in a salt/ice water bath (−8 °C) in a cold room (5 °C). If performed at higher temperatures, increased rates of isocyanate hydrolysis result in lower yields. Furthermore, non-specific labeling has been observed at 37 °C [43]. Analytical-scale reactions can be performed using the following procedure, provided reaction amounts are scaled down in such a way that all concentrations of reactants and buffer remain constant.

Preparative scale reactions of 2'-amino-modified RNA with aliphatic isocyanates

(1) Dissolve the crude (i.e. not yet gel- or HPLC-purified), deprotected 2'-amino-containing RNA (one-quarter of a 1-μmol synthesis) in 100 μl 70 mM boric acid buffer, pH 8.6.
(2) Cool the solution in a salt/ice water bath (−8 °C) in a cold room (5 °C).
(3) Treat the solution sequentially with pre-cooled solutions of formamide (60 μl, 0 °C) and freshly prepared isocyanate in anhydrous DMF (75 mM, 40 μl, −8 °C). Final concentrations: 1 mM 2'-amino RNA, 15 mM isocyanate 4; 50% aqueous borate buffer, 30% formamide, 20% DMF, v/v/v.
(4) Incubate for 1 h at −8 °C.
(5) Treat the oligoribonucleotide solution with a second aliquot of freshly prepared isocyanate (40 μl, 75 mM in DMF).
(6) Incubate for 1 h at −8 °C.
(7) Wash the solution with $CHCl_3$ (2 × 300 μl) at room temperature.
(8) Add sodium acetate (3.0 M, 20 μl, pH 5.3).
(9) Add absolute ethanol (−20 °C, 1.3 ml).
(10) Precipitate the RNA by storage at −20 °C for 4 h.
(11) Centrifuge the sample (12 000 g, 15 min, 5 °C).
(12) Remove the supernatant.
(13) Wash the pellet with cold absolute ethanol (2 × 50 μl).
(14) Dry the pellet *in vacuo*.
(15) Dissolve the pellet in water (50 μl).
(16) Dilute with aqueous urea (8 M, 150 μl).
(17) Purify the RNA by 20% denaturing PAGE (20-cm gel for short oligos up to 20 nt in length, 20 h at 400–600 V or three-quarters the length of the gel; 40-cm gel for longer oligos up to 50 nt in length, up to 72 h at 600 V or less time if higher voltage, e.g. 1200–1800 V, is used).
(18) Yields typically range from 100–170 nmol for one-quarter of a 1.0-μmol synthesis, depending on the length and quality of the RNA synthesis.

To monitor the extent of labeling:

(1) Remove an aliquot (1.0 μl) of the reaction mixture from step 5 of the above protocol.
(2) Dilute with water (19 μl).
(3) Wash with chloroform (2 × 75 μl) to remove excess labeling reagent.
(4) Analyze the reaction by one of the following three methods:
 (a) Reversed-phase HPLC on an analytical column (C_{18}, 4.6 × 250 mm, 5-μm column) at 1.5 ml/min using the following protocol: solvent A, 50 mM Et_3NHOAc (pH 7.0); solvent B, 70% CH_3CN/30% of 50 mM Et_3NHOAc (pH 7.0); 15-min linear gradient from 0 to 23% B, 5-min linear gradient to 100% B, isocratic for 10 min, 3-min linear gradient to initial conditions, 15 min equilibration time between runs. A representative example is given

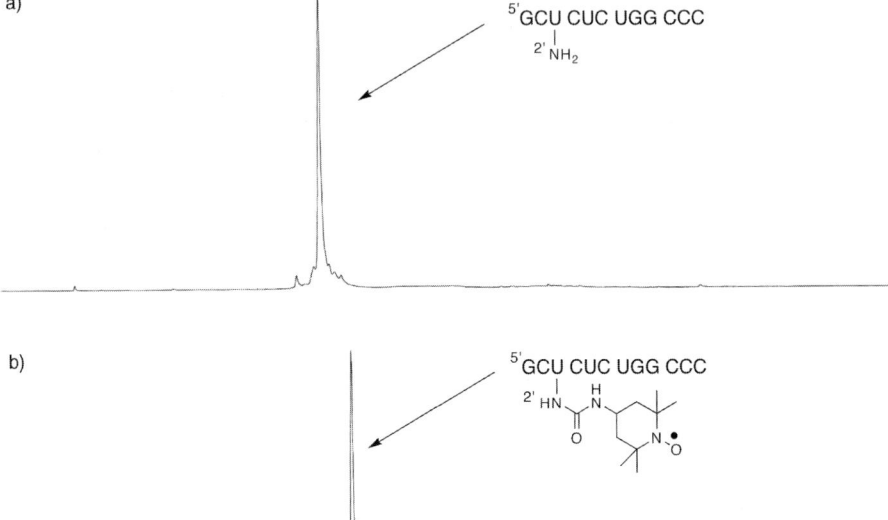

Fig. 8.4. HPLC analysis of 5′-GC(2′-NH₂ U) CUC UGG CCC before (a) and after (b) reaction with **4** which shows 95% conversion to the labeled RNA with increased retention time. The asterisks correspond to **4** and its hydrolysis products. HPLC chromatograms were obtained at 260 nm using an analytical column (C18, 4.6 × 250 mm, 5-μm column) run at 1.5 ml/min according to the following protocol: solvent A, 50 mM Et₃NHOAc (pH 7.0); solvent B, 70% CH₃CN/30% of 50 mM Et₃NHOAc (pH 7.0); 15-min linear gradient from 0 to 23% B, 5-min linear gradient to 100% B, isocratic for 10 min, 3-min linear gradient to initial conditions, 15 min equilibration time between runs.

in Fig. 8.4, which shows the reaction of **4** with 5′-GC(2′-NH₂ U) CUC UGG CCC.

(b) 20% denaturing PAGE (20-cm gel, 400 V for 3.5 h) by UV shadow visualization.

(c) Analytical ion exchange (IE) HPLC on a Dionex DNA Pac PA-100 4 × 250 mm analytical column heated at 50 °C by a column warmer. Separation will not be achieved without heating the column. Solvent gradients for analytical IE-HPLC were run at 1.0 ml/min as follows: solvent A, 25 mM Tris–HCl, pH 8.0; Solvent B, 1.0 M NaCl, 25 mM Tris–HCl, pH 8.0; 35-min linear gradient from 10% B to 80% B, 5-min linear gradient to 10% B.

Short labeled RNAs (up to 20 nt in length) can also be purified utilizing these HPLC protocols, although we recommend 20% denaturing PAGE purification, since the hydrolysis products of some isocyanates may co-elute with the labeled RNA using RP-HPLC.

8.3.3
Post-synthetic Labeling of 4-Thiouridine-modified RNA

If one knows *a priori* that modification of the 2′-hydroxyl group will likely interfere with biological function (e.g. 2′-OH is involved in an essential hydrogen bond), it may be necessary to label using an alternative post-synthetic labeling strategy. In this case, another simple, straightforward method is the labeling of 4-thiouridine residues with iodoacetamides [69, 70] or sulfur-based compounds [71]. One of the advantages of this method is that the labeling reaction can be followed by monitoring the consumption of UV signal at 320 nm, which corresponds to the thiocarbonyl. This labeling strategy changes the base-pairing properties of this residue. However, UV thermal denaturation melting temperature and hypochromicity data as well as NMR structural data indicate that 4-thiouridine residues can be labeled in this manner without disruption of helical stacking [71]. Labeling of the 4-amine group of cytidine with a crudely analogous modification, however, resulted in severe thermal instability of DNA [72]. Therefore, caution should be exercised when choosing such a labeling strategy for base-pairing residues.

Labeling of 4-thiouridine with the iodoacetamide spin-labeling reagent 3-(2-iodoacetamido)-proxyl (modified procedure from that reported in [70])

(1) Dissolve 4-thiouridine-modified RNA (one-quarter of a 1-µmol synthesis) in 166 µl of buffer (100 mM sodium phosphate, pH 8).
(2) Acquire UV spectrum of an aliquot of the above mixture, monitoring at 260 and 320 nm.
(3) Dissolve 6 mg of 3-(2-iodoacetamido)-proxyl, **5** (Fig. 8.3c, Sigma) into 14 µl of ethanol and 20 µl of anhydrous DMF (0.5 M labeling reagent).
(4) Mix **5** and 4-thiouridine-modified RNA; final concentrations: around 1 mM RNA, 85 mM **5**, 83% phosphate buffer/7% ethanol/10% DMF (v/v/v).
(5) Due to light sensitivity of 4-thiouridine residues, cover samples with aluminum foil.
(6) Vortex vigorously until absorbance at 320 nm disappears (typically 18–28 h).
(7) Once the reaction is complete, precipitate and purify RNA as described above.

8.3.4
Verification of Label Incorporation

Whenever a modification is introduced into RNA, either by solid-phase chemical synthesis using a phosphoramidite or by post-synthetic modification, several steps are necessary to verify that the incorporation was successful. Not all modifications are incorporated as intended. For example, the 5-trifluoromethyl-2′-deoxyuridine phosphoramidite was prepared for the purpose of ^{19}F-NMR spectroscopy of nucleic acids; however, standard oligonucleotide deprotection conditions converted the 5-trifluoromethyl group to a 5-cyano group, prompting the use of alternate mild deprotection conditions [73]. Incorporation of the modified nucleoside should

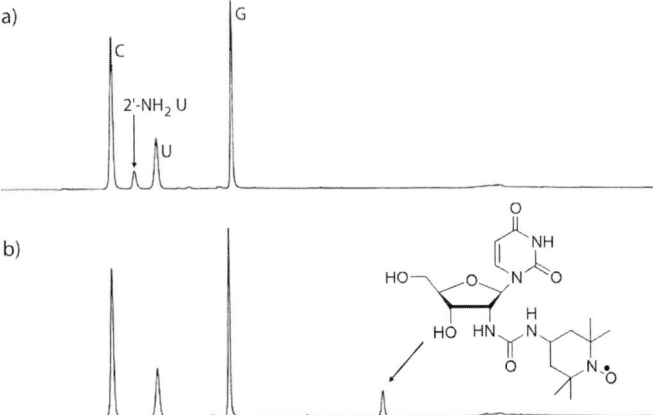

Fig. 8.5. HPLC analysis of enzymatically digested RNA.
(a) Enzymatic digestion of 5′-GCU C(2′-NH$_2$ U)C UGG CCC;
(b) enzymatic digestion of the product of the 2′-NH$_2$-modified oligonucleotide from a after reaction with isocyanate **4**. HPLC chromatograms were obtained as in Fig. 8.4.

be verified by mass spectrometry and enzymatic digestion in conjunction with RP-HPLC analysis. The RNAs (2.0 nmol or around 0.2 OD$_{260}$) should be digested with snake venom phosphodiesterase (0.5 U) and calf intestinal alkaline phosphatase (8 U) at 37 °C for 5 h in 5 mM Tris–HCl, pH 7.4 (20 µl) and then analyzed by analytical RP-HPLC using the same protocol as that listed above for monitoring the extent of labeling. For example, HPLC analysis of the enzymatic digestion of 5′-GCU C(2′-NH$_2$ U)C UGG CCC resulted in peaks corresponding to C, 2′-NH$_2$ U, U and G (Fig. 8.5a), whereas after labeling with the spin-label isocyanate **4** HPLC analysis revealed the absence of the 2′-NH$_2$ U peak and the presence of a new peak (Fig. 8.5b) that co-eluted with the expected modified spin-labeled nucleoside prepared by chemical synthesis [37].

8.3.5
Potential Problems and Troubleshooting

It is always important to determine if the modification interferes (intentionally or unintentionally) with the structure and function of the molecule using a standard structural (e.g. UV thermal denaturation and/or other biophysical spectroscopy or crystallography) and functional (binding or enzymatic) assay. For example, the effect of incorporation of nitroxide spin-labels at the 2′-position on RNA has been investigated by UV thermal denaturation [37], whereas their effect on RNA–protein complex formation was investigated by electrophoretic mobility shift analysis [38].

If the labeling reaction does not work or the yields of the labeling reactions are low, this is generally a result of one of four problems.

(1) The isocyanate may be hydrolyzed or not prepared properly. The quality of the isocyanate can be determined by spectroscopic methods (e.g. NMR) and/or tested by reaction with a simple aliphatic amine such as benzylamine (30 min in CH_2Cl_2 at room temperature).
(2) The 2′-trifluoroacetyl protecting group may not have been fully removed, which may not be readily apparent because 2′-trifluoroacetamido- and 2′-amino-modified RNAs often have similar mobility on HPLC or in gels. However, this can be readily investigated by enzymatic digestion of the RNA, followed by HPLC analysis as described above. For example, if the 2′-trifluoroacetyl group is not fully removed a new peak will be observed by HPLC analysis with a retention time of around 5 min corresponding to the 2′-trifluoroacetamido uridine nucleoside (e.g. in the order of C, 2′-NH_2 U, U, 2′-$NHCOCF_3$ U, G, A).
(3) If the temperature of the reaction is not low enough, the yields are lower, presumably because of the competing hydrolysis of the isocyanate. Therefore, it is important to monitor the temperature of the ice/salt bath with a thermometer.
(4) Lower yields will be obtained if the RNA is not completely dissolved at the beginning of the reaction.

Note added in proof. Recently, a paper published by Pham et al. (Nucleic Acids Res. 2004, 32, 3446–3455) showed that 2′-ureido-modified RNAs are significantly more stable than analogous 2′-amido-modified RNAs.

References

1 R. F. GESTELAND, T. R. CECH, J. F. ATKINS (eds), *The RNA World*, Cold Springs Harbor Laboratory Press, Cold Springs Harbor, NY, **1999**.
2 M. MENGER, F. ECKSTEIN, D. PORSCHKE, *Nucleic Acids Res.* **2000**, *29*, 4428–4434.
3 T. TUSCHL, C. GOHLKE, T. M. JOVIN, E. WESTHOF, F. ECKSTEIN, *Science* **1994**, *266*, 785–788.
4 G. S. BASSI, A. I. H. MURCHIE, F. WALTER, R. M. CLEGG, D. M. J. LILLEY, *EMBO J.* **1997**, *16*, 7481–7489.
5 D. A. LAFONTAINE, D. G. NORMAN, D. M. J. LILLEY, *EMBO J.* **2002**, *21*, 2461–2471.
6 S. T. SIGURDSSON, T. TUSCHL, F. ECKSTEIN, *RNA* **1995**, *1*, 575–583.
7 N. VENKATESAN, S. J. KIM, B. H. KIM, *Curr. Med. Chem.* **2003**, *10*, 1973–1991.
8 B. S. SPROAT, *J. Biotech.* **1995**, *41*, 221–238.
9 D. J. EARNSHAW, M. J. GAIT, *Biopolymers (Nucleic Acid Sci.)* **1998**, *48*, 39–55.
10 S. VERMA, F. ECKSTEIN, *Annu. Rev. Biochem.* **1998**, *67*, 99–134.
11 F. ECKSTEIN, *Biochemie* **2002**, *84*, 841–848.
12 H. J. LENOX, C. P. MCCOY, T. L. SHEPPARD, *Org. Lett.* **2001**, *3*, 2415–2418.
13 S. DEY, T. L. SHEPPARD, *Org. Lett.* **2001**, *3*, 3983–3986.
14 J. S. SUMMERS, B. R. SHAW, *Curr. Med. Chem.* **2001**, *8*, 1147–1155.
15 G. A. HOLLOWAY, C. PAVOT, S. A. SCARINGE, Y. LU, T. B. RAUCHFUSS, *ChemBioChem* **2002**, *3*, 1061–1065.
16 B. CHEN, E. R. JAMIESON, T. D. TULLIUS, *Bioorg. Med. Chem. Lett.* **2002**, *12*, 3093–3096.
17 M. E. DOUGLAS, B. BEIJER, B. S. SPROAT, *Bioorg. Med. Chem. Lett.* **1994**, *4*, 995–1000.
18 J. T. HWANG, M. M. GREENBERG, *J. Org. Chem.* **2001**, *66*, 363–369.
19 L. BEIGELMAN, A. KARPEISKY, N. USMAN, *Nucleosides, Nucleotides* **1995**, *14*, 901–905.

20 J. B. Murray, D. P. Terwey, L. Maloney, A. Karpeisky, N. Usman, L. Beigelman, W. G. Scott, Cell **1998**, *92*, 665–673.

21 A. P. Massey, S. T. Sigurdsson, Nucleic Acids Res. **2004**, *32*, 2017–2022.

22 A. Földesi, A. Trifonova, Z. Dinya, J. Chattopadhyaya, J. Org. Chem. **2001**, *66*, 6560–6570.

23 K. Yamana, R. Iwase, S. Furutani, H. Tschida, H. Zako, T. Yamaoka, A. Murakami, Nucleic Acids Res. **1999**, *27*, 2387–2392.

24 V. A. Korshun, D. A. Stetsenko, M. J. Gait, J. Chem. Soc., Perkin Trans. 1 **2002**, *8*, 1092–1104.

25 N. N. Dioubankova, A. D. Malakhov, D. A. Stetsenko, M. J. Gait, P. E. Volynsky, R. G. Efremov, V. A. Korshun, ChemBioChem **2003**, *4*, 841–847.

26 K. Yamana, T. Mitsui, H. Nakano, Tetrahedron **1999**, *55*, 9143–9150.

27 P. B. Rupert, A. P. Massey, S. T. Sigurdsson, A. R. Ferré-D'Amaré, Science **2002**, *298*, 1421–1424.

28 S. I. Khan, M. W. Grinstaff, J. Am. Chem. Soc. **1999**, *121*, 4704–4705.

29 K. Musier-Forsyth, P. Schimmel, Biochemistry **1994**, *33*, 773–779.

30 M. M. Konarska, Methods **1999**, *18*, 22–28.

31 P. Z. Qin, S. E. Butcher, J. Feigon, W. L. Hubbell, Biochemistry **2001**, *40*, 6929–6936.

32 J. A. Fidanza, L. W. McLaughlin, J. Org. Chem. **1992**, *57*, 2340–2346.

33 S. B. Cohen, T. R. Cech, J. Am. Chem. Soc. **1997**, *119*, 6259–6268.

34 D. J. Earnshaw, B. Masquida, S. Muller, S. T. Sigurdsson, F. Eckstein, E. Westhof, M. J. Gait, J. Mol. Biol. **1997**, *274*, 197–212.

35 H. Aurup, T. Tuschl, F. Benseler, J. Ludwig, F. Eckstein, Nucleic Acids Res. **1994**, *22*, 20–24.

36 S. K. Silverman, T. R. Cech, Biochemistry **1999**, *38*, 14224–14237.

37 T. E. Edwards, T. M. Okonogi, B. H. Robinson, S. T. Sigurdsson, J. Am. Chem. Soc. **2001**, *123*, 1527–1528.

38 T. E. Edwards, T. M. Okonogi, S. T. Sigurdsson, Chem. Biol. **2002**, *9*, 699–706.

39 T. E. Edwards, S. T. Sigurdsson, Biochemistry **2002**, *41*, 14843–14847.

40 T. E. Edwards, S. T. Sigurdsson, Biochem. Biophys. Res. Commun. **2003**, *303*, 721–725.

41 O. Schiemann, A. Weber, T. E. Edwards, T. F. Prisner, S. T. Sigurdsson, J. Am. Chem. Soc. **2003**, *125*, 3434–3435.

42 S. T. Sigurdsson, Methods **1999**, *18*, 71–77.

43 S. T. Sigurdsson, F. Eckstein, Nucleic Acids Res. **1996**, *24*, 3129–3133.

44 J. T. Hwang, M. M. Greenberg, Org. Lett. **1999**, *1*, 2021–2024.

45 T. J. Meade, J. F. Kayyem, Angew. Chem. Int. Ed. **1995**, *34*, 352–353.

46 E. S. Krider, J. E. Miller, T. J. Meade, Bioconjugate Chem. **2002**, *13*, 155–162.

47 E. J. Merino, K. M. Weeks, J. Am. Chem. Soc. **2003**, *125*, 12370–12371.

48 T. S. Zatsepin, D. A. Stetsenko, A. Arzumanov, E. Romanova, M. J. Gait, T. S. Oretskaya, Bioconjugate Chem. **2002**, *13*, 822–830.

49 K. F. Blount, O. C. Uhlenbeck, Biochemistry **2002**, *41*, 6834–6841.

50 T. K. Stage-Zimmerman, O. C. Uhlenbeck, Nat. Struct. Biol. **2001**, *8*, 863–867.

51 K. L. Buchmueller, B. T. Hill, M. S. Platz, K. M. Weeks, J. Am. Chem. Soc. **2003**, *125*, 10850–10861.

52 D. M. John, K. M. Weeks, Chem. Biol. **2000**, *7*, 405–410.

53 D. M. John, K. M. Weeks, Biochemistry **2002**, *41*, 6866–6874.

54 S. K. Silverman, M. L. Deras, S. A. Woodson, S. A. Scaringe, T. R. Cech, Biochemistry **2000**, *39*, 12465–12475.

55 S. K. Silverman, T. R. Cech, RNA **2001**, *7*, 161–66.

56 K. F. Blount, Y. Tor, Nucleic Acids Res. **2003**, *31*, 5490–5500.

57 K. F. Blount, N. L. Grover, V. Mokler, L. Beigelman, O. C. Uhlenbeck, Chem. Biol. **2002**, *9*, 1009–1016.

58 S. I. Chamberlin, E. J. Merino, K.

M. Weeks, *Proc. Natl. Acad. Sci. USA* **2002**, *99*, 14688–14693.
59 C. Hendrix, B. Devreese, J. Rozenski, A. van Aerschot, A. De Bruyn, J. Van Beeumen, P. Herdewijn, *Nucleic Acids Res.* **1995**, *23*, 51–57.
60 C. Hendrix, M. Mahieu, J. Anne, S. V. Calenbergh, A. Van Aerschot, J. Content, P. Herdewijn, *Biochem. Biophys. Res. Comm.* **1995**, *210*, 67–73.
61 O. P. Kryatova, W. H. Connors, C. F. Bleczinski, A. A. Mokhir, C. Richert, *Org. Lett.* **2001**, *3*, 987–990.
62 P. Sergiev, S. Dokudovskaya, E. Romanova, A. Topin, A. Bogdanov, R. Brimacombe, O. Dontsova, *Nucleic Acids Res.* **1998**, *26*, 2519–2525.
63 S. T. Sigurdsson, F. Eckstein, *Trends Biotech.* **1995**, *13*, 286–289.
64 I. Huq, N. Tamilarasu, T. M. Rana, *Nucleic Acids Res.* **1999**, *27*, 1084–1093.
65 L. Huang, L. L. Chappell, O. Iranzo, B. F. Baker, J. R. Morrow, *J. Biol. Inorg. Chem.* **2000**, *5*, 85–92.
66 S. Alefelder, S. T. Sigurdsson, *Bioorg. Med. Chem.* **2000**, *8*, 269–273.
67 B. Bramlage, S. Alefelder, P. Marschall, F. Eckstein, *Nucleic Acids Res.* **1999**, *27*, 3159–3167.
68 S. T. Sigurdsson, B. Seeger, U. Kutzke, F. Eckstein, *J. Org. Chem.* **1996**, *61*, 3883–3884.
69 H. Hara, T. Horiuchi, M. Saneyoshi, S. Nishimura, *Biochem. Biophys. Res. Comm.* **1970**, *38*, 305–311.
70 A. Ramos, G. Varani, *J. Am. Chem. Soc.* **1998**, *120*, 10992–10993.
71 P. Z. Qin, K. Hideg, J. Feigon, W. L. Hubbell, *Biochemistry* **2003**, *42*, 6772–6783.
72 W. Bannwarth, D. Schmidt, *Bioorg. Med. Chem. Lett.* **1994**, *4*, 977–980.
73 J. C. Markley, P. Chirakul, D. Sologub, S. T. Sigurdsson, *Bioorg. Med. Chem. Lett.* **2001**, *11*, 2453–2455.
74 Y. Kinoshita, K. Nishigaki, Y. Husimi, *Nucleic Acids Res.* **1997**, *25*, 3747–3748.
75 J. T. Rodgers, P. Patel, J. L. Hennes, S. L. Bolognia, D. P. Mascotti, *Anal. Biochem.* **2000**, *277*, 254–259.
76 A. J. Harwood, *Methods Mol. Biol.* **2002**, *187*, 17–22.
77 D. L. McMinn, M. M. Greenberg, *J. Am. Chem. Soc.* **1998**, *120*, 3289–3294.
78 P. S. Chockalingam, L. A. Jurado, H. W. Jarrett, *Mol. Biotechnol.* **2001**, *19*, 189–99.
79 R. W. Roberts, J. W. Szostak, *Proc. Natl. Acad. Sci. USA* **1997**, *94*, 12297–12302.
80 M. Teplova, C. J. Wilds, Z. Wawrzak, V. Tereshko, Q. Du, N. Carrasco, Z. Huang, M. Egli, *Biochemie* **2002**, *84*, 849–858.
81 A. A. Koshkin, J. Fensholdt, H. M. Pfundheller, C. Lamholt, *J. Org. Chem.* **2001**, *66*, 8504–8512.
82 A. Okamoto, T. Taiji, K. Tainaka, I. Saito, *Bioorg. Med. Chem. Lett.* **2002**, *12*, 1895–1896.
83 C. R. Allerson, S. L. Chen, G. L. Verdine, *J. Am. Chem. Soc.* **1997**, *119*, 7423–7433.
84 L. Bellon, C. Workman, J. Scherrer, N. Usman, F. Wincott, *J. Am. Chem. Soc.* **1996**, *118*, 3771–3772.

Part II
Structure Determination

II.1
Molecular Biology Methods

9
Direct Determination of RNA Sequence and Modification by Radiolabeling Methods

Olaf Gimple and Astrid Schön

9.1
Introduction

The large numbers of genome sequences now available have allowed the identification of many novel RNA species, simply by deduction from the published DNA sequences [1]. Even though, direct sequencing of RNA molecules is still an indispensable method for a number of reasons. The most important reason is that even nowadays, novel RNA species may be discovered following a "functional assay". If no hint to the sequence can be obtained by genomic data mining, if RNA editing may occur in this organism or, simply, if the RNA is derived from an organism where no genomic sequences are available, the RNA has to be purified prior to direct sequence determination. The second, and probably even more intriguing, rationale is the observation that a large number of RNAs, such as tRNAs, snRNAs, snoRNAs and others, contain modified nucleobases, which in many cases play crucial roles in the function of these RNAs [2–7]. Although in many instances the plain RNA sequence can be extracted from the genomic sequence, the type and position of the modified bases have to be determined by direct analysis of the purified RNA. In this chapter, we will describe methods for the rigorous purification of single small RNA species from the bulk of cellular RNA, their sequence determination and the identification of modified nucleotides by post-labeling methods.

9.2
Methods

It is anticipated that the reader is familiar with standard biochemical and molecular biology practice, such as gel electrophoresis, chromatography and handling of radioactive materials.

In order to avoid RNase contamination, all aqueous solutions and plasticware should be sterilized by autoclaving or prepared from RNase-free [diethylpyrocarbonate (DEPC)-treated] water. Glassware should be baked at 150 °C for 4 h. Centrifugations are performed in a microcentrifuge at 10 000 g, if not stated otherwise.

Handbook of RNA Biochemistry. Edited by R. K. Hartmann, A. Bindereif, A. Schön, E. Westhof
Copyright © 2005 WILEY-VCH Verlag GmbH & Co. KGaA, Weinheim
ISBN: 3-527-30826-1

9.2.1
Isolation of Pure RNA Species from Biological Material

9.2.1.1 Preparation of Size-fractionated RNA

Numerous methods for the preparation of crude RNA from tissues can be found in standard molecular biology reference works, e.g. [8]. They all consist of a cell disruption step under denaturing conditions, followed by separation of the nucleic acids from protein and cell debris. For purification of a single RNA species from this total RNA population, it is preferable to perform a crude size selection prior to further manipulations. The simplest procedure consists of a fractionated precipitation with NaCl, where large RNAs and polysaccharides are separated from the "soluble" (i.e. small) RNAs by centrifugation [9]. A more elaborate scheme for large-scale purification of tRNAs (and other small RNAs) from human and animal tissue has been described by Roe [10]. This procedure can be easily scaled down and adapted to other tissues. The DEAE anion-exchange chromatography described in that paper can be conveniently replaced by ready-to-use columns for small-scale nucleic acid preparations, following the manufacturer's instructions for RNA preparation (e.g. Macherey-Nagel Nucleobond AX). Alternatively, a ribosome-free cell extract (S100) can be used for isolation of non-rRNAs [11]. The RNAs obtained by any of these purification schemes are ready for further purification and functional assays.

9.2.1.2 Isolation of Single Unknown RNA Species Following a Functional Assay

If a functional assay such as aminoacylation, ribozyme activity or similar is available, any RNA of interest can be purified and identified, regardless of prior sequence information. Although chromatographic procedures such as anion-exchange, gel filtration or reverse-phase chromatography can be used in any combination [2, 3], isolation of a single species usually requires preparative separation by one- or two-dimensional gel electrophoresis. To achieve the best possible separation by length, base composition, nucleotide modifications and structure, the first dimension gel should be run at acidic pH under semi-denaturing conditions, and the second dimension at slightly basic pH and fully denaturing conditions as described [12, 13].

Materials for Staining and Elution of RNAs after Electrophoresis

- Staining solution: 0.4% toluidine blue O (w/v) in 50% MeOH (v/v), 10% glacial acetic acid (v/v).
- Destaining solution: 50% MeOH (v/v), 10% glacial acetic acid (v/v).
- Elution buffer: 0.5 M Tris–HCl, pH 7, 0.1% SDS (w/v), 0.1 mM Na$_2$EDTA, 1 mM MgCl$_2$.

Comments on the Electrophoretic Purification and Elution of RNA Species

To obtain optimal resolution, not more than 50 µg of a pre-fractionated "small RNA" preparation should be loaded onto each 1 cm wide lane of a first dimension gel (10% PAA; 0.5 mm thick; 40 cm total gel length). After electrophoresis, a 16 cm long part of the lane is cut out. The vertical position of this strip depends on the anticipated migration distance of the desired RNA. The strip is then polymerized into the second dimension gel such that the direction of electrophoresis is turned by 90° compared to the first dimension. Following electrophoresis, the RNA species are visualized by toluidine blue staining for 20 min and destaining until a clear background is achieved. The isolated spots are cut out with a scalpel and 150 μl of elution buffer is added to each reaction tube containing a gel piece of 4 mm maximum diameter (if pieces are larger, adjust buffer volume accordingly). After quick-freezing the contents in dry ice, the RNA is eluted by vigorous shaking at room temperature for at least 8 h. The buffer is collected, another 150 μl is added to each tube to wash the gel piece, and the RNA is precipitated from the combined buffer fractions by addition of 750 μl EtOH, overnight incubation at −20 °C and centrifugation at 10 000 g. After washing with ice-cold 70% ethanol and vacuum drying for 10 min, the RNAs can be dissolved in H_2O or the desired buffer for functional analysis.

9.2.1.3 Isolation of Single RNA Species with Partially Known Sequence

If the primary sequence of an RNA species is known, e.g. from genomic analysis or direct sequencing, a hybrid selection method can be applied to obtain the desired species in sufficient quantities for further studies. This protocol has been optimized following published procedures [14, 15].

Materials for Hybrid Selection of Single RNA Species

Buffers
- 20 × SSC, 6 × SSC, 1 × SSC.
- TE: 10 mM Tris–HCl, pH 7.5, 1 mM Na_2EDTA.
- 2 M NaOAc, pH 5.
- Urea gel loading buffer: 8 M urea, 0.03% (w/v) bromophenol blue (BPB), 0.03% (w/v) xylene cyanol FF (XC), 0.03% (w/v) Sigma brilliant blue (SBB); make up from dye stock (Section 9.2.3.1).

Affinity matrix
- Ultra-Link Streptavidin Plus-Beads (Pierce; 53117); capacity 66.7 pmol biotin/μl according to the manufacturer's information.

Nucleic acids
- 3′-Biotinylated deoxyoligonucleotide, complementary to the desired RNA. Most manufacturers (e.g. IBA, Germany or Eurogentec, Belgium) offer the biotin-coupled oligos at excellent quality. Note that T_m of the oligo should be about 70 °C. If possible, chose a single-stranded variable RNA region as target. The working solution of the oligonucleotide is adjusted to 1 nmol/μl H_2O.
- Crude RNA preparation (see Section 9.2.1.1) in TE or H_2O. Hybrid selection is most efficient with size-fractionated RNA preparations.

Equipment
- Two to three thermostated shakers holding 1.5 ml reaction tubes; alternatively, put standard mixers/shakers in an incubator with the desired temperature. Shaking speed should be adjusted such that the beads are just kept from settling, but do not move vigorously.

Procedure for the Purification of a Single RNA Species from 1 mg Crude Small RNAs

Coupling of biotinylated oligonucleotide to streptavidin beads
The streptavidin beads are supplied as a suspension and tend to settle down fast. Before removing the required amount, shake the suspension well until homogeneous and use a pipette tip with a larger opening (cut with a scalpel) to avoid clogging. Per 1 mg total RNA, take 15 µl of bead suspension. Wash beads twice in 200 µl TE and resuspend in 200 µl TE ("Washing" of beads means: suspend *thoroughly but carefully* by vortexing, collect by centrifugation at 2000 r.p.m. for 5 min in a microcentrifuge, remove supernatant). Add 1 µl biotin–oligonucleotide solution; shake for 15 min at room temperature and wash twice in 6 × SSC.

Hybridization of RNA to coupled oligonucleotides
Pre-heat one shaker at 65 °C and a second one at 90 °C. Adjust RNA volume with TE to 100 µl (final concentration: 10 mg/ml). Denature by heating for 2 min at 90 °C and snap-cool in ice water. Adjust RNA to 6 × SSC by adding 43 µl of 20 × SSC and add to beads; shake for 10 min at 65 °C, turn off heater and continue shaking while the block cools down to room temperature.

Removal of undesired (contaminating) RNAs
After hybridization, collect beads at 2000 r.p.m., wash at room temperature 3 times with 6 × SSC followed by 3 times with 1 × SSC. Save the beads for elution, keep the supernatants in case the hybridization needs further optimization.

Elution of desired RNA species
Pre-heat two shakers at 60 and 75 °C, respectively. Elution of the desired RNA from the beads is achieved in three steps; the supernatant of each step is retained for further analysis. First, the beads are resuspended in 100 µl TE and shaken at room temperature for 5 min; after collecting the beads, the elution is repeated at 60 and 75 °C with the same amount of TE buffer, pre-warmed at the respective temperature. If no RNA is recovered, elution may be repeated at 90 °C.

Electrophoresis of affinity purified RNAs
RNA is precipitated from the fractions by addition of 0.1 volume of 2 M NaOAc (pH 5) and 2.5 volumes of EtOH, followed by incubation at −20 °C for at least 30 min and centrifugation. The precipitate is washed with 100 µl ice-cold 70% EtOH, vacuum-dried and dissolved in 10–20 µl urea gel loading buffer. After denaturation for 2 min at 95 °C, the samples are separated on a denaturing PAA gel (40 cm long, 0.5 mm thick). The gel is stained with toluidine blue (Section 9.2.1.2) and

RNA bands are cut out and eluted (Section 9.2.1.2). Because in some cases, co-purification of RNA species with similar sequence cannot be totally avoided, this step is required to ensure absolute purity of the desired RNA. Alternatively, for the determination of RNA purity on an analytical scale, 2–5 µl of the column eluates are mixed with 4 volumes of loading buffer, electrophoresed as above and detected by silver staining [16].

Comments on electrophoresis and elution of RNAs
Details on preparation-scale electrophoresis of RNA can be found in Chapter 1. The polymer concentration of analytical or preparative gels should be adjusted to the expected size of the RNA of interest, e.g. 15% PAA should be used for an expected length between 75 and 90 nt. In this case, electrophoresis should proceed until XC (the second dye marker) has reached the bottom of the gel. Note that the efficiency of gel elution is also dependent on the gel concentration – whereas a 350 nt long RNA is easily recovered from an 8 or 10% PAA gel, yield is low from a 15% gel. If very small amounts of RNA have to be recovered from large volumes of elution buffer, 1–10 µg glycogen may be added as a co-precipitant if it does not interfere with later analysis.

9.2.2
Radioactive Labeling of RNA Termini

End-labeling of RNA with ^{32}P is a prerequisite for direct sequence analysis and free 5′- or 3′-OH-groups are required for most labeling reactions. The removal of 5′-cap structures using tobacco pyrophosphatase has been described [17]. 5′- and 3′-phosphate residues can be easily removed by calf intestine alkaline phosphatase (CIP) prior to the labeling reaction. Since the different enzymes have different substrate preferences, not all reactions will work equally well with all types of RNA. For example, the "hidden" 5′ end of tRNAs is often hard to dephosphorylate; thus, labeling by phosphate exchange (Section 9.2.2.1) is the preferred method in these cases.

9.2.2.1   5′ Labeling of RNAs

Material Required for 5′-end-labeling of RNAs

Enzymes
- Calf intestine alkaline phosphatase (CIP; 10 mU/µl) and T4 polynucleotide kinase (PNK; 5 U/µl) are from Roche Biochemicals.

Buffers and reagent
- 50 mM nitrilo-tri-acetic acid (NTA), pH 8.
- IMID mix: 250 mM imidazole, 25 mM DTT, 0.5 mM spermine, 0.5 mM Na_2EDTA, 50 mM $MgCl_2$, pH 6.6 with HCl.

- MgCl$_2$/spermine solution: 0.2 M MgCl$_2$, 32 mM spermine.
- 100 mM DTT.

Radioactive materials
- [γ-^{32}P]ATP (10 µCi/µl; 3000 Ci/mmol).

5′ Labeling of RNA after Dephosphorylation

For dephosphorylation, 10–100 pmol of purified RNA is vacuum dried, dissolved in 8 µl H$_2$O, denatured for 2 min at 90 °C and snap-cooled in ice water. Then, 1 µl 1 M Tris–HCl, pH 8 and 10 mU CIP are added and the mixture is incubated at 50 °C for 1 h. The enzyme is inactivated by addition of 3.3 µl NTA solution followed by incubation at 50 °C for 20 min.

For radioactive labeling, prepare one 1.5-ml reaction tube containing 200 µCi [γ-^{32}P]ATP and one with 10 µl of urea gel loading buffer (Section 9.2.1.3) and dry in a desiccator. For phosphorylation, the [γ-^{32}P]ATP is dissolved by adding 6.5 µl of the above RNA preparation. Then, 1 µl each of the MgCl$_2$/spermine solution, the 100 mM DTT solution and 1 µl PNK are added, and incubated at 37 °C for 30 min. The reaction is terminated by transferring the whole mixture into the tube containing dry loading buffer. After denaturation (2 min, 90 °C) the labeled RNA is separated on a denaturing PAA gel, localized by autoradiography and recovered by elution (Sections 9.2.1.2). To increase recovery of the labeled RNA during precipitation, 10 µg yeast tRNA per 300 µl elution buffer may be added as a co-precipitant.

5′ Labeling by Phosphate Exchange

Since many small RNAs are highly structured and have a recessed 5′ end difficult to access by the phosphatase, the exchange reaction first introduced by Berkner and Folk [18] is the labeling method of choice for these RNAs.

Between 10 and 100 pmol dry RNA is dissolved in 2 µl IMID mix, 1.25 µl 0.5 mM ADP and 5.75 µl H$_2$O, and transferred into a reaction tube containing 200 µCi dry [γ-^{32}P]ATP. The reaction is initiated by addition of 1 µl PNK, run for 30 min at 37 °C, terminated by pipetting onto loading buffer and separated by electrophoresis as described above.

9.2.2.2 3′ Labeling of RNAs

All intact RNAs possess a 3′-OH end and can thus be directly labeled at this terminus.

The most popular method is the ligation of radioactive pCp to the RNA [19]. The method has been described in detail [12, 20]; an abbreviated and efficient variation including the synthesis of pCp is presented here.

3′ labeling of RNAs is usually more efficient than 5′ labeling and requires less material. However, larger RNAs are often poor substrates for the ligation reaction and may preferably be labeled by poly(A) polymerase, using 3′-deoxyadenosine (Cordycepin) to prevent chain elongation [21]. A special method to label RNAs with (at least partially) known sequence is "splint labeling" of the 3' end with DNA polymerase [22].

Materials Required for 3'-end-labeling of RNAs

Enzymes
- PNK (5 U/µl) and T4 RNA ligase (1.5–3 U/µl) are from Roche Biochemicals; yeast poly(A) polymerase and T7 DNA polymerase are from United States Biochemicals.

Buffers and reagents
- 1 mM 3'-cytidine monophosphate (3'-Cp).
- pCp mix: 175 mM Tris–HCl, pH 8, 15 mM MgCl$_2$, 12 mM DTT, 2.4 mM spermine.
- D mix: 10 mM HEPES–KOH, pH 8.3 in 33% (v/v) DMSO.
- Ligase mix: 120 mM HEPES–KOH, pH 8.3, 10 mM DTT, 30 mM MgCl$_2$, 30 µg/ml RNase-free BSA or gelatin, 3 mM ATP in 25% (v/v) DMSO. Note that the DMSO should be deionized freshly before preparation of these solutions; aliquots of DMSO or the reaction mixes can be stored frozen for several months.

Radioactive materials
- [γ-^{32}P]ATP (10 µCi/µl; 3000 Ci/mmol); [α-^{32}P]Cordycepin triphosphate; [α-^{32}P]dATP (each of highest specific activity available).

3' Labeling of RNA by Ligation of [5'-^{32}P]pCp

Preparation of [5'-^{32}P]pCp
For one labeling reaction, dry down 100 µCi (3.7 MBq) [γ-^{32}P]ATP and dissolve in 2 µl pCp-mix. Add 1 µl 1 mM 3'-Cp and 5 U PNK, incubate for 1 h at 37 °C, and vacuum dry. Alternatively, [5'-^{32}P]pCp can be purchased from several suppliers and used directly for ligation.

Ligation of [5'-^{32}P]pCp to RNA
The dry RNA (4–10 pmol) is dissolved in 2 µl D mix, denatured for 2 min at 90 °C and snap-cooled in ice. Then, 2 µl ligase mix is added and the mixture transferred into the tube containing the dry [^{32}P]pCp. The reaction is started by addition of 2 µl T4 RNA ligase and incubated at 4 °C for 16–30 h. Purification is performed by one-dimensional gel electrophoresis as described (Sections 9.2.1.2).

3' Labeling of RNA with Poly(A) Polymerase and Cordycepin

For this highly sensitive labeling method, 2–10 nmol of RNA 3' ends is sufficient. The purified RNA is mixed with 2 µl 5 × reaction buffer (supplied by the manufacturer), 2 µl (20 µCi) [α-^{32}P]Cordycepin triphosphate and H$_2$O to a final volume of 9 µl. Then, 1 µl poly(A) polymerase is added and the reaction incubated at 30 °C for 20 min. Because the enzyme tends to bind to the RNA, a phenol extraction/precipitation step is advisable before proceeding to one-dimensional gel electrophoresis (Section 9.2.1.2).

3′ Labeling of RNA with DNA Polymerase

This "splint labeling" method is a special case because a single RNA species may be labeled specifically within a mixture of other nucleic acids, provided that at least the immediate 3′ sequence of that RNA is known. A DNA oligonucleotide should be obtained which is complementary to the immediate 3′ RNA end (T_m about 40 °C) and has one extra T residue at its 5′ end, providing a "5′-T overhang" after annealing. An RNA preparation containing approximately 10–50 pmol of the desired species is annealed with 50 pmol of this oligonucleotide in a total volume of 17 µl H_2O by incubating for 5 min at 70 °C and cooling down to 50 °C over about 30 min. The annealing process is terminated on ice, 5 µl of 5 × reaction buffer (supplied by the manufacturer), 2 µl (20 µCi) [α-^{32}P]dATP and 1 µl T7 DNA polymerase (about 20 U/µl) are added and the labeling reaction is incubated for 30 min at 37 °C. The reaction products are further analyzed and purified as described (Section 9.2.1.2).

9.2.3
Sequencing of End-labeled RNA

Genomic sequencing has opened the view on a large number of putative novel RNA species. Their existence and primary sequence can easily be verified by a number of indirect techniques, including RT-PCR for the known part of the RNA sequence and variations of the RACE method to determine the initiation and termination points of transcription, or of processing sites during maturation [8]. However, if no hint to the sequence of an interesting functional RNA can be obtained by data mining or if nucleotide modifications are suspected to play a role in RNA function, direct sequence analysis of the RNA should be performed. Base specific enzymatic and chemical cleavages of end-labeled RNA provide hints on the identity of many modified nucleotides [23–28]; detailed working protocols for both methods will be presented here. The enzymatic as well as the chemical sequencing method rely on cleavage reactions that are not completely specific for all of the four major nucleobases; thus, the sequence has to be deduced from a partly ambiguous cleavage pattern in both cases. The advantage of the enzymatic over the chemical method is that sequence can be obtained from either labeled end, that more information on the nature of modifications can be deduced and that the procedure is straightforward and fast. The main disadvantage is that strong secondary structure of an RNA may inhibit cleavage by certain enzymes and that it may be difficult to obtain the required enzymes at sufficient quality. In contrast, the chemical modifications and subsequent cleavage reactions require only a small number of chemicals and are mostly insensitive to secondary structure under the conditions used.

If the exact nature and position of the modification is to be determined, a position-specific nucleotide analysis by thin-layer chromatography (TLC) has to be performed [20, 29]. This post-labeling method avoids end-labeling of the RNA and gives the best information regarding modified nucleotides, but is quite laborious and time-consuming to perform. Also, it is quite often prone to secondary cleavages and thus not absolutely reliable for primary sequence determination. A major disadvantage of all three approaches discussed here is that none of them will allow unambiguous reading of the terminal few nucleotides; the mobility shift method used to solve this problem has been described in detail elsewhere [12, 20, 30] and will not be presented here. If desired, the labeled 5'- or 3'-terminal nucleotide can be determined by cleavage with nuclease P1 or RNase T2, respectively, and subsequent TLC analysis, as described below (Section 9.2.4.1). In conclusion, the inherent advantages and disadvantages of the aforementioned methods demand a careful evaluation of the specific goals of each sequencing project in order to determine in which combination they should be used.

9.2.3.1 Sequencing by Base-specific Enzymatic Hydrolysis of End-labeled RNA

For enzymatic sequencing, either 5'- or 3'-labeled RNA may be used. Although 10 000 c.p.m. per reaction is optimal, as little as 1000 c.p.m. may be used if sufficient exposure time is allowed. To resolve problems arising from secondary structure, all reactions are performed under denaturing conditions (8 M urea, 50 °C).

Materials Required for Enzymatic Sequencing

RNA
- 5'- or 3'-^{32}P-labeled, gel purified RNA (see Sections 9.2.2.1 and 9.2.2.2), minimum amount 10 000 c.p.m. total; 1 µg/µl yeast tRNA (Roche Biochemicals).

Enzymes
- The sequencing nucleases (RNases T1, CL 3, *Staphylococcus aureus* nuclease and RNase U2) are available from Calbiochem, BRL, Roche Biochemicals, Worthington and Pharmacia, respectively. Because the quality differs between production batches, each lot should be tested separately using the protocol below. Working solutions of 100 mU/µl (RNases T1 and U2), 3 U/µl for *Staphylococcus* nuclease and 13 mU/µl for RNase CL3 should be made up fresh in H$_2$O before use.

Reaction mixes
- T1 mix and H$^+$ mix should be made up fresh from appropriate stock solutions; all others may be prepared in advance and stored at −20 °C.
- T1 mix: 20 mM Na citrate, pH 3.5, 1 mM Na$_2$EDTA, 0.03% (v/v) dye stock, 8.1 M urea.
- S7 mix: 20 mM Tris–HCl, pH 7.5, 10 mM CaCl$_2$, 0.03% (v/v) dye stock, 8.1 M urea.
- CL3 mix: 20 mM Tris–HCl, pH 7.5, 0.03% (v/v) dye stock, 8.1 M urea.
- H$^+$ mix: 0.22 N H$_2$SO$_4$, 0.03% (v/v) dye stock, 6.8 M urea.
- Dye stock: 1% (w/v) BPB, 1% (w/v) XC, 2% (w/v) SBB in 10 mM Tris, pH 7.5.

Tab. 9.1. Cleavage reactions for enzymatic sequencing

Reaction	Yeast RNA (μg)	Mix	Enzyme	Incubation	Specificity
–E	1	T1 mix	–	15 min, 50 °C	–
T1			RNase T1		G
U2			RNase U2		A ≫ ms^2i^6A, G
S7		S7 mix	*Staphylococcus* nuclease		A, U > T, s^4U, ms^2i^6A, m^2A
CL3		CL3 mix	RNase CL3		C ≫ A, T
H^+	5	H^+ mix	–	3 min, 100 °C	All except 2'-O-methyl

Working Procedure for Enzymatic Sequencing

Prepare 10 Eppendorf-type reaction tubes with equal aliquots of your end-labeled, gel-purified RNA (between 1000 and 10 000 c.p.m. per reaction). Label one tube as control (–E), one for the ladder (H^+) and two for each of the four enzymes (see Table 9.1). Calculate the amount of carrier RNA (from the elution and precipitation) in each aliquot, and adjust to 1 μg/tube for the control and enzymatic reactions. For the acid ladder adjust carrier to 5 μg. Dry down the contents of all tubes at room temperature (this takes about 30 min in a Speed Vac or 2 h in an evacuated desiccator over fresh desiccant). Prepare an ice box with wet ice and one with finely crushed dry ice, both large enough to hold the 10 tubes deeply immersed. Pre-heat one water bath at 50, 65 and 95–100 °C, respectively. Alternatively, metal-based heating blocks may be used, but heat transfer is faster and more efficient in water.

Add 4 μl of the respective enzyme reaction mix to each labeled tube (except H^+; see Table 9.1), spin down shortly, denature for 5 min at 65 °C and quick chill on ice. Add 1 μl of the respective enzyme working solution to the first of your two tubes for the same enzyme (e.g. T1). Mix by pipetting in and out, and transfer exactly 1 μl to the second tube. Immediately put the two tubes in the 50 °C bath and incubate for exactly 15 min; stop the reactions in dry ice. The remaining tubes are treated the same way. For the acid ladder, 4 μl of H^+ mix are added to the respective tube, incubated in a boiling water bath for exactly 3 min and quenched in dry ice. The samples may be stored overnight at –80 °C at this stage.

After a short spin, the samples are directly loaded onto a sequencing gel (40 × 20 cm, 0.4 mm thick, 12 lanes per RNA). Details on the composition of RNA sequencing gels may be found in Chapters 10 and 11. For RNAs of 70–90 nt, use 20% PAA; for longer RNAs, use 15% PAA. To read over the whole sequence length, it is advisable to prepare enough material for two runs: a short run (BPB just leaving the 20% gel) and a long run (XC at the edge of the gel). If the sequence should be read up to the labeled end, precipitate the cleavage reactions with EtOH and omit BPB from the loading mix.

Interpretation and Troubleshooting

From the counting ladder (H^+) and the highly base specific RNases T1 and U2 (and CL3 if good quality is available) it is straightforward to deduce a large part of the sequence. To read the band pattern created by *Staphylococcus* nuclease, recall that this enzyme cleaves 5' of the respective nucleotide, leading to a band shift (see Fig. 9.1). A gap in the counting ladder (H^+) indicates a 2'-ribose methylation of the corresponding nucleotide. Single weak bands may result from base modifi-

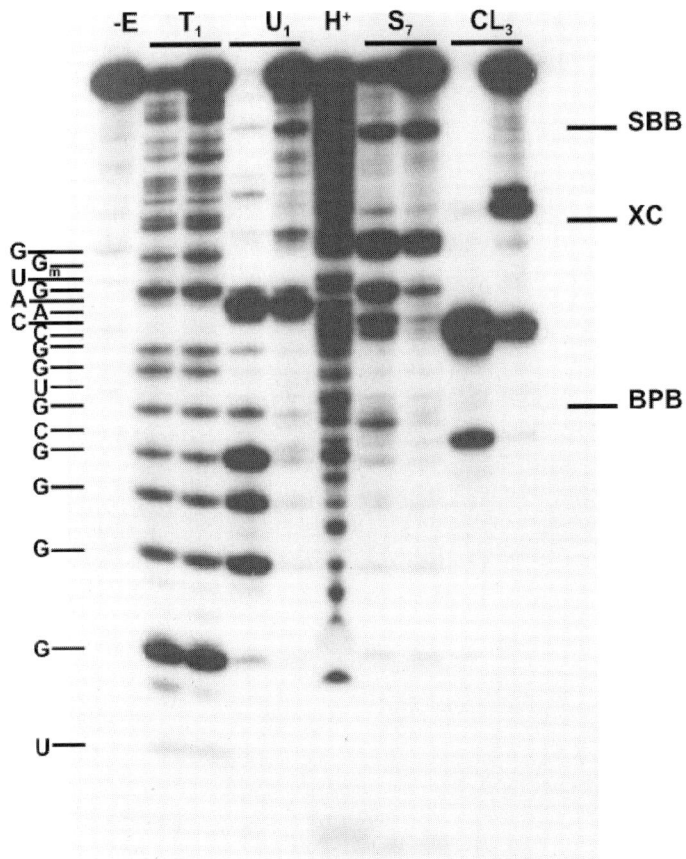

Fig. 9.1. Sequence analysis by enzymatic cleavage of 5'-^{32}P-labeled tRNA^{Gln} from barley chloroplasts. Enzymatic cleavages by RNase T1, U2, CL3 and *Staphylococcus* nuclease (S7), and acid hydrolysis (H^+) were performed as described in Section 9.2.3.1; the two lanes with the same specificity differ by a factor of 5 in the amount of enzyme used. A 25% PAA sequencing gel (29:1) was used to allow reading from the second nucleotide (bottom of the gel). The position of the dyes (SBB, XC, BPB) is given on the right, and the sequence of the first 18 nt on the left side of the panel (the terminal U was determined by end group analysis). Note the gap in the ladder (H^+) at the position of 2'-O-methyl-guanosine (G_m,). For details of the sequence and its interpretation, see [5].

cations; see Table 9.1 for an incomplete overview and [12, 25] for a full discussion of this phenomenon. If a high-quality preparation of RNase CL3 is not commercially available, the enzyme may also be prepared according to the procedure described in [26].

If parts of the gel show weak bands in all lanes, strong secondary structure may hinder efficient cleavage; in this case, denature the RNA at 90 °C and run the reactions at 65 °C (you may have to use more enzyme). If bands are compressed on the gel, insufficient denaturation during electrophoresis is the reason. Make sure that the gel is run at 40 W (for a 20 × 40-cm gel) or use a thermostated electrophoresis apparatus at 60–65 °C.

9.2.3.2 Sequencing by Base-specific Chemical Modification and Cleavage

Chemical sequencing gives clear results only for 3′-labeled RNA, because of the inhomogeneous cleavage products 5′ of the attacked nucleotide [24]. The precipitations required to stop the reactions and to remove the aniline prior to electrophoresis lead to some loss of material; thus, a higher amount of radioactive starting material should be used than for enzymatic sequencing. The following protocol follows a simplified and slightly modified version of the original, which should be consulted for full details [24, 27].

Materials Required for Chemical Sequencing

RNA
- 3′-^{32}P-labeled, gel purified RNA (see Section 9.2.2.2), minimum amount 25 000 c.p.m. total; 10 µg/µl yeast tRNA (Roche Biochemicals).

Chemicals
- Hydrazine, DEPC, aniline, NaBH$_4$, dimethylsulfate (DMS) and EtOH should be of the highest purity available and stored dry at 4 °C.

Buffers and reagents
- 50 mM NaOAc, pH 4.
- 1 M aniline acetate, pH 4.5 (mix H$_2$O:HOAc:aniline at a 7:3:1 volume ratio; spin out precipitate, check pH of an aliquot, store frozen in aliquots).
- 0.5 M NaBH$_4$ (make fresh before use).
- 0.3% (w/v) DMS in NaOAc (mix directly before use).
- Hydrazine/H$_2$O: mix equal volumes and keep on ice until use.
- 3 M NaCl in hydrazine: dry NaCl in a 120 °C oven for 2 h, store in a desiccator. Dissolve in hydrazine and keep on ice until use.

Waste disposal
- DMS is a carcinogen; all solutions containing it (e.g. reagents and EtOH supernatants from the first precipitation) should be disposed into a bottle containing 5 M NaOH. Hydrazine waste is inactivated with 3 M FeCl$_3$, aniline, DMS and hydrazine bottles should be opened only under a flow hood.

Tab. 9.2. Working table for chemical sequence analysis of RNA

Specificity	G	A	U	C
Modification		150 µl NaOAc + 1 µl DEPC	10 µl hydrazine/ H_2O	10 µl NaCl/ hydrazine
reagent	10 µl 0.3% DMS in NaOAc	150 µl NaOAc + 1 µl DEPC	10 µl hydrazine/ H_2O	10 µl NaCl/ hydrazine
incubation	40 s, 90 °C	10 min, 90 °C	10 min, 0 °C	10 min, 90 °C
first precipitation	150 µl NaOAc, 650 µl EtOH	400 µl EtOH	150 µl NaOAc, 550 µl EtOH	500 µl 80% EtOH
second precipitation	–	150 µl NaOAc, 450 µl EtOH	150 µl NaOAc, 450 µl EtOH	150 µl NaOAc, 450 µl EtOH
wash	800 µl EtOH	800 µl EtOH	800 µl EtOH	800 µl EtOH
Reduction		–	–	–
reagent	10 µl 0.5 M $NaBH_4$	–	–	–
incubation	10 min, 0 °C (dark)	–	–	–
precipitation	150 µl NaOAc, 650 µl EtOH	–	–	–
wash	800 µl EtOH	–	–	–

Starting material is dried 3′-end-labeled RNA containing 20 µg yeast tRNA per reaction tube. All centrifugations are performed in a microcentrifuge at 4 °C (10 000 g); all precipitations are done in crushed dry ice for 10 min.

Working Procedure for Chemical RNA Sequencing

Prepare five reaction tubes with equal aliquots of your end-labeled, gel purified RNA (between 5000 and 20 000 c.p.m. per reaction) and 20 µg yeast tRNA. Label one tube as control (–E) and for each of the four reactions (A, C, G and U; see Table 9.2). Dry down the contents of all tubes at room temperature (this takes about 30 min in a Speed Vac or 2 h in an evacuated desiccator over fresh desiccant). Prepare an ice box with wet ice, and pre-heat one water bath at 60 and 90 °C. All modifications are done according to the flow sheet (Table 9.2); note that all precipitations are on ice (or at −20 °C) for 5 min, all centrifugations are at 4 °C and 10 000 g for 5 min (precipitations) or 1 min (wash), respectively. Be careful to remove all of the supernatants to avoid a smear on the sequencing gel.

For aniline cleavage, add 10 µl of aniline acetate to all tubes including the control tube. Incubate for 20 min at 60 °C, stop on ice and precipitate with 150 µl NaOAc and 650 µl EtOH. After 2 washes with 800 µl EtOH to completely remove residual aniline, the RNA is dried, dissolved in urea gel loading buffer, denatured for 3 min at 95 °C and analyzed on a sequencing gel (Section 9.2.3.1).

If desired, a counting ladder may be prepared by acid hydrolysis (Section 9.2.3.1) and run in parallel. Note that the resulting banding pattern is shifted about one nucleotide away from the 3′ end if compared to the corresponding band obtained by chemical cleavage.

Interpretation and Troubleshooting
From the counting ladder (H$^+$) and the highly base-specific cleavages for G, A and U, a large part of the sequence can be easily deduced. Because the C reaction modifies both pyrimidines (although with lower efficiency for U residues), a "subtractive reading" of the U and C lanes is required to unambiguously identify both bases. Some modified bases can be identified very easily: m^{7}G is sensitive towards aniline without any further modification and thus appears as an extremely strong band in all lanes including the control. All uridine derivatives except pseudouridine are weakly reactive towards the U reaction; ac^{4}C can be recognized as a band in all lanes, but weaker than the appearance of m^{7}G. For a more complete overview, see [12, 25]. Band compression due to strong secondary structure of the RNA can be avoided as described above (Section 9.2.3.1).

9.2.4
Determination of Modified Nucleotides by Post-labeling Methods

In many cases, it is desirable to obtain an overview of the modified nucleotides present in a purified RNA species or in an RNA population obtained from a certain organism. If the RNA material can be easily obtained, HPLC analysis is the method of choice because UV spectra provide additional information on the nature of the nucleobase. Coupled HPLC-MS will even identify unknown or novel nucleotides [31]. However, the required apparative infrastructure is not readily available for most laboratories and, even though the sensitivity of the methods is impressive, availability of the biological samples may be limiting. A reliable alternative to determine the nucleotide content of subpicomolar amounts of RNA is the post-labeling of an RNA hydrolysate, followed by two-dimensional TLC analysis of the products [20, 32–34]. The determination of sequence and base modification at the same time has been made possible by the coupling of limited RNA fragmentation and end-group identification of the terminally labeled, separated fragments [29].

9.2.4.1 Analysis of Total Nucleotide Content
The first step of this procedure is the hydrolysis by a mixture of RNases T2 and A. The resulting nucleoside 3′-phosphates are then radioactively labeled at the 5′ end by PNK; these 5′-^{32}P-labeled 3′,5′-nucleoside diphosphates are then converted to the corresponding nucleoside 5′-phosphates by nuclease P1. After elimination of residual ATP by Apyrase, the mixture is subjected to two-dimensional TLC, with an excess of non-labeled nucleoside 5′-phosphates co-migrating as standards.

Materials Required for RNA Nucleotide Analysis

Enzymes
- RNase T2 (Invitrogen); pancreatic ribonuclease (RNase A) and T4 polynucleotide kinase (Roche Biochemicals). Prepare a working solution (T2/A mix) containing 50 mU/μl RNase T2 and 0.1 μg/μl RNase A in H$_2$O (can be stored frozen).

- Apyrase (5 U/ml; Sigma).
- Nuclease P1 (Gibco/BRL; prepare a working solution of 10 ng/µl in 50 mM ammonium acetate, pH 5.3).

Solvents and plates for TLC
- Solvent A: isobutyric acid:concentrated ammonia:H_2O [57.7:3.8:38.5, (v/v)]; pH 4.3.
- Solvent B: isopropanol:concentrated HCl:H_2O (70:15:15).
- Cellulose TLC plates (plastic or glass backed, non-fluorescent, analytical scale), 20 × 20 cm (Macherey-Nagel or Merck).

Radioactive materials
- [γ-^{32}P]ATP (10 µCi/µl; 3000 Ci/mmol).

Preparation of 5'-^{32}P-labeled Nucleoside Monophosphates

Purified RNA (2–20 pmol) is vacuum dried and dissolved in 2 µl 50 mM ammonium acetate, pH 4.5 and 6 µl H_2O (include a control sample without RNA). 1 µl of RNase T2/A mixture is added and the sample is incubated for 5 h at 37 °C. To the resulting hydrolysate, add 1 µl 10 × concentrated PNK buffer (provided by the manufacturer), 0.5 µl 0.1 mM ATP, 25 µCi [γ-^{32}P]ATP and 5 U PNK; incubate for 30 min at 37 °C. Add 1 µl Apyrase, incubate for another 30 min at 37 °C and proceed with half of the mixture (save the other half at −20 °C). Vacuum-dry this aliquot, add 10 µl of nuclease P1 solution and incubate for 3 h at 37 °C.

Two-dimensional TLC of Nucleoside Monophosphates

For analytical TLC, 1 µl of above hydrolysate is mixed with 1 µl pN marker mix (5 mg/ml each of pA, pG, pC and pU). The start point is marked with a soft pencil in the lower left corner of a cellulose plate, 1.5 cm from each edge. The sample is applied in a small spot (preferably with a drawn-out capillary) and dried. The first dimension is developed in solvent A until the front has reached the top edge; the plate is dried thoroughly under a hood. For chromatography in the second dimension, the plate is turned by 90° compared to the first dimension and developed in solvent B. After drying, the plates can be exposed to X-ray film or a Phospho-Imager. The marker nucleotides are visualized as dark blue spots under a UV lamp at 254 nm; their position is marked as an aid in the identification of unknown nucleotides. If a specific nucleotide has to be prepared for secondary analysis, the whole reaction mix (Section 9.2.4.1) is applied to the plate. The corresponding spot is then localized by positioning the plate on top of the X-ray film, scraped off the plate, and eluted with H_2O [12].

Interpretation and Troubleshooting

The four marker nucleotides should appear under UV as clearly separated spots (see reference patterns in [13, 32–34]; the $^{32}P_i$ (resulting from hydrolysis of unused ATP) should be visible as a prominent spot on the X-ray film in the center of the right edge. If separation of nucleotides is unacceptable, check the pH of the solvents and replace them if necessary; make sure that the sample was dried completely after application (use an infrared lamp or hot air if necessary) and check that the lids of the tanks close tightly.

If the starting point is streaked out in the first dimension, the problem might be the amount of protein in your sample; try to reduce the amount of enzyme used (extend the incubation times instead). If comparison to the standard pattern reveals that many dinucleotides are present in your sample, you should first analyze an aliquot of your sample before P1 digestion. This reveals whether you should increase the amount of T2/A mix and/or P1 nuclease, or the respective incubation time.

9.2.4.2 Determination of Position and Identity of Modified Nucleotides

In this case, limited RNA hydrolysis (ideally, one cut per RNA molecule) is performed non-enzymatically and the resulting fragments are radioactively labeled [20, 29]. After electrophoretic separation, the 5′-terminal nucleotide of each isolated fragment is determined by TLC.

The material required is mostly identical to that specified in Section 9.2.4.1; in addition, a sterile glass capillary (5 or 10 µl size) and a gas burner is needed.

Generation and Separation of 5′-labeled Random RNA Fragments

In separate reaction tubes, dry down 20–40 pmol purified RNA, 50–100 µCi [γ-^{32}P]ATP and 10 µl of urea gel loading buffer (see Section 9.2.1.3); pre-heat a water bath to 95 °C. Dissolve the RNA in 1.5 µl H$_2$O, transfer it into the center of the capillary by aspiration and seal the ends with the flame. Hydrolysis is performed for exactly 30 s in the boiling water bath and stopped in ice water. Cut open the ends of the capillary, transfer the contents back to the original tube and rinse the capillary with 5 µl H$_2$O. Transfer the whole contents to the tube containing the dry [γ-^{32}P]ATP and proceed with 5′-labeling and electrophoresis on a 15% PAA gel as described in Section 9.2.2.1. For best separation of the RNA fragments, it is advisable to use gels of 60 or 80 cm length and run them until BPB has reached the bottom; if this is not available, a short and long run should be performed on a 40 cm gel. Fragments are localized by autoradiography, cut out and eluted, including 10 µg yeast tRNA per band (see Sections 9.2.1.2).

Identification of the 5′-end Group of the RNA Fragments

Each sample is dissolved in 10 µl of nuclease P1 solution (Section 9.2.4.1), incubated for 2 h at 50 °C, and an aliquot (1–5 µl, depending on labeling efficiency) is removed and dried (the rest may be stored frozen). The dry samples are then dissolved in 2 µl of pN marker mix (Section 9.2.4.1) and equal amounts applied to each of two TLC plates. The cellulose plates are prepared such that 12–16 samples can be applied as thin streaks, 1.5 cm from the bottom edge; they are then analyzed by one-dimensional separation in solvent A and B, respectively (Section 9.2.4.1). The RNA sequence can then be directly read from the TLC plate. If a 2′-O-methylated dinucleotide is detected, an aliquot of the corresponding sample is digested with 1–10 µg of P1 nuclease (5 h at 65 °C) and analyzed as before or by two-dimensional TLC (Section 9.2.4.1).

Interpretation and Troubleshooting

Ideally, the band pattern visible after electrophoresis should show an even distribution, reaching up to the penultimate nucleotide. If the ladder is shifted significantly towards the smaller fragments, try to increase the amount of RNA or reduce the hydrolysis time. A specific problem may arise if a labile modified nucleotide or a C–A bond in a single-stranded region is present in the RNA. In this case, near-quantitative hydrolysis of the corresponding phosphodiester bond may even lead to a complete lack of bands above this point, and the nucleotide 3′ of the cleaved bond will be visible in all other samples [4]. Most problems arising from the TLC systems have been discussed in Section 9.2.4.1. In some cases, it may be necessary to re-analyze specific modified nucleotides in a different solvent system. A two-dimensional chromatography system with slightly different separation properties has been described in [34]. ac^4C and m^5C are not separated in solvents A and B (Section 9.2.4.1), but can be readily distinguished by chromatography on PEI plates [35]. Thionucleotides can be identified after modification with CNBr and separation of the products on Cellulose [36].

9.3
Conclusions and Outlook

The increasing number of genomic sequences has led to the detection of numerous novel RNAs with mostly unknown functions. In many cases, modified nucleotides may play a role in increasing their structural stability, or facilitating specific interactions with proteins or other RNAs; in some cases, editing may even change the primary sequence and coding potential of an RNA. The methods presented here do not only allow the rigorous purification of any desired RNA from biological samples, but also permit the identification of type and position of modified nucleotides. They may thus help in elucidating the function of these RNAs by identifying novel interaction points with other macromolecules. We anticipate an increasing application of direct RNA sequencing methods, specifically in context with the future investigation of novel RNA species.

Acknowledgments

Research on RNase P in my laboratory is supported by grants from the Deutsche Forschungsgemeinschaft (Scho515/7-3).

References

1 W. FILIPOWICZ, *Proc. Natl. Acad. Sci. USA* **2000**, *97*, 14035–14037.
2 H. BEIER, M. BARCISZEWSKA, G. KRUPP, R. MITNACHT, H. J. GROSS, *EMBO J.* **1984**, *3*, 351–356.
3 H. BEIER, M. BARCISZEWSKA, H. D.

1 Sickinger, *EMBO J.* **1984**, *3*, 1091–1096.
4 A. Schön, G. Krupp, S. Gough, S. Berry-Lowe, C. G. Kannangara, D. Söll, *Nature* **1986**, *322*, 281–284.
5 A. Schön, C. G. Kannangara, S. Gough, D. Söll, *Nature* **1988**, *331*, 187–190.
6 T. Yasukawa, T. Suzuki, N. Ishii, S. Ohta, K. Watanabe, *EMBO J.* **2001**, *20*, 4794–4802.
7 I. Behm-Ansmant, A. Urban, X. Ma, Y. T. Yu, Y. Motorin, C. Branlant, *RNA* **2003**, *9*, 1371–1382.
8 J. Sambrook, E. F. Fritsch, T. Maniatis, *Molecular Cloning: A Laboratory Manual*, Cold Spring Harbor Laboratory Press, Cold Spring Harbor, NY, **2001**.
9 G. Zubay, *J. Mol. Biol.* **1962**, *4*, 247–362.
10 B. Roe, *Nucleic Acids Res.* **1975**, *2*, 21–42.
11 C. Heubeck, A. Schön, *Methods Enzymol.* **2001**, *342*, 118–134.
12 G. Krupp, H. J. Gross, in: *The Modified Nucleotides of Transfer RNA II*, P. F. Agris, R. A. Kopper (eds), Alan R. Liss, New York, **1983**.
13 Y. Kuchino, N. Hanyu, S. Nishimura, *Methods Enzymol.* **1987**, *155*, 379–396.
14 K. Wakita, Y.-I. Watanabe, T. Yokogawa, Y. Kumazawa, S. Nakamura, T. Ueda, K. Watanabe, K. Nishikawa, *Nucleic Acids Res.* **1994**, *22*, 347–353.
15 T. Suzuki, T. Ueda, K. Watanabe, *FEBS Lett.* **1996**, *381*, 195–198.
16 H. Blum, H. Beier, H. J. Gross, *Electrophoresis* **1987**, *8*, 93–99.
17 J. M. d'Alessio, in: *Gel Electrophoresis of Nucleic Acids*, D. Rickwood, B. D. Hames (eds), IRL Press, Oxford, UK, **1982**.
18 L. Berkner, W. R. Folk, *J. Biol. Chem.* **1977**, *252*, 3176–3184.
19 T. E. England, A. G. Bruce, O. C. Uhlenbeck, *Methods Enzymol.* **1980**, *65*, 65–74.
20 H. Beier, H. J. Gross, in: *Essential Molecular Biology II*, T. A. Brown (ed.), IRL Press, Oxford, UK, **1991**.
21 J. Lingner, W. Keller, *Nucleic Acids Res.* **1993**, *21*, 2917–2920.
22 T. P. Hausner, L. M. Giglio, A. M. Weiner, *Genes Dev.* **1990**, *4*, 2146–2156.
23 H. Donis-Keller, A. M. Maxam, W. Gilbert, *Nucleic Acids Res.* **1977**, *4*, 2527–2538.
24 D. A. Peattie, *Proc. Natl. Acad. Sci. USA* **1979**, *76*, 1760–1764.
25 B. Lankat-Buttgereit, H. J. Gross, G. Krupp, *Nucleic Acids Res.* **1987**, *15*, 7649.
26 C. C. Levy, T. P. Karpetsky, *J. Biol. Chem.* **1980**, *255*, 2153–2159.
27 R. Waldmann, H. J. Gross, G. Krupp, *Nucleic Acids Res.* **1987**, *15*, 7209.
28 Y. Kuchino, S. Nishimura, *Methods Enzymol.* **1989**, *180*, 154–163.
29 J. Stanley, S. Vassilenko, *Nature* **1978**, *274*, 87–89.
30 M. Silberklang, A. M. Gillum, U. L. RajBhandary, *Methods Enzymol.* **1979**, *59*, 58–109.
31 J. A. McCloskey, A. B. Whitehill, J. Rozenski, F. Qiu, P. F. Crain, *Nucleosides Nucleotides* **1999**, *18*, 1549–1553.
32 S. Nishimura, in: *Transfer RNA: Structure, Properties and Recognition*, P. Schimmel, D. Söll, J. Abelson (eds), Cold Spring Harbor Laboratory Press, Cold Spring Harbor, NY, **1979**.
33 S. Nishimura, Y. Kuchino, in: *Methods of DNA and RNA Sequencing*, S. Weisman (ed.), Praeger, New York, **1983**.
34 G. Keith, *Biochimie* **1995**, *77*, 142–144.
35 R. C. Gupta, E. Randerath, K. Randerath, *Nucleic Acids Res.* **1976**, *3*, 2915–2921.
36 M. Saneyoshi, S. Nishimura, *Biochim. Biophys. Acta* **1970**, *204*, 389–399.

10
Probing RNA Structures with Enzymes and Chemicals *In Vitro* and *In Vivo*

Eric Huntzinger, Maria Possedko, Flore Winter, Hervé Moine,
Chantal Ehresmann and Pascale Romby

10.1
Introduction

A renewal of interest in RNA was brought about the recent discovery of a large number of new regulatory RNA molecules in bacteria and in eukaryotes (for reviews, see [1, 2]). Many studies in bacteria have also now confirmed that mRNA can adopt highly structured domains that serve genetic switches in response to ligand binding, ranging from proteins to RNA and even metabolites (e.g. [3–5]). Thus, the structural features of a RNA most often are of key importance for its biological function and consequently there is an increasing interest in studies on the structure of RNAs either free or in interaction with ligands.

Chemical and enzymatic probing has become one of the most popular approaches for mapping the conformation of RNA molecules of any size under defined experimental conditions. The method maps the reactivity of each nucleotide towards enzymes or chemicals, which reflects its environment within the RNA molecule. The elaboration of a secondary structure model requires the use of probes with different and complementary specificities. For long RNA molecules, the interpretation of the data can be facilitated with the help of computer programs that predict secondary structure from the sequence. One powerful method is to combine energy minimization with co-variation while other programs tend to simulate the kinetics of RNA folding during transcription (for a review, see [6]). Since the probing approach defines unambiguously the unpaired regions of the RNA, they can be given as constraints in the computer folding programs. The resulting secondary structure model can be further validated by a site-directed mutagenesis study coupled to the probing approach to analyze the effect of the mutation on the RNA structure. For instance, compensatory base changes validate the existence of Watson–Crick base pairs and appropriate deletion may help to define independent structural domains.

Probing the structure with chemicals and enzymes may also provide information on the tertiary folding of large RNA molecules. The tertiary structure of large RNAs is composed of stable secondary structure elements that are brought

Handbook of RNA Biochemistry. Edited by R. K. Hartmann, A. Bindereif, A. Schön, E. Westhof
Copyright © 2005 WILEY-VCH Verlag GmbH & Co. KGaA, Weinheim
ISBN: 3-527-30826-1

Tab. 10.1. Structure-specific probes for RNA

Probes	MW	Target	Product	Detection			Special considerations Buffers, pH, temperature, etc.
				direct	RT	in vivo	
Chemicals and divalent ion							
DMS	126	$A(N^1)$	N^1-CH_3	–	+	+	reactive at pH ranging from 4.5 to 10 and temperature from 4 to 90 °C; tris buffers should be avoided as DMS reacts with amine groups
		$C(N^3)$	N^3-CH_3	s	+	+	idem
		$G(N^7)$	N^7-CH_3	s	s	+	idem
DEPC	174	$A(N^7)$	$N^7-CO_2H_5$	s	+	–	reactive at pH ranging from 4.5 to 10 and temperature from 4 to 90 °C; tris buffers should be avoided as DEPC reacts with amine groups
Kethoxal	148	$G(N^1-N^2)$	N^1-CHOH | N^2-CROH	+ (a)	+	–	borate ions are required to stabilize the guanine–kethoxal adduct
CMCT	424	$G(N^1)$	$N^1-C=N-R$ | $NH-R'$	–	+	–	optimal reactivity at pH 8 and over a wide range of temperature; CMCT still soluble up to 300 mg/ml in water
		$U(N^3)$	$N^3-C=N-R$ | $NH-R'$	–	+	–	idem
Pb(II) acetate	207	specific binding sites dynamic regions	...Np (3'p)	+	+	+	buffers with chlorure ions should be avoided as Pb(II) can form precipitates with it. Pb(II) acetate must be dissolved in water just before use
Biological nucleases							
T1 RNase	11 000	unpaired G	....Gp (3'p)	+	+	–	active under a wide range of conditions (e.g. temperature between 4 and 55 °C, with or without magnesium ion and salt, in urea)
T2 RNase	36 000	unpaired A > C, U, G	....Ap (3'p)	+	+	–	active under a wide range of conditions (e.g. temperature between 4 and 55 °C, with or without magnesium ion and salt)
V1 RNase	15 900	paired or stacked N	pN.... (5'p)	+	+	–	absolutely requires divalent cations; active under a wide range of temperature (4–50 °C)

DMS, dimethylsulfate; DEPC, diethylpyrocarbonate; kethoxal, β-ethoxy-α-ketobutyraldehyde; CMCT, 1-cyclohexyl-3-(2-morpholinoethyl) carbodiimide metho-p-toluene sulfonate. Detection method: (direct) detection of cleavages on end-labeled RNA molecule; (RT) detection by primer extension with reverse transcriptase. (+) the corresponding detection method can be used; (s) a chemical treatment is necessary to cleave the ribose-phosphate chain prior to the detection; (a) RNase T1 hydrolysis can be used after kethoxal modification with end-labeled RNA. Modification of guanine at N1, N2 will prevent RNase T1 hydrolysis [46]. *In vivo* mapping: probes which diffuse efficiently across membranes and walls (+), the other probes can be used only after permeabilization of the cell (–). Molecular weight, specificity, and products generated by the probe action are indicated. The table is adapted from Brunel & Romby [23].

Position N^7 of purines, involved in Hoogsteen or reverse Hoogsteen interactions, can also be probed by diethylpyrocarbonate (DEPC) for adenines and by DMS or nickel complex for guanines. Nickel complex [24] and DEPC [25] are very sensitive to the stacking of the base rings, and therefore N^7 of purines within a helix are never reactive except if the deep groove of the helix is widened.

10.2.3
Lead(II)

Divalent metal ions such as Mg^{2+} are required for stabilization of the RNA structure, but under special conditions can promote cleavage in RNA. This catalytic activity was first discovered with Pb^{2+} ions, and latter with many other di- and trivalent cations (for a review, see [26]). Strong cleavage was first described as the consequence of a tight divalent metal ion-binding site and of an appropriate stereochemistry of the cleaved phosphodiester bond. Lead(II) is also considered as a single-strand-specific probe since weaker cleavages at several sites occurred mainly in unpaired and flexible regions (interhelical or loop regions and bulged nucleotides). In contrast to RNases, lead(II) is not sensitive to steric hindrance, but detects subtle conformational changes that can occur upon ligand binding. Lead(II) was also successfully used *in vivo* to map the structure of mRNA and regulatory RNAs [14]. In contrast to DMS modification, lead(II) is less sequence dependent, and thus can be used to assess single- and double-stranded regions of RNA. The cleavage patterns obtained on three different RNAs indicated that similar conformations were observed *in vivo* and *in vitro* [14].

10.3
Methods

Probing the conformation of RNAs with different enzymes and chemicals requires the use of defined buffer conditions (pH, ionic strength, magnesium concentration, temperature). Indeed, the optimal conditions vary slightly with the different probes and the possibility exists that subtle conformational changes may occur under different incubation conditions (Table 10.1). The probe:RNA ratio must also be adapted so that the experiments are conducted under limited and statistical conditions. For the first experiment, different concentrations of the probes and a time-scale dependence should be performed. This is also required when the commercial source of the probes has been changed.

10.3.1
Equipment and Reagents

Equipment and Material
Electrophoresis instrument for sequencing gels. Eppendorf tubes, tips and buffers should be sterilized before used.

Chemicals and Enzymes

CMCT and lead(II) acetate can be purchased from Merck; DMS from Acros Organics (ref. 11682-0100); calf intestinal phosphatase and T4 RNA ligase from P-L Biochemicals; avian myeloblastosis virus reverse transcriptase from Q biogene (France) or Life Sciences (USA); T4 polynucleotide kinase, [γ-^{32}P]ATP (3200 Ci/mmol) and [5'-^{32}P]pCp (3000 Ci/mmol) from Amersham. RNase T1 was from Industrial Research Limited (IRL, New Zealand) or from Fermentas; RNase T2 from Invitrogen (ref. 18031-013) and RNase V1 from Pierce (ref. MB092701).

Safety Rules using Chemicals

Most of the chemical reagents are potential carcinogens and therefore chemical modifications until the removal of the first ethanol supernatant (see below) are carried out under a fume hood while wearing protective gloves. DMS and kethoxal solutions are discarded in 1 M sodium hydroxide waste and CMCT in 10% acetic acid waste.

Buffers

The buffer conditions given here are indicative and can be modulated according to the system used and the nature of the ligand. *buffer N1*: 50 mM sodium HEPES, pH 7.5, 5 mM MgAc, 100 mM KAc; *buffer N2*: 50 mM sodium cacodylate, pH 7.5, 5 mM MgCl$_2$, 100 mM KCl; *buffer D2*: 50 mM sodium cacodylate, pH 7.5, 1 mM EDTA; *buffer N3*: 50 mM sodium borate, pH 8, 5 mM MgCl$_2$, 100 mM KCl; *buffer D3*: 50 mM sodium borate, pH 8, 1 mM EDTA; *buffer N4*: 50 mM sodium borate, pH 7.5, 5 mM MgCl$_2$, 100 mM KCl; *buffer D4*: 50 mM sodium borate, pH 7.5, 1 mM EDTA; *Buffer $\Delta T1$*: 20 mM sodium citrate, pH 4.5, 1 mM EDTA, 7 M urea, 0.02% xylene cyanol, 0.02% bromophenol blue; *Ladder buffer*: Na$_2$CO$_3$ 0.1 M/ NaHCO$_3$ 0.1 M pH 9; *RNA loading buffer*: 0.02% xylene cyanol, 0.02% bromophenol blue in 8 M urea; *DNA loading buffer*: 0.02% xylene cyanol, 0.02% bromophenol blue in formamide; *RTB buffer*: 50 mM Tris–HCl, pH 7.5, 20 mM MgCl$_2$ and 50 mM KCl; *TBE buffer*: 0.09 M Tris–borate, pH 8.3, 1 mM EDTA. All buffers are given 1 × concentrated.

10.3.2
RNA Preparation and Renaturation Step

The RNA is usually transcribed *in vitro* from a DNA template using T7 RNA polymerase. The RNA is then purified from shorter RNA fragments, DNA template and the excess of NTP by using either a gel-filtration column [27], monoQ column [28] or denaturing polyacrylamide–urea gel electrophoresis [29]. More recently, ion-pairing reversed-phase high-performance liquid chromatography (IP-RPLC) was used for the fractionation of short RNA fragments [30].

For 5'-end-labeling, the RNA is first dephosphorylated at its 5' end, and then labeled using [γ-^{32}P]ATP and T4 polynucleotide kinase [31]. The dephosphorylation step can be avoided if transcription is carried out in the presence of GMP or with ApG. The 3'-end-labeling is performed with [5'-^{32}P]pCp and T4 RNA ligase [32].

The labeled RNAs are purified by electrophoresis on 8% polyacrylamide (0.5% bis)–8 M urea slab gels. Before each experiment, the RNA is eluted from the gel slice in 100 μl of 500 mM ammonium acetate/1 mM EDTA, precipitated with 2.5 volumes of cold ethanol in the presence of 1 μg of glycogen. After two washing steps with 200 μl of 80% cold ethanol, the pellet is then dissolved in sterile H_2O (to obtain about 50 000 c.p.m./μl).

Since the RNA is often in contact with denaturing reagents during its purification, it is worth spending effort to carry out a renaturation process before the probing experiments. One possible renaturation protocol is as follows: the RNA is pre-incubated 1 min at 90 °C in H_2O, quickly cooled on ice (2 min) and brought back slowly (20 min) at 20 or 37 °C in the appropriate buffer containing 5 mM $MgCl_2$.

10.3.3
Enzymatic and Lead(II)-induced Cleavage Using End-labeled RNA

This direct method which uses end-labeled RNA is restricted to the detection of cleavage in the RNA after RNase hydrolysis or after chemical modifications that allow subsequent strand scission by an appropriate treatment (see Table 10.1).

Enzymatic probing and lead-induced cleavages were adapted for the *thrS* mRNA regulatory region (around 250 nt). Some of the experiments are illustrated in Fig. 10.1(A–C). All reactions were conducted in a total volume of 10 μl. Appropriate dilutions of enzymes and of lead(II) acetate were done in sterile water just before use. Incubation controls in the absence of the probes were performed in order to detect non-specific cleavage in RNA. In these controls, the enzyme or lead(II) was replaced by sterile H_2O. For RNA–ligand footprinting experiments, the complex was pre-formed before the enzymatic or chemical reaction in an appropriate buffer optimal for binding.

Labeled RNA (50 000 c.p.m./μl) sufficient for the planned experiments was first denatured in sterile H_2O at 90 °C for 1 min then cooled on ice for 1 min.

RNase T1
Labeled mRNA (1 μl, 50 000 c.p.m.) was renatured in the presence of 5 μl of H_2O and 2 μl of buffer N1 (5 × concentrated) at 20 °C for 20 min. Then, 1 μl of total tRNA (2 μg/μl) was added and reaction was performed with 1 μl of RNase T1 (0.1 U from IRL, or 0.2 U from Fermentas) for 5 min at 20 °C.

RNase T2
The same protocol as for RNase T1 except that reaction was performed with 1 μl of RNase T2 (0.05 U) for 5 min at 20 °C.

RNase V1
The same protocol as for RNase T1 except that reaction was performed with 1 μl of RNase V1 (0.05 U) for 5 min at 20 °C.

In order to define the best conditions for the hydrolysis, try for the first time three different concentrations of the enzymes: RNase T1 (0.05–0.1–0.5 U from

158 | 10 Probing RNA Structures with Enzymes and Chemicals In Vitro and In Vivo

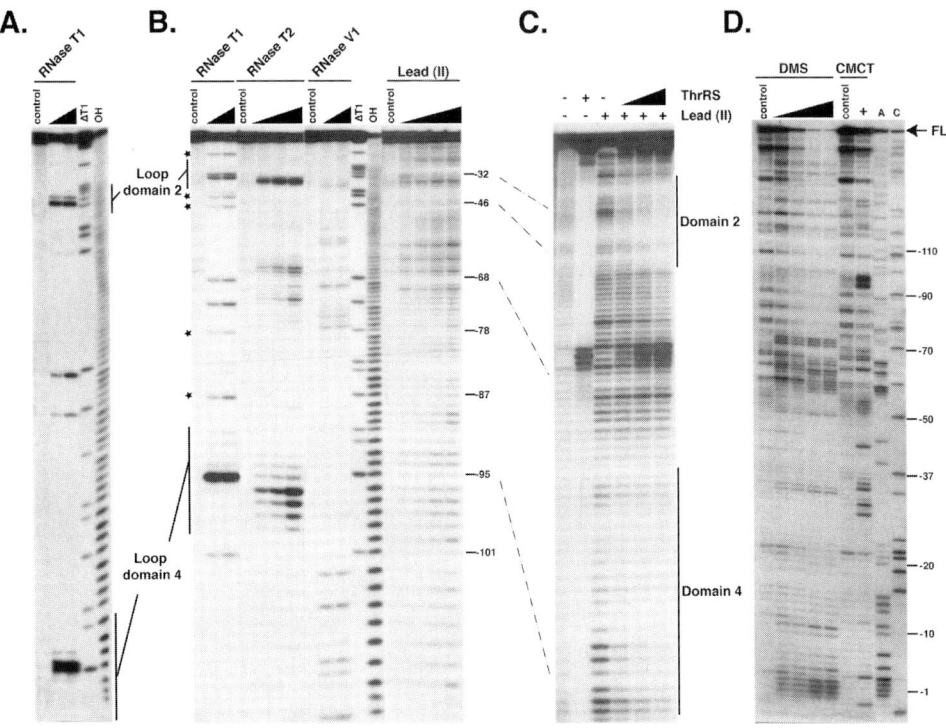

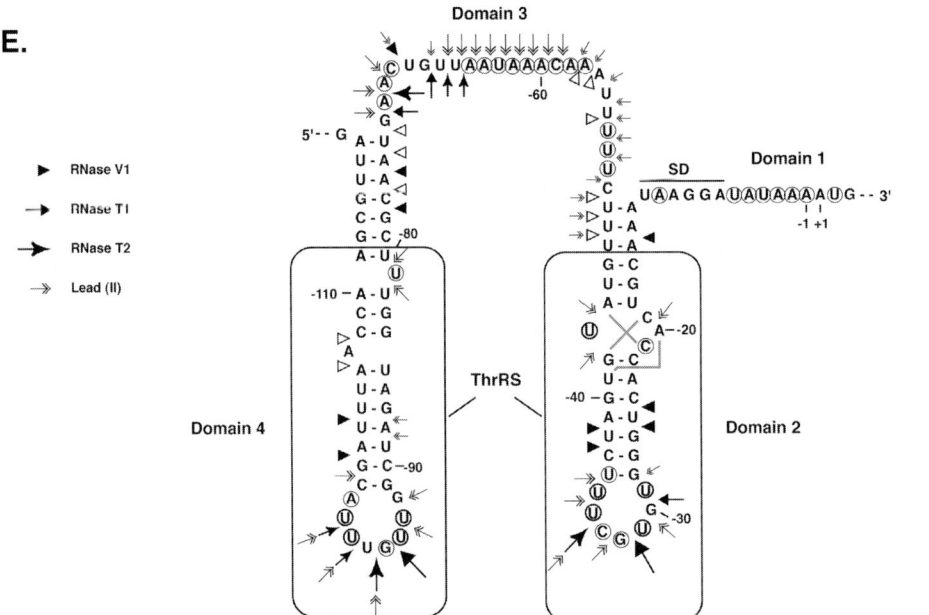

IRL or 0.1–0.2–0.5 U from Fermentas), RNase T2 (0.01–0.05–0.1 U) and RNase V1 (0.01–0.05–0.1 U).

Lead(II) Acetate

Labeled mRNA (1 µl, 50 000 c.p.m.) was renatured in the presence of 3.5 µl of H_2O and 2 µl of buffer N1 (5×) at 20 °C for 20 min. Hydrolysis was initiated with 2.5 µl of different concentrations of lead(II) acetate from 12, 40, 80 to 120 mM for 5 min at 20 °C in the presence of 1 µl of total tRNA (2 µg/µl). Then, 5 µl of 0.1 M EDTA were added to stop the reaction. The best conditions for *thrS* mRNA was 40 mM.

Reaction Stop

Enzymatic hydrolysis were stopped by phenol extraction while the RNA treated with lead(II) was directly precipitated.

- To all samples, 40 µl of 0.3 M sodium acetate, pH 6 and 50 µl of phenol saturated with chloroform (v/v) were added. The samples were mixed for 1 min and then centrifuged 1 min at high speed.
- The aqueous phase was removed carefully and transferred into a new sterile 1.5 ml micro tube and 2.5 volumes of cold ethanol (150 µl) was added to precipitate the RNA. After mixing, the samples were left in a dry ice/ethanol bath for 10 min and centrifuged (13 000 r.p.m. at 4 °C for 15 min).
- The supernatant was discarded and the pellet was washed twice with 200 µl of 80% cold ethanol. After a short centrifugation (13 000 r.p.m. for 5 min at 4 °C), the supernatant was removed and the pellets were vacuum-dried (no more than 5 min) and dissolved in 6 µl of RNA loading buffer.

Fig. 10.1. Enzymatic and chemical probing on *thrS* mRNA. (A–C) Enzymatic hydrolysis and lead(II)-induced cleavages performed on 5′-end-labeled *thrS* mRNA either free (A and B) or in the presence of increasing concentrations of threonyl-tRNA synthetase (C). (A and B) The conditions for RNase and lead(II) concentrations were as follows: (A) RNase T1 from IRL (0.05 and 0.1 U), (B) RNase T1 from IRL (0.2 and 0.5 U), RNase T2 (0.01, 0.05 and 0.1 U) and RNase V1 (0.05 and 0.1 U), and lead(II) (12, 40, 80 and 120 mM). (C) Hydrolysis was performed with 40 mM lead(II). ThrRS concentrations were as follows: 0.01, 0.05 and 0.1 µM. (control) incubation control in the absence of the probes. (ΔT1, OH) RNase T1 under denaturing conditions and alkaline ladder, respectively. RNase T1 cleavages not reproducibly found are noted by an asterisk. (D) Gel electrophoresis fractionation of products resulting from DMS ($N^1A \gg N^3C$) and CMCT ($N^3U \gg N^1G$) modification followed by primer extension analysis. Reactions have been performed on free mRNA under native conditions in the absence (control) or in the presence of increasing concentrations of DMS (see text for details) or in the presence of 4 µl of CMCT 40 mg/ml (+). FL = full-length product. (Lanes A and C) The two sequencing ladders correspond to the RNA sequence. (E) Reactivity of Watson–Crick positions, enzymatic and lead(II)-induced cleavages reported on the secondary structure of the *thrS* mRNA adapted from Caillet et al. [20]. Circled nucleotides are reactive towards DMS (N^1A, N^3C) and CMCT (N^3U, N^1G) modifications. The two domains of the RNA protected by ThrRS are squared.

Fractionation of End-labeled RNA Fragments

RNase T1 and alkaline ladders were required to identify the cleavage positions.

- RNase T1 ladder. Labeled mRNA (0.5 μl, 25 000 c.p.m.) was preincubated at 50 °C for 5 min in 5 μl of buffer ΔT1 containing 1 μg total tRNA. The reaction is then performed at 50 °C for 10 min in the presence of 1 μl of RNase T1 (0.1 U for IRL or 0.5 U for Fermentas).
- Alkaline ladder. Labeled mRNA (2 μl, 100 000 c.p.m.) was incubated at 90 °C for 3 min in the presence of total tRNA (2 μg) in 5 μl of ladder buffer.

The end-labeled RNA fragments were sized by electrophoresis on 12 or 15% polyacrylamide (0.5% bis)–8 M urea slab gels (0.5 mm × 30 cm × 40 cm) in 1 × TBE. Gels should be pre-run (30 min at 75 W) and run warm (75 W) to avoid band compression. The migration conditions must be adapted to the length of the RNA, knowing that on 15% gel, xylene cyanol migrates as a 39-nt RNA and bromophenol blue as 9 nt. The 15% polyacrylamide gel is convenient to collect the data on small-size fragments (1–50 nt RNA fragments). For a 250-nt long RNA, two migrations are necessary to interpret correctly the reactivity of nucleotides of the whole RNA molecule. At the end of the run, the 12% gel is fixed for 5 min in a 10% ethanol/6% acetic acid solution, transferred to Whatman 3 MM paper and dried. The 15% gel was transferred without drying on a plastic support and wrapped with a plastic film. Exposure is done at −80 °C using an intensifying screen.

10.3.4
Chemical Modifications

Examples of chemical modifications performed on *thrS* mRNA regulatory region are shown in Fig. 10.1(D). Reactions were carried out on 1 pmol of unlabeled *thrS* mRNA in a total volume of 20 μl. For enzymatic and lead(II)-induced cleavages, the same experimental conditions could be used as described above except that the end-labeled RNA is replaced by 1 pmol of cold RNA. Control of an unmodified RNA was done in parallel, in order to detect pauses of reverse transcriptase due to stable secondary structures and/or non-specific cleavage. The reactions were conducted either in the presence of mono- and divalent ion ("native conditions") or in the absence of ions ("semi-denaturing" conditions). Unlabeled mRNA was first heated in sterile H_2O at 90 °C for 1 min and then cooled on ice for 1 min.

DMS modification

- Native conditions: 1 μl of mRNA (1 pmol) was first renatured by incubation at 20 °C for 20 min in the presence of 4 μl of buffer N2 (5×) and 13 μl of sterile H_2O. The reaction was performed at 20 °C for 5 min in the presence of 1 μl of tRNA (2 μg/μl) and 1 μl of pure DMS or diluted freshly into ethanol 1:2, 1:5 and 1:10. The optimal chemical modification was obtained with DMS diluted 1:10 (Fig. 10.1D).

- Semi-denaturing conditions: same procedure as for native conditions, but the reaction was performed in buffer D2.

CMCT modification

- Native conditions: 1 µl of mRNA (1 pmol) was first incubated at 20 °C for 20 min in the presence of 4 µl of buffer N3 (5×) and 10 µl of sterile H_2O. The reaction was done at 20 °C during 20 min in the presence of 1 µl of tRNA (2 µg/µl) and 4 µl of CMCT (40 or 60 mg/ml in water just before use). The optimal chemical modification for *thrS* mRNA was obtained with CMCT at 40 mg/ml (Fig. 10.1D).
- Semi-denaturing conditions: same procedure as for native conditions, but in buffer D3.

Kethoxal modifications

- Native conditions: 1 µl of mRNA (1 pmol) was first incubated at 20 °C for 20 min in the presence of 4 µl of buffer N4 (5×) and 12 µl of sterile H_2O. The reaction was done at 20 °C for 10 min in the presence of 1 µl of tRNA (2 µg/µl) and 2 µl of kethoxal (at 10 or 20 mg/ml diluted in 20% ethanol).
- Semi-denaturing conditions: same procedure as for native conditions but reaction was done in buffer D4 for 5 min.

All these steps have to be conducted under a fume hood, and DMS and CMCT solutions should be destroyed in 1 M NaOH and 10% acetic acid waste, respectively.

Reaction stops

All the reactions were stopped by ethanol precipitation of the RNA.

- To all samples, 50 µl of 0.3 M sodium acetate, pH 6 and 250 µl of cold ethanol were added. For RNA–protein footprinting experiments, the protein was removed by phenol extraction. The samples were then mixed, placed in a dry-ice/ethanol bath for 15 min and centrifuged (13 000 r.p.m. at 4 °C for 15 min).
- The supernatants were removed carefully (do not touch the pellet) and 200 µl of 80% cold ethanol added to the pellets. The samples were centrifuged (13 000 r.p.m. at 4 °C for 5 min) and the supernatants were removed with the same caution.
- The pellets were vacuum-dried (no more than 5 min) and resuspended in 4 µl of sterile H_2O.

10.3.5
Primer Extension Analysis

Primer extension with reverse transcriptase was originally developed by HuQu et al. [33] for probing the structure of large RNA molecules. Reverse transcriptase stops its incorporation of dNTP at the residue preceding a cleavage or a modification at a

Watson–Crick position. While carbethoxylation of $A(N^7)$ by DEPC is sufficient to stop reverse transcriptase, a subsequent treatment is necessary to induce a cleavage at $G(N^7)$ after DMS modification (Table 10.1 [23]).

The length of the primer varies usually from 12 to 18 nt. This provides sufficient specificity even if the primers are used on a mixture of RNAs. For long RNA, primers are selected every 200 nt due to the gel resolution limitation. Before probing the RNA structure, assays should be performed to define the best concentration of the RNA, the choice of the primer sequence and the hybridization conditions in order to get an efficient primer extension. For *thrS* mRNA, primer annealing conditions were selected in order to maximize the unfolding of the probed RNA and to minimize RNA degradation. The primer TTACAGCGTGATCGT, complementary to nucleotides +47 to +61 of *thrS* mRNA, was used (Fig. 10.1D).

Hybridization

To the 4 μl of modified mRNA (1 pmol), 1 μl of 5′-end-labeled DNA primer (around 100 000 c.p.m.) was added. The samples were then heated at 90 °C for 1 min and then quickly cooled on ice. Then, 1 μl of 5 × RTB buffer was added and the samples were incubated for 15 min at 20 °C.

Primer Extension

The reaction was done in 15 μl of total volume. To the hybridization mix were added 2 μl of 5 × RTB, 2 μl of dNTP mix (2.5 mM of each dNTP), 4 μl sterile H_2O and 1 μl of reverse transcriptase (2 U/μl diluted freshly into sterile H_2O). Incubate the samples for 30 min at 37 °C.

For kethoxal modification, 2 μl of 50 mM sodium borate, pH 7.0 was added in the extension reaction to stabilize the adduct.

- To all samples, 50 μl of 0.3 M sodium acetate, pH 6 and 200 μl of cold ethanol were added. After precipitation, the pellets were washed twice with 80% ethanol and vacuum-dried as described above. The end-labeled DNA fragments were resuspended in 6 μl of the DNA loading buffer.
- To improve the resolution of the gels, the RNA template can be subjected to alkaline hydrolysis.
- Just after primer extension, 20 μl of the buffer containing 50 mM Tris–HCl, pH 7.5, 7.5 mM EDTA, 0.5% SDS and 3.5 μl of 3 M KOH were added. The samples were incubated at 90 °C for 3 min then at 37 °C for at least 1 h.
- To all samples, 6 μl of 3 M acetic acid, 100 μl 0.3 M sodium acetate, pH 6 and 300 μl of cold ethanol were added. The following procedure is as described above.

Gel Fractionation

The DNA fragments were denatured by incubation at 90 °C for 3 min and were fractionated on 8% polyacrylamide (0.4% bis)–8 M urea slab gels in 1 × TBE. As described above, gels should be pre-run (30 min at 75 W) and run warm (75 W). The migration conditions must be adapted to the size of the fragments to be analyzed, knowing that on 8% gel, xylene cyanol migrates as 81 nt and bromophenol blue as 19 nt.

The modification or cleavage positions were identified by running in parallel a sequencing reaction. The elongation step was performed as described above except in the presence of one of the didesoxyribonucleotide ddXTP (2.5 µM), the corresponding desoxyribonucleotide dXTP (25 µM) and the three other desoxyribonucleotides (100 µM).

After migration, the gels were dried, and autoradiographed at −80 °C with an intensifying screen overnight.

10.3.6
In Vivo RNA Structure Mapping

10.3.6.1 *In Vivo* DMS Modification

DMS has been successfully used to probe several RNA species from a variety of cells, including Gram-negative [12, 34] and Gram-positive bacteria [11, 35], yeast [36], protozoa [37], plants [38], and fibroblast cells [39]. This reagent is capable to diffuse efficiently across cell wall and membrane, and to modify unpaired adenines (N^1) and cytosines (N^3). Occasionally, modifications of uridines at N^3 have been identified during *in vivo* DMS modification [13]. Information on the accessibility of guanines at N^7 can also be obtained. The protocol given below was adapted for bacteria and one typical experiment performed on *Staphylococcus aureus* RNAIII is illustrated in Fig. 10.2.

The bacterial strain was grown to mid-log phase and then treated with DMS. As for the *in vitro* experiment, it is important to verify that the reaction occurred under limited conditions such as less than one modification per molecule was statistically induced. Thus, a range of DMS quantities (100 µl of DMS diluted 1:10, 1:5, 1:2 or pure) and time intervals for incubation (2–15 min) were initially tested. After treatment, the reaction was stopped just before the extraction of the total RNA. Sites of DMS modification were detected by primer extension on total RNA extracts (5–20 µg), using end-labeled primers specific for a chosen region of interest of the tested RNA.

- Bacteria (20 ml of culture) were grown in LB medium in a 50-ml sterile tube to mid-logarithmic phase at 37 °C (until an OD_{600} of 0.5 was reached).
- 100 µl of DMS (diluted 1:2 in ethanol) were added and the culture was incubated for an additional 5 min at 37 °C after gentle shaking.
- The reaction was stopped by adding 10 ml of cold stop buffer containing 100 mM Tris–HCl, pH 8, 200 mM β-mercaptoethanol, 5 mM EDTA.
- The cells were then pelleted (3 000 r.p.m. for 15 min at 4 °C), and were resuspended in 1.5 ml of cold buffer 10 mM Tris–HCl, pH 8, 100 mM NaCl, 1 mM EDTA. The cells were transferred in a 1.5-ml micro tube and centrifuged at 13 000 r.p.m. for 15 min at 4 °C.
- The cells were then disrupted by adding 200 µl of buffer containing 50 mM Tris–HCl, pH 8.0, 8% sucrose, 0.5% Triton, 10 mM EDTA, 4 mg/ml lysozyme, and

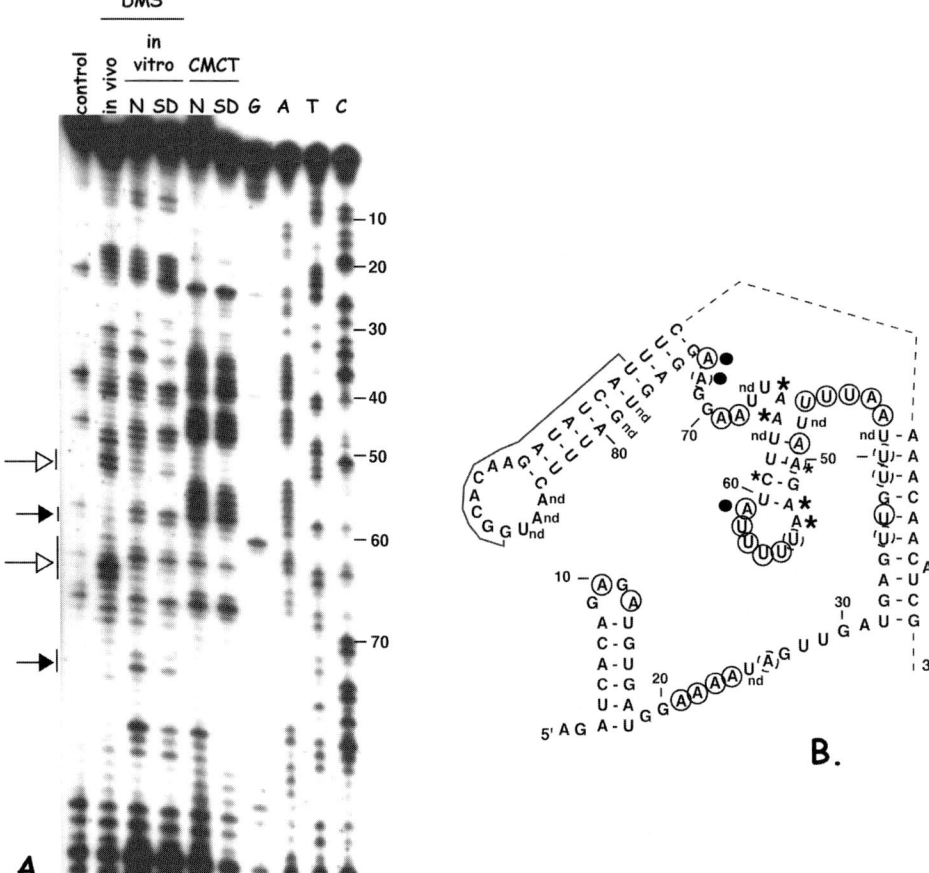

Fig. 10.2. Chemical probing on *S. aureus* RNAIII performed *in vivo* and *in vitro* (A), and part of the secondary structure of *S. aureus* RNAIII (B). (A) DMS and CMCT modifications were performed *in vitro* under native conditions (N) and semi-denaturing conditions (SD), and *in vivo* for DMS. Experimental details are given in the text. (Control) incubation control; (lanes G, A, T, C) DNA sequencing ladders. (B) Circled nucleotides are reactive towards DMS ($N^1A \gg N^3C$) and CMCT ($N^3U \gg N^1G$) modifications. (A and B) Reactivity differences are shown by empty arrows or by stars for nucleotides, which are only reactive *in vivo* and by dark arrows or black circles for nucleotides reactive only *in vitro*. nd = not determined. The position of the primer, complementary to nucleotides G88 to U102, is given. Adapted from Benito et al. [35].

incubated 5 min in ice. For *S. aureus*, the cells were treated with lysostaphine (50 µg/ml) in the presence of 1% SDS.
- Then, 200 µl of phenol saturated with 0.1 M sodium acetate, pH 5.5 and 10 mM EDTA were added. Cells were vortexed for 30 s at high speed. The samples were heated at 65 °C during 15 min and mixed every 5 min.
- The mixture was cooled on ice and centrifuged 10 min at 13 000 r.p.m. The

aqueous phases were carefully collected, and the phenol and interface were re-extracted by vortexing the samples with 100 µl 0.1 M sodium acetate, pH 5.5.
- After centrifugation, the aqueous phases were pooled and extracted once with phenol/chloroform previously saturated in sodium acetate 0.1 M, pH 5.5 and once with chloroform.
- The RNA was then precipitated twice with three volumes of cold ethanol in the presence of 0.3 M sodium acetate (final concentration).
- The pellets were washed twice with 200 µl of 80% ethanol, vacuum-dried (no more than 5 min) and dissolved in a small volume of sterile H_2O. The RNA concentration was measured.
- 10 µg of material was used for primer extension.
- Primer hybridization and elongation by reverse transcriptase were as described above, except that elongation was conducted at 45 °C for 30 min with 5 U of reverse transcriptase. Do not forget the sequencing reactions, which help to identify the position of modifications.
- Incubation control is performed on cells grown and treated in the same conditions as above but in the absence of DMS.
- A stop control was done in order to verify that little or no DMS modification occurred during the RNA extraction. In that control, DMS was added after the addition of the stop buffer.

10.3.6.2 *In Vivo* Lead(II)-induced RNA Cleavages

To avoid secondary cleavages, it is of prime importance to perform lead(II)-induced cleavage under conditions where less than one cleavage per molecule is induced. Thus, a range of lead(II) acetate concentrations (25–200 mM final concentration) and time intervals for hydrolysis (2–15 min) should be tested. After treatment, the reaction is stopped using an excess of EDTA and total RNA can be isolated. Incubation control should be performed under the same experimental conditions, except that lead(II) acetate is avoided. The cleavage positions were detected by primer extension analysis and assigned using in parallel sequencing reactions. The protocol described below was adapted for *Escherichia coli* to map the accessible regions of several non-coding RNAs and of mRNA [14].

- Bacteria (20 ml of culture) were grown in LB medium to mid-logarithmic phase at 37 °C (until an OD_{600} of 0.5 was reached) in a 50 ml conical tube.
- Make up a fresh solution of 1 M lead(II) acetate in sterile water [lead(II) acetate precipitates at high concentration in LB medium]. Then, 2.8 ml of this solution is mixed with 3.2 ml of sterile water and 2 ml of pre-warmed 4 × concentrated LB (at 37 °C) to give 8 ml of lead(II) acetate/LB solution at 350 mM. This step should be done just before use [This step is essential for reproducibility. Some lead(II) acetate precipitation was always observed in LB medium. Consequently, the intracellular concentration must be lower than the nominal concentration in the medium].
- 8 ml of this lead(II) acetate/LB solution (350 mM) was then added to 20 ml of cells at mid-logarithmic phase. This gave a final concentration of lead(II) acetate

100 mM. For the first trials, different concentrations of lead(II) acetate (50, 100, 150 and 200 mM final concentration) should be used.
- Cultures were incubated with gentle shaking for 7 min at 37 °C.
- Reactions were stopped by addition of 10 ml of cold 0.5 M EDTA (1.5-fold molar excess) and immediately put on ice.
- The cells were pelleted (3 000 r.p.m. for 15 min at 4 °C) and resuspended in 1.5 ml of cold buffer 10 mM Tris–HCl, pH 8, 100 mM NaCl, 1 mM EDTA. The cells were transferred in a 1.5-ml micro tube and centrifuged at 13 000 r.p.m. for 15 min at 4 °C.
- RNA extraction and analysis was carried out as described for the *in vivo* DMS mapping.

Lead(II)-induced cleavage *in vivo* can be performed under different growth conditions. However, lead(II) acetate has some tendency to precipitate, in particular when chloride ions are present within the medium. In that case, it is essential to test different concentrations of the lead(II) acetate. One simple and reliable method for evaluating cleavage conditions is to fractionate total RNA on agarose gels [14]. Upon increases of lead(II) acetate concentration (25–200 mM), the intensities of 16S and 23S rRNA, the predominant RNA species, significantly decreased and optical inspection of the patterns could be used to calibrate conditions.

10.4
Commentary

10.4.1
Critical Parameters

RNA Preparation
The RNA is usually synthesized by *in vitro* transcription using T7 RNA polymerase. Therefore, the 3′ or 5′ end of the RNA might be heterogeneous, and several abortive transcription products might also accumulate. To obtain homogenous size RNA, the method of choice remains the fractionation of the transcript by electrophoresis on denaturing urea gel. For long RNA molecules, electro-elution might help to increase the elution efficiency.

Homogeneous RNA Conformation
Since during the purification, the RNA can be partially denatured, it is essential to design renaturation protocols in order to have a conformationally homogeneous RNA population and to test whether this conformation is biologically relevant (enzymatic activity for ribozyme, efficient ligand binding, etc.). Alternative RNA conformations may co-exist, and can be revealed by the simultaneous presence of single-stranded and double-stranded specific cleavages or modifications at the same position. If the conformers have different electrophoretic mobilities on native

polyacrylamide gel, chemical probing can be used to distinguish them [40]. After chemical modification, the co-existing structures are separated on a native polyacrylamide gel and the modification sites for each conformer are then identified by primer extension.

Chemical and Enzymatic Probing *In Vitro*

The chemical reactions and RNase T1 hydrolysis can be conducted under a variety of experimental conditions. For instance, the influence of divalent ion (such as magnesium) can be tested on the folding of the RNA and by varying the temperature (between 4 and 90 °C) one can follow thermal transition of RNA molecules [41]. It is essential, however, to adapt for each condition the chemical reactions or the enzymatic hydrolysis in order to have less than one cut or modification per molecule, i.e. more than 80% of the RNA should not be modified or cleaved. For example, for DMS modification, reaction at 4 °C is for 20 min in the presence of 1 µl of DMS, whereas at 50 °C, reaction is for 5 min with 1 µl of DMS diluted 1:16.

- Kethoxal might have a partially denaturing effect on RNA structure even if the reaction was not too strong [41]. Concentration of kethoxal or the incubation time should be reduced in order to get only modifications at guanines present in single-stranded regions.
- The RNase cleavages in the RNA molecule can induce conformational rearrangements potentially able to provide new targets for secondary cleavages. Usually these secondary cleavages occur when the reaction is too strong; they also are of weak intensity and are not reproducibly found in all experiments. These cleavages can be distinguished from primary cuts by comparing the hydrolysis patterns obtained from the 5'- or 3'-end-labeled RNA.
- RNase V1 hydrolysis generates RNA fragments which end up with a 3'-OH group in contrast to alkali and most of the RNases. Therefore, 5'-end-labeled fragments generated by alkali will migrate faster than the RNase V1 fragments. This difference is only observed for the shortest RNA fragments (see Fig. 10.1B).
- Appropriate incubation controls are required to identify cleavages that are induced during the incubation treatments and pauses of reverse transcriptase that are due to stable secondary structures. Nucleotides for which strong bands are visible in the control lanes are not considered for the interpretation.
- Each experiment should be repeated several times, and only the reproducible cleavages and modifications will be considered for the interpretation. As mentioned previously, the elaboration of a secondary structure RNA model requires to collect data from enzymes and chemicals with different and complementary specificities. Only this combination will allow to define helical and loop regions. The presence of RNase V1 cleavages and nucleotides not reactive at Watson–Crick position is a strong indication for the existence of a helical region.
- Footprinting assays. The experiments should be done in the presence of increasing concentrations of ligand (Fig. 10.1C). Lead(II)-induced cleavages and hydroxyl radicals are appropriate probes to map the ligand binding site due to their small size and their specificity. Results should be interpreted with care because

decreased reactivity does not necessarily result from a direct shielding effect, but could be due to a steric hindrance effect (particularly observed with the bulky RNases) or to a conformational change of the RNA.

10.4.2
In Vivo Mapping

- Due to the inability to diffuse within the cells, only a few probes have been used to map the RNA structure within the cells. DMS- and lead(II)-induced cleavages are to date the most commonly used probes. Other probes like RNase T1 [42], kethoxal [13] and CMCT [43] have been used *in vivo* after permeabilization of the cells. However, due to this additional treatment, particular caution has to be taken to ensure that the cells remain intact during the time of incubation. It is also essential to verify that the reaction was efficiently stopped before the RNA extraction procedure.
- Alternative to phenol extraction, other protocols used to extract total RNA can be used. Reagents combining phenol and guanidine thiocyanate enable a straightforward isolation of total RNA from samples of human, yeast, bacterial and viral origin.
- Data from *in vivo* probing may be more complex to interpret than the *in vitro* probing. One of the main reasons is that the studied RNA may be involved simultaneously in several complexes (e.g. regulatory RNAs). However, *in vivo* mapping becomes powerful when it is used in a comparative manner. For example, conformational changes of mRNA induced by a *trans*-acting ligand have been identified upon repression or activation of translation (e.g. [11, 34]). DMS- and lead(II)-induced cleavages can also be used to monitor the conformational changes of mRNA under different growth conditions and under different environmental cues such as temperature.

10.5
Troubleshooting

10.5.1
In Vitro Mapping

- RNase T1 cleaves all guanines (Fig. 10.1B). Significant unfolded RNA molecules were present (improve the renaturation protocol) or the hydrolysis was too strong. As shown in Fig. 10.1(A), the cleavage pattern significantly changed by reducing the concentration of RNase T1.
- Compression of bands due to stable secondary structure (in general rich in GC base pairs). Heat the end-labeled RNA samples at 90 °C for 3 min before gel loading. Never heat the alkaline ladder and the RNase T1 ladder.
- Cleavages of end-labeled RNA are doubled: the T7 RNA transcript is not homogenous in size (purify the RNA on polyacrylamide–urea gel).

- Too many bands in the incubation controls of end-labeled RNA: RNase contamination, repurify the RNA and prepare new sterile buffers.
- Aggregation of end-labeled RNA in the gel pockets; only fragments of small size can be visualized. The data cannot be taken into account. The pellets were not correctly dried after ethanol precipitation. Heat the samples before loading on the gel.
- Samples do not migrate correctly during electrophoresis. This is probably due to the presence of salt. Add several washing steps with ethanol 80% at the end of the procedure.
- No full-length RNA after DMS modification (Fig. 10.1D): adapt the conditions by reducing the amount of DMS in order to get more than 80% of RNA molecules unmodified.
- Absence of signal after primer extension: the modified RNA did not efficiently precipitate. Since the modified RNA is not labeled, particular caution should be taken to prevent the loss of the pellet.
- To keep high resolution of the gels, acrylamide, urea solutions and, in particular, ammonium persulfate should be prepared freshly.

10.5.2
In Vivo Probing

- Low yield of total RNA: incomplete homogenization or lysis of samples, degradation of the RNA.
- Strong stops in the control lanes: degradation of RNA, pauses of reverse transcription due to stable secondary structures (perform elongation at higher temperature, increase the concentration of the enzyme and dNTP or change the primer sequence). Many RNA molecules carry post-transcriptional modifications that may interfere with reverse transcriptase elongation (primer should be changed in order to cover the modified base).
- No more full-length RNA product after modification: reaction was too strong (reduce either the quantity of the reagent or/and the time of incubation). Check that the reaction was efficiently stopped before the extraction of total RNA.
- Weak or smearing signal after primer extension: increase the concentration of total RNA, check that the primer hybridization protocol is efficient. Optimal conditions for primer hybridization should be established in a series of pilot experiments, another protocol was described by Sambrook et al. [44]. The optimal temperature for annealing varies from RNA to RNA, depending on the $G+C$ content, the propensity of the RNA to form secondary structure and the length of the primer. The aim is also to minimize the formation of mismatched DNA primer–RNA hybrids.

Acknowledgments

We are grateful to B. Ehresmann for his constant support and for critical reading of the manuscript, and we thank C. Brunel, J. C. Paillart and other members from the laboratory for discussions. This work was supported by the Centre National de Recherche (CNRS), the Ministère délégué à la Recherche et aux Nouvelles Technologies (ACI "Microbiologie et Pathologie Infectieuses"), la Ligue de la Recherche sur le Cancer and Région Alsace (M. P.).

References

1 STORZ, G. *Science* 2002, *296*, 1260–1263.
2 CARRINGTON, J. C., V. AMBROS. *Science* 2003, *301*, 336–338.
3 GOTTESMAN, S. *Genes Dev.* 2002, *16*, 2829–2842.
4 JOHANSSON, J., P. COSSART. *Trends Microbiol.* 2003, *11*, 280–285.
5 STORMO, G. G. *Mol. Cell* 2003, *11*, 1419–1420.
6 ZUKER, M. *Curr. Opin. Struct. Biol.* 2000, *10*, 303–310.
7 LEONTIS, N. B., E. WESTHOF. *Curr. Opin. Struct. Biol.* 2003, *13*, 300–308.
8 LATHAM, J. A., T. R. CECH. *Science* 1989, *245*, 276–282.
9 RALSTON, C. Y., B. SCLAVI, M. SULLIVAN, M. L. DERAS, S. A. WOODSON, M. R. CHANCE, M. BRENOWITZ. *Methods Enzymol.* 2000, *317*, 353–358.
10 DAS, R., L. W. KWOK, I. S. MILLETT, Y. BAI, T. T. MILLS, J. JACOB, G. S. MASKEL, S. SEIFERT, S. G. MOCHRIE, P. THIYAGARAJAN, S. DONIACH, L. POLLACK, D. HERSCHLAG. *J. Mol. Biol.* 2003, *332*, 311–319.
11 MAYFORD, M., B. WEISBLUM. *EMBO J.* 1989, *6*, 4307–4314.
12 ALTUVIA, S., D. WEINSTEIN-FISCHER, A. ZHANG, L. POSTOW, G. STORZ. *Cell* 1997, *90*, 43–53.
13 BALZER, M., R. WAGNER. *Anal. Biochemistry* 1998, *256*, 240–242.
14 LINDELL, M., P. ROMBY, E. G. H. WAGNER. 2002, *RNA 8*, 534–541.
15 MOINE, H., B. EHRESMANN, C. EHRESMANN, P. ROMBY. Probing the RNA structure and function in solution, in: *RNA structure and function*, SIMONS, R. W., GRUNBERG-MANAGO, M. (eds), Cold Spring Harbor Laboratory Press, Cold Spring Harbor Laboratory, NY, 1998, pp. 77–115.

16 JOSEPH, S., H. F. NOLLER. *Methods Enzymol.* 2000, *318*, 175–190.
17 WILSON, K. S., H. F. NOLLER. *Cell.* 1998, *92*, 131–139.
18 EHRESMANN, C., F. BAUDIN, M. MOUGEL, P. ROMBY, J. P. EBEL, B. EHRESMANN. *Nucleic Acids Res.* 1987, *15*, 9109–9128.
19 GIEGÉ, R., M. HELM, C. FLORENTZ. *Comp. Natural Prod. Chem.* 1998, *6*, 63–80.
20 CAILLET, J., T. NOGUEIRA, B. MASQUIDA, F. WINTER, M. GRAFFE, A. C. DOCK-BRÉGEON, A. TORRES-LARIOS, R. SANKARANARAYANAN, E. WESTHOF, D. MORAS, B. EHRESMANN, C. EHRESMANN, P. ROMBY, M. SPRINGER. *Mol. Microbiol.* 2003, *47*, 961–974.
21 CHRISTIANSEN, C., J. EGEBJERG, N. LARSEN, R. A. GARRETT. *Methods Enzymol.* 1989, *164*, 456–472.
22 KROL, A., P. CARBON. *Methods Enzymol.* 1989, *180*, 212–227.
23 BRUNEL, C., P. ROMBY. *Methods Enzymol.* 2000, *318*, 3–21.
24 CHEN, C., S. A. WOODSON, C. J. BURROWS, S. E. ROKITA. *Biochemistry* 1993, *32*, 7610–7616.
25 WEEKS, K. M., D. M. CROTHERS. *Science* 1993, *261*, 1574–1577.
26 PAN, T., D. M. LONG, O. C. UHLENBECK. Divalent metal ions in RNA folding and catalysis, in: *The RNA world*, GESTELAND, R. F., ATKINS, J. F. (eds), Cold Spring Harbor Laboratory Press, Cold Spring Harbor Laboratory, NY, 1993, pp. 271–302.
27 ROMANIUK, P. J., I. L. DE STEVENSON, H. H. WONG. *Nucleic Acids Res.* 1987, *15*, 2737–2755.
28 JAHN, M. J., D. JAHN, A. M. KUMAR, D. SOLL. *Nucleic Acids Res.* 1991, *19*, 2786.

29 MILLIGAN, J. F., D. R. GROEBE, G. W. WITHERELL, O. C. UHLENBECK. *Nucleic Acids Res.* **1987**, *15*, 8783–8798.
30 GELHAUS, S. L., W. R. LACOURSE, N. A. HAGAN, G. K. AMARASINGHE, D. FABRIS. *Nucleic Acids Res.* **2003**, *31*, e135–e140.
31 SILBERKLANG, M., A. M. GILLUM, U. L. RAJBHANDARY. *Methods Enzymol.* **1979**, *59*, 58–109.
32 ENGLAND, T. E., A. G. BRUCE, O. C. UHLENBECK. *Methods Enzymol.* **1980**, *65*, 65–74.
33 HUQU, L., B. MICHOT, J. P. BACHELLERIE. *Nucleic Acids Res.* **1983**, *11*, 5903.
34 WULCZYN, F. G., R. KAHMANN. *Cell* **1991**, *65*, 259–269.
35 BENITO, Y., F. A. KOLB, P. ROMBY, G. LINA, J. ETIENNE, F. VANDENESCH. *RNA* **2000**, *6*, 668–679.
36 MÉREAU, A., R. FOURNIER, R. GRÉGOIRE, A. MOUGIN, P. FABRIZIO, R. LÜHRMANN, C. BRANLANT. *J. Mol. Biol.* **1997**, *273*, 552–571.
37 ZAUG, A. J., T. R. CECH. *RNA* **1995**, *1*, 363–374.
38 SENECOFF, J. F., R. B. MEAGHER. *Plant Mol. Biol.* **1992**, *18*, 219–234.
39 GRANGER, S. W., H. FAN. *J. Virol.* **1998**, *72*, 8961–8970.
40 SCHRÖDER, A. R. W., T. BAUMSTARK, D. RIESNER. *Nucleic Acids Res.* **1998**, *26*, 3449.
41 JAEGER, L., E. WESTHOF, F. MICHEL. *J. Mol. Biol.* **1993**, *234*, 331–346.
42 BERTRAND, E., M. FROMONT-RACINE, R. PICTET, T. GRANGE. *Proc. Natl Acad. Sci. USA* **1993**, *90*, 3496–3500.
43 DROZDZ, M., C. CLAYTON. *RNA* **1999**, *5*, 1632–1644.
44 SAMBROOK, J., E. F. FRITSCH, T. MANIATIS. Extraction, purification and analysis of mRNA from eukaryotic cells, in: *Molecular Cloning: A Laboratory Manual*, Cold Spring Harbor Laboratory Press, Cold Spring Harbor, NY, 1989, pp. 7.79–7.81.
45 SWERDLOW, H., C. GUTHRIE. *J. Biol. Chem.* **1984**, *259*, 5197–5207.

11
Study of RNA–Protein Interactions and RNA Structure in Ribonucleoprotein Particles

Virginie Marchand, Annie Mougin, Agnès Méreau and Christiane Branlant

11.1
Introduction

In cells, RNAs almost invariably function in association with proteins and form ribonucleoprotein particles (RNP). For most of the characterized RNP, one or more proteins with RNA-binding properties first associate with the RNA. Subsequently, other protein components may associate to the complex by protein–protein interactions or both protein–protein and RNA–protein interactions. Cellular RNA molecules can be classified into various groups according to their function or localization and different classes of proteins are associated with each of these groups [1, 2]. The RNA-associated proteins have diverse functions. They can stabilize, protect, package or transport RNAs, or participate in their subcellular localization. They can also mediate RNA interactions with other macromolecules or be catalysts [3].

In contrast to DNA, RNA can adopt a large variety of three-dimensional (3-D) structures. RNA–protein interactions may involve a defined nucleotide sequence and/or a specific 2- or 3-D RNA structure. Several protein domains were selected in the course of evolution for their ability to bind peculiar RNA motifs with either a narrow or a broad specificity. One of the best-studied examples is the RNA recognition motif (RRM) [4] that was first discovered in spliceosomal particles. It turned to be a very general RRM present in many different proteins [5]. Members of the RRM family include proteins that bind mRNAs, snRNAs or rRNAs. The KH domain was first identified in the human hnRNP K protein [6] and is also found in a large variety of RNA-binding proteins [1]. RRM and KH domains have well-defined and conserved 3-D structures [2, 7, 8]. Another protein RNA-binding motif with a defined 3-D structure was first described in the ribosomal L30 protein (L30 motif) [9]. It was later found to bind RNAs that form peculiar "K-turn" structures [10, 11]. Several "K-turn" structures were discovered in rRNAs [12, 13], and in small nuclear and nucleolar RNAs (snRNAs and snoRNAs) [10, 11] and they were found to bind "L30 type" protein domains [10, 11, 14–16]. Another characterized RNA-binding domain, which was first found in the *Escherichia coli* RNase III [17], binds specifically double-stranded RNA (dsRBD). This domain is limited to interactions

Handbook of RNA Biochemistry. Edited by R. K. Hartmann, A. Bindereif, A. Schön, E. Westhof
Copyright © 2005 WILEY-VCH Verlag GmbH & Co. KGaA, Weinheim
ISBN: 3-527-30826-1

with the A-form RNA helix. Other conserved RNA-binding domains are frequently encountered in proteins that bind RNAs, but their 3-D structures are not well characterized. This is the case for the RGG motif that was initially identified in the hnRNP U protein and is often found in combination with RRM motifs [18]. Other types of RNA-binding motifs have also been described such as zinc fingers, arginine rich and cold shock domains [2, 19–22].

There are numerous approaches to characterize RNA–protein interactions. First of all, one has to identify the proteins which are associated with the studied RNA. This implies the purification to homogeneity of the authentic RNP complex. Classical immunoselection approaches have been used for a long time as a first step in RNP purifications. They were based on the use of antibodies directed against one of the protein component of the complex or against the specific cap structure of the RNA [23, 24]. This first immunoselection step was followed by MonoQ/FPLC chromatography and/or fractionation by glycerol gradient centrifugation. During the last few years, new approaches have been developed for purification of RNP by two successive immunoselection steps [10, 25, 26]. They are based on the expression in cells of tagged components of the RNP (either two proteins or one protein and the RNA). The tagged components are included in the RNA–protein complexes *in vivo*, which allow RNP purification by successive affinity chromatography steps [10, 27, 28]. Identification of protein components of the isolated RNP is done by mass spectrometry.

At this stage the proteins associated with the RNA molecules are identified. The next question is to know which protein(s) bind(s) directly to the RNA and which ones are associated to the complex either only by protein–protein interactions or by protein–protein interactions together with interactions of low stability with the RNA. One way to identify the primary binding proteins is to produce them in a recombinant form in *E. coli*, yeast or animal cells and then to test their *in vitro* capacity to bind RNA by electrophoretic mobility shift assays (EMSA). This approach can produce information on the affinity of the RNA and protein partners, and also on the conditions that favor the interaction (ionic strength, pH, temperature, etc.). Sometimes, it is difficult to produce recombinant protein because of solubility problems. In this case, another approach can be used if specific antibodies directed against the proteins of the RNP are available. This approach is based on the formation of covalent bonds between the RNA and proteins, which are in a very close contact, by UV irradiation at 254 nm. Crosslinking is followed by RNA digestion. The free and crosslinked protein is immunoselected by using antibodies coated on Sepharose beads. After electrophoresis, only the proteins that were covalently linked to the RNA are labeled by the residual crosslinked nucleotides, so that they can be visualized by autoradiography. Measurement of the radioactivity in the gel can give an indication of the affinity of the two partners. Crosslinking experiments may be performed either with cellular extracts or with recombinant proteins. Doing crosslinking experiments in cellular extract is informative because under these conditions all the proteins in the extract are in competition with each other, as is the case in cells. Sometimes, it may happen that two proteins in the extract are in competition with one another for the same site on the RNA. In this case, crosslinking experiments can be performed with the labeled RNA and various relative amounts of the two recombinant proteins. Using this approach, one can obtain important information on the relative affinities of these two proteins for the RNA site.

After identification of the proteins, which directly interact with the RNA, a further step consists of the precise mapping of their binding sites. One possibility is to produce different pieces of the RNA and to test for their ability to bind the proteins. However, pieces of the RNA may fold into structures different from those present in the entire molecule. One more direct method consists of mapping the protein-binding sites by the use of chemical and enzymatic probes in solution. The bound RNA regions are protected against the action of the probes, and are identified by comparison of the cleavages and modifications obtained under the same conditions in the RNP and naked RNA. Such footprinting experiments can be performed on authentic native RNP, and on complexes reconstituted from recombinant proteins and *in vitro* transcribed RNAs or on complexes formed upon incubation of an *in vitro* transcribed RNA in a cellular or nuclear extract. However, as described below, when the experiment is performed in an extract, special digestion and modification conditions have to be used due to the presence of a large amount of protein in the extracts. As sites of cleavages and modifications are identified by primer extension analysis by the use of specific oligodeoxynucleotide primers, the footprinting analyses can be made in the presence of the endogenous RNAs from the extract. Thus, this approach is extremely powerful, since purification of the reconstituted complexes is not required.

Footprinting analysis of RNA–protein complexes formed in cellular extracts can be performed without knowing the identity of the bound proteins. These data allow the delineation of the RNA fragments that are free or bound to proteins within the extract. Then, for rapid identification of the proteins that are bound to the RNA in extracts, without a purification step, supershift experiments can be performed. The principle is to form complexes between the RNA and proteins from the cellular extract, and to incubate the mixture with specific antibodies directed against a protein expected to be bound to the RNA. An EMSA is performed with or without incubation with the antibody. If the antibody binds the protein without dissociation of the complex, the mobility of the RNP complex in the gel is decreased (supershift). If binding of the antibody to the protein dissociates the complex, the RNP band disappears.

If the secondary structure of the studied RNA target is not known, we recommend to study it in parallel with the footprinting experiments, since the same series of chemical and enzymatic probes are used (see Chapter 10). Knowledge of the RNA secondary structure and, if possible, RNA tertiary structure allows a better delineation of the RNA-binding domain. However, the fact that binding of proteins may alter the RNA structure should be taken into consideration for interpretation of these data.

When the binding site of a given protein has been delineated by this approach, confirmation of its functional role can be done *in vitro* and *in vivo*. The absence of protein binding after disruption of the RNA-binding site can be tested *in vitro* by reconstitution and EMSA experiments after site-directed mutagenesis of the RNA [16]. The effect on RNP activity of the disruption of the RNA-binding site can be also studied *in vivo* [16].

More generally, the biological relevance of the *in vitro* probing data can be tested *in vivo*, since one of the chemical probes, dimethylsulfate (DMS), can be used *in vivo* [29, 30] (see Chapter 10). As only a limited number of probes can be used *in vivo*, the best strategy is to perform a deep analysis of the RNP *in vitro* and then, by using DMS as a probe, verify that both the RNA secondary structure and protected areas are identical *in vivo* and *in vitro*.

When an RNA–protein binding site has been identified, the details of the RNA–protein interaction and the mechanism of its formation can be studied at the atomic level. To this end, the 3-D structure of the free RNA and proteins partners and of the complex that they form have to be determined by X-Ray or NMR analysis [11, 31–33].

Finally, for a more precise definition of the RNA-binding specificity of the studied protein, Systematic Evolution of Ligands by EXponential enrichment (SELEX) experiments can be performed [34, 35].

11.2
Methods

11.2.1
RNP Purification

RNP complexes contained in cytoplasmic or nuclear cell extracts are usually purified using immunoaffinity chromatography. The specific antibodies used can be directed against endogenous tagged or untagged proteins expressed from a modified gene (Tap-Tag technique; for experimental details, see [25]). The occurrence of a particular cap structure or the insertion of a tag sequence in the RNA can also be used for RNP immunoselection, for experimental details see [10, 26, 36, 37]. For instance, the presence of a m^{3}G cap structure at the 5′ extremity of the spliceosomal UsnRNAs and of some snoRNAs was largely used for purification of the spliceosomal UsnRNP and snoRNP using immobilized anti-m^{3}G cap antibodies [24, 38, 39]. The spliceosomal 25S [U4/U6.U5] tri-snRNP, 20S U5 snRNP, 17S U2 snRNP and 12S U1 snRNP contained in the RNP mixture that is retained on the anti-m^{3}G antibodies can then be separated by glycerol gradient centrifugation. The importance of the salt concentration in these purification steps is evidenced by the fact that at KCl concentrations above 250 mM, the [U4/U6.U5] tri-snRNP is disrupted into 12S U4/U6 and 20S U5 snRNP, and the 17S U2 snRNP is converted into a 15S or 12S particle [40]. Very powerful methods were recently developed for spliceosomal complex purification [26, 41]. They are based on the addition of an aptamer that binds the tobramycin aminoglycoside at one extremity of the RNA.

11.2.2
RNP Reconstitution

11.2.2.1 Equipment, Materials and Reagents

Equipment

- Electrophoresis instruments for small vertical slab gels. Localization of the RNP complexes in gels is performed either by autoradiography using X-Ray films (Fuji or Kodak) processed in an X-ray film developer or by use of a PhosphorImager.
- Temperature controlled baths (96, 65, 30 and 20 °C).

Materials

Eppendorf tubes, tips, buffers and MilliQ water should be sterilized before use. Wearing gloves is strongly recommended to avoid contamination of the samples by RNases.

Reagents

RNP

Nuclear or cytoplasmic extracts from HeLa cells or other cell lines can be purchased from CilBiotech, Belgium (around 13 mg/ml of total protein) or prepared according to the method developed by Dignam [42]. Before use, the extracts are incubated for 10 min at 30 °C in order to consume the endogenous ATP and kept on ice.

Proteins

RNP proteins can be either extracted from the purified native RNP particles [39] or produced as recombinant proteins in *E. coli* or using other expression systems [43–45].

Antibodies

Some of the primary antibodies used in the described examples can be purchased from Immuquest.

Chemicals and enzymes

SP6 RNA polymerase, T4 polynucleotide kinase and T4 RNA ligase are purchased from MBI Fermentas (Lithuania); T7 RNA polymerase is from Ambion; calf intestine phosphatase, glycogen (10 mg/ml) and RNase-free DNase I from MBI Fermentas; yeast total tRNA (20 mg/ml) from Roche Diagnostics; heparin sodium salt from porcine intestinal mucosa (H3393) from Sigma; Hybond C nitrocellulose membrane and ECL detection system are purchased from Amersham Pharmacia Biotech.

Radiochemicals
[^{32}P]pCp (3000 Ci/mmol), [γ-^{32}P]ATP (3000 Ci/mmol) and [α-^{32}P]UTP (800 Ci/mmol) are purchased from Amersham Biosciences or ICN.

Buffers
Some of the buffers used here are identical to buffers used in probing experiments.

- 1 × buffer D: 20 mM HEPES–KOH (pH 7.9), 0.2 mM EDTA, 100 mM KCl, 20% glycerol (w/v). Add freshly prepared 0.5 mM dithiothreitol (DTT) and 0.5 mM phenylmethylsulfonyl fluoride (PMSF) (dissolved in 96% ethanol).
- 1 × Tris buffer: 50 mM Tris–HCl (pH 7.5), 100 mM KCl, 2.5 mM MgCl$_2$.
- 1 × binding buffer A: 20 mM HEPES–KOH (pH 7.9), 0.2 mM EDTA, 150 mM KCl, 10% glycerol (w/v), 1.5 mM MgCl$_2$.
- 1 × binding buffer B: 20 mM HEPES–KOH (pH 7.9), 0.2 mM EDTA, 100 mM KCl, 20% glycerol (w/v), 3.125 mM MgCl$_2$.
- CSB loading buffer: 20 mM HEPES–KOH (pH 7.9), 40% glycerol (w/v), 0.05% bromophenol blue, 0.03% xylene cyanol.
- DNA loading buffer: 0.02% bromophenol blue, 0.02% xylene cyanol in formamide.
- SDS–PAGE loading buffer: 62.5 mM Tris–HCl (pH 6.8), 2% SDS, 100 mM β-mercaptoethanol, 10% glycerol, 0.01% bromophenol blue.
- 1 × TBE buffer: 90 mM Tris-borate (pH 8.0), 2 mM Na$_2$EDTA.
- 1 × elution buffer: 500 mM sodium acetate (pH 5.2), 1 mM EDTA.
- PBS-TM: 1 × PBS containing 0.1% Tween 20 and 5% dry non-fat milk powder.

11.2.2.2 RNA Preparation and Renaturation Step

Production of Labeled and Unlabeled RNAs by *In Vitro* Transcription
RNAs are generated by run-off transcription from a DNA template (usually 0.5–2 pmol of a linearized plasmid or PCR product) using the SP6 or T7 RNA polymerase [45–49]. Efficiency of transcription is usually higher for T7 RNA polymerase than for SP6 RNA polymerase. However, efficient transcription with T7 RNA polymerase requires the presence of at least one G residue at the initiation site. The presence of the GGG, GAG or GGA sequence strongly reinforces the transcription yield [50]. However, addition of these residues at the 5′ extremity of the RNA may modify its RNA secondary structure and/or protein-binding capacity. For instance, the presence of a UAGGGA/U sequence at the 5′ extremity of the transcript often generates a hnRNP A1 binding site. Noticeably, it is not easy to get small RNA transcripts (less than 50 nt) in high yield. For small RNA production, we recommend the use of the MEGAscript® or MEGAshortscript™ kit provided by Ambion (catalog reference 1330, 1333 or 1354). Several factors affecting the transcription yield must also be taken into account, such as the quantity of DNA template (generally 0.5–2 pmol), the incubation time (2–4 h), the Mg^{2+}/NTP ratio (usually

1/1.75), the pH of NTP stocks, and the preparation in a defined order and at room temperature of the transcription reaction mixture.

Uniformly labeled transcripts are produced by incorporation of an [α-^{32}P]NTP during transcription. After transcription, the DNA template is degraded by RNase-free DNase I (10 U). The RNA transcript is purified from shorter RNA fragments and excess of NTPs by polyacrylamide–urea gel electrophoresis in 1 × TBE buffer (for other protocols, see Chapters 10 and R. Hartmann). RNAs are eluted from the gel slices in 100 µl of 1 × elution buffer and are precipitated by the addition of 3 volumes of 96% EtOH, in the presence of 1 µg of glycogen. After centrifugation, the RNA pellet is washed with 70% EtOH, vacuum-dried and dissolved in MilliQ water.

For 5′-end-labeling, RNA (10–100 pmol) is first dephosphorylated at the 5′ end with the calf intestine phosphatase, and then labeled with [γ-^{32}P]ATP (3000 Ci/mmol) and the T4 polynucleotide kinase [51]. For 3′-end-labeling, [^{32}P]pCp (3000 Ci/mmol) is ligated to the RNA transcript in the presence of the T4 RNA ligase [52]. Labeled RNAs are purified by denaturing PAGE and eluted as described above.

Unlabeled RNA transcripts used RNP reconstitution and 2-D structure analyses are produced by similar methods, except that, in order to improve the transcription efficiency, the concentration of NTPs and Mg^{2+} is higher (up to 5 mM of each NTP).

RNA Transcript Renaturation

A renaturation process is required to produce a homogeneous population of RNA molecules in terms of RNA secondary structure. Before probing of naked RNA or reconstitution of RNP complexes, the RNA transcript dissolved in 1 × buffer D or 1 × Tris buffer is incubated for 10 min at 65 °C and then slowly cooled down to room temperature. Then, addition of divalent ions such as Mg^{2+} (at a concentration between 1.5 and 10 mM) is required to favor RNA 2- and 3-D structure formation and its stabilization during the probing and reconstitution experiments. After Mg^{2+} addition, the RNA is incubated for 10 min at room temperature. RNA should not be heated at 65 °C for a too long time, since phosphodiester bonds can be cleaved under these conditions, especially when Mg^{2+} ions are present in the incubation buffer (1 × Tris buffer) [53]. In spite the frequent use of 1 × Tris buffer described in literature, we recommend the use of buffer D and the addition of Mg^{2+} ions after cooling down to room temperature.

11.2.3
EMSA

EMSA can be used for several purposes. EMSA experiments can be performed with an *in vitro* transcribed RNA and purified proteins or an appropriate cellular extract. For determination of dissociation constant values (K_D), a fixed concentration of labeled RNA and increasing concentrations of the protein or cell extract are used. For estimation of the RNA/protein ratio, which has to be used to form RNP complexes for footprinting analysis, unlabeled RNA is added to the labeled

RNA. Under these conditions the same RNA concentration can be used in the EMSA and probing assays.

To perform EMSA with nuclear extract, the buffer, previously defined for pre-mRNA *in vitro* splicing assays [54] is generally used, except that ATP and creatine phosphate are omitted [48, 55, 56]. We recommend the following incubation mixture: HeLa cell nuclear extract 40–50% of total assay volume (dialyzed against buffer D), 2.5 mM $MgCl_2$ and buffer D. Other conditions can also be used for snRNP reconstitution; however, for the biological relevance of the data, it is important to select *in vitro* conditions, as close as possible, to the *in vivo* conditions.

11.2.3.1 EMSA Method

The reactions are performed in a total volume of 10–20 µl.

An amount of labeled RNA between 1 and 50 fmol, with or without 1–10 pmol of cold RNA, can be used for the assay. The use of 3'- or 5'-end-labeled RNA is recommended for K_D determination, since uniformly labeled RNAs give less-defined bands on the EMSA gels, especially in the case of RNAs longer than 60 nt. However, uniformly labeled RNAs (0.2 fmol) mixed with cold RNAs (2–10 pmol) are convenient for verification of RNP reconstitution or determination of RNP formation conditions. Competitor tRNAs (usually total tRNAs from yeast *Saccharomyces cerevisiae*) (10- to 1000-fold mass excess as referred to the tested RNA) can be added to the mixture in order to prevent non-specific RNA–protein interactions.

RNPs are generally reconstituted in the binding buffers (1 × Tris buffer or 1 × buffer D) in the presence of Mg^{2+} ions at concentrations varying from 1 to 10 mM. Nuclear extract, or another appropriate extract, or purified proteins are used as protein sources. A large range of protein concentration can be used: generally from 10 nM to 10 µM for purified proteins and from 1 to 100 µg of total proteins for nuclear extract. Pre-incubation of the protein or nuclear extract with the total tRNA mixture is recommended in order to limit the formation of non-specific RNA–protein interactions. A control experiment is performed in the absence of protein extract or purified protein (replaced by 1 × buffer D).

At the end of the incubation, 7 µl of CSB loading buffer is added. Note that in order to limit the non-specific electrostatic interactions between RNA and proteins, heparin, a negatively charged polyelectrolyte, can be added (at a concentration of 5 µg/µl). Heparin addition is followed by a 10-min incubation at room temperature. The presence of heparin usually improves the selection of specific RNP complexes. Addition of heparin is recommended in the case of nuclear extracts that contain a large amount of positively charged proteins. Depending on the electrostatic properties of the protein, heparin may be omitted when purified proteins are used.

For all types of EMSA, the CSB loading buffer is used. Electrophoresis is performed on 5–10% non-denaturing polyacrylamide gels containing 0.5 × TBE buffer and 5% (v/v) of glycerol (acrylamide:bisacrylamide ratio, 19:1). In order to limit RNP disruption and to obtain reproducible results, electrophoresis is generally performed at 4 °C and 10 V/cm. Conditions of electrophoresis, the acrylamide: bisacrylamide ratio as well as the type of running buffer can be modified if required [57].

At the end of the electrophoresis, the gel is transferred on a sheet of Whatman 3MM paper and dried. Exposure is overnight at −80 °C, using intensifying screens, or at room temperature in a PhosphorImager cassette. Free and bound RNAs are visualized by autoradiography.

For K_D determination, the amounts of radioactivity in the bands of gel are estimated by PhosphorImager measurement and SigmaPlot Software (SPSS Science Software) can be used for K_D calculations using the measured radioactivity [16].

Example: Experimental protocol used for K_D determination (Fig. 11.1)

Figure 11.1 illustrates the determination of the K_D values for the complexes formed between the recombinant S. cerevisiae Snu13 protein and one of its wild-type or mutated target RNA motifs. This target motif is one of the "K-turn" structures present in the nucleolar snoRNA U3 (U3B/C WT). In the variant RNA designated as U3B/C G.C → G.G, 1 bp pair of the motif has been disrupted [33] (Fig. 11.1a). RNA–protein complexes were produced under the following conditions: about 50 fmol of uniformly labeled RNA, mixed with 1 μg of yeast tRNA, was renatured in 6.5 μl of 1 × binding buffer A. The recombinant L7Ae protein was added at various concentrations ranging from 50 to 2000 nM and the mixtures were incubated for 30 min at 4 °C. After addition of 7 μl of CSB loading buffer, the RNA–protein complexes formed were fractionated by electrophoresis on a 6% (19:1) non-denaturing polyacrylamide gel, as described above (Fig. 11.1b).

Example: EMSA protocols used to define optimal conditions for RNP reconstitution in view to probe the RNA structure in the reconstituted RNP (Fig. 11.2)

In Fig. 11.2(a), EMSA was used to define the optimal conditions for formation of an RNP complex. The studied RNA was an HIV-1 BRU RNA fragment designated as SLS2,3, that corresponds to positions 7970–8068 in the entire RNA. It was produced by *in vitro* transcription with the SP6 RNA polymerase and was incubated in a HeLa cell nuclear extract [45]. In Fig. 11.2(b), the EMSA is used for the same purpose, but in this case the RNP complexes are formed between the SLS2,3 RNA and the recombinant hnRNP A1 protein [45].

The following experimental conditions were used: about 2.5 fmol of the 3′-end-labeled RNA was mixed with 2.5 μg of yeast total tRNA and renatured for 10 min at 65 °C in 6.5 μl of 1 × binding buffer B. After cooling for 10 min, HeLa cell nuclear extract purchased from CilBiotech (26 μg of total protein) (in Fig. 11.2a) or increasing amounts of the purified recombinant hnRNP A1 protein (0, 12.5, 25 and 50 pmol dialyzed against buffer D) [45] (in Fig. 11.2b) were added and the mixture was incubated for 15 min at 4 °C. The RNA–protein complexes formed were subsequently fractionated by electrophoresis at room temperature on a 6% (19:1) non-denaturing polyacrylamide gel for 90 min at 10 V/cm and 4 °C.

11.2.3.2 Supershift Method
Supershift experiments are a variant of EMSA. RNA–protein complex formation is performed in the presence or absence of antibodies. Binding of antibodies to their

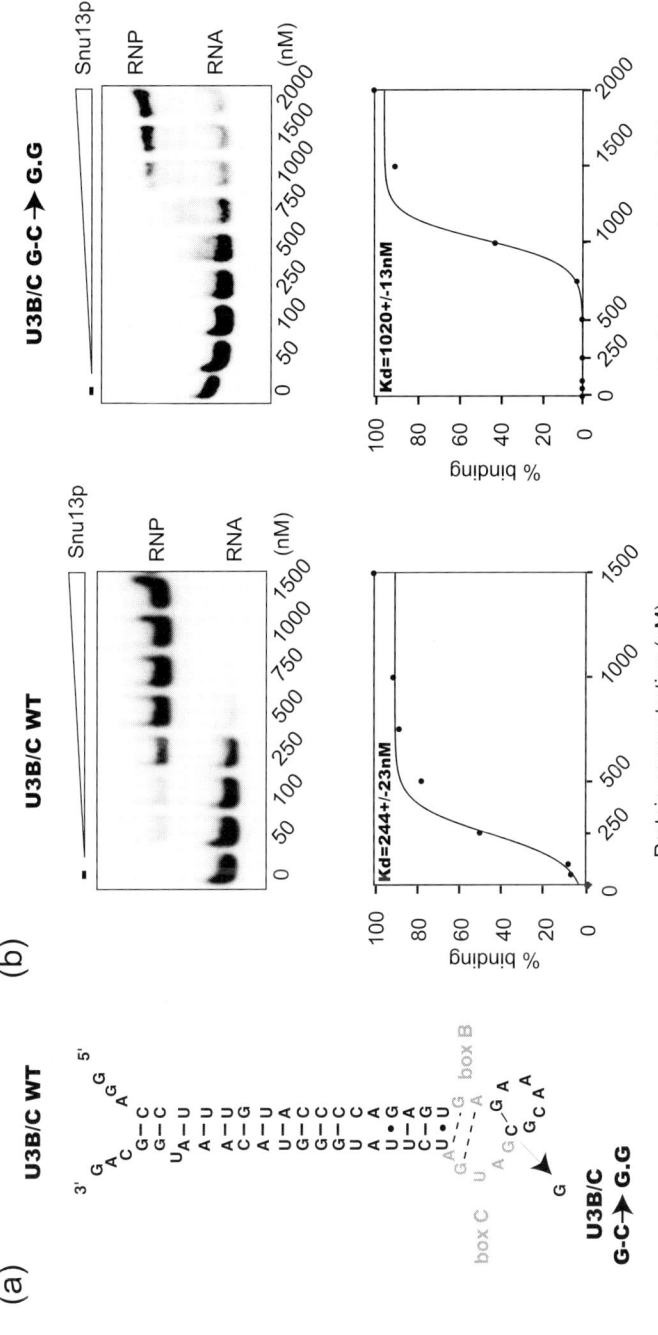

Fig. 11.1. Use of EMSA to study the binding of the recombinant protein Snu13p to the wild-type or mutated B/C motif of yeast U3 snoRNA [33]. The secondary structure of the wild-type U3B/C RNA and position of the C → G mutation in the variant RNA are shown (a). The 3′-end-labeled RNA was incubated in the presence of increasing amounts of the Snu13 protein, as described above. The auto- radiograms obtained after gel electrophoresis are shown (b). The protein concentration (in nM) is given below each lane. The K_D values of the RNA–protein complexes (protein concentration for which 50% of the input RNA is shifted to an RNP complex) were calculated for each experiment with the SigmaPlot Software (SPSS Science Software).

(a)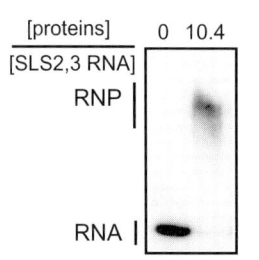

(b)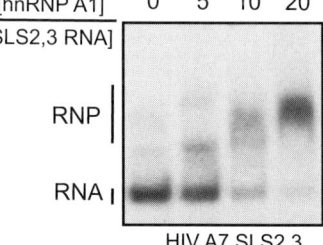

Fig. 11.2. Use of EMSA to study formation of complexes upon incubation of the HIV-1 A7 SLS2,3 RNA with an HeLa cell nuclear extract or the purified recombinant hnRNP A1 protein [45]. (a) and (b) The 3′-end-labeled RNA was incubated in the presence of either a nuclear extract (a) or increasing amounts of the purified recombinant hnRNP A1 protein (b). The [protein] (μg)/[RNA] (pmol) (a) or [hnRNP A1] (pmol)/[RNA] (pmol) (b) ratios (P/R) used in each assays are given above the lanes. Positions of the RNP complexes (RNP) and free RNA (RNA) are indicated on the left of the autoradiograms. The production of complexes with decreasing electrophoretic mobility upon increasing the hnRNP A1 concentration (b) is explained by the multimerization of protein hnRNP A1 along the RNA.

target protein in the RNP complex increases the size of the RNP complex and, thus, lowers its electrophoretic mobility on the gel (so-called "supershift"). In some cases, binding of the antibodies may lead to RNP complex disruption and loss of the shifted RNA band.

For supershift experiments, 0.5–2 μl of antibodies of interest is added before the heparin treatment and incubation is continued for 10 min at 4 °C. The amount of antibody added depends upon the antibody specificity and concentration; both monoclonal and polyclonal antibodies can be used for this type of experiments.

RNP complexes formed with or without antibodies are fractionated in parallel on a non-denaturing polyacrylamide gel (5–10% concentration can be used) and the electrophoresis is performed in 0.5 × TBE buffer containing 5% of glycerol at 4 °C. The gel is dried and the presence of radioactivity is detected as described above.

Example: "Supershift" experimental protocol (Fig. 11.3a)

RNP reconstitution experiments are performed with two different fragments of HIV-1 BRU RNA. The biological components used for the assays are: the SLS2,3 wild-type RNA (positions 7970–8068) described above, the C3 wild-type HIV RNA fragment (positions 5359–5408 in the entire molecule) [58], HeLa cell nuclear extract, and specific anti-hnRNP A1 and hnRNP H antibodies, that were provided by G. Dreyfuss (University of Pennsylvania School of Medicine, Philadelphia, USA) and D. Black (University of California, Los Angeles, USA), respectively. The RNA

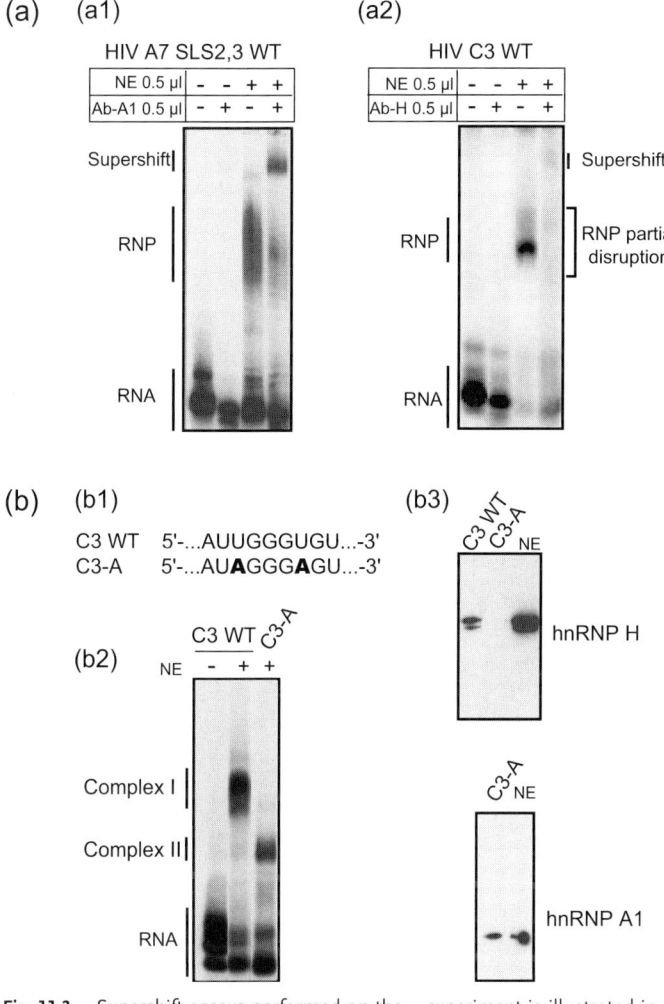

Fig. 11.3. Supershift assays performed on the RNP complexes formed by incubation in a HeLa cell nuclear extract (NE) of the A7 SLS2,3 wild-type (a1) and C3 wild-type (a2) RNAs [45, 58]. (a) RNAs were 3′-end-labeled, and anti-hnRNP A1 (Ab-A1) or anti-hnRNP H (Ab-H) antibodies were used in (a1) and (a2), respectively. Positions of the RNAs, RNP and the supershifted RNP are indicated on the left of the autoradiogram. (b) EMSA experiments coupled with a second gel electrophoresis and Western Blot analysis, performed on complexes formed by the HIV wild-type or mutated C3 RNAs (C3 wild-type and C3-A) in HeLa cell nuclear extract [58]. The mutations present in the C3-A RNA are shown in (b1), the EMSA experiment is illustrated in (b2). The positions of the free RNAs and RNP complexes I and II obtained with the HIV C3 wild-type and C3A RNAs, respectively, are indicated on the left of the autoradiogram. The bands containing complexes I (C3 wild-type) and II (C3 A) were included in an SDS–polyacrylamide gel. After electrophoresis proteins were transferred on a nitrocellulose membrane, that was used for Western blot analysis first with hnRNP H antibodies (upper panel) and then with anti-hnRNP A1 antibodies (lower panel). Proteins from nuclear extract (NE) were loaded on the gel for a control. Complex I contained hnRNP H protein and complex II contained hnRNP A1 protein.

fragments transcribed with SP6 RNA polymerase were 3′-end-labeled. With the HIV A7 SLS2,3 wild-type RNA, a supershift is obtained, demonstrating the presence of protein hnRNP A1 in the complex (Fig. 11.3a1). The Cl complex formed with the C3 wild-type RNA is dissociated by binding of the anti-hnRNP H antibody (Fig. 11.3a2). This suggests the presence of protein hnRNP H in the complex and an essential role of this protein for complex stability.

The following experimental conditions were used: the HIV C3 and A7 SLS2,3 RNAs (2.5 and 6 pmol, respectively) were 3′-end-labeled by using [^{32}P]pCp. Labeled RNAs were incubated at 4 °C in the presence of 5 μg of yeast total tRNA, with 0.5 and 4 μl of nuclear extract, respectively. After a 15-min incubation, 0.5 μl of anti-hnRNP A1 antibodies [59] (Fig. 11.3a1) or 1 μl of the monoclonal anti-hnRNP H antibodies [60] (Fig. 11.3a2) were added, respectively. The incubation was continued for 10 min on ice and was followed by gel electrophoresis. As a control, the two RNAs were incubated under the same conditions in the absence of antibodies.

11.2.3.3 Identification of Proteins Contained in RNP by EMSA Experiments Coupled to a Second Gel Electrophoresis and Western Blot Analysis

This method also allows the identification of proteins present in RNP complexes fractionated on EMSA gels. The EMSA are performed as described above, and the band containing RNP complexes is cut out from the gel, soaked in a SDS–PAGE loading buffer and included in a 5–10% SDS–PAGE. Then, the fractionated proteins are electrotransferred on a nitrocellulose membrane. The search for the presence of defined proteins is done by immunoblotting using specific antibodies directed against these proteins [58].

Example 4: Protocol for an EMSA experiment coupled to a second gel electrophoresis (Fig. 11.3b)

Figure 11.3(b) illustrates the search for the presence of proteins hnRNP H and A1 in RNP complexes I and II by Western blot analysis (Fig. 11.3b). Complexes I and II were formed by incubation in a HeLa cell nuclear extract of the HIV C3 wild-type and C3-A mutant RNAs (Fig. 11.3b2), respectively. The HIV C3-A RNA is a variant of the HIV wild-type C3 RNA (2U were substituted by 2A) (Fig. 11.3b1) [58]. The bands of gel containing each RNP complex were cut out. They were soaked in 10 μl of SDS–PAGE loading buffer for 1.5 h at 37 °C, and the band and buffer were heated for 5 min at 96 °C before their inclusion at the top of a 10% SDS–polyacrylamide gel (1.5 mM thickness). After 2 h electrophoresis at 20 V/cm, the fractionated proteins were electrotransferred onto a Hybond C nitrocellulose membrane (Pharmacia Amersham Biotech) (for 1 h at 100 V). Then, the membrane was blocked with 20 ml of PBS-TM buffer overnight at 4 °C or 2 h at room temperature with gentle shaking. It was then probed with the anti-hnRNP A1 (0.5 μl) or hnRNP H (2 μl) antibodies [58]. The bound antibodies were detected with peroxidase-conjugated anti-mouse and anti-rabbit IgG antibodies, respectively, and visualized by the ECL detection system (Fig. 11.3b3) [58].

11.2.4
Probing of RNA Structure

11.2.4.1 Properties of the Probes Used

Conditions used for RNP probing with chemical reagents or enzymes are chosen as mild as possible in order to preserve the structural integrity of RNP particles: probing reactions are performed in the buffer used for the RNP purification or reconstitution. Incubations are performed at moderate temperature for short times. A defined amount of yeast total tRNA is added, in order to control the [RNA]/[probe] ratio. Modification and enzymatic digestion conditions should be selected in order that less than one modification or cleavage statistically occurs per RNA molecule. The chemical and enzymatic probes used for footprinting are also used for determination of RNA secondary structure in solution (see Chapter 10). When the probing experiments are performed on purified natural complexes or complexes reconstituted between an RNA transcript and recombinant proteins, almost all the probes used for RNA secondary structure analysis can be used. Only the S1 nuclease, which has an optimum pH of action of 4.5, cannot be used. However, when cellular extracts are used for RNP formation without further purification of the complexes, some of the chemical probes, especially 1-cyclohexyl-3(2-morpholinoethyl)carbodiimide metho-p-toluene sulfonate (CMCT), cannot be used. The method employed for the identification of cleavage and modification positions depends on the labeled state of the RNA. For unlabeled RNAs (authentic purified RNP or RNP reconstituted with unlabeled RNAs), primer extension analyses with reverse transcriptase are performed: stops of extension occur at the cleavage site, or one residue before the cleaved (depending on the enzyme used) or modified nucleotide. When 3'- or 5'-end-labeled RNAs are used for RNP reconstitution, only enzymatic probes are used and cleavages are directly localized by gel electrophoresis without the reverse transcriptase step.

Chemical Probes

DMS methylates RNAs at the N^7-G, N^1-A and N^3-C positions of the bases. CMCT alkylates RNAs at the N^3-U and N^1-G positions, and kethoxal reacts at the N^1-G and N^2-G positions. Only N^7-G methylation by DMS can occur in double-stranded RNAs – all the other modifications are impaired.

Enzymes

RNase T1 cleaves the phosphodiester bonds 3' to G residues, whereas RNase T2 cleaves after any residue. Both enzymes are used in conditions such that they preferentially cleave single-stranded RNA regions. RNase V1 is used to cleave double-stranded or stacked RNA regions. More details on these probes are available in Chapter 10.

To identify the positions that are protected by the proteins in an RNP, the naked RNA and the RNP are subjected to the same chemical and enzymatic treatments, and the reactive positions in RNA and RNP are compared. It should be noticed that in addition to RNA protection, RNP probing may detect some RNA conformational changes occurring upon protein binding [61].

Safety Rules
DMS is a potential carcinogen and special care should be taken when using it (see Chapters 9 and 10 for the safety rules).

11.2.4.2 Equipment, Material and Reagents

Equipment

- Sequencing gels for primer extension analysis.
- Exposure with X-Ray films (Fuji or Kodak) using an X-Ray film developer is recommended. However visualization with a PhosphorImager equipment can also be used.
- Temperature controlled baths (96, 65, 30 and 20 °C).

Reagents

Probes
RNase T1 is purchased from Roche Diagnostics, RNase T2 from Invitrogen, RNase V1 from Ambion, DMS from Aldrich, CMCT from Fluka and kethoxal from Amersham.

Chemicals and enzymes
The avian myeloblastosis virus (AMV) reverse transcriptase is purchased from Q-Bio Gene; glycogen (10 mg/ml), dNTPs (100 mM of each) and ddNTPs are from MBI Fermentas. The yeast total tRNA (20 mg/ml) is from Roche Diagnostics; cacodylic acid and boric acid are from Sigma.

Radiochemicals
[γ-^{32}P]ATP (3000 Ci/mmol) is purchased from Amersham Biosciences or ICN.

Materials
See Section 11.2.2.1.

Buffers
- 1 × buffer D: see Section 11.2.2.1.
- 1 × Tris buffer: see Section 11.2.2.1.
- 1 × TBE: see Section 11.2.2.1.
- 1 × DMS/Ke buffer: 50 mM sodium cacodylate (pH 7.5), 100 mM KCl, 2.5 mM MgCl$_2$.
- 1 × CMCT buffer: 50 mM sodium borate (pH 8.0), 100 mM KCl, 2.5 mM MgCl$_2$.
- DMS stop buffer: 1 M Tris–acetate (pH 7.5), 1.5 M sodium acetate, 1 M β-mercaptoethanol.

- Buffer A: 10 mM Tris–HCl (pH7.5), 10 mM MgCl$_2$, 3 mM CaCl$_2$, 250 mM sucrose, 0.7 M β-mercaptoethanol.
- 0.7 M ice cold β-mercaptoethanol.
- 0.1 M EDTA (pH 8.0).
- 0.5 M potassium borate (pH 7.0).
- 0.5 M sodium cacodylate (pH 7.0).
- 3 M sodium acetate (pH 6.0).
- 10 × RT buffer: 500 mM Tris–HCl (pH 8.3), 60 mM MgCl$_2$, 400 mM KCl (provided with the reverse transcriptase purchased from Q-BioGene).

11.2.4.3 Probing Method

Enzymatic and Chemical Probing of Native Purified RNP Particles
Modifications and cleavages are performed in the purification or storage buffer. To ensure statistical modifications and cleavages, all the reactions are performed in the presence of 1.25 µg of yeast total tRNA.

Chemical Modifications

DMS
DMS modifications are performed for 6 min at 20 °C in 50 µl of 1 × DMS/Ke buffer with 1 µl of a DMS/EtOH solution (1/1, v/v).

Kethoxal
Same protocol as for DMS, except that modifications are performed for 10 min at 0 °C at a kethoxal concentration of 10 mg/ml.

CMCT
CMCT modifications are performed for 6 min at 20 °C in 50 µl of 1 × CMCT buffer and at CMCT concentrations of 30–60 mM.

Enzymatic Cleavages
T1, T2 and V1 RNase cleavages are performed for 6 min at 20 °C in 40 µl of 1 × Tris buffer with 5×10^{-3} U/µl of RNase.

Reaction Stop
DMS modification is quenched by addition of 10 µl of DMS stop buffer (20% of the reaction mixture), followed by phenol extraction. CMCT modification is stopped by phenol extraction, followed by ethanol precipitation. Kethoxal modification is stopped by addition of 0.5 M potassium borate (pH 7.0) to stabilize the kethoxal–guanine adduct (25% of the reaction mixture volume), followed by phenol extraction and ethanol precipitation. RNase V1 digestion is stopped by the addition of 5 µl of 100 mM EDTA (pH 8.0) before phenol extraction. RNase T1 and T2 digestions are stopped by the addition of an excess of yeast total tRNA (10 µg), followed

by rapid phenol extraction. To avoid reaction of the enzymes or chemical on the free RNA, phenol extractions should be quickly performed on ice.

Modification and digestion products are ethanol precipitated, washed with 70% (v/v) ethanol, dried and dissolved in MilliQ water (except kethoxal-modified RNA pellets, that are dissolved in potassium borate 25 mM, pH 7.0, in order to stabilize the chemical adducts [62]).

Primer Extension Analysis
For primer extension analysis, 5′-end-labeled primers are annealed to chemically modified or digested RNAs. As the RNA length that can be examined with one primer ranges between 100 and 200 nt, different primers (generally 12–20 nt) complementary to regions that are spaced by 100–200 nt have to be used. As each primer has its own efficiency for reverse priming, preliminary assays should be performed for each primer to define the amount of RNA which is suitable for the analysis. The 5′-end-labeling (with [γ-^{32}P]ATP, 3000 Ci/mmol) is described in [63]. Extension is achieved with the AMV reverse transcriptase in the presence of the four dNTPs, the conditions are described in [64].

Hybridization
The RNA sample (1–10 pmol in 1 μl) is mixed with the 5′-end-labeled primer (100 000 c.p.m.) and 10 × RT buffer, in a total volume of 2.5 μl. The mixture is incubated for 10 min at 65 °C and quickly cooled on ice for 10 min.

Primer Extension
The primer extension reaction is performed in a final volume of 5 μl. The elongation mixture is prepared as follows: 0.1 μl of dNTP mixture (5 mM of each dNTP), 0.25 μl of 10 × RT buffer, 0.25 μl of reverse transcriptase (2 U/μl extemporaneous dilution) and 1.9 μl of H$_2$O. The hybridization mixture is mixed with 2.5 μl of elongation mixture and samples are incubated for 45 min at 42 °C. The primer extension is stopped by addition of 4 μl of the DNA loading buffer. To prepare the sequencing ladder, the unmodified RNA is used as a template. The elongation mixture contains a ddNTP at a 0.5 mM concentration and dNTPs with a dNTP:ddNTP ratio of 2.

Gel Fractionation
The elongation mixture is denatured for 2 min at 96 °C and 2 μl aliquots are fractionated on a 7% denaturing (8 M urea) polyacrylamide (19:1 ratio acrylamide:bisacrylamide) sequencing gel in 1 × TBE. The gel is preheated for 30 min at 50 V/cm and electrophoresis is performed at 50 °C using the same voltage. The migration time is adjusted depending on the sequence to be read. After migration, gels are transferred on sheets of Whatman 3MM paper, dried and autoradiographed (X-Ray films from Kodak or Fuji) overnight at −80 °C with an intensifying screen (Amersham, Biosciences).

Example: Native RNP probing protocol (Fig. 11.4)

The yeast U5 snRNA was probed as free RNA (RNA) and in the yeast [U4/U6.U5] tri-snRNP (25S) (Fig. 11.4a). About 200 ng of RNP or 100 ng of free renatured RNA was incubated for 10 min at 20 °C in the presence of 1.25 µg of yeast total tRNA, in 50 µl of the digestion or modification buffer (1 × DMS/Ke buffer or 1 × Tris buffer). For kethoxal, the mixture was then put on ice for 10 min, as the reaction is performed at 0 °C. The reactions with DMS, kethoxal, RNase T2 and V1 were performed and stopped in the conditions described above. Aliquots of the treated RNAs were reverse transcribed with the 5′-labeled specific oligodeoxynuclotide primer O-335, complementary to positions 69–81 of the *S. cerevisiae* U5 snRNA (Fig. 11.4b). The cDNA fragments obtained were fractionated on a sequencing gel, using as a reference a sequencing ladder performed with the unmodified RNA and the same 5′-end-labeled oligodeoxynucleotide (Fig. 11.4a).

Enzymatic and Chemical Probing of RNP Particles Formed in Nuclear Extract or with purified Proteins

As for EMSA experiments, the RNA is renatured in 1 × buffer D, before complex formation, as described in Section 11.2.2.2.

1. Enzymatic Reactions

RNase T1
RNA (200 ng, 1 pmol) is mixed with 1 µl of 62.5 mM $MgCl_2$, 3.6 µl of 1 × buffer D, 5 µg of total tRNA and the final volume is adjusted to 14 µl with water. An adequate amount (based on EMSA) of HeLa cell nuclear extract, as defined by the EMSA experiments, is added and the reaction mixture is incubated at 30 °C for 10 min. The cleavage reaction is performed with 1 µl of RNase T1 (0.025–2 U) at 30 °C for 10 min.

RNase T2
Same protocol as for RNase T1, except that the reaction is performed with 1 µl of RNase T2 (1–3 U) at 30 °C for 10 min.

RNase V1
Same protocol as for RNase T1, except that the reaction is performed with 1 µl of RNase V1 (0.02–0.2 U) at 30 °C for 10 min.

2. Chemical Modification by DMS

The RNA (200 ng, 1 pmol) is mixed with 1 µl of 62.5 mM $MgCl_2$, 3.6 µl of 1 × buffer D, 5 µg of total tRNA are added and the final volume is adjusted with water to 14 µl. An adequate amount of HeLa cell nuclear extract dialyzed against buffer D, as deduced from the EMSA experiment, is added and the reaction mixture is incubated at 30 °C for 10 min. To improve the efficiency of DMS modification, sodium cacodylate at a final concentration of 50 mM (pH 7.5) is added in

190 *11 Study of RNA–Protein Interactions and RNA Structure in Ribonucleoprotein Particles*

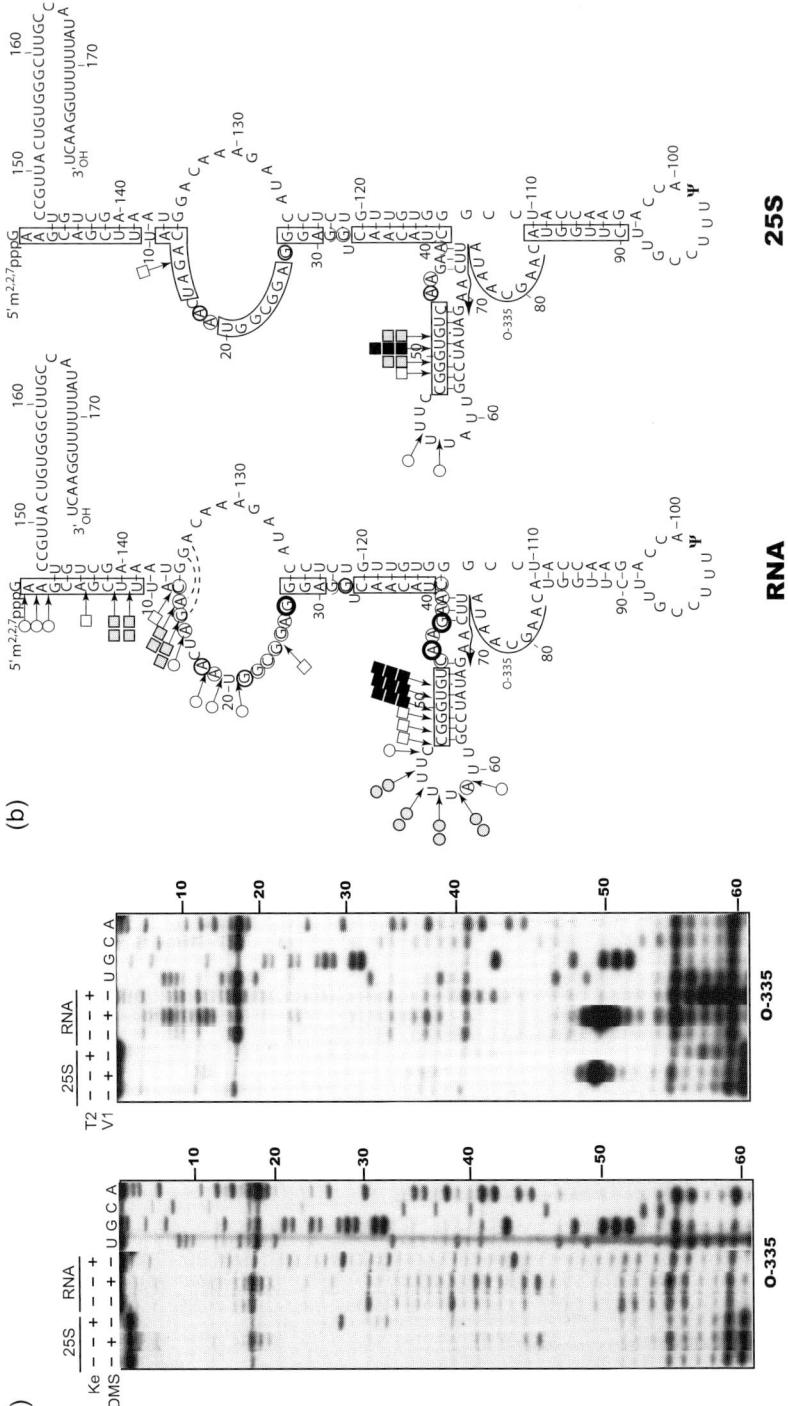

Fig. 11.4

buffer D before the reaction. The modification reaction is performed at 30 °C for 10 min with between 1 and 5 μl of a 1/1 (v/v) DMS/EtOH solution.

3. Reaction Stop

Same protocol as described in Section 11.2.4.3. The hydrolyses by RNases T1, T2 and V1 are stopped by addition of 20 μg of total yeast tRNA; in addition, 1 μl of 100 mM EDTA is added for the RNase V1. These additions are followed by phenol extraction on ice. DMS modifications are stopped by addition of 10 μl of DMS stop buffer before phenol extraction and ethanol precipitation.

Ethanol precipitation of all samples is done by addition of 10 μl of 3 M sodium acetate (pH 6.0), 1 μg of glycogen and at least 3 volumes of 96% EtOH, followed by a 15 min incubation at −80 °C. After centrifugation at 13 000 r.p.m. for 15 min and at 4 °C, the supernatants are discarded and the RNA pellets are washed with 200 μl of 70% EtOH. A second centrifugation is performed for 5 min at 13 000 r.p.m. and 4 °C. The RNA pellets are vacuum-dried for 2 min and dissolved in 4 μl of MilliQ water before primer extension analysis (see Section 11.2.4.3).

Example: Protocol used for probing of a reconstituted RNP complex (Fig. 11.5)

The complexes formed upon incubation of the HIV-1 A7 SLS1,2,3 RNA fragment (positions 7903–8170) in a HeLa cell nuclear extract or with the purified hnRNP A1 protein were analyzed by chemical and enzymatic probing of the RNA structure and accessibility (Fig. 11.5). The following conditions were used: 1.12 pmol of cold HIV-1 A7 SLS1,2,3 RNA was incubated in the presence of 5 μg of yeast total tRNA with 1 μl of 62.5 mM MgCl$_2$ and 3.6 μl of 1 × buffer D in a total volume of 14 μl. Assays were performed in the presence (+) of 4 μl of nuclear extract ([Protein]/[RNA] (P/R) = 46) (Fig. 11.5a) or 50 fmol of purified hnRNP A1 protein

Fig. 11.4. Probing of U5 snRNA in the spliceosomal [U4/U6.U5] tri-snRNP purified from S. cerevisiae by use of chemical and enzymatic probes [49]. The yeast [U4/U6.U5] 25S tri-snRNP (25S, a) and the natural free U5 snRNA (RNA, a) were probed with kethoxal, DMS, RNase T2 and RNase V1 in conditions described in Section 11.2.4.3. (a) Primer extension analyses performed with the primer O-335. For each probe, a control experiment in the absence of the probe was performed (−). Lanes U, G, C and Λ correspond to the sequencing ladder. Positions of nucleotides in U5 snRNA are indicated on the right side of the panels. (b) The probing data illustrated in (a) are schematically represented on the secondary structure of U5 snRNA (left: 2-D structure results for the naked RNA; right: 2-D structure results for the 25S tri-snRNP). Nucleotides modified by DMS or kethoxal are circled; the thickness of the circles indicates the levels of modification (weak, medium and strong). RNase V1 or T2 cleavages are indicated by arrows linked to squares or circles, respectively. The color and number of symbols indicate the yield of cleavage. Boxed nucleotides are not modified. In the U5 snRNA region that is analyzed with primer O-335, the tri-snRNP components generate a strong protection, except for the lateral stem–loop structure formed by residues 40–75. This stem–loop structure is additional in the yeast U5 snRNA compared to vertebrate U5 snRNA [49].

192 | 11 Study of RNA–Protein Interactions and RNA Structure in Ribonucleoprotein Particles

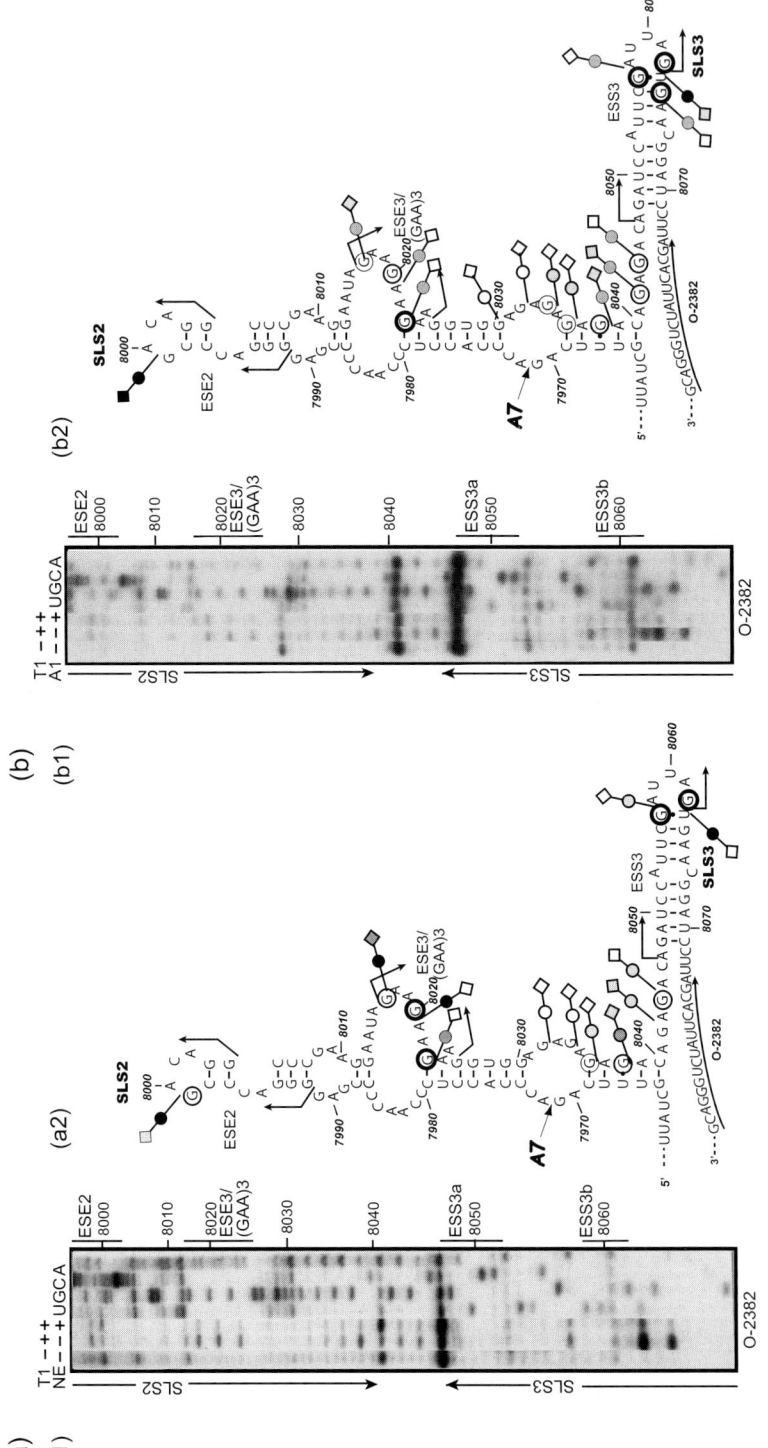

Fig. 11.5

(P/R = 45) (Fig. 11.5b) or in the absence of added extract or protein (−) (Fig. 11.5a and b). After a 10-min incubation at 30 °C, 0.025 U of RNase T1 was added. For identification of reverse transcriptase pauses which are not due to RNA cleavage, a primer extension was performed on the naked RNA incubated in buffer D without RNase T1. The reactions were stopped by addition of 20 µg of tRNAs, followed by phenol extraction as described in Section 11.2.4.3.

Probing of Yeast RNAs Modified *in vivo* by DMS Treatment
DMS is able to penetrate efficiently in bacterial, yeast and animal cells, and can thus be used to probe RNA structure and accessibility in these cells. However, to get interpretable results the experiments should be performed on RNAs that form well-defined homogeneous particles in cells. Otherwise, information on RNA molecules having different structures and accessibilities would be superimposed in the primer extension analysis. The *in vivo* analysis of the *S. cerevisiae* U3 snoRNA structure and accessibility in the U3 snoRNP illustrated in Fig. 11.6 was a very successful example of an RNP analysis *in vivo*. The methylation sites detected *in vivo* were compared to those detected *in vitro* by treatment of a partially purified U3 snoRNP and the naked *in vitro* transcribed U3 snoRNA [29]. The results obtained validated the protein binding sites identified by *in vitro* analysis and demonstrated the interaction of U3 snoRNA with the pre-ribosomal RNA [29].

Example: Protocol used for *S. cerevisiae* U3 snoRNP probing by DMS *in vivo* (Fig. 11.6)

Fig. 11.5. Probing of the RNA structure and accessibility in RNP formed by the HIV-1 A7 SLS2,3 RNA and a HeLa cell nuclear extract (NE) or the purified hnRNP A1 protein (A1) [45]. Primer extension analyses of the A7 SLS1,2,3 RNA cleaved by RNase T1 in buffer D in the absence (−) or presence (+) of nuclear extract (a1) or in the presence of the recombinant hnRNP A1 protein (b1) are shown. As a control, a primer extension was performed using the intact RNA transcript incubated without RNase T1 as the template (left lane of the autoradiogram). Lanes U, G, C and A correspond to the sequencing ladder. Positions of nucleotides in the HIV-1 BRU RNA and of the RNA secondary structure elements described for the HIV-1 A7 RNA region are indicated on the right side of the autoradiograms. (a2 and b2) Schematic representations of the probing data illustrated in (a1) and (b1). Positions of RNase T1 cleavages are represented on the RNA secondary structure established for the A7 SLS1,2,3 RNA [45]. Cleavages 3′ to the G residues are indicated by thin lines. They are surmounted with circles when the cleavages occurred in the naked RNA (a2 and b2) Cleavages occurring in the presence of nuclear extract (a2) or hnRNP A1 (b2) are indicated by the presence of a square. The colors of circles and squares indicate the level of cleavage observed in the naked RNA and the RNP complex, respectively (grey, dark gray and black represent increasing intensities of cleavages, respectively). The G residues protected either in the presence of nuclear extract (a2) or with hnRNP A1 protein (b2) are circled; the intensity of the circle corresponds to the yield of the protection. The hybridization site of the oligodeoxynucleotide O-2382 is indicated. The *cis* regulatory elements of splicing, acting at site A7 (ESE2, ESE3/(GAA)3, ESS3) are delimited by two opposite broken arrows and the name of the element is given.

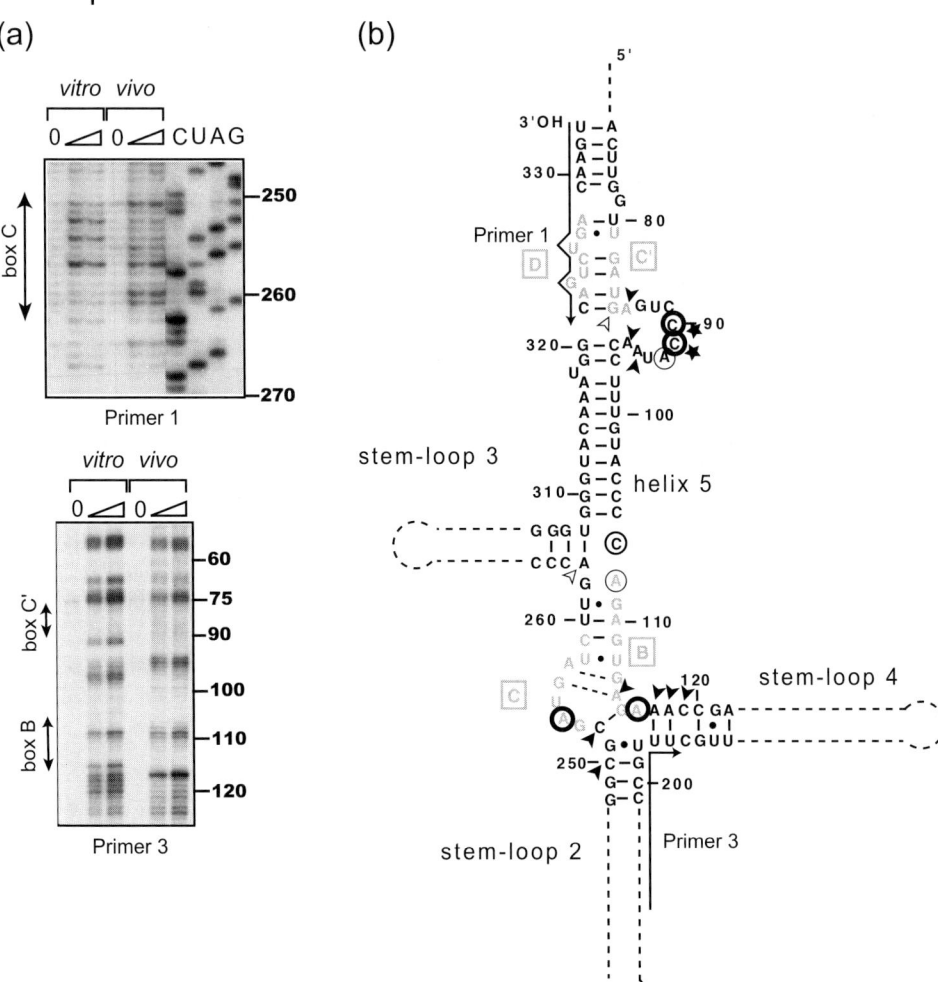

Fig. 11.6. Probing of the structure and the accessibility of the S. cerevisiae U3 snoRNA *in vivo* [29]. Primer extension analyses of the U3 snoRNA modified by DMS *in vitro* or *in vivo* are presented in (a). As a control, primer extensions were performed with an untreated U3 snoRNA transcript (lanes 0). Lanes U, G, C and A correspond to the sequencing ladders. Positions of nucleotides in the *S. cerevisiae* U3 snoRNA sequence and the phylogenetically conserved RNA segments of U3 snoRNA [29] are shown on the right and left sides of the autoradiogram, respectively. (b) Experimental data shown in (a) are represented on the secondary structure proposed for the *S. cerevisiae* U3 snoRNA in interaction with the pre-rRNA [16, 29]. Nucleotides modified *in vitro* and protected *in vivo* are indicated by black triangles, nucleotides modified *in vivo* and *in vitro* are circled; the thickness of the circles reflects the yield of modification *in vivo*. Asterisks mark the nucleotides with an increased reactivity *in vivo* compared to *in vitro*.

About 15-ml aliquots of a *S. cerevisiae* culture, grown in YEPD medium at 30 °C to an A_{600} between 0.5 and 1.0, were gently rocked at room temperature for 2 min in the presence of DMS at a concentration between 40 and 160 mM. The reaction was quenched by addition of 0.7 M ice cold β-mercaptoethanol and 5 ml of cold water-saturated isoamyl alcohol, followed by shaking and centrifugation. Cell pellets were washed with Buffer A (10 mM Tris–HCl, pH 7.5, 10 mM $MgCl_2$, 3 mM $CaCl_2$, containing 250 mM sucrose and 0.7 M β-mercaptoethanol). Cells were centrifuged for 10 min at 2500 r.p.m. and 4 °C. Total RNA was extracted by the method described by Domdey et al. [65]. Sites of RNA methylation were mapped by primer extension using the 5'-end-labeled oligodeoxynucleotides complementary to two distinct regions of the yeast U3 snoRNA [29] and 10 μg of total RNA as the template.

11.2.5
UV Crosslinking and Immunoselection

Formation of covalent bonds between RNA and proteins can be established by incorporation of photoactivable residues such as 4-thiouridine in the course of reverse transcription [66]. However, the easiest way to test for a very near proximity between RNA and proteins is to use UV irradiation at 254 nm [45, 58]. Although the yield of crosslinking is low, it is sufficient to detect RNA–protein contacts by using RNA molecules with a high specific radioactive labeling.

11.2.5.1 Equipment, Materials and Reagents

Equipment

- Electrophoresis instruments for small vertical slab gels. Exposure of EMSA gels can be done with X-Ray films (Fuji or Kodak) or a PhosphorImager.
- Temperature controlled baths (65, 50, 30, 20 °C).

Materials
The short wavelength (254 nm) UV lamp and 96-well ELISA plates are purchased from VWR.

Reagents

Antibodies
Some of the primary antibodies used in the described examples can be purchased from Immuquest.

Chemicals and enzymes
Protein G–Sepharose, Protein A–Sepharose, Hybond C nitrocellulose membrane and the ECL detection system are purchased from Amersham Pharmacia Biotech; bovine serum albumin (B8894) is from Sigma; RNase T1 and RNase A are from Roche Diagnostics.

The Protein (G or A)–Sepharose beads are coated with the antibodies. The digested crosslinked products are then incubated with the coated beads in the presence of the immunoselection buffer. Subsequently, the beads are washed with the immunoselection buffer containing 0.25% Nonidet P-40, suspended in the SDS–PAGE loading buffer and heated 5 min at 96 °C for elution of the bound proteins. The proteins are further fractionated by a SDS–polyacrylamide gel (8–10% polyacrylamide concentration).

Each sample of eluted proteins is divided in two parts. One part is fractionated on a gel analyzed by autoradiography. The amount of radiolabeled protein is estimated by a PhosphorImager. The second half of the eluted proteins is fractionated on a gel used for immunoblotting with the antibody that was coated on the beads. This allows an estimation of the total amount of the protein which was retained on the beads. The ratio between the amount of protein estimated by immunoblotting and the radioactivity detected in the gel gives an indication on the level of crosslinking.

Example: Immunoselection protocol (Fig. 11.7)

About 20 µl of a Protein (G or A)–Sepharose beads suspension was coated for 2 h at 4 °C with 1 µl of anti-hnRNP A1 (4B10) or 2 µl anti-ASF/SF2 antibodies provided by G. Dreyfuss (University of Pennsylvania School of Medicine, Philadelphia, USA) and J. Stevenin (IGBMC, Strasbourg, France), respectively. The digested crosslinked products (see Section 11.2.5.2) were incubated with the coated beads by end-over-end rotation for 2 h at 4 °C in 400 µl of immunoselection buffer containing 0.1 mg/ml bovine serum albumin. Subsequently, the beads were washed 3 times by incubation with 500 µl of the immunoselection buffer containing 0.25% Nonidet P-40, followed by centrifugation (2 min at 3000 r.p.m. at room temperature). At this stage, the beads were suspended in 20 µl of SDS–PAGE loading buffer and boiled for 5 min to elute the bound proteins. The proteins are further fractionated by 10% SDS–Polyacrylamide gel for 1–2 h at 20 V/cm.

Each sample of the eluted proteins was divided in two parts: one part was fractionated by electrophoresis and the amount of radiolabeled protein was estimated by PhosphorImager scanning (Fig. 11.7b and c). The other part was fractionated on a gel and subjected to immunoblotting (Fig. 11.7b).

11.3
Commentaries and Pitfalls

11.3.1
RNP Purification and Reconstitution

11.3.1.1 RNA Purification and Renaturation
In order to obtain an RNA transcript of homogeneous size, a step of purification on denaturing PAGE (8 M urea) is advisable. Indeed, minor RNA degradation or abortive transcription products can be generated during transcription.

A homogeneous conformation of the RNA molecules is required. Thus, the RNA renaturation procedure prior to complex formation and probing experiment is a critical step. Alternative conformations lead to the simultaneous detection of single-stranded and double-stranded specific cleavages at some of the positions.

11.3.1.2 EMSA

Some naked RNAs are resolved in two bands in EMSA gels due to the occurrence of two distinct conformations or dimerizations. If the RNA used corresponds to a fragment of a larger RNA, it is advisable in this case to prepare another fragment by choosing other 5′ and 3′ extremities. Templates for production of small RNA may be produced by use of partially complementary DNA oligodeoxynucleotide primers and DNA polymerase to form a full-length double-stranded DNA.

For probing experiments, it is very important to get a complete and homogeneous formation of the RNPs. Otherwise, heterogeneous probing data will be obtained that will be difficult to interpret.

11.3.2
Probing Conditions

The probing conditions (pH, ionic strength, Mg^{2+} ions concentration, temperature, probe concentration) have to be defined by several preliminary tests. The stability of the RNP in these various conditions has to be tested by electrophoresis on native gel [49].

11.3.2.1 Choice of the Probes Used

Some of the widely used probes for RNA 2-D structure investigation [like CMCT, Pb^{2+}, Fe-EDTA(OH) and S1 nuclease] cannot be used when the assays are directly performed in nuclear or cellular extracts. The reasons for this are given below.

- The presence of a nuclear extract inhibits RNA modification by CMCT, probably due to its accelerated hydrolysis or interactions with other components of the extract. Note that CMCT can be used on purified RNP complexes.
- Pb(II) (Pb^{2+}) cannot be used because the presence of chloride ions in the RNP buffers induces precipitation and inhibits the reaction. However, it can be used for purified complex using a suitable buffer and also *in vivo*, using defined conditions (see Chapter 10).
- Fe-EDTA(OH) cannot be used since hydroxylated compounds, such as glycerol present in the nuclear extract and buffer D, are known to be ·OH scavengers and can inhibit ·OH-mediated cleavage. In addition, Tris or HEPES buffers also reduce RNA cleavage by hydroxyl radicals, presumably by acting as free-radical scavengers. Again, Fe-EDTA(OH) can be used on purified RNP.
- S1 nuclease cannot be used for any kind of RNP probing as its optimal pH of action is 4.5, which is generally deleterious for RNP complexes [48].

It should also be taken into account that the overall efficiency of a given enzymatic or chemical probe may be considerably diminished in the cellular extract and this apparent decrease in efficiency of cleavage or modification does not necessarily represent RNA protection due to protein binding. For instance, DMS reacts with both RNA and proteins; thus, DMS has a large number of substrates when modifications are directly performed in extracts, which may decrease its activity on RNA.

To get informative data by RNP probing, the probes should be selected in order that reactive residues or sensitive phosphodiester bonds are present all along the RNA molecule. It is sometimes difficult to fulfill these conditions in the case of highly structured RNAs. The reactivity of the naked RNA with the selected probes should be sufficient to see clear variations upon protein binding. Hence, in order to define conditions suitable for RNP analyses, one has first to test different conditions of modifications and cleavages on the naked RNA. The effects of different parameters can be examined in these preliminary tests, like the [RNA]/[probe] ratio, time of incubation, temperature of incubation and Mg^{2+} concentration. The conditions given in this paper were found to be suitable for analyses of several RNP. However, when they turned out to be unsuitable, they could be use as a starting point to look for other more favorable conditions.

11.3.2.2 Ratio of RNA/Probes

The added exogenous tRNA (usually total yeast tRNA mixture) minimizes the nonspecific interactions between RNA and proteins, and is also used to get a defined ratio of [RNA]/[Probe].

Another difficulty which may be encountered in the course of RNP probing is a very strong protection generated by the proteins. They may mask very large parts of the RNA. This may be the case for probing with extracts containing a large number of proteins or with proteins, like hnRNP A1, that are able to multimerize along the RNA [45]. To get convincing probing data one needs to be sure that the protections observed are not simply due to a global inhibition of the activity of the probes. To this end, it is necessary to obtain nearly unaltered modifications and cleavages, together with strongly diminished ones, in the same experiment. Here, again, conditions of reaction often have to be adapted to obtain such contrasted modifications of the reactivities along the RNA molecule. When proteins of the RNP generate very strong protections we recommend the use of chemical probes which are less sensitive to steric hindrance.

Sometimes, new cleavage sites are observed in RNP compared to naked RNA. This may reflect an RNA conformational change. However, one has to verify that these additional or reinforced cleavages are not due to the presence of an RNase activity present in the extract or the purified protein.

11.3.3
UV Crosslinking

11.3.3.1 Photoreactivity of Individual Amino Acids and Nucleotide Bases

Crosslink formation depends on the photoreactivity of both individual amino acids and nucleotide bases. Pyrimidines residues form covalent bonds with protein more efficiently compared to purine residues. Thus, upon UV irradiation a protein that binds a purine-rich sequence may be undetectable by UV crosslinking.

11.3.3.2 Labeled Nucleotide in RNA
Due to the low level of crosslinking established by UV irradiation, the specific activity of the labeled RNA must be high enough for detection by autoradiography of the crosslinked residues bound to the protein. The choice of the labeled nucleotide and the RNase used for the digestion should be made taking into consideration the nucleotide sequence of the expected binding site.

11.3.4
Immunoprecipitations

The amount of serum, or antibodies, required for complete precipitation of a particular protein has to be determined for each individual batch of serum.

11.3.4.1 Efficiency of Immunoadsorbents for Antibody Binding
Antibodies from humans, rabbits or guinea-pigs have a stronger affinity for Protein A, than those from mouse or rat. Binding to Protein G provides a convenient alternative for the use of mouse and rabbit antibodies [67]. Poor binding of antibodies to Protein A and G can be circumvented by the use of secondary antibodies (e.g. anti-mouse immunoglobulin raised in rabbits) that do bind to Protein A. Alternatively, the secondary antibodies can be directly coupled to CNBr-activated Sepharose. These coupled secondary antibodies will then serve as efficient adsorbents.

11.4
Troubleshooting

11.4.1
RNP Reconstitution

- No RNP complex is observed: first modify the [RNA]/[protein] ratio used for the reconstitution assay, then, if there is still a problem decrease the quantity of competitor tRNA used.
- The RNP complex does not penetrate in the gel: use a lower polyacrylamide concentration.

11.4.2
RNA Probing

- High smearing in the gel: decrease the amount of loaded material or digest the RNA with a DNase-free RNase.
- No elongation stop signal is detected in primer extension analysis of the modified or cleaved RNA: decrease the amount of probe used.

- No protection is observed: check if the RNP complex is formed and stable under the conditions used.
- Too much protections of the RNA: decrease the quantity of purified protein or cellular extract added.

11.4.3
UV Crosslinking

- No crosslinked proteins are obtained: verify that the UV wavelength is correct and that the UV light is still working.
- No crosslinked proteins are immunoselected: if the Western Blot analyses indicate that immunoselection occurred, try to use another labeled nucleotide. If labeled proteins are still not immunoselected, you have to use multiple approaches to understand what are the parameters that govern the binding of the protein to the RNA target.

11.4.4
Immunoprecipitations

- High background in the membrane: it is recommended to use less serum or antibodies.
- No signal is detected on the membrane: check that primary and/or secondary antibodies are still active.

Acknowledgments

S. Jacquenet and A. Clery are acknowledged for providing materials for illustrations of EMSA experiments. Y. Motorin is thanked for helpful discussion and critical reading of the manuscript. V. M. was a fellow of the French Ministère de la Jeunesse, de l'Education Nationale et de la Recherche. C. B., A. M. and A. M. are Staff Scientists from the Centre National de la Recherche Scientifique.

References

1 C. G. BURD, G. DREYFUSS, *Science* **1994**, *265*, 615–621.
2 K. B. HALL, *Curr. Opin. Struct. Biol.* **2002**, *12*, 283–288.
3 T. A. STEITZ, P. B. MOORE, *Trends Biochem. Sci.* **2003**, *28*, 411–418.
4 C. C. QUERY, R. C. BENTLEY, J. D. KEENE, *Cell* **1989**, *57*, 89–101.
5 D. J. KENAN, C. C. QUERY, J. D. KEENE, *Trends Biochem. Sci.* **1991**, *16*, 214–220.
6 H. SIOMI, M. J. MATUNIS, W. M. MICHAEL, G. DREYFUSS, *Nucleic Acids Res.* **1993**, *21*, 1193–1198.
7 C. OUBRIDGE, N. ITO, P. R. EVANS, C. H. TEO, K. NAGAI, *Nature* **1994**, *372*, 432–438.
8 N. V. GRISHIN, *Nucleic Acids Res.* **2001**, *29*, 638–643.
9 H. MAO, S. A. WHITE, J. R. WILLIAMSON, *Nat. Struct. Biol.* **1999**, *6*, 1139–1147.

10 N. J. Watkins, V. Segault, B. Charpentier, S. Nottrott, P. Fabrizio, A. Bachi, M. Wilm, M. Rosbash, C. Branlant, R. Luhrmann, Cell **2000**, *103*, 457–466.
11 I. Vidovic, S. Nottrott, K. Hartmuth, R. Luhrmann, R. Ficner, Mol. Cell **2000**, *6*, 1331–1342.
12 D. J. Klein, T. M. Schmeing, P. B. Moore, T. A. Steitz, EMBO J. **2001**, *20*, 4214–4221.
13 N. Ban, P. Nissen, J. Hansen, P. B. Moore, T. A. Steitz, Science **2000**, *289*, 905–920.
14 E. J. Tran, X. Zhang, E. S. Maxwell, EMBO J. **2003**, *22*, 3930–3940.
15 T. S. Rozhdestvensky, T. H. Tang, I. V. Tchirkova, J. Brosius, J. P. Bachellerie, A. Huttenhofer, Nucleic Acids Res. **2003**, *31*, 869–877.
16 N. Marmier-Gourrier, A. Clery, V. Senty-Segault, B. Charpentier, F. Schlotter, F. Leclerc, R. Fournier, C. Branlant, RNA **2003**, *9*, 821–838.
17 I. Fierro-Monti, M. B. Mathews, Trends Biochem. Sci. **2000**, *25*, 241–246.
18 M. Kiledjian, G. Dreyfuss, EMBO J. **1992**, *11*, 2655–2664.
19 O. Theunissen, F. Rudt, U. Guddat, H. Mentzel, T. Pieler, Cell **1992**, *71*, 679–690.
20 J. D. Puglisi, L. Chen, S. Blanchard, A. D. Frankel, Science **1995**, *270*, 1200–1203.
21 S. Cusack, Curr. Opin. Struct. Biol. **1999**, *9*, 66–73.
22 M. Ranjan, S. R. Tafuri, A. P. Wolffe, Genes Dev. **1993**, *7*, 1725–1736.
23 P. Fabrizio, S. Esser, B. Kastner, R. Luhrmann, Science **1994**, *264*, 261–265.
24 M. Bach, P. Bringmann, R. Luhrmann, Methods Enzymol. **1990**, *181*, 232–257.
25 O. Puig, F. Caspary, G. Rigaut, B. Rutz, E. Bouveret, E. Bragado-Nilsson, M. Wilm, B. Seraphin, Methods **2001**, *24*, 218–229.
26 K. Hartmuth, H. Urlaub, H. P. Vornlocher, C. L. Will, M. Gentzel, M. Wilm, R. Luhrmann, Proc. Natl. Acad. Sci. USA **2002**, *99*, 16719–16724.
27 G. Neubauer, A. Gottschalk, P. Fabrizio, B. Seraphin, R. Luhrmann, M. Mann, Proc. Natl. Acad. Sci. USA **1997**, *94*, 385–390.
28 F. Dragon, J. E. Gallagher, P. A. Compagnone-Post, B. M. Mitchell, K. A. Porwancher, K. A. Wehner, S. Wormsley, R. E. Settlage, J. Shabanowitz, Y. Osheim, A. L. Beyer, D. F. Hunt, S. Baserga, Nature **2002**, *417*, 967–970.
29 A. Mereau, R. Fournier, A. Gregoire, A. Mougin, P. Fabrizio, R. Luhrmann, C. Branlant, J. Mol. Biol. **1997**, *273*, 552–571.
30 M. Balzer, R. Wagner, Anal. Biochem. **1998**, *256*, 240–242.
31 T. Hamma, A. R. Ferre-D'Amare, Struct. **2004**, *12*, 893–903.
32 T. Moore, Y. Zhang, M. O. Fenley, H. Li, Struct. **2004**, *12*, 807–818.
33 C. Charron, X. Manival, A. Cléry, V. Senty-Ségault, B. Charpentier, N. Marmier-Gourrier, C. Branlant, A. Aubry, J. Mol. Biol. **2004**, in press.
34 D. Irvine, C. Tuerk, L. Gold, J. Mol. Biol. **1991**, *222*, 739–761.
35 C. Tuerk, L. Gold, Science **1990**, *249*, 505–510.
36 H. Urlaub, K. Hartmuth, R. Luhrmann, Methods **2002**, *26*, 170–181.
37 M. Bachler, R. Schroeder, U. von Ahsen, RNA **1999**, *5*, 1509–1516.
38 C. L. Will, S. E. Behrens, R. Luhrmann, Mol. Biol. Rep. **1993**, *18*, 121–126.
39 B. Kastner, R. Luhrmann, Methods Mol. Biol. **1999**, *118*, 289–298.
40 A. Gottschalk, G. Neubauer, J. Banroques, M. Mann, R. Luhrmann, P. Fabrizio, EMBO J. **1999**, *18*, 4535–4548.
41 K. Hartmuth, H. P. Vornlocher, R. Luhrmann, Methods Mol. Biol. **2004**, *257*, 47–64.
42 J. D. Dignam, R. M. Lebovitz, R. G. Roeder, Nucleic Acids Res. **1983**, *11*, 1475–1489.
43 F. Baneyx, Curr. Opin. Biotechnol. **1999**, *10*, 411–421.
44 C. M. Griffiths, M. J. Page, Methods Mol. Biol. **1997**, *75*, 427–440.
45 V. Marchand, A. Mereau, S.

JACQUENET, D. THOMAS, A. MOUGIN, R. GATTONI, J. STEVENIN, C. BRANLANT, *J. Mol. Biol.* **2002**, *323*, 629–652.

46 V. V. GUREVICH, I. D. POKROVSKAYA, T. A. OBUKHOVA, S. A. ZOZULYA, *Anal. Biochem.* **1991**, *195*, 207–213.

47 J. F. MILLIGAN, D. R. GROEBE, G. W. WITHERELL, O. C. UHLENBECK, *Nucleic Acids Res.* **1987**, *15*, 8783–8798.

48 S. JACQUENET, D. ROPERS, P. S. BILODEAU, L. DAMIER, A. MOUGIN, C. M. STOLTZFUS, C. BRANLANT, *Nucleic Acids Res.* **2001**, *29*, 464–478.

49 A. MOUGIN, A. GOTTSCHALK, P. FABRIZIO, R. LUHRMANN, C. BRANLANT, *J. Mol. Biol.* **2002**, *317*, 631–649.

50 J. F. MILLIGAN, O. C. UHLENBECK, *Methods Enzymol.* **1989**, *180*, 51–62.

51 M. SILBERKLANG, A. M. GILLUM, U. L. RAJBHANDARY, *Methods Enzymol.* **1979**, *59*, 58–109.

52 T. E. ENGLAND, O. C. UHLENBECK, *Nature* **1978**, *275*, 560–561.

53 T. PAN, D. M. LONG, O. C. UHLENBECK, in *The RNA World*, R. F. GESTELAND, J. F. ATKINS (eds), Cold Spring Harbor Laboratory Press, Cold Spring Harbor Laboratory, NY, **1993**.

54 R. GATTONI, P. SCHMITT, J. STEVENIN, *Nucleic Acids Res.* **1988**, *16*, 2389–2409.

55 A. MAYEDA, A. R. KRAINER, *Methods Mol. Biol.* **1999**, *118*, 315–321.

56 A. M. ZAHLER, C. K. DAMGAARD, J. KJEMS, M. CAPUTI, *J. Biol. Chem.* **2004**, *279*, 10077–10084.

57 M. M. KONARSKA, *Methods Enzymol.* **1989**, *180*, 442–453.

58 S. JACQUENET, A. MEREAU, P. S. BILODEAU, L. DAMIER, C. M. STOLTZFUS, C. BRANLANT, *J. Biol. Chem.* **2001**, *276*, 40464–40475.

59 S. PINOL-ROMA, Y. D. CHOI, M. J. MATUNIS, G. DREYFUSS, *Genes Dev.* **1988**, *2*, 215–227.

60 M. Y. CHOU, N. ROOKE, C. W. TURCK, D. L. BLACK, *Mol. Cell. Biol.* **1999**, *19*, 69–77.

61 J. R. WILLIAMSON, *Nat. Struct. Biol.* **2000**, *7*, 834–837.

62 D. MOAZED, H. F. NOLLER, *Cell* **1986**, *47*, 985–994.

63 J. SAMBROOK, E. F. FRITSCH, T. MANIATIS, *Molecular Cloning: A Laboratory Manual*, Cold Spring Harbor Laboratory Press, Cold Spring Harbor, NY, **1989**.

64 A. MOUGIN, A. GREGOIRE, J. BANROQUES, V. SEGAULT, R. FOURNIER, F. BRULE, M. CHEVRIER-MILLER, C. BRANLANT, *RNA* **1996**, *2*, 1079–1093.

65 H. DOMDEY, H. APOSTOL, R. J. LIN, A. NEWMAN, E. BRODY, J. ABELSON, *Cell* **1984**, *39*, 611–621.

66 A. D. BRANCH, B. J. BENENFELD, C. P. PAUL, H. D. ROBERTSON, *Methods Enzymol.* **1989**, *180*, 418–442.

67 D. LANE, E. HARLOW, *Nature* **1982**, *298*, 517.

12
Terbium(III) Footprinting as a Probe of RNA Structure and Metal-binding Sites

Dinari A. Harris and Nils G. Walter

12.1
Introduction

Cations play a pivotal role in RNA structure and function. A functional RNA tertiary structure is stabilized by metal ions that neutralize and, in the case of multivalents, bridge the negatively charged phosphoribose backbone [1]. The current chapter describes the use of the trivalent lanthanide metal ion terbium(III) as a versatile probe of high-affinity metal ion binding sites, as well as of RNA secondary and tertiary structure. Terbium(III) has a similar ion radius (0.92 Å) as magnesium(II) (0.72 Å) and a similar preference for oxygen and nitrogen ligands over softer ones. Thus, terbium(III) binds to similar sites on RNA as magnesium(II); however, with 2–4 orders of magnitude higher affinity. Unlike magnesium(II), the low pK_a of the aqueous terbium(III) complex (around 7.9) generates enough Tb(OH)$_{(aq)}^{2+}$ to make it capable of hydrolyzing the RNA backbone around neutral pH via deprotonation of the 2′-hydroxyl group and nucleophilic attack of the resulting oxyanion on the adjacent 3′,5′-phosphodiester to form 2′,3′-cyclic phosphate and 5′-hydroxyl termini [2]. Under physiological conditions, therefore, low (micromolar) concentrations of Tb^{3+} ions readily displace medium (millimolar) concentrations of Mg^{2+} ions from high-affinity binding sites (both specific and non-specific ones) and promote slow phosphodiester backbone cleavage, revealing their location on the RNA. Higher (millimolar) concentrations of Tb^{3+} ions bind less specifically to RNA and result in backbone cleavage in a sequence-independent manner, preferentially cutting solvent accessible, single-stranded or non-Watson–Crick base-paired regions, thus providing a footprint of the RNA's secondary and tertiary structure at nucleotide resolution.

Terbium(III) can be a very straightforward and useful probe of metal binding and tertiary structure formation in RNA. However, there are several precautions that need to be considered in order to obtain a reliable and reproducible RNA footprinting pattern. Since low (micromolar) concentrations of terbium(III) bind to high-affinity metal binding sites within a folded RNA, while high (millimolar) concentrations produce a footprinting pattern of solvent accessible regions, it is critical to perform terbium(III) induced cleavage reactions over a wide range of Tb^{3+} con-

Handbook of RNA Biochemistry. Edited by R. K. Hartmann, A. Bindereif, A. Schön, E. Westhof
Copyright © 2005 WILEY-VCH Verlag GmbH & Co. KGaA, Weinheim
ISBN: 3-527-30826-1

centrations. To ensure conformational homogeneity, pre-folding the RNA under optimized buffer conditions and magnesium concentrations is necessary. This is especially important when trying to identify metal-binding sites, since there will be relatively few cleavage events at low Tb^{3+} concentrations. All cleavage reactions should be performed near physiological pH (7.0–7.5) to allow for the accumulation of the cleavage active $Tb(OH)_{(aq)}^{2+}$ species [3]. Insoluble polynuclear hydroxo aggregates of terbium(III) can form at pH 7.5 and above [4, 5], which should be avoided. Another parameter that needs to be empirically optimized is the temperature and duration of the metal-ion-induced cleavage reactions. Higher temperatures result in faster cleavage rates, but also increase the amount of background degradation. Therefore, typical reaction temperatures range from 25 to 45 °C over a period of 0.5–2 h. All of these parameters need to be well established prior to carrying out terbium(III) footprinting experiments in earnest.

12.2
Protocol Description

12.2.1
Materials

Reagents and buffers
Appropriate buffers to fold RNA (usually Tris, MES and/or HEPES of desired pH)
1 M $MgCl_2$ solution.
100 mM $TbCl_3$ in 5 mM sodium cacodylate buffer, pH 5.5 (store in small aliquots at -20 °C).
0.5 M Na_2EDTA, pH 8.0.
Urea loading buffer: 8 M urea, 50 mM EDTA, pH 8.0, 0.01% bromophenol blue, 0.01% xylene cyanol.

Equipment
Heating block at 90 °C.
Water bath.
Phosphor screens and phosphorImager with appropriate software [e.g. PhosphorImager Storm 840 with ImageQuant software (Molecular Dynamics)].

(1) Prior to performing terbium(III) mediated footprinting of an RNA molecule, the RNA should be end-labeled (typically with ^{32}P at either the 5′ or 3′ end), purified by denaturing gel electrophoresis, and stored in water (or an appropriate buffer) at -20 °C.

(2) Prepare a single pool with 250 000–500 000 c.p.m. (typically 0.5–2 pmol) of end-labeled RNA per reaction aliquot under appropriate buffer conditions and heat denature at 90 °C for 2 min. The total pool volume should be sufficient for single or duplicate reaction aliquots at each desired Tb^{3+} concentration.

(3) Pre-fold the RNA, by incubating the pool at an optimized temperature (typically 25–45 °C) for approximately 10 min to ensure structural homogeneity. Some RNAs fold best when a slow-cooling procedure is used or when certain cations are already added at this stage.

(4) To obtain the desired Mg^{2+} concentration, add an aliquot of $MgCl_2$ from an appropriately diluted stock solution and equilibrate at the optimized temperature for an additional 5–10 min. At this point, the total volume of the reaction mixture should be 8 µl per reaction aliquot.

(5) From the 100 mM $TbCl_3$ stock solution, make a serial set of $TbCl_3$ dilutions in water, ranging from micromolar to millimolar concentrations (5× over the final reaction concentration). This wide range of $TbCl_3$ concentrations should be sufficient to probe for both high-affinity metal binding sites and secondary/tertiary structure formation. *Note: The 100 mM $TbCl_3$ stock solution is dissolved in a 5 mM sodium cacodylate buffer at pH 5.5 to prevent formation of terbium(III) hydroxide precipitates at higher pH. The $TbCl_3$ dilutions in water should be made immediately prior to use. A serial set of dilutions is recommended to ensure consistency in cleavage band intensity between gel lanes. Use a fresh aliquot of 100 mM $TbCl_3$ stock solution each time. Final $TbCl_3$ concentrations used in the cleavage reactions should be optimized together with other experimental conditions for the specific RNA and experimental goal.*

(6) To initiate terbium(III) mediated cleavage, mix an 8-µl aliquot from the pool containing the end-labeled RNA, buffer components and Mg^{2+} with 2 µl of an appropriate dilution of $TbCl_3$ to achieve the desired final Tb^{3+} concentration (typically ranging from 5 µM to 5 mM , in addition to a 0 mM Tb^{3+} control). Continue to incubate at the optimized temperature for an optimized amount of time (typically 30 min to 2 h). *Note: The incubation times should be chosen to generate a partial digestion pattern of end-labeled RNA under single-hit conditions. Extended incubation times will increase secondary hits that may reflect structural distortions of the RNA.*

(7) Quench the cleavage reaction by adding EDTA, pH 8.0, to a final concentration of 50 mM (or at least a 2-fold excess over the total concentration of multivalent metal ions in the reaction aliquot).

(8) Perform an ethanol precipitation of the RNA by adding Na(OAc) to a final concentration of 0.3 M and 2–2.5 volumes of 100% ethanol, and precipitate at −20 °C overnight. Centrifuge 30 min at 12 000 g, 4 °C. Decant supernatant, wash with 80% ethanol, decant supernatant, and dry RNA in a Speed Vac evaporator. Re-dissolve samples in 10–20 µl of urea loading buffer.

(9) Partial alkaline hydrolysis and RNase T1 digestion reactions of the same RNA should be performed as calibration standards by incubating the end-labeled RNA in the appropriate buffers.

(10) Heat samples at 90 °C for 5 min and place on ice water to snap cool. Analyze the cleavage products on a high-resolution denaturing (8 M urea) polyacrylamide sequencing gel, using the partial alkaline hydrolysis and RNase T1 digestion reactions as size markers to identify the specific terbium(III) cleavage products at nucleotide resolution. *Note: In the example cited below, a wedged,*

8 M urea, 20% polyacrylamide *gel was run at a constant power of 80 W for separating the reaction products of a radiolabeled 39mer RNA. Identical samples can be loaded at different times on the same gel to resolve different regions of longer RNA.*

(11) Product bands are directly visualized by exposing the gel to a phosphor screen. *Note: The exposure can take several hours to overnight, depending on the level of radioactivity of the bands in the gel.*

(12) Quantify the full-length RNA and cleavage product bands using a volume count method. (For a more qualitative evaluation, a line scan method can be used.) At every Tb^{3+} concentration, calculate a normalized extent of cleavage (Π) by substituting the peak intensities in the equation:

$$\Pi = \frac{\left(\frac{\text{band intensity at nucleotide } x}{\sum_i \text{band intensity at nucleotide } i}\right)_{y[Tb^{3+}]}}{\left(\frac{\text{band intensity at nucleotide } x}{\sum_i \text{band intensity at nucleotide } i}\right)_{0 \text{ mM } [Tb^{3+}]}}$$

where y is the terbium(III) concentration in a particular cleavage reaction and x the analyzed nucleotide position of the RNA. 0 mM $[Tb^{3+}]$ signifies a control reaction incubated in the same fashion as the terbium(III) containing ones except that no terbium(III) is added. A Π value of ≥ 2 indicates significant cleavage over background degradation. *Note: By dividing the ratio of a*

Fig. 12.1. Terbium(III) footprinting of the *trans*-acting HDV ribozyme. (A) Synthetic HDV ribozyme construct D1. The ribozyme portion is shown in bold, and consists of two separate RNA strands A and B. 3′ Product (3′P) is shown outlined. The substrate variant S3 contains eight additional nucleotides (gray) 5′ of the cleavage site (arrow). To generate non-cleavable substrate analogs, the 2′-OH of the underlined nucleotide immediately 5′ of the cleavage site was modified to 2′-methoxy and the suffix "nc" added to their name. Dashed lines, tertiary structure hydrogen bonds of C75 and the ribose zipper of A77 and A78 in joiner J4/2. (B) Terbium(III)- and magnesium(II)-mediated footprint of the 5′-^{32}P-labeled HDV ribozyme strand A upon incubation with terbium(III) for 2 h in 40 mM Tris–HCl, pH 7.5, 11 mM MgCl$_2$ at 25 °C. From left to right as indicated: strand A fresh after radiolabeling; incubated in buffer without Tb^{3+}; incubated with excess strand B in buffer without Tb^{3+}; incubated in buffer without Tb^{3+}; incubated with excess strand B and non-cleavable substrate analog ncS3 in buffer without Tb^{3+}; RNase T1 digest; alkali (OH$^-$) ladder; footprint with increasing Tb^{3+} concentrations in the presence of excess strand B and ncS3; incubated in buffer without Tb^{3+}; incubated with excess strand B in buffer without Tb^{3+}; footprint with increasing Tb^{3+} concentrations in the presence of excess strand B and 3′ product (3′P). As the terbium(III) concentration increases, backbone scission becomes more intense. The 5′ and 3′ segments of P1.1 (boxed) footprint very differently in the precursor and product complexes. Far right, magnesium(II)-induced cleavage at pH 9.5 and 37 °C, for comparison; from left to right: precursor (ncS3) complex, control incubated at pH 7.5; precursor complex, footprinted at pH 9.5; product complex, control incubated at pH 7.5; product complex, footprinted at pH 9.5. Reprinted with permission from [6].

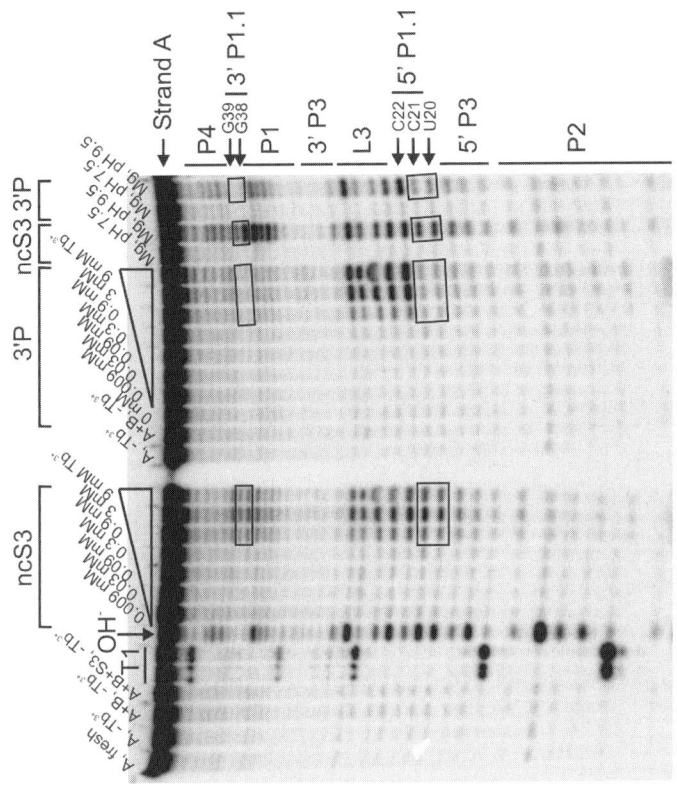

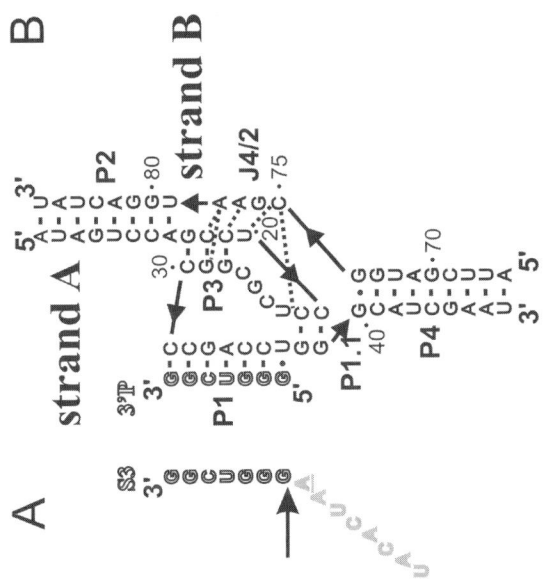

Fig. 12.1

single band intensity over total RNA in the presence of terbium(III) by the ratio of a single band intensity over total RNA in the absence of terbium(III), one normalizes for the effect of non-specific background degradation.

12.3
Application Example

Terbium(III) has been successfully used on a number of RNAs as probe for high-affinity metal-binding sites and tertiary structure formation. For example, taking advantage of its luminescent, as well as RNA footprinting properties, terbium(III) has recently revealed subtle structural differences between the precursor and product forms of the hepatitis delta virus (HDV) ribozyme [6]. The HDV ribozyme is a unique RNA motif found in the human HDV, a satellite of the hepatitis B virus that leads to frequent progression towards liver cirrhosis in millions of patients worldwide [7]. There is a strong interest, both for medical and fundamental reasons, in understanding structure and function relationships in this catalytic RNA. We found that the terbium(III) mediated footprinting pattern of the 3′ product (3′P) complex of a *trans*-acting version of the HDV ribozyme (Fig. 12.1A), obtained in the presence of millimolar Tb^{3+} concentrations, is consistent with a post-cleavage crystal structure. In particular, protection is observed in all five Watson–Crick base-paired stems, P1 through P4 and P1.1, while the backbone of the L3 loop region and that of the J4/2 joining segment are strongly cut (Fig. 12.1) [6]. Cuts in the J4/2 joiner are particularly relevant since it encompasses the catalytic residue C75 and its neighboring G76, and the strong terbium(III) hits implicate it as a region of high negative charge density with high affinity for metal ions.

Strikingly, terbium(III) footprinting reveals the precursor (ncS3) structure as distinct; while P1, P2, P3 and P4 remain protected, both the 5′ and 3′ segments of the P1.1 stem (as well as U20, immediately upstream) are strongly hit, suggesting that this helix in the catalytic core is formed to a lesser extent than in the product complex. In addition, scission in J4/2 extends to A77 and A78, implying that the ribose zipper motif involving these nucleotides may not be fully formed in the precursor complex (Fig. 12.1B). These differences in extent of backbone scission in the precursor versus the 3′ product complexes show that a significant conformational

Fig. 12.2. Sites of backbone scission mediated by 3 mM terbium(III) in 40 mM Tris–HCl, pH 7.5, 11 mM $MgCl_2$ at 25 °C and superimposed onto two-dimensional representations of the precursor and product HDV ribozyme secondary structures. Only the catalytic core residues are explicitly shown. Relative scission intensities were calculated as described in Section 12.2 and are represented by the symbol code. Scission is located 3′ of the indicated nucleotides. Only the product structure is likely to fully form P1.1 and the ribose zipper of A77 and A78 in J4/2, as suggested by solid and dashed lines, respectively. Reprinted with permission from [6].

12.3 Application Example | 211

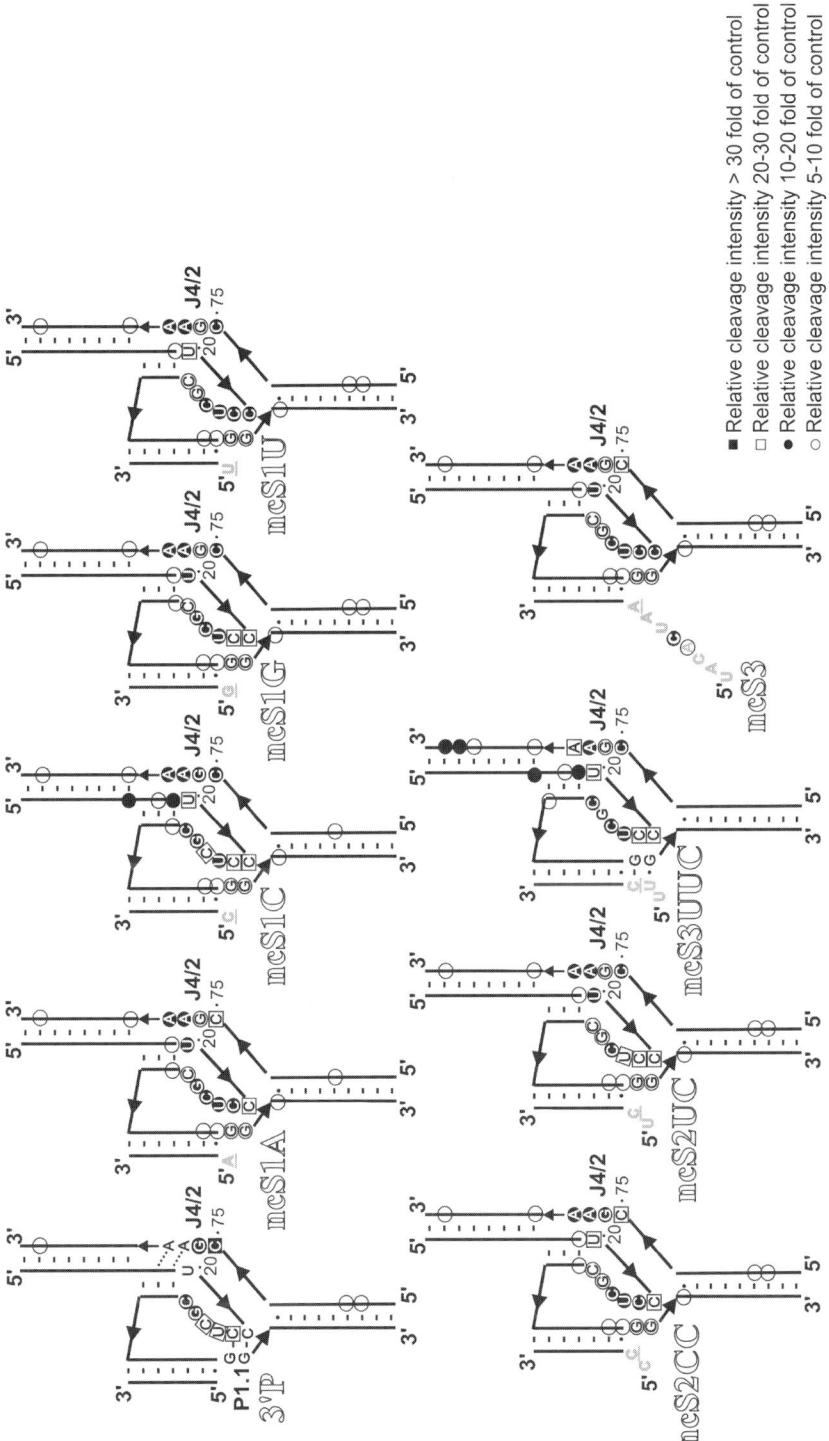

change occurs upon HDV ribozyme catalysis and 5′ product dissociation from the 3′ product [6].

While previous evidence from fluorescence resonance energy transfer [8], 2-aminopurine fluorescence quenching [9] and NMR spectroscopy [10, 11] had already hinted at structural differences between the precursor and 3′ product forms of the *trans*-acting HDV ribozyme, terbium(III)-mediated footprinting complements these techniques by providing specifics of these rearrangements at nucleotide resolution. Particularly intriguing are the differences in the catalytic core structure around C75 and P1.1 that may control access to the cleavage transition state and may therefore explain differences in the catalytic rate constants for substrates with different 5′ sequences (Fig. 12.2) [6]. In fact, the 5′ substrate sequence subtly modulates the terbium(III) footprinting pattern, but all the substrates consistently show strong cuts in the P1.1 stem and the ribose zipper motif of J4/2 (Fig. 12.2). This implies that in the precursor these tertiary structure interactions are not fully formed, in contrast to the 3′P complex. Interestingly, these subtle differences in the catalytic core structure of the various precursor complexes translate into significant changes in fluorescence resonance energy transfer (FRET) efficiency between fluorophores attached to the termini of P4 and P2 stems [6]. Taken together, these results indicate that the various precursor complexes differ in structure both locally (in the catalytic core) and globally (as measured by FRET), providing an explanation for the wide range of catalytic activities of substrates with varying 5′ extensions [6, 12].

Several other labs have also found terbium(III) to be a useful probe of high-affinity metal binding sites and tertiary structure in RNA. Musier-Forsyth and coworkers were able to show that terbium substitutes for several well known metal binding sites in human tRNA^{Lys,3} and works as a sensitive probe of tertiary structure. At low Tb^{3+} concentrations, cleavage of tRNA^{Lys,3} is restricted to nucleotides that were previously identified from X-ray crystallography as specific metal-binding pockets [13]. The use of higher Tb^{3+} concentrations resulted in an overall footprint of the L-shaped tRNA structure, showing increased cleavage in the loop regions (D and anticodon loops). Binding of HIV nucleocapsid protein could then be shown to result in the disruption of the tRNA's metal binding pockets and, at higher concentrations, to induce subtle structural changes in, for example, the tRNA acceptor–TψC stem minihelix [14]. Other RNAs that have similarly been studied by terbium(III)-mediated footprinting include the hammerhead [15], aminoacyl-transferase [16, 17], RNase P [18] and group II intron ribozymes [19].

12.4
Troubleshooting

Initial titration experiments will be necessary to obtain the optimal Tb^{3+} concentration(s) to use for structure probing of any individual RNA [typical terbium(III) and RNA concentrations for determining tertiary structural features are around 1–5 mM and 1 µM, respectively]. The trivalent terbium(III) has been shown to induce slight perturbations in the RNA structure [13], but careful titration will reveal the optimal terbium(III):RNA ratio needed for detecting unbiased secondary and tertiary structure features in a given RNA molecule.

To verify a high-affinity metal-ion-binding site, it is advisable to first decrease the Tb^{3+} concentration until a very narrow cleavage pattern is observed (typically at 10–100 μM Tb^{3+}) and then to perform a competition experiment with increasing concentrations of Mg^{2+}. The intensity (or fraction of RNA cleaved at a particular nucleotide position) should decrease as the Mg^{2+} concentration increases. Quantifying the intensities of cleaved bands at each nucleotide position directly relates to the structure of the RNA. It is critical to keep the extent of total cleavage lower than 20% of the uncleaved or full-length band. This will ensure that the RNA is undergoing a single cleavage event. Finally, it is important to keep in mind that, while terbium(III) footprinting will reveal many high-affinity metal ion binding sites, it may not reveal all. This is due to the fact that there is a steric requirement of Tb^{3+} to bind close to the 2′-hydroxyl group on the ribose for inducing backbone cleavage. This restraint is very unfavorable in A-type RNA helices and, therefore, strong metal sites that occur in RNA helical regions may be underestimated or go undetected by Tb^{3+} cleavage, as may binding sites that are highly specific for a particular metal ion [19].

References

1 Pyle, A. M. *J. Biol. Inorg. Chem.* **2002**, 7, 679–690.
2 Ciesiolka, J., T. Marciniec, W. Krzyzosiak. *Eur. J. Biochem.* **1989**, 182, 445–450.
3 Walter, N. G., N. Yang, J. M. Burke. *J. Mol. Biol.* **2000**, 298, 539–555.
4 Baes, C. F., R. E. Mesmer. *The Hydrolysis of Cations*, Wiley Interscience, New York, **1976**.
5 Matsumura, K., M. Komiyama. *J. Biochem.* **1997**, 122, 387–394.
6 Jeong, S., J. Sefcikova, R. A. Tinsley, D. Rueda, N. G. Walter. *Biochemistry* **2003**, 42, 7727–7740.
7 Hadziyannis, S. J. *J. Gastroenterol. Hepatol.* **1997**, 12, 289–298.
8 Pereira, M. J., D. A. Harris, D. Rueda, N. G. Walter. *Biochemistry* **2002**, 41, 730–740.
9 Harris, D. A., D. Rueda, N. G. Walter. *Biochemistry* **2002**, 41, 12051–12061.
10 Luptak, A., A. R. Ferre-D'Amare, K. Zhou, K. W. Zilm, J. A. Doudna. *J. Am. Chem. Soc.* **2001**, 123, 8447–8852.
11 Tanaka, Y., M. Tagaya, T. Hori, T. Sakamoto, Y. Kurihara, M. Katahira, S. Uesugi. *Genes Cells* **2002**, 7, 567–579.
12 Shih, I., M. D. Been. *EMBO J.* **2001**, 20, 4884–4891.
13 Hargittai, M. R., K. Musier-Forsyth. *RNA* **2000**, 6, 1672–1680.
14 Hargittai, M. R., A. T. Mangla, R. J. Gorelick, K. Musier-Forsyth. *J. Mol. Biol.* **2001**, 312, 985–997.
15 Feig, A. L., M. Panek, W. D. Horrocks, Jr, O. C. Uhlenbeck. *Chem. Biol.* **1999**, 6, 801–810.
16 Flynn-Charlebois, A., N. Lee, H. Suga. *Biochemistry* **2001**, 40, 13623–13632.
17 Vaidya, A., H. Suga. *Biochemistry* **2001**, 40, 7200–7210.
18 Kaye, N. M., N. H. Zahler, E. L. Christian, M. E. Harris. *J. Mol. Biol.* **2002**, 324, 429–442.
19 Sigel, R. K., A. Vaidya, A. M. Pyle. *Nat. Struct. Biol.* **2000**, 7, 1111–1116.

13
Pb^{2+}-induced Cleavage of RNA

Leif A. Kirsebom and Jerzy Ciesiolka

13.1
Introduction

Certain metal ions induce degradation of RNA in a non-oxidative manner, and in some RNA molecules this process is exceptionally efficient and specific. The best-known example, yeast tRNA^{Phe}, undergoes specific fragmentation in the D-loop in the presence of Pb^{2+} [1–3] and other ions, e.g. Eu^{3+} [4, 5], Mn^{2+} [6] and Mg^{2+} [7, 8]. Based on X-ray analysis of yeast tRNA^{Phe} crystals it was suggested that in order to promote cleavage, Pb^{2+} has to be positioned at an optimal distance from the 2'-OH that acts as the nucleophile [9, 10]. These findings gave rise to an experimental approach that uses Pb^{2+} and other ions to localize high-affinity metal ion binding sites as well as to probe the structure of RNA molecules.

Highly efficient and specific Pb^{2+}-induced cleavages are rather rarely observed. The majority of cleavages are weaker and usually comprises several consecutive phosphodiester bonds. Most information on the specificity of Pb^{2+}-induced RNA fragmentation has been obtained from studies on ribosomal 16S RNA [11] and 5S RNAs [12–14]. Cleavages occur preferentially in bulges, loops and other single-stranded RNA regions except those involved in stacking or other higher-order interactions. Double-stranded RNA segments are essentially resistant to breakage. Cleavages are also observed in paired regions destabilized by the presence of non-canonical interactions, bulges or other structural distortions. In general, it seems that flexibility of the polynucleotide chain determines its sensitivity to Pb^{2+}-induced cleavage [11–15].

It has been suggested [16] that the mechanism proposed for the specific, Pb^{2+}-induced fragmentation of yeast tRNA^{Phe} [9, 10, 17] might account for all types of cleavage induced by metal ions. The simplified mechanism shown in Fig. 13.1 is helpful for understanding the relation between RNA structure and sensitivity of a particular RNA region to cleavage.

In the first step, the ionized metal ion hydrate acts as a Brönsted base and abstracts a proton from the 2'-OH group of the ribose. Subsequently, the activated anionic 2'-O$^-$ attacks the phosphorus atom and a penta-coordinated intermediate is formed. The phosphodiester bond is cleaved generating 2',3'-cyclic phosphate

Handbook of RNA Biochemistry. Edited by R. K. Hartmann, A. Bindereif, A. Schön, E. Westhof
Copyright © 2005 WILEY-VCH Verlag GmbH & Co. KGaA, Weinheim
ISBN: 3-527-30826-1

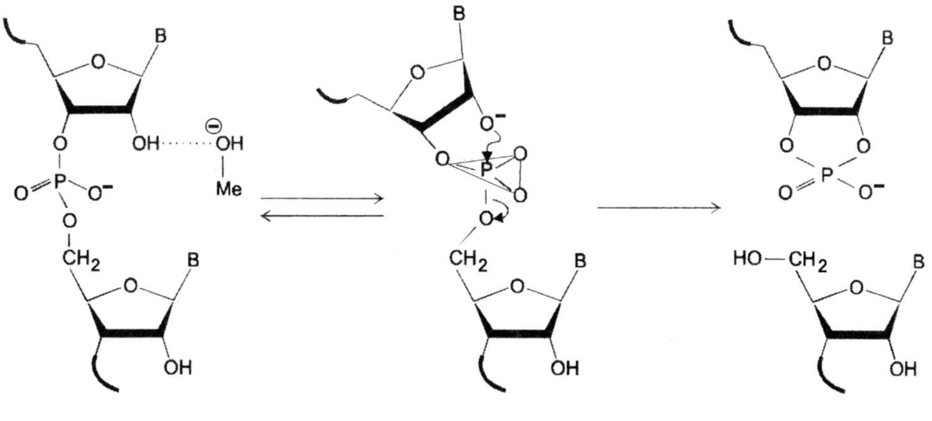

Fig. 13.1. Mechanism of metal ion-induced cleavage of RNA (see text for details).

and 5'-hydroxyl groups as cleavage products. However, based on the discussion about the role of metal ions in, for example, the hammerhead cleavage reaction [18, 19], one has also to consider the possibility that the metal ion acts as a Lewis acid by accepting electrons from the 2'-oxygen, thereby facilitating a nucleophilic attack on the phosphorus atom. Irrespective of mechanism, metal ion interaction with the 2'-OH results in a nucleophilic attack on the phosphorus atom and subsequent cleavage of the phosphodiester bond. Experimentally, the data suggests an inverse correlation between the pK_a values for different metal ion hydrates and cleavage rates: Pb^{2+} with a pK_a of 7.2 induces cleavage more efficiently than, for example, Mg^{2+} (pK_a = 11.4). This would be in keeping with the suggestions that the metal ion either acts as a Lewis acid or Brönsted base. For metal ions with higher pK_a values, such as Eu^{3+}, Zn^{2+}, Mn^{2+}, Mg^{2+} and Ca^{2+} (pK_a = 8.5, 9.6, 10.6, 11.4 and 12.6, respectively), the reaction pH, time or temperature have to be increased and/or, for example, ethanol has to be added to detect substantial cleavage.

The cleavage efficiency of a particular phosphodiester bond in an RNA molecule depends on: (1) proper localization of the metal ion hydrate facilitating deprotonation of the 2'-OH group (Fig. 13.1, transition A to B), and (2) sufficient conformational flexibility of the analyzed region allowing formation of the penta-coordinated intermediate/transition state and subsequent breaking of the phosphodiester chain (transition B to C). Optimal distance and correct orientation of the bound metal ion hydrate seems to be of primary importance when RNAs undergo efficient and highly specific fragmentation. The cleavage at these sites occurs at relatively low concentrations of Pb^{2+} (below 0.1 mM) – conditions under which breakage of other phosphodiester bonds takes place only at significantly reduced rates. Cleavages with lower efficiencies are most likely induced by ions acting from the solution, from weak binding sites and/or from sites at which the Pb^{2+} ion(s) is positioned suboptimally. Moreover, metal hydrates interact equally well with all accessible 2′-hydroxyl groups. Thus, differences in rigidity/flexibility of the phosphates, hindering or facilitating conformational transitions necessary for the reaction to occur, influence cleavage efficiency at individual phosphodiester bonds [16]. The contribution of rigidity/flexibility to the cleavage reaction is difficult to assess. However, the value of the potential rate enhancement derived from constraining a free RNA linkage to an optimal orientation for nucleophilic attack has recently been estimated not to be greater than 50- to 100-fold [20, 21].

13.2
Pb^{2+}-induced Cleavage to Probe Metal Ion Binding Sites, RNA Structure and RNA–Ligand Interactions

The Pb^{2+} cleavage approach has been used in structural analysis of several RNAs and RNA complexes in various ways as summarized in Table 13.1. The information in the table can be classified into three groups: (1) high-affinity metal ion binding sites, (2) RNA structure and (3) RNA–ligand interactions. Figure 13.2 also shows Pb^{2+}-induced cleavage of RNase P RNA in the presence of various divalent metal ions as a typical example.

13.2.1
Probing High-affinity Metal Ion Binding Sites

A strong, highly specific metal ion-induced cleavage suggests the presence of a tight metal ion binding site in the RNA. Cleavage occurring in a particular RNA region does not implicate, however, the direct involvement of that region in coordination of the metal ion. Also, tightly bound metal ions may not induce cleavage at all due to unfavorable distance constraints and/or high rigidity of the polynucleotide chain. For instance, in yeast tRNA^{Phe} a Pb^{2+} ion induces cleavage in the D-loop, but is bound primarily in the TΨC-loop, while the ion positioned in the anticodon loop does not induce specific cleavage [9, 10, 17]. Furthermore, a metal ion binding pocket can usually accommodate different ions, thereby acting as a "general" metal ion binding site. To probe for a "general" metal ion binding site the following two experimental approaches can be and have been used (see Table 13.1 and Fig. 13.2).

The first approach takes advantage of the fact that metal ion-induced cleavage is suppressed if the reaction is performed in the presence of other ions competing for a common metal ion binding site. Thus, addition of metal ions, such as Mg^{2+}, Mn^{2+} or Ca^{2+}, results in suppression of Pb^{2+}-induced cleavage. Quantitative analysis of the inhibition data can also give the K_D value for binding of Mg^{2+} and information about the relative binding affinity of different metal ions for metal ion binding sites in RNA [22, 23].

The second approach relies on the observation that Pb^{2+}, Mg^{2+}, Mn^{2+} and Eu^{3+} ions, bound in the D–TΨC region of yeast tRNA^{Phe}, induce strong cleavage at the same site (and/or at neighboring sites) in the RNA chain. Thus, it seems likely that

Tab. 13.1. Examples of structural analysis of RNAs and RNA complexes by means of the metal ion-induced cleavage approach[1]

RNA or RNA complex	Structural probe	Type of analysis	References
In vitro selected RNAs, aptamers and model oligonucleotides	Pb^{2+}	ion binding sites RNA structure	24, 41, 54, 70 16, 30, 41, 66, 71–73
tRNAs and mutants, in vitro transcripts, and fragments thereof	Pb^{2+} various Me^{2+}	ion binding sites RNA structure ion binding sites RNA structure	2, 3, 17, 22, 37, 50, 74 3, 53, 75, 76 5, 6, 8, 42–45, 77, 78 42–45
HDV ribozyme	various Me^{2+} Pb^{2+}	ion binding sites RNA structure	38, 46, 47 47
4.5S RNA	Pb^{2+}	RNA structure	55
5S rRNA	Pb^{2+}	RNA structure	12–14, 27
U1 snRNA	Pb^{2+}	RNA structure	79
RNase P RNA	Pb^{2+} various Me^{2+}	ion binding sites ion binding sites	28, 29, 51, 52, 59 23, 25, 35
Group I and II intron RNAs	Pb^{2+} various Me^{2+}	ion binding sites ion binding sites	39 26, 48, 49
10Sa RNA (tmRNA)	Pb^{2+}	RNA structure	80
mRNA fragments with trinucleotide repeats	Pb^{2+}	RNA structure	56–58
TfR mRNA fragment	Pb^{2+}	RNA structure	60
SECIS mRNA fragment	Pb^{2+}	RNA structure	81, 82
BRCA1 mRNA fragment	Pb^{2+}	RNA structure	83
CaMV 35S RNA leader	Pb^{2+}	RNA structure	84
16S rRNA fragment in 30S subunit and 70S ribosome	Pb^{2+}	RNA structure RNA–protein interaction RNA–RNA interaction	11
16S and 23S rRNA in 70S ribosome	Pb^{2+} various Me^{2+}	ion binding sites ion binding sites	40 85
RNA aptamer–citrulline complex	Pb^{2+}	RNA–amino acid interaction	65
RNA aptamer–antibiotic complex	Pb^{2+}	RNA–antibiotic interaction	32, 66, 67
HDV ribozyme-antibiotic complex	Pb^{2+}	RNA–antibiotic interaction	38
tRNA–neomycin complex	Pb^{2+}	RNA–antibiotic interaction	31, 69
Phe-tRNA^{Phe}-EF-Tu:GTP complex	Pb^{2+}	RNA–protein interaction	61
4.5S RNA–P48 protein complex	Pb^{2+}	RNA–protein interaction	55
5S rRNA–L18 protein complex	Pb^{2+}	RNA–protein interaction	13
RNase P RNA–neomycin complex	Pb^{2+}	RNA–antibiotic interaction	68
RNase P RNA–tRNA complex	Pb^{2+}	RNA–RNA interaction	52, 63, 64
TfR mRNA fragment–IRP1 complex	Pb^{2+}	RNA–protein interaction	60
3′ end of HEV RNA–viral polymerase complex	Pb^{2+}	RNA–protein interaction	62

[1] Reproduced and modified from Table 1 in [36].

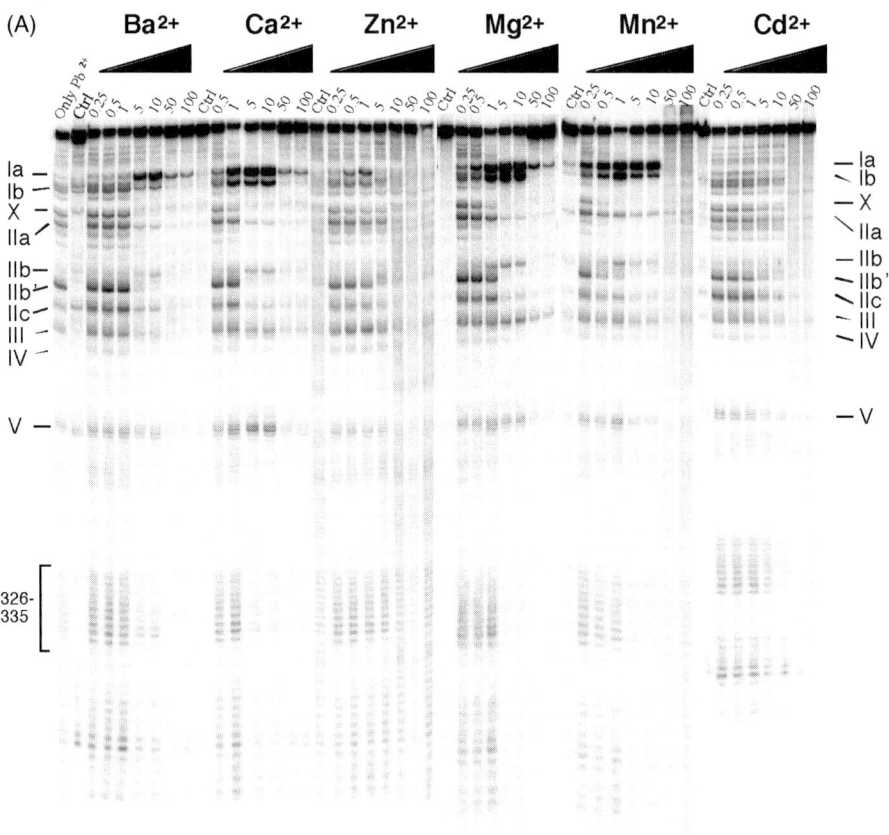

Fig. 13.2. Pb^{2+}-induced cleavage patterns of E. coli RNase P RNA. (A) Pb^{2+}-induced cleavage pattern in the presence of increasing concentrations of Ba^{2+}, Ca^{2+}, Zn^{2+}, Mg^{2+}, Mn^{2+} and Cd^{2+}. Cleavage was performed at specified concentrations and 37 °C as outlined. Lanes: Only Pb^{2+}, incubation in the presence of only 0.5 mM Pb^{2+}; Ctrl, incubation in the absence of Pb^{2+}, but in the presence of Ba^{2+}, Ca^{2+}, Zn^{2+}, Mg^{2+}, Mn^{2+} or Cd^{2+}, 10 mM (final concentration), as indicated; the band denoted X was only observed in the presence of Pb^{2+} and at low concentrations of the other metal ions except Cd^{2+} (see also [23]); reprinted with permission of Nucleic Acids Research. (B) Secondary structure of E. coli RNase P RNA; roman numerals refer to the sites of Pb^{2+} cleavage shown in (A). Roman numerals in italic refer to sites where Mg^{2+}-induced cleavage has also been observed [35]. In (A) it is apparent that increasing the concentration of different divalent metal ions results in suppression of Pb^{2+}-induced cleavage at specific sites, although to different degrees (e.g. compare the effects of different metal ions on the cleavage site IIa). This suggests that these metal ions and Pb^{2+} bind to at least overlapping sites (see text for details). Moreover, these data indicate that different divalent metal ions bind with different affinities to RNase P RNA as well as that the conformation of E. coli RNase P RNA is very similar in the presence of Ba^{2+}, Ca^{2+}, Mg^{2+} and Mn^{2+}, while it is changed compared to the Mg^{2+}-induced conformation in the presence of others, e.g. Cd^{2+}. For further details, see Brännvall et al. [23]; region 326–335 represents an example of a flexible single-stranded region with cleavage at several successive phosphodiester bonds [52]. In the case of cleavage sites Ia and Ib, the second divalent metal ion, such as Ba^{2+}, Ca^{2+}, Mg^{2+} and Mn^{2+}, enhances Pb^{2+}-induced cleavage at lower concentrations due to supporting RNA structure formation, but displaces Pb^{2+} at these sites at higher concentrations.

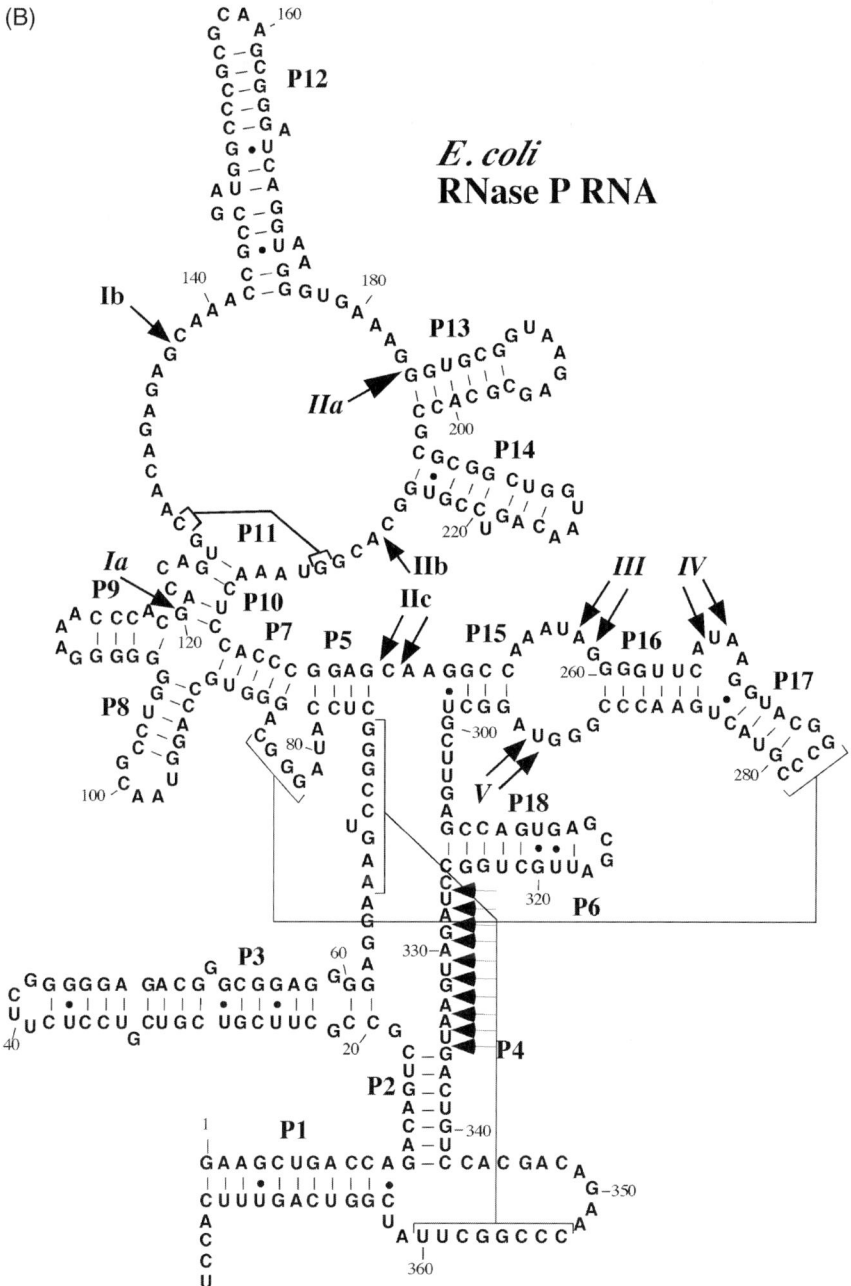

Fig. 13.2.B

different metal ions occupying at least overlapping sites can also induce specific cleavage in the same RNA region in other RNA molecules. However, cleavage induced by different metal ions does not necessarily occur at exactly the same site. Rather, the cleavage sites usually differ by 1 or 2 nt. This is rationalized considering the different coordination preferences and sizes of various metal ions, resulting in a slightly different arrangement of their hydrates in metal ion binding pockets. Note that the cleavage reactions are performed at conditions where the different pK_a values of metal ion hydrates have been taken into account (discussed above, see also below). In addition to Pb^{2+}, typical metal ions used in these experiments are Mg^{2+}, Mn^{2+}, Ca^{2+} and Eu^{3+}.

An additional approach using Tb^{3+}, which has the same coordination geometry as Mg^{2+}, has been used to probe for metal ion binding sites in RNA. Cleavage of RNA with Tb^{3+} gives different cleavage patterns compared to cleavage with, for example, Pb^{2+} [24–26]. Thus, Tb^{3+}-induced cleavage can be used in combination with cleavage induced by other metal ions and thereby more information concerning metal ion binding to RNA can be obtained.

13.2.2
Pb^{2+}-induced Cleavage and RNA Structure

Pb^{2+}-induced cleavage of several RNA molecules has been studied and the cleavage patterns have been used in analysis of RNA structures (see Table 13.1). Moreover, metal ion-induced cleavage allows identification of similarities and differences in related RNA molecules in the regions involved in metal ion binding. However, note that Pb^{2+} patterns do not always correspond precisely to RNA secondary structure models. Experimental results are most consistent with cleavage occurring preferentially in "flexible regions" of an RNA polynucleotide chain. Taking into account that our knowledge of RNA conformational dynamics is still insufficient, the term "flexible regions of RNA" should be used cautiously in the interpretation of experimental data.

Recently, Pb^{2+}-promoted cleavage of several well-defined RNA secondary structure motifs, such as bulges, hairpin loops and single-stranded RNA, has been characterized [16]. These studies show that the cleavage patterns of single nucleotide bulge regions depend on the structural context provided by adjacent base pairs. In general, a pyrimidine flanking the bulged nucleotide, particularly at its 5′ side, facilitates cleavage, while a purine makes the bulge more resistant to cleavage. This effect seems to correlate with the ability of the bulge to form stacking interactions with its neighbors. Cleavage of 2- and 3-nt bulges depends only slightly on their nucleotide composition. In case of terminal loops, cleavage usually increases with the loop size and strongly depends on its nucleotide composition. Particularly resistant are stable tetraloops, most likely due to their high conformational rigidity. Most single-stranded RNA regions are highly susceptible to Pb^{2+}-induced cleavage. However, clusters of G residues and, in most cases, also phosphodiester bonds at the junction of paired and unpaired RNA regions are more resistant. This can be attributed to extensive stacking interactions and increased conformational rigidity

[16]. For some RNAs, however, efficient Pb^{2+}-induced cleavage at the junction between unpaired and double-stranded regions can be seen. It might be that the enhanced reactivity in those cases results from increased "breathing" of the base pair at the junction between single- and double-stranded RNA.

The Pb^{2+} cleavage method is very sensitive for the detection of conformational changes in RNA molecules and useful in determining alternative hairpin structures formed by transcripts composed of trinucleotide repeats (Table 13.1; see also Fig. 13.2 where it is shown that the Pb^{2+}-induced cleavage pattern changes with increasing concentration of other divalent metal ions indicating differences in folding). Several studies have further demonstrated that cleavage of the same structural motifs present in different RNAs results in essentially identical patterns. This raises the interesting possibility to use the Pb^{2+}-induced cleavage approach to identify certain RNA structural motifs in RNA molecules of unknown structure ([16, 27]; Ciesiolka et al., unpublished results).

Lastly, Pb^{2+}-induced cleavage in combination with genetics, i.e. introduction of point mutations, has been used to provide support for the existence of long-range interactions in RNA [3, 50, 51, 59]. Note that point mutations may either result in increased/decreased cleavage at specific positions or in the appearance of cleavage at new positions [28, 29].

13.2.3
Pb^{2+}-induced Cleavage to Study RNA–Ligand Interactions

Remarkable reduction of Pb^{2+}-induced cleavage intensities has been observed upon the formation of RNA complexes with proteins, other RNAs or low-molecular-weight ligands – amino acids, antibiotics and other divalent metal ions (see Table 13.1 and above; also see e.g. [23, 30]).

In RNA–protein complexes, the shielding effect of a bound protein is most likely responsible for changes in cleavage intensities. However, it is not excluded that the RNA changes its conformation due to interaction with protein(s) such that the positioning of Pb^{2+} is altered or that the ion is displaced. In the case of RNA–aminoglycoside interaction, structural studies have provided evidence for displacement of Pb^{2+} as a result of aminoglycoside binding [31]. Furthermore, there are several examples of enhanced Pb^{2+} cleavage upon complex formation (for references, see Table 13.1). Here, moderately enhanced cleavage may suggest increased flexibility of the analyzed RNA regions. The appearance of a very strong cleavage may indicate formation of a new strong metal ion binding site or that a previously inactive metal ion has been repositioned in such a way that efficient metal ion-induced cleavage becomes feasible. In both cases the presence of a tightly bound ion needs to be verified by other methods, since it is conceivable that strand breakage may also occur due to transient, low affinity binding of a metal ion. Needless to say, this mapping method is obviously restricted to RNA regions susceptible to Pb^{2+}-induced cleavage, mainly bulges, loops and other single-stranded RNA stretches.

In some cases, changes in the Pb^{2+} cleavage pattern due to ligand binding may include unexpectedly large regions of the polynucleotide chain [30, 32]. This is probably caused by the loss of flexibility, i.e. formation of a more rigid conformation of a large RNA fragment, which is unstructured in the absence of the ligand.

13.3
Protocols for Metal Ion-induced Cleavage of RNA

The information that can be extracted using the metal ion-induced cleavage approach suggests that a divalent metal ion(s) is positioned close to the metal ion-induced cleavage site, but this does not give any information about how the metal ion is coordinated. However, since the 2′-hydroxyl immediately 5′ of the scissile bond is actively involved in the chemistry of cleavage some structural constraints for the positioning of the divalent metal ion(s) can be derived. These aspects have to be kept in mind when interpreting the data. Here, we describe three protocols used to cleave RNA with divalent metal ions, *in vitro* using Pb^{2+} and Mg^{2+}, and *in vivo* using Pb^{2+} [33]. We will use RNase P RNA, the catalytic subunit of the endoribonuclease P, as an example of an RNA that has been studied using these protocols (see e.g. [34]; unpublished data). Further protocols on Pb^{2+}-induced cleavage of RNA *in vitro* and *in vivo* can be found in Chapter 10.

Protocol 1: Pb^{2+}-induced Cleavage of RNA

(1) The RNA is ^{32}P-labeled at the 3′ end with [5′-^{32}P]pCp or at the 5′ end with [γ-^{32}P]ATP using standard procedures.
(2) The RNA is purified on a denaturing polyacrylamide gel containing 7 M urea in TBE buffer (90 mM Tris–borate, pH 8.5, 2.5 mM EDTA) and eluted (see Chapters 1 and 3). The RNA is renatured by incubation for 5 min at 55 °C in water or buffer of choice.
(3) Pb^{2+}-induced cleavage of, for example, RNase P RNA. Typically, approximately 20 000–30 000 Cerenkov c.p.m. of labeled RNA is mixed with around 2.5 pmol of unlabeled RNA and pre-incubated in 50 mM Tris–HCl, pH 7.5, 100 mM NH_4Cl and 10 mM $MgCl_2$ for 10 min at 37 °C. Note when analyzing metal ion binding and/or structure of an RNA, you have to adjust the conditions such that the RNA adopts a conformation relevant to what you would like to investigate.
(4) Cleavage is initiated by the addition of freshly prepared $Pb(OAc)_2$ to a final concentration of 0.5 mM. Depending on the nature of the experiment you can use other concentrations of $Pb(OAc)_2$, but usually not higher than 2 mM (see also below). Chloride buffer salts can be used, but for higher concentrations of Pb^{2+} (above 2 mM) acetate instead of chloride salts should be used to avoid precipitation of Pb^{2+} ions. The final volume of the reaction is 10 µl.
(5) The reaction is terminated after 10–15 min by the addition of 2 volumes of stop solution (9 M urea, 25 mM EDTA, 0.1% bromophenol blue). The time

of incubation in the presence of Pb^{2+} has to be adjusted experimentally, and depends on the nature of the RNA, cleavage conditions and the question you address.

(6) The cleavage products are separated on denaturing gels, where the percentage depends on the size of the RNA under study (6–12% polyacrylamide gels are generally used).

(7) The Pb^{2+}-induced cleavage sites are mapped using size markers and [OH$^-$] ladders (see Chapter 10). It is also possible to map cleavage sites by primer extension analysis (see Protocol 4) using primers complementary to specific positions in the RNA. In the case of RNase P RNA, we use 15- to 20-nt primers.

We emphasize that an increase in the concentration of Mg^{2+} (or some other divalent metal ion such as Mn^{2+}) results in suppression of Pb^{2+}-induced cleavage as illustrated for cleavage of RNase P RNA (see Fig. 13.2A), suggesting that Mg^{2+} and Pb^{2+} bind, if not to the same, at least to overlapping sites (see e.g. [23]). In combination with the use of genetics (i.e. by using in our example RNase P RNA variants) or by studying cleavage of the RNase P RNA–substrate complex, it is also possible to use the Pb^{2+}-induced cleavage to probe for structural changes in RNase P RNA. Note that the formation of RNase P RNA–substrate complexes requires a higher concentration of Mg^{2+} (20 mM or above). Therefore, an increased concentration of Pb^{2+} is needed to detect cleavage. This might also apply when other RNA molecules are studied and hence the conditions have to be adjusted accordingly.

Protocol 2: Cleavage of RNA in the presence of Mg^{2+}
RNase P RNA is also cleaved by other divalent metal ions such as Mg^{2+}, first described by Kazakov and Altman [35]. However, Mg^{2+}-induced cleavage of RNase P RNA is less efficient compared to Pb^{2+}-induced cleavage (see above); in order to detect cleavage, the reaction has to be performed at a higher pH and in the presence of 10% ethanol. Hence, Steps 3 and 4 of Protocol 1 are modified.

(3) In our studies of RNase P RNA we have used the following conditions: 50 mM CHES buffer, pH 9.5, 100 mM NH_4Cl, 10 mM $MgCl_2$ (higher concentrations of Mg^{2+} can be used) and 10% ethanol [35].
(4) The reaction is incubated at 37 °C for 6 h.
(5) The reaction is terminated and the cleavage products are separated and characterized as described in Protocol 1.

Protocol 3: Pb^{2+}-induced cleavage of RNA *in vivo*
Here the protocol is adapted to study cleavage in growing bacteria, e.g. *Escherichia coli* [33].

(1) Typically, *E. coli* cells are grown in Luria-Broth media (LB) overnight at 37 °C (or temperature of choice).

(2) The culture is diluted 400-fold in LB and allowed to grow to an $OD_{600} \sim 0.5$ (mid-log phase).

(3) Freshly prepared $Pb(OAc)_2$ solutions of appropriate concentrations are prepared by diluting pre-heated (37 °C) 4 × LB media [3 volumes of $Pb(OAc)_2$ and 1 volume of 4 × LB]. Hence, to give a final (theoretical) concentration of 100 mM in the *E. coli* cell suspension (see Step 4), that typically has been used, the freshly prepared $Pb(OAc)_2$ solution should be 467 mM. For reproducibility mixing has to be performed rapidly. Note that when LB and $Pb(OAc)_2$ are mixed, there is always some precipitation, and hence the final concentration of $Pb(OAc)_2$ in solution is lower. Moreover, replacing LB with minimal media results in substantial precipitation and poor RNA cleavage.

(4) Then 1 volume of the $Pb(OAc)_2$/LB solution ($[Pb(OAc)_2] = 350$ mM) is added to 2.5 volumes of cell culture ($OD_{600} \sim 0.5$, see above) and incubated for 7 min at 37 °C under vigorous shaking (the total volume will be 3.5 volumes = V_{tot}). The final concentration of $Pb(OAc)_2 = 100$ mM (not taking the precipitation into account).

(5) The reaction is stopped by adding excess EDTA to a final concentration of 1.5 times the $Pb(OAc)_2$ concentration, typically 150 mM final concentration of EDTA.

(6) The solution is put on ice.

(7) The cells are harvested by centrifugation and the cell pellet is snap frozen in liquid nitrogen and stored at −70 °C.

(8) The cell pellet is re-suspended in a volume of $0.5 \times V_{tot}$ pre-heated (65 °C) lysis buffer (100 mM Tris–HCl, pH 7.5, 40 mM EDTA, 200 mM NaCl, 0.5% SDS w/v) and incubated at 65 °C for 3–5 min.

(9) This is followed by addition of pre-heated (65 °C) phenol solution (volume: $0.5 \times V_{tot}$). To prepare the phenol solution, 1 volume phenol is saturated with 1 volume 10 mM Tris–HCl, pH 8.0, 10 mM EDTA. The RNA is extracted at 65 °C.

(10) The phenol extraction is followed by chloroform/isoamylalcohol (24:1) extraction at room temperature and ethanol precipitation in the presence of 0.3 M sodium acetate, pH 6.0.

(11) The RNA is dissolved in DNase buffer (40 mM Tris–HCl, pH 7.9, 100 mM NaCl, 60 mM $MgCl_2$, 1 mM $CaCl_2$) and 80 U of DNase I (RNase-free) are added. This mixture is incubated for 15 min at 37 °C followed by standard phenol extraction and ethanol precipitation in the presence of 0.3 M sodium acetate, pH 6.0.

(12) The RNA is stored at −70 °C.

(13) The Pb^{2+}-induced cleavage sites are mapped by primer extension analysis using appropriate oligodeoxyribonucleotides as primers (see Protocol 4).

Note that RNA samples prepared from untreated cells (no Pb^{2+} added) have to be analyzed in parallel to RNA prepared from Pb^{2+}-treated cells. Hence, $Pb(OAc)_2$ is omitted in Step 4 by replacing $Pb(OAc)_2$ with RNase-free water and the RNA is prepared following the same procedure as outlined above. This is an essential control to be able to discriminate signals (stops) in the primer extension analysis that are related to Pb^{2+}-induced cleavage from those that are due to pausing (pre-termination) in the reverse transcription reaction.

Protocol 4: Primer Extension

(1) An appropriate 5′-^{32}P-end-labeled oligodeoxyribonucleotide is mixed with 10 μg of total cellular RNA from step 11 (Protocol 3) in RNase-free water and incubated for 1 min at 90 °C.
(2) The mixture is put on ice for 1 min followed by warming at 20 °C for 5 min.
(3) The actual primer extension is performed in a total volume of 15 μl in 50 mM Tris–HCl, pH 8.5, 6 mM $MgCl_2$, 40 mM KCl and dNTPs (1.0 mM each) and 200 U of reverse transcriptase (e.g. Superscript II; Life Technologies). The primer extension mixture is incubated at 45 °C for 30 min.
(4) The reaction is terminated by the addition of 20 μl stop buffer (50 mM Tris–HCl, pH 7.5, 0.1% SDS) and 3.5 μl of 3 M KOH. This solution is incubated for 3 min at 90 °C followed by incubation at 37 °C for 3 h.
(5) Add 6 μl of 3 M acetic acid and ethanol precipitate the cDNA in the presence of 0.3 M sodium acetate, pH 6.0. The products are separated on denaturing gels (see Step 6, Protocol 1).

13.4 Troubleshooting

13.4.1
No Pb^{2+}-induced Cleavage Detected

- Old solution of Pb^{2+}. The action is to prepare a new solution.
- Your crystalline $Pb(OAc)_2$ is old or has moistened. The action is to buy a new bottle of solid $Pb(OAc)_2$.
- Cleavage conditions are not optimized with respect to: time of incubation, concentration of other divalent metal ions, e.g. Mg^{2+}, concentration of $Pb(OAc)_2$. The action is to optimize the conditions: increase/decrease concentration of Mg^{2+} and/or Pb^{2+}, increase the time of incubation.
- Poor quality of RNA and/or the RNA solution contains metal ions that interfere with Pb^{2+}-induced cleavage or is contaminated with metal ion chelators, e.g. EDTA. The action is to prepare a new batch of RNA.
- The pH is too low. The action is to increase the pH.

13.4.2
Complete Degradation of the RNA

- Too high concentration of Pb^{2+}. The action is to decrease $[Pb^{2+}]$ and/or time of incubation.

- Time of incubation is too long. The action is to decrease the time and/or to decrease $[Pb^{2+}]$.
- The pH of the reaction is too high. The action is to lower pH.
- Contamination of your solutions with RNase. The action is to change all solutions (from experience, the RNase-free water is usually the candidate that is most often contaminated).

Acknowledgments

We thank our colleagues and, in particular, Mr M. Lindell for critical reading of the section of the chapter concerning Pb^{2+}-induced cleavage *in vivo*. The ongoing research in the laboratories of L. A. K. and J. C. are supported by the Swedish Research Council (L. A. K.), the Wallenberg Foundation (L. A. K.), the Swedish Foundation for Strategic Research (L. A. K.) and the Polish Committee for Scientific Research grant no 6P04B 01720 (J. C.).

References

1 C. WERNER, B. KREBS, G. KEITH, G. DIRHEIMER, *Biochim. Biophys. Acta* **1976**, *432*, 161–175.
2 J. R. SAMPSON, F. X. SULLIVAN, L. S. BEHLEN, A. B. DIRENZO, O. C. UHLENBECK, *Cold Spring Harbor Symp. Quant. Biol.* **1987**, *52*, 267–275.
3 W. J. KRZYŻOSIAK, T. MARCINIEC, M. WIEWIÓROWSKI, P. ROMBY, J. EBEL, R. GIEGÉ, *Biochemistry* **1988**, *27*, 5771–5777.
4 B. F. RORDORF, D. R. KEARNS, *Biopolymers* **1976**, *15*, 1491–1504.
5 J. CIESIOŁKA, T. MARCINIEC, W. J. KRZYŻOSIAK, *Eur. J. Biochem.* **1989**, *182*, 445–450.
6 J. WRZESINSKI, D. MICHAŁOWSKI, J. CIESIOŁKA, W. J. KRZYŻOSIAK, *FEBS Lett.* **1995**, *374*, 62–68.
7 W. WINTERMEYER, G. ZACHAU, *Biochim. Biophys. Acta* **1973**, *299*, 82–90.
8 T. MARCINIEC, J. CIESIOŁKA, J. WRZESINSKI, W. J. KRZYŻOSIAK, *Acta Biochim. Polon.* **1989**, *36*, 115–122.
9 R. S. BROWN, B. E. HINGERTY, J. C. DEWAN, A. KLUG, *Nature* **1983**, *303*, 543–546.
10 J. R. RUBIN, M. SUNDARALINGAM, *J. Biomol. Struct. Dyn.* **1983**, *1*, 639–646.
11 P. GÓRNICKI, F. BAUDIN, P. ROMBY, M. WIEWIÓROWSKI, W. J. KRZYŻOSIAK, J. P. EBEL, C. EHRESMANN, B. EHRESMANN, *J. Biomol. Struct. Dyn.* **1989**, *6*, 971–984.
12 C. BRUNEL, P. ROMBY, E. WESTHOF, C. EHRESMANN, B. EHRESMANN, *J. Mol. Biol.* **1991**, *221*, 293–308.
13 J. CIESIOŁKA, S. LORENZ, V. A. ERDMANN, *Eur. J. Biochem.* **1992**, *204*, 575–581.
14 J. CIESIOŁKA, S. LORENZ, V. A. ERDMANN, *Eur. J. Biochem.* **1992**, *204*, 583–589.
15 L. JOVINE, S. DJORDJEVIC, D. RHODES, *J. Mol. Biol.* **2000**, *301*, 401–414.
16 J. CIESIOŁKA, D. MICHAŁOWSKI, J. WRZESINSKI, J. KRAJEWSKI, W. J. KRZYŻOSIAK, *J. Mol. Biol.* **1998**, *275*, 211–220.
17 R. S. BROWN, J. C. DEWAN, A. KLUG, *Biochemistry* **1985**, *24*, 4785–4801.
18 T. A. STEITZ, J. A. STEITZ, *Proc. Natl. Acad. Sci. USA* **1993**, *90*, 6498–6502.
19 B. W. PONTIUS, W. B. LOTT, P. H. VON HIPPEL, *Proc. Natl. Acad. Sci. USA* **1997**, *94*, 2290–2294.
20 S. MIKKOLA, U. KAUKINEN, H. LÖNNBERG, *Cell Biochem. Biophys.* **2001**, *34*, 95–119.

21 G. M. EMILSSON, S. NAKAMURA, A. ROTH, R. R. BREAKER, *RNA* **2003**, *9*, 907–918.
22 D. LABUDA, K. NICOGHOSIAN, R. CEDERGREN, *J. Biol. Chem.* **1985**, *260*, 1103–1107.
23 M. BRÄNNVALL, N. E. MIKKELSEN, L. A. KIRSEBOM, *Nucleic Acids Res.* **2001**, *29*, 1426–1432.
24 A. FLYNN-CHARLEBOIS, N. LEE, H. SUGA, *Biochemistry* **2001**, *40*, 13623–13632.
25 N. M. KAYE, N. H. ZAHLER, E. L. CHRISTIAN, M. E. HARRIS, *J. Mol. Biol.* **2002**, *324*, 429–442.
26 R. K. SIGEL, A. VAIDYA, A. M. PYLE, *Nat. Struct. Biol.* **2000**, *7*, 1111–1116.
27 J. CIESIOŁKA, W. J. KRZYŻOSIAK, *Biochem. Mol. Biol. Int.* **1996**, *39*, 319–328.
28 J. G. MATTSSON, S. G. SVÄRD, L. A. KIRSEBOM, *J. Mol. Biol.* **1994**, *241*, 1–6.
29 S. G. SVÄRD, J. G. MATTSSON, K. E. JOHANSSON, L. A. KIRSEBOM, *Mol. Microbiol.* **1994**, *11*, 849–859.
30 J. CIESIOLKA, M. YARUS, *RNA* **1996**, *2*, 785–793.
31 N. E. MIKKELSEN, K. JOHANSSON, A. VIRTANEN, L. A. KIRSEBOM, *Nat. Struct. Biol.* **2001**, *8*, 510–514.
32 S. T. WALLACE, R. SCHROEDER, *RNA* **1998**, *4*, 112–123.
33 M. LINDELL, P. ROMBY, E. G. WAGNER, *RNA* **2002**, *8*, 534–541.
34 S. ALTMAN, L. A. KIRSEBOM, RNase P, in: *The RNA World*, 2nd edn, R. F. GESTELAND, T. R. CECH, J. F. ATKINS (eds), Cold Spring Harbor Laboratory Press, Cold Spring Harbor, NY, **1999**.
35 S. KAZAKOV, S. ALTMAN, *Proc. Natl. Acad. Sci. USA* **1991**, *88*, 9193–9197.
36 J. CIESIOŁKA, Metal ion-induced cleavages in probing of RNA structure, in: *RNA Biochemistry and Biotechnology*, J. BARCISZEWSKI, B. F. C. CLARK (eds), Kluwer, Dordrecht, **1999**.
37 E. J. MAGLOTT, G. D. GLICK, *Nucleic Acids Res.* **1998**, *26*, 301–308.
38 J. ROGERS, A. H. CHANG, U. VON AHSEN, R. SCHROEDER, J. DAVIES, *J. Mol. Biol.* **1996**, *259*, 916–925.
39 B. STREICHER, U. VON AHSEN, R. SCHROEDER, *Nucleic Acids Res.* **1993**, *21*, 311–317.
40 D. WINTER, N. POLACEK, I. HALAMA, B. STREICHER, A. BARTA, *Nucleic Acids Res.* **1997**, *25*, 1817–1824.
41 J. CIESIOLKA, J. GORSKI, M. YARUS, *RNA* **1995**, *1*, 538–550.
42 J. CIESIOŁKA, J. WRZESINSKI, P. GÓRNICKI, J. PODKOWINSKI, W. J. KRZYŻOSIAK, *Eur. J. Biochem.* **1989**, *186*, 71–77.
43 T. MARCINIEC, J. CIESIOŁKA, J. WRZESINSKI, W. J. KRZYŻOSIAK, *FEBS Lett.* **1989**, *243*, 293–298.
44 D. MICHAŁOWSKI, J. WRZESINSKI, J. CIESIOŁKA, W. J. KRZYŻOSIAK, *Biochimie* **1996**, *78*, 131–138.
45 D. MICHAŁOWSKI, J. WRZESINSKI, W. J. KRZYŻOSIAK, *Biochemistry* **1996**, *35*, 10727–10734.
46 D. A. LAFONTAINE, S. ANANVORANICH, J. P. PERREAULT, *Nucleic Acids Res.* **1999**, *27*, 3236–3243.
47 M. MATYSIAK, J. WRZESINSKI, J. CIESIOŁKA, *J. Mol. Biol.* **1999**, *291*, 282–294.
48 B. STREICHER, E. WESTHOF, R. SCHROEDER, *EMBO J.* **1996**, *15*, 2556–2564.
49 M. HERTWECK, M. W. MUELLER, *Eur. J. Biochem.* **2001**, *268*, 4610–4620.
50 L. BEHLEN, J. R. SAMPSON, A. B. DIRENZO, O. C. UHLENBECK, *Biochemistry* **1990**, *29*, 2515–2523.
51 K. ZITO, A. HÜTTENHOFER, N. R. PACE, *Nucleic Acids Res.* **1993**, *21*, 5916–5920.
52 J. CIESIOLKA, W.-D. HARDT, J. SCHLEGL, V. A. ERDMANN, R. K. HARTMANN, *Eur. J. Biochem.* **1994**, *219*, 49–56.
53 K. N. NOBLES, C. S. YARIAN, G. LIU, R. H. GUENTHER, P. F. AGRIS, *Nucleic Acids Res.* **2002**, *30*, 4751–4760.
54 T. PAN, O. C. UHLENBECK, *Biochemistry* **1992**, *31*, 3887–3895.
55 G. LENTZEN, H. MOINE, C. H. EHRESMANN, B. EHRESMANN, W. WINTERMEYER, *RNA* **1996**, *2*, 244–253.
56 M. NAPIERAŁA, W. J. KRZYŻOSIAK, *J. Biol. Chem.* **1997**, *272*, 31079–31085.
57 A. JASINSKA, G. MICHLEWSKI, M. DE MEZER, K. SOBCZAK, P. KOZŁOWSKI, M. NAPIERAŁA, W. J. KRZYŻOSIAK, *Nucleic Acids Res.* **2003**, *31*, 5463–5468.
58 K. SOBCZAK, M. DE MEZER, G.

Michlewski, J. Krol, W. J. Krzyżosiak, *Nucleic Acids Res.* **2003**, *31*, 5469–5482.

59 A. Tallsjö, S. G. Svärd, J. Kufel, L. A. Kirsebom, *Nucleic Acids Res.* **1993**, *21*, 3927–3933.

60 J. Schlegl, V. Gegout, B. Schläger, M. W. Hentze, E. Westhof, C. H. Ehresmann, B. Ehresmann, P. Romby, *RNA* **1997**, *3*, 1159–1172.

61 D. E. Otzen, J. Barciszewski, B. F. C. Clark, *Biochem. Mol. Biol. Int.* **1993**, *31*, 95–103.

62 S. Agrawal, D. Gupta, S. K. Panda, *Virology* **2001**, *282*, 87–101.

63 W.-D. Hardt, J. Schlegl, V. A. Erdmann, R. K. Hartmann, *Biochemistry* **1993**, *32*, 13046–13053.

64 W.-D. Hardt, J. Schlegl, V. A. Erdmann, R. K. Hartmann, *J. Mol. Biol.* **1995**, *247*, 161–172.

65 P. Burgstaller, M. Kochoyan, M. Famulok, *Nucleic Acids Res.* **1995**, *23*, 4769–4776.

66 M. G. Wallis, B. Streicher, H. Wank, U. von Ahsen, E. Clodi, S. T. Wallace, M. Famulok, R. Schroeder, *Chem. Biol.* **1997**, *4*, 357–366.

67 C. Berens, A. Thain, R. Schroeder, *Bioorg. Med. Chem.* **2001**, *9*, 2549–2556.

68 N. E. Mikkelsen, M. Brännvall, A. Virtanen, L. A. Kirsebom, *Proc. Natl. Acad. Sci. USA* **1999**, *96*, 6155–6160.

69 S. R. Kirk, Y. Tor, *Bioorg. Med. Chem.* **1999**, *7*, 1979–1991.

70 M. Olejniczak, Z. Gdaniec, A. Fischer, T. Grabarkiewicz, L. Bielecki, R. W. Adamiak, *Nucleic Acids Res.* **2002**, *30*, 4241–4249.

71 Z. Szweykowska-Kulinska, J. Krajewski, K. Wypijewski, *Biochim. Biophys. Acta* **1995**, *1264*, 87–92.

72 I. Majerfeld, M. Yarus, *RNA* **1998**, *4*, 471–478.

73 M. Welch, I. Majerfeld, M. Yarus, *Biochemistry* **1998**, *36*, 6614–6623.

74 H. Y. Deng, J. Termini, *Biochemistry* **1992**, *31*, 10518–10528.

75 C. Baron, E. Westhof, A. Böck, R. Giegé, *J. Mol. Biol.* **1993**, *231*, 247–292.

76 V. M. Perreau, G. Keith, W. M. Holmes, A. Przykorska, M. A. Santos, M. F. Tuite, *J. Mol. Biol.* **1999**, *293*, 1039–1053.

77 J. Ciesiołka, T. Marciniec, P. Dziedzic, W. J. Krzyżosiak, M. Wiewiórowski, in: *Biophosphates and their Analogues: Synthesis, Structure, Metabolism and Activity*, K. S. Bruzik, W. J. Stec (eds), Elsevier, Amsterdam, **1987**.

78 M. Matsuo, T. Yokogawa, K. Nishikawa, K. Watanabe, N. Okada, *J. Biol. Chem.* **1995**, *270*, 10097–10104.

79 E. Zietkiewicz, J. Ciesiołka, W. J. Krzyżosiak, R. Słomski, in: *Nuclear Structure and Function*, J. R. Harris, J. B. Zbarsky (eds), Plenum Press, New York, **1990**.

80 B. Felden, H. Himeno, A. Muto, J. P. McCutcheon, J. F. Atkins, R. F. Gesteland, *RNA* **1997**, *3*, 89–103.

81 R. Walczak, E. Westhof, P. Carbon, A. Krol, *RNA* **1996**, *2*, 367–379.

82 R. Walczak, P. Carbon, A. Krol, *RNA* **1998**, *4*, 74–84.

83 K. Sobczak, W. J. Krzyżosiak, *J. Biol. Chem.* **2002**, *277*, 17349–17358.

84 M. Hemmings-Mieszczak, G. Steger, T. Hohn, *J. Mol. Biol.* **1997**, *267*, 1075–1088.

85 S. Dorner, A. Barta, *Biol. Chem.* **1999**, *380*, 243–251.

14
In Vivo Determination of RNA Structure by Dimethylsulfate

Christina Waldsich and Renée Schroeder

14.1
Introduction

Considerable progress has been made in the past years in elucidating RNA structure, its folding pathways and the functional roles RNA molecules play in cellular processes. Despite the wealth of insights we have obtained about structure and folding, there is a significant drawback as our knowledge is so far predominately based on, and limited to, biochemical and biophysical analyses of RNA molecules performed *in vitro* [1]. However, the *in vitro* folding conditions significantly contrast the intracellular environment. For example, it is well known that many catalytic RNAs, which usually function at non-physiological reaction conditions *in vitro*, associate with protein cofactors *in vivo*, which are thought to stabilize them [2]. It is therefore essential to extend our understanding of RNA structure and function by studying those molecules within cells [3].

The methodologies useful and suitable for studying RNA structure *in vivo* are limited. Structural probing with dimethylsulfate (DMS), which proved to be a powerful tool *in vitro* [4], has so far been the main chemical used to analyze RNA *in vivo*. DMS is a base-specific probe that modifies, in addition to the N^7 position of guanines, the Watson–Crick positions N^1 of adenines and N^3 of cytidines. The modified sites can be mapped by primer extension and subsequently compiled into a pattern profile of nucleotides protected from or accessible to DMS [4–6]. Notably, certain uridines and guanines are occasionally stabilized in an enol-tautomer due to a specific local environment, and are therefore reactive to DMS at their N^3 or N^1 positions. Also, it has to be kept in mind that naturally modified nucleotides like m^7G in rRNAs could occur in your RNA of interest. A protection from DMS modification can result from base pairing, but also from an interaction with a protein, while accessibility indicates that those residues (at least their N^1, N^3 or N^7 positions) do not participate in any intra- or intermolecular contacts. Thus, DMS modification can be employed to determine RNA structure and folding as well as to study RNA–protein interactions and their associated conformational changes in living cells. As DMS easily and rapidly penetrates cells, this method can be applied

Handbook of RNA Biochemistry. Edited by R. K. Hartmann, A. Bindereif, A. Schön, E. Westhof
Copyright © 2005 WILEY-VCH Verlag GmbH & Co. KGaA, Weinheim
ISBN: 3-527-30826-1

to various organisms such as bacteria both Gram-negative [7, 8] and Gram-positive [9], protozoa [10, 11], yeast [12], and plants [13].

However, there are significant limitations to this method that have to be considered to allow a correct interpretation of the results. First, the method hinges on a few out of many functional groups of the RNA nucleotide that can participate in interactions. Thus, you can only determine RNA structure and conformational changes or interactions with proteins that involve those base-functional groups. Additionally, protection from DMS modification observed due to the presence of a protein does not necessarily indicate a physical interaction. Specific binding of a protein to a RNA often leads to a structural stabilization and, as a consequence, to a protection from DMS modification [14]. However, the major "problem" is that the obtained DMS modification pattern is averaged over the entire RNA population and over time. The lack of time resolution is especially problematic when it comes to studying *in vivo* folding and conformational changes, which are time-related events, and thus these questions are rendered more difficult to be addressed within cells. Secondly, your RNA of interest does not necessarily represent a single population, but might be partitioned among distinct species leading to mixed RNA populations. In other words, you have to be aware of what you are looking at. It is important to be able to differentiate, for example, between folded versus unfolded molecules, naked RNA versus RNA–protein particles or spliced versus precursor RNA in order to assign the modification pattern and its concomitant interpretation to a specific population.

In order to prevent fundamental pitfalls it is of great importance to check whether the modified RNA is still functional. Thus, we recommend incorporating the *in vivo* DMS modification step into a well-established experimental procedure that in the end allows you to test the activity of your RNA. For example, we incorporated the modification step into our *in vivo* splicing analysis procedure. As the splicing efficiency and RNA levels were not affected by DMS treatment of cells (no change compared to non-treated cells), we were confident that the RNA we were analyzing was in good condition [15].

14.2
Description of Method

The methodology of *in vivo* DMS structural probing of RNA described in here has been optimized for *Escherichia coli* and has mainly been adjusted from [11].

14.2.1
Cell Growth and *In Vivo* DMS Modification

For the successful application of this method it is important to standardize the way of growing the bacterial cell cultures as well as their DMS treatment [14]. Note that DMS is very toxic (for information, see http://www.state.nj.us/health/eoh/rtkweb/

0768.pdf), therefore you have to take precautions, such as working in a hood as well as wearing gloves and a lab coat, when working with this harmful reagent.

(1) Grow a cell culture of at least 100 ml to an $OD_{600\,nm}$ of 0.2–0.3. Then harvest the cells (2 × 50 ml) by centrifuging at 5000 r.p.m. in SS34 Sorval tubes at 4 °C for 5 min. Discard the supernatant and resuspend the cell pellet in 1 ml ice cold TM buffer (10 mM Tris–HCl, pH 7.5, 10 mM $MgCl_2$).

(2) Partition the pellet into two equal samples and add DMS to a final concentration of 150 mM to one of the two samples and vortex briefly. Incubate the cells with DMS for one minute and afterwards add β-mercaptoethanol to a final concentration of 0.7 M in order to quench the DMS. Vortex strongly! Centrifuge the tubes immediately at 6000 r.p.m. for 2 min in an Eppendorf tube. After careful removal of the supernatant, freeze the pellet at −80 °C until you proceed with the RNA preparation, but not longer than overnight. As a control, treat the second sample equally but without adding DMS.

After the modification step it is absolutely necessary to get rid of the DMS, because it will interfere with subsequent steps such as RNA extraction (degradation) and it can lead to modification of RNA during its isolation. Making a stop control, in which you add β-mercaptoethanol before DMS, will help to determine whether DMS is sufficiently quenched by β-mercaptoethanol [5]. This will provide confidence that the observed modification did not occur during RNA extraction, but did indeed occur *in vivo*. If you have trouble in removing DMS completely then you can solubilize DMS with isoamylalcohol [5].

14.2.2
RNA Preparation

(1) The frozen cell pellet is resuspended carefully in 157 μl solution A [150 μl TE, 1.5 μl 1 M DTT, 0.75 μl RNasin (35.5 U/μl), 4 μl lysozyme (10 mg/ml), 1 μl ddH_2O]. The cell suspension is frozen in liquid nitrogen and then thawed in a room temperature water bath. These steps (freeze and thaw) have to be done 3 times.
(2) Add 20 μl Solution B [4 μl 1 M $MgOAc_2$, 3.5 μl DNase I (RNase-free) (10 U/μl), 0.1 μl RNasin (40 U/μl), 12.4 μl ddH_2O], mix gently and incubate the samples on ice for 1 h.
(3) Add 20 μl Solution C (10 μl 0.2 M acetic acid, 10 μl 10% SDS), mix gently and incubate the samples at room temperature for 5 min.
(4) Perform phenol extraction: once with 1 volume of phenol, then with PCI (phenol:chloroform:isoamylalcohol = 25:24:1) and finally with CI; for each step vortex the samples and then centrifuge at 15.000 r.p.m. at 4 °C for 5 min. Collect the upper (aqueous) phase and proceed with the next extraction step.
(5) Precipitate the RNA with 1/100 volumes 0.5 M EDTA, pH 8.0, 1/10 volumes 3 M NaOAc, pH 5.0 and 2.5 volumes ethanol abs. Freeze the samples at −80 °C for 1 h and then centrifuge at 15 000 r.p.m. at 4 °C for 30 min. Discard the

supernatant, wash the RNA pellet with 70% ethanol and dry it carefully, but briefly (1 min at 65 °C). Resuspend the pellet in an appropriate volume of ddH$_2$O.

We gained sufficient structural information from analyzing the accessibility of adenine and cytidine residues [15]. However, if you wish to determine whether the N^7 positions of G residues are modified as well, then you need to perform aniline cleavage before reverse transcription [5]. Alternatively, you can study the accessibility of guanine nucleotides using kethoxal *in vivo* [6].

14.2.3
Reverse Transcription

Primer kinase reaction

(1) Set up the reaction in a total volume of 20 µl as follows: 4 µl 5 × PNK buffer (500 mM Tris–HCl, pH 8.0, 50 mM MgCl$_2$, 35 mM DTT), 8 pmol primer with 10 pmol [γ-^{32}P]ATP and 1 µl T4 polynucleotide kinase (10 U/µl).
(2) Incubate the sample at 37 °C for 45 min.
(3) Add 1 µl 500 mM EDTA, pH 8.0. Incubate the sample at 95 °C for 1 min, put the sample immediately on ice.
(4) Precipitate the primer with 1 µl glycogen (10 µg/µl) and 2.5 volumes ethanol p.a. and 1/10 volumes 3 M NaOAc, pH 5.0. Freeze the sample at −20 °C for 30 min and then centrifuge the sample at 4 °C at 15 000 r.p.m. for 30 min. Discard the supernatant and resuspend the washed and dried pellet in an appropriate amount of ddH$_2$O.

Depending on the primer and on the assay it might be necessary to purify the oligonucleotide before labeling it and/or after the labeling reaction with [γ-^{32}P]ATP. In general, newly synthesized oligonucleotides should always be purified prior to use.

Annealing reaction

(1) Combine 2.5 µl of *in vivo* isolated RNA (16 µg/µl) with 1 µl labeled primer (50 000 c.p.m.) and 1 µl 4.5 × hybridization buffer (225 mM K-HEPES, pH 7.0, 450 mM KCl).
(2) Incubate the sample at 90 °C for 1 min.
(3) Subsequently transfer the hot water into another glass box and let the sample cool down to 42 °C.

Extension of the primer

(1) Add 2.2 µl extension mix (0.6 µl 10 × extension buffer (1.3 M Tris–HCl, pH 8.0, 100 mM MgCl$_2$, 100 mM DTT), 0.3 µl nucleotide mix (2.5 mM each dNTP),

1 μl AMV reverse transcriptase (4 U/μl) and 1 μl ddH$_2$O to the 4.5 μl annealing reaction.
(2) Incubate the samples in a water bath at 42 °C for 1 h.
(3) Add 1.5 μl 1 M NaOH in order to degrade the RNA and incubate the sample for another hour at 42 °C. Then add 1.5 μl 1 M HCl in order to neutralize the pH.
(4) Precipitate with 1/10 volumes 500 mM EDTA, pH 8.0, 1 volume 3 M NaOAc, pH 5.0, 3 volumes ethanol p.a. Freeze the sample at −20 °C for 1 h and then centrifuge the sample at 4 °C at full speed for 30 min. Discard the supernatant and resuspend the dried pellet in 10 μl loading buffer [7 M urea, 0.25% bromphenol blue, 0.25% xylene cyanol in 1 × TBE (0.089 M Tris base, 0.089 M boric acid, 2 mM EDTA)]. Separate the extension products on an 8% PAA gel.

For obtaining sequencing ladders proceed as described above, but add in addition to the extension mix 1.5 μl of the appropriate 1 mM ddNTP solution to the reaction. If you have difficulties in generating satisfactory sequencing ladders, optimize the ddNTP concentrations. Usually the A and C lanes are sufficient for orientation along the molecule.

14.3
Evidence for Protein-induced Conformational Changes within RNA *In Vivo*

The *in vivo* DMS modification method described above was used to study the mode of action of the StpA protein in *E. coli* cells. StpA was shown to promote folding of the group I intron containing pre-mRNA of the thymidylate synthase (*td*) gene [16, 17]. First, the modification pattern of the *in vivo* DMS treated intron RNA was used to visualize the *in vivo* folding state of the *td* intron (Fig. 14.1). Importantly, we concluded from this DMS modification pattern that the secondary structure model, which was derived from phylogeny and biochemical data obtained *in vitro*, accurately describes the structure of the *td* group I intron *in vivo* [15]. We then addressed the question how the RNA chaperone StpA rescues folding of the *td* pre-mRNA. For this purpose we determined the DMS modification pattern of *td* RNA in the absence and presence of StpA, and compared it to the *td* RNA in the presence of the specific RNA-binding protein Cyt-18 (Fig. 14.2). In the presence of StpA, residues belonging to tertiary structure elements become more accessible to DMS. In contrast, the presence of Cyt-18 leads to a protection of bases involved in tertiary structure formation. Thus, StpA, a protein with RNA chaperone activity, and Cyt-18, a specific RNA binding protein, have opposite effects on the intron RNA structure *in vivo*.

From these results we concluded that StpA leads to a general loosening of the *td* group I intron RNA structure, while Cyt-18 contributes to the overall compactness of the RNA. In brief, using *in vivo* DMS modification we have been able to provide first evidence for protein induced conformational changes within a catalytic RNA *in vivo* and gained first insights into the mechanism of action of an RNA chaperone.

234 | *14 In Vivo Determination of RNA Structure by Dimethylsulfate*

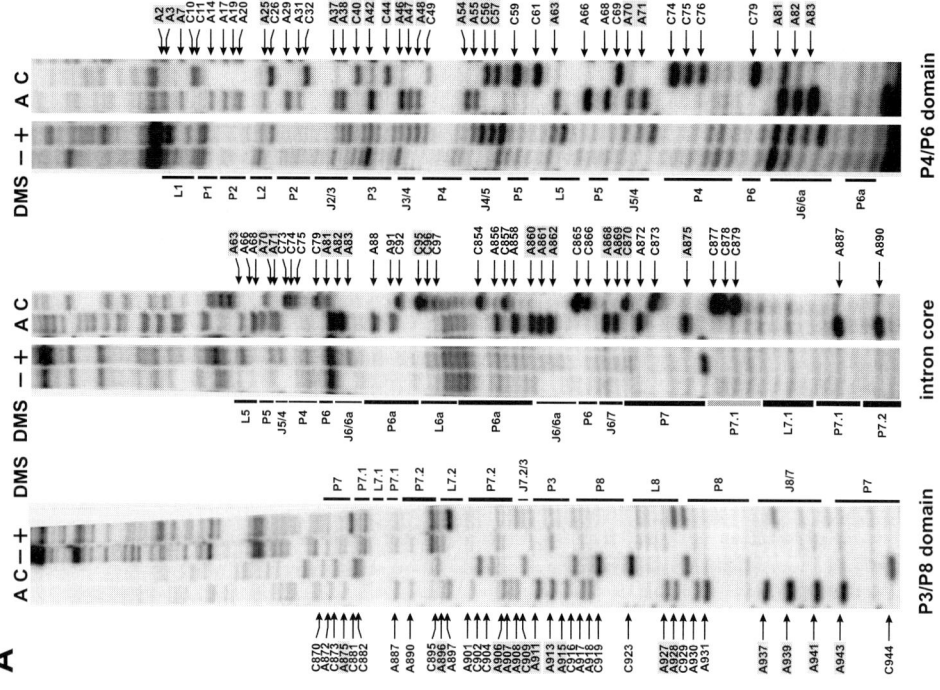

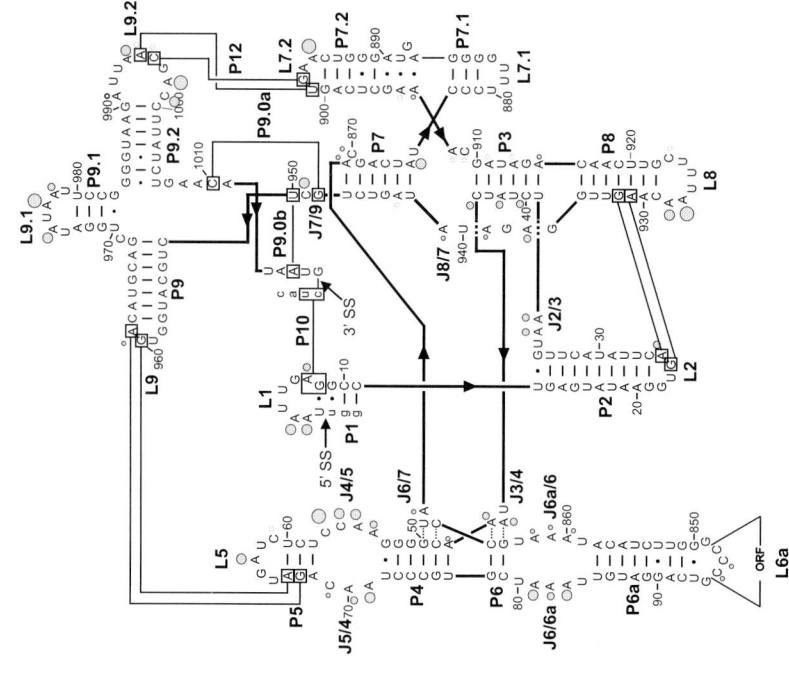

Fig. 14.1

14.4 Troubleshooting

The most likely problems that will occur when performing *in vivo* DMS probing of your RNA of interest is over-modification and RNA degradation. For this reason, we recommend that you carefully determine the optimal concentration of DMS and the incubation time. Both parameters highly depend on the intracellular levels of RNA. The aim is to achieve about one modification per molecule. Thus, reverse transcription from an untreated control as well as from DMS-modified RNA should result in a comparable amount of full-length extension products. If primer extension of DMS-treated RNA runs off a lot earlier than the untreated control, the RNA is probably over-modified. On the other hand it might happen that you observe poor modification. Thus, it is advisable to buy fresh DMS every couple of months (about every 6 months to 1 year). DMS is usually a colorless solution, which becomes more and more yellow the older the solution is. As far as RNA degradation is concerned, there are typically two main reasons for it. First, if DMS is not removed completely before RNA extraction is performed, this results in a very low yield of isolated RNA (proceed as described in the method description). Second, strong stops in the reverse transcription control, which is obtained from untreated RNA, are indicative for contamination with the pancreatic nuclease RNase A that prefers UpA sites for cleavage. In this case you should try to raise the amount of RNase inhibitor. In addition, it is important to note that *in vivo* isolated RNA is not very stable and thus the best results for reverse transcription are obtained within the first 2–3 weeks after the extraction. Potential pitfalls for reverse transcription are the choice of reverse transcriptase, since every reverse transcriptase does not recognize and stop at methylated N^1 of As and N^3 of Cs. We highly recommend using AMV reverse transcriptase. Good and specific priming is typically observed for primers of 18 nt in length. Nevertheless, it is advisable to check the T_m of the primers, which should be above the primer extension temperature (42 °C). If primers do not label or prime efficiently, this might be due to the formation of competing structures within the primer. In that case you should redesign the primer. If the primer extension is not satisfactory, you should try to increase the levels of target RNA and sometimes it is also helpful to optimize the KCl concentration for annealing reaction. A very common phenomenon is that a primer is not significantly extended but there occur very strong reverse transcrip-

Fig. 14.1. DMS modification of the *td* intron *in vivo*. (a) Intron residues accessible to DMS are displayed in these representative gels. Boxed nucleotides correspond to positions within the intron, which are modified by DMS. The P3–P8 domain of the intron core (left panel), the center of the intron core covering the P7 stem as well as the P6–P6a element (middle panel) and the P4–P6 domain of the intron core and the stem–loops P1–P2 (right panel) are shown. A and C denote the sequencing lanes. (B) Summary of the *td* intron residues modified by DMS *in vivo*. Modified sites are indicated by dots. The size of the dots correlates with the relative modification intensities. The largest dot corresponds to the highest modification intensity.

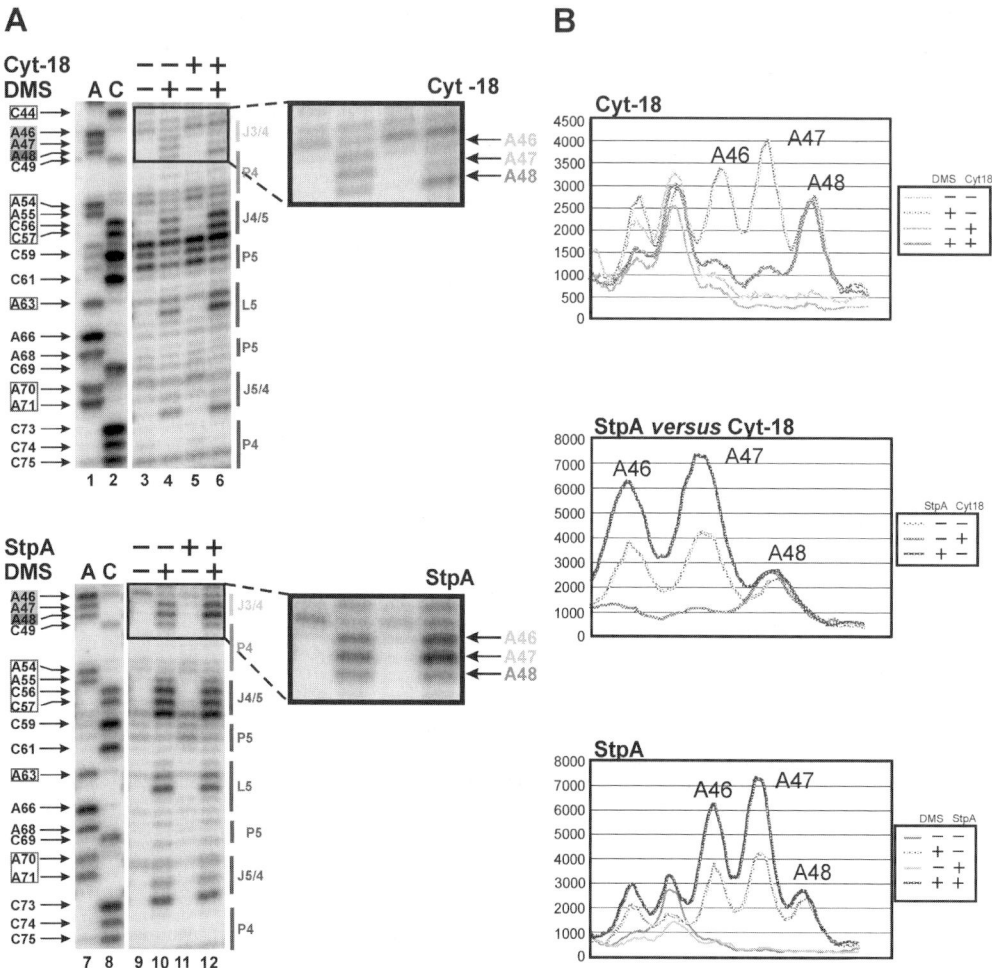

Fig. 14.2. StpA and Cyt-18 induce structural changes in the base triple interactions between adenines in J3/4 and stem P6. Changes in the DMS modification pattern of the td intron in vivo due to the presence of Cyt-18 is shown in the upper panel or due to the presence of StpA in the lower panel. (A) The P4/P5 domain is shown in these representative gels. Numbered nucleotides, which are highlighted by boxes at the left of the gel, are modified by DMS. The gel part boxed in black is outlined to point out the different effect of Cyt-18 versus StpA on the residues A46 and A47 in J3/4. The sequencing lanes are labeled with A and C. In the presence of Cyt-18 the amount of td RNA is increased in the cells as reflected by the increase of non-specific stops in untreated samples (cf. lanes 3 and 5) as well as by the increased modification intensity of residues A55, C56, C57 and A63 in lanes 4 and 6. These differences in the td RNA amount were normalized. (B) PhosphoImager quantification (right panel) of the outlined gel segments in the presence of Cyt-18 or StpA. The opposite effects of these proteins on the accessibility of the two adenines in J3/4 to DMS are summarized in the middle panel.

tion stops at the beginning of the extension. Usually it is sufficient to overcome this problem by setting the primer a few nucleotides more upstream or downstream of the original hybridization site. In summary, the best results for reverse transcription are obtained using clean and freshly labeled primers in conjunction with newly prepared RNA.

References

1 BRION, P., E. WESTHOF. *Annu. Rev. Biophys. Biomol. Struct.* **1997**, *26*, 113–137.
2 WEEKS, K. M. *Curr. Opin. Struct. Biol.* **1997**, *7*, 336–342.
3 SCHROEDER, R., R. GROSSBERGER, A. PICHLER, C. WALDSICH. *Curr. Opin. Struct. Biol.* **2002**, *12*, 296–300.
4 EHRESMANN, C., F. BAUDIN, M. MOUGEL, P. ROMBY, J. P. EBEL, B. EHRESMANN. *Nucleic Acids Res.* **1987**, *15*, 9109–9128.
5 WELLS, S. E., J. M. HUGHES, A. H. IGEL, M. ARES, JR. *Methods Enzymol.* **2000**, *318*, 479–493.
6 BALZER, M., R. WAGNER. *Anal. Biochem.* **1998**, *256*, 240–242.
7 CLIMIE, S. C., J. D. FRIESEN. *J. Biol. Chem.* **1988**, *263*, 15166–15175.
8 MOAZED, D., J. M. ROBERTSON, H. F. NOLLER. *Nature* **1988**, *334*, 362–364.
9 MAYFORD, M., B. WEISBLUM. *EMBO J.* **1989**, *8*, 4307–4314.
10 HARRIS, K. A., JR, D. M. CROTHERS, E. ULLU. *RNA* **1995**, *1*, 351–362.
11 ZAUG, A. J., T. R. CECH. *RNA* **1995**, *1*, 363–374.
12 ARES, M., JR, A. H. IGEL. *Genes Dev.* **1990**, *4*, 2132–2145.
13 SENECOFF, J. F., R. B. MEAGHER. *Plant Mol. Biol.* **1992**, *18*, 219–234.
14 WALDSICH, C., R. GROSSBERGER, R. SCHROEDER. *Genes Dev.* **2002**, *16*, 2300–2312.
15 WALDSICH, C., B. MASQUIDA, E. WESTHOF, R. SCHROEDER. *EMBO J.* **2002**, *19*, 5281–5291.
16 SEMRAD, K., R. SCHROEDER. *Genes Dev.* **1998**, *12*, 1327–1337.
17 CLODI, E., K. SEMRAD, R. SCHROEDER. *EMBO J.* **1999**, *18*, 3776–3782.

15
Probing Structure and Binding Sites on RNA by Fenton Cleavage

Gesine Bauer and Christian Berens

15.1
Introduction

In the 20 years that have followed the discovery of RNA-based catalysis, many novel biological functions and catalytic activities of RNA have been either discovered *in vivo* or obtained through *in vitro* selection techniques [1]. Roughly, these can be divided into two groups. While the first, exemplified by mRNA, snoRNA, guide RNA and siRNA, utilizes sequence-dependent Watson–Crick interactions, the activities exerted by the second group, containing ribozymes, aptamers and riboswitches, are based on their three-dimensional structures. Knowledge of these structures and how they are formed is a prerequisite to understanding how these RNAs function mechanistically.

As X-ray crystallography of RNA molecules has proven difficult and since many interesting RNA molecules are still too large for NMR analysis, a profusion of RNA probing methods have been developed for structural analysis [2–4]. One very versatile method for analyzing RNA is probing with hydroxyl radicals. They are the smallest molecule species used for chemical probing, cleave nucleic acids with little or no sequence specificity [5, 6], and a significant secondary structure preference has not been observed in radical-induced cleavage of single- and double-stranded forms of RNA and DNA [7].

Hydroxyl radicals are generated physically by radiolysis of water using synchrotron X-ray beams or, more often, chemically by the reduction of peroxo-groups with Fe^{2+} in the so-called Fenton reaction [8]. Like most transition metals, iron has more than one oxidation state besides the ground state and its valence electrons may be unpaired allowing one-electron redox reactions [9]. As such, Fe^{2+} reacts with H_2O_2 (or other peroxo molecules like peroxonitrous acid) to generate short-lived, highly reactive hydroxyl radicals. These cleave the bases of a nucleic acid, its phosphodiester backbone and also peptide bonds in spatial proximity of Fe^{2+}. Sodium ascorbate is often added to the reaction mixture in order to reduce Fe^{3+} to Fe^{2+}, thereby establishing a catalytic cycle and permitting low, micromolar, concentrations of Fe^{2+} to be effective in cleaving the substrates. Consequently,

Handbook of RNA Biochemistry. Edited by R. K. Hartmann, A. Bindereif, A. Schön, E. Westhof
Copyright © 2005 WILEY-VCH Verlag GmbH & Co. KGaA, Weinheim
ISBN: 3-527-30826-1

hydroxyl-radical-based probing methods have been widely used for structural analysis of RNA and also as a tool to study interactions of nucleic acids with proteins or other ligands.

The classical interaction study is a footprinting experiment. The presence of the interaction partner protects the nucleic acid at the site bound from cleavage by the hydroxyl radicals. Detailed protocols for footprinting protein–DNA and protein–RNA complexes are given in [10, 11]. However, this method can also be used to identify contact sites of 16S rRNA in 30S subunits with 50S subunits [12] or to determine the structural elements of an internal ribosomal entry site that interact with a 40S ribosomal subunit [13].

The versatility of this approach was greatly extended by tethering Fe^{2+} to defined sites on proteins and RNA using the reagent 1-(p-bromoacetamidobenzyl)-EDTA (BABE), originally synthesized by Meares et al. [14, 15]. Hydroxyl radical footprints with Fe^{2+} tethered either to various ribosomal proteins or to rRNA gave important insights into the three-dimensional organization of the ribosome [16–18] that were later confirmed by the crystal structure of the 70S ribosome (reviewed in [19]). Detailed protocols for interaction studies with hydroxyl radicals generated by Fe^{2+} either tethered to proteins or RNA have been published [20, 21].

Hydroxyl radical cleavage is also used for RNA structure analysis. Similar to the interaction analysis method described above, Fe^{2+} can be tethered to RNA to induce intramolecular self-cleavage as was shown by Newcomb and Noller [22] who determined the RNA neighborhoods of specific nucleotides in the rRNA of 70S ribosomes or by Huq et al. [23] who obtained structural information on the three-dimensional fold of the HIV-1 *trans*-activation responsive region RNA.

In addition to tethered Fe^{2+}/EDTA, free Fe^{2+}/EDTA is used to identify solvent-accessible and solvent-excluded sugar moieties and, thus, aids in modeling the three-dimensional structure of an RNA [24]. Protection of tRNA bound to the ribosomal P-site from hydroxyl radical cleavage gave important hints for the mechanism of tRNA-ribosome interaction [25]. Swisher et al. presented hydroxyl radical footprints [26] demonstrating that a group II intron ribozyme has a tightly packed, solvent-inaccessible core like other large ribozymes [27, 28]. In addition, hydroxyl radical footprinting allows us to determine the relative stabilities of individual structural motifs by examining the protection pattern as a function of added Mg^{2+} or urea. Experiments like this have been done with RNase P [29] or the *Tetrahymena* LSU group I ribozyme [30]. Furthermore, synchrotron generated hydroxyl radicals have been employed successfully for time-resolved footprinting of RNA folding [31] (a detailed methods protocol is presented in [32]).

Fe^{2+} is similar to Mg^{2+} in size and coordination geometry [33] and has been used to replace the latter in proteins to map metal ion binding sites [34–36]. Catalytic RNAs either require divalent cations for achieving a stable tertiary structure and for catalysis or their activity is greatly enhanced by the presence of divalent metal ions [37]. The identification of metal ion binding sites is therefore essential for a thorough structure–function analysis of catalytic RNA. In addition to NMR studies (summarized in [38]), hydroxyl-radical-induced cleavage based on limited replacement of Mg^{2+} with Fe^{2+} provides a powerful method for the analysis of RNA–metal ion interactions [39].

Here we present two different protocols for hydroxyl radical probing of RNA. The first describes structural probing of large RNAs. Comparison of cleavage patterns obtained with Fe^{2+} in the presence and in the absence of EDTA allows to distinguish between solvent-exposed and solvent-occluded regions of the RNA and to identify metal-ion-binding sites.

The second protocol describes an interaction study that exploits the ability of Fe^{2+} to replace the Mg^{2+} ion chelated to tetracycline. A subsequent hydroxyl radical digestion can then identify the residues in proximity of the [Fe^{2+}-tetracycline] chelate. This has already been done for the tetracycline proton-antiporter TetA [40] and the tetracycline-dependent regulatory protein TetR [41]. We used hydroxyl radical cleavage of 16S rRNA to identify tetracycline-binding sites in the 70S ribosome [42].

15.2
Description of Methods

15.2.1
Fe^{2+}-mediated Cleavage of Native Group I Intron RNA

The method presented here was used to detect Mg^{2+}-binding sites in the *Tetrahymena* LSU group I intron [39]. For a successful reaction, it is important to prepare the solutions of $FeCl_2$ and H_2O_2 freshly. Sodium ascorbate can be prepared as a 10-fold stock solution and stored at $-20\ °C$. In order to mix the reagents accurately after 1 min, we recommend using a small table Microfuge (Qualitron). The appropriate reagents are added subsequently to the wall of each Eppendorf cap and then mixed simultaneously by briefly applying the centrifuge.

Hydroxyl radical cleavage

- For experiments with native RNA, take 1 µl RNA (5 pmol cold RNA, spiked with approximately 50 000 c.p.m. of the RNA labeled with ^{32}P at either the 5′ or the 3′ end) and add 1 µl of 5 × native cleavage buffer (1 × NCB: 25 mM MOPS–KOH, pH 7.0; 3 mM $MgCl_2$; 400 µM spermidine; 200 mM NaCl). Incubate the RNA for 2 min at 56 °C, followed by 3 min incubation at room temperature.
- Add 1 µl 1.25 mM $FeCl_2$ to the reaction tube, mix by centrifugation and incubate for 1 min before adding 1 µl 12.5 mM sodium ascorbate.
- After 1 min, add 1 µl of 12.5 mM H_2O_2 and mix rapidly to initiate the reaction. The final concentrations are 250 µM for Fe^{2+} and 2.5 mM for both sodium ascorbate and H_2O_2.
- Stop the cleavage reaction after 1 min by adding 1 µl 1 M thio-urea. The RNA is then precipitated with 1 µl glycogen (10 µg/µl), 1 µl sodium acetate (3 M) and 30 µl 96% (v/v) ethanol.

Gel electrophoresis

- After precipitation, resuspend the RNA in gel loading buffer [7 M urea; 0.01% (w/v) bromophenol blue and xylenecyanol each].
- Separate the cleavage products on 6–20% denaturing polyacrylamide sequencing gels.
- For obtaining sequencing ladders, carry out limited hydrolysis with RNase T1 and NaHCO$_3$ [43].

Mg^{2+} competition of Fe^{2+}-mediated cleavage

- Mix equal volumes of a 2.5 mM FeCl$_2$ solution and a Mg^{2+} stock solution. Notice that to achieve the desired Mg^{2+} concentration for competition, you have to take into account the Mg^{2+} already present in the reaction when calculating the Mg^{2+} concentration of the stock solution. For a final MgCl$_2$ concentration of 10 mM, e.g. mix 50 µl of 2.5 mM FeCl$_2$ and 50 µl of 70 mM MgCl$_2$. Then 1 µl of the Fe^{2+}/Mg^{2+} mixture is pipetted into the reaction tube and the cleavage reaction continued as above. The total reaction volume is 5 µl and a combined Mg^{2+} concentration of 10 mM is obtained by 3 mM originating from the native cleavage buffer and 7 mM resulting from the added Fe^{2+}/Mg^{2+} mixture.

Visualization of metal-ion binding sites in group I introns by Fe^{2+}-mediated Fenton cleavage

Cleavage by Fe^{2+} is observed in distinct regions of the group I intron RNA and only with native RNA (cf. lanes 7 and 8 in Fig. 15.1). It is competed by Mg^{2+} (cf. lanes 8 and 10 in Fig. 15.1) indicating that both ions interact with the same or overlapping binding sites. Lanes 8 and 12 show a comparison of the cleavage sites obtained using Fe^{2+} or Fe^{2+}/EDTA. Most of the sites cleaved by Fe^{2+} are embedded in regions protected from cleavage by Fe^{2+}/EDTA. They are, thus, located in the interior of the ribozyme where they would be expected to be if the metal ions they reflect bury phosphate oxygens [44]. In a three-dimensional model of the bacteriophage T4-derived *td* intron and in the crystal structure of the P4P6 domain of the *Tetrahymena* LSU intron [45], cleavage sites separated in secondary structure come together in three-dimensional space to form several distinct pockets (see figures 8 and 9 in [39]). There is also very good agreement between nucleotides cleaved by Fe^{2+} and nucleotides close to metal ions determined by phosphorothioate substitution [46–48] metal-hydroxyl cleavage [49], or X-ray crystallography [45]. Figure 15.2 shows cleavage sites in the hinge region of the P4P6 domain coincide nicely with a diffusely bound metal ion that was predicted from microenvironment analysis [50], but not observed in the crystal structure [45].

15.2.2
Fe^{2+}-mediated Tetracycline-directed Hydroxyl Radical Cleavage Reactions

This method describes the identification of tetracycline-binding sites on rRNA in 70S ribosomes of *Escherichia coli* [42]. Hydroxyl radical cleavage of the RNA in the

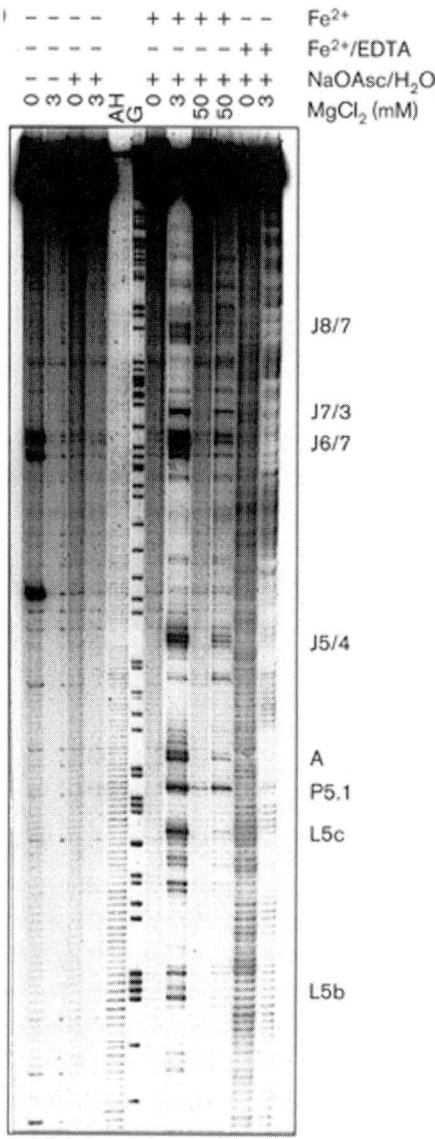

Fig. 15.1. Mapping the Fe^{2+} cleavage sites in the *Tetrahymena* LSU group I intron. Autoradiogram of a 6% denaturing polyacrylamide gel with 5'-end-labeled *Tetrahymena* L-21 RNA cleaved with 10 µM Fe^{2+} (lanes 7 and 9), 250 µM Fe^{2+} (lane 8 and 10) or with 250 µM Fe^{2+}/500 µM EDTA (lanes 11 and 12). Controls with untreated RNA (lanes 1 and 2) and in which Fe^{2+} was omitted (lanes 3 and 4), as well as competition of Fe^{2+} cleavage by 50 mM Mg^{2+} (lanes 9 and 10) are also shown. The respective final concentrations of Mg^{2+}, as well as the presence (+) or absence (−) of Fe^{2+}, sodium ascorbate and H_2O_2 are displayed above each lane. Renatured RNA is in lanes 2, 4, 8, 10 and 12; denatured RNA in lanes 1, 3, 7, 9 and 11. Secondary structure elements cleaved by Fe^{2+} are marked on the right. Sequencing markers are AH (alkaline hydrolysis) and G (RNase T1). Reprinted from [39].

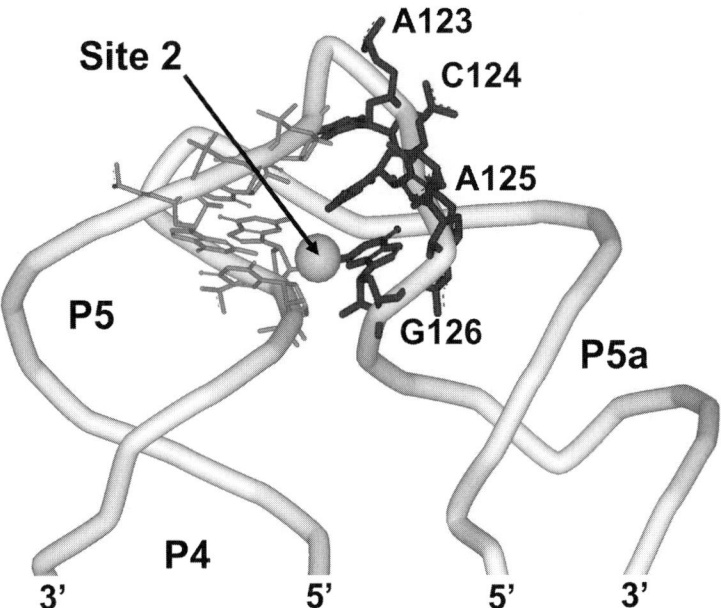

Fig. 15.2. Correlation between a computationally predicted Mg^{2+}-binding site and Fe^{2+} cleavage sites in the *Tetrahymena* P4P6 domain. The phosphodiester backbones of the two RNA strands in the hinge region of the P4P6 domain of the *Tetrahymena* LSU group I intron are shown as closed white ribbons with their polarity and the secondary structure elements indicated. Site 2 (gray sphere) is a potential diffusely bound Mg^{2+} site [50] and possible coordinating residues are displayed as thin gray sticks. Residues that are cleaved by Fe^{2+} are shown as thick black sticks. Coordinates were taken from the RCSB entry 1hr2 [60] and table 3 of [50].

vicinity of bound tetracyclines is detected by primer extension. Fe^{2+}-mediated hydroxyl radical cleavage of the 70S ribosome is carried out similar to the method described above. Tetracycline solutions have to be prepared freshly.

Hydroxyl radical cleavage

- Add 4 µl of a ribosome solution (5 pmol ribosomes in a buffer containing 5 mM $MgCl_2$) to 1 µl of a 10 × tetracycline stock solution of the final tetracycline concentration.
- After addition of 2 µl 5 × cleavage buffer (1 × CB: 25 mM MOPS–KOH, pH 7.0; 3 mM $MgCl_2$; 400 µM spermidine), incubate for 30 min at 37 °C followed by 10 min incubation at room temperature.
- Add 1 µl of 1.25 mM $FeCl_2$ to the reaction tube and mix by centrifugation.
- Incubate for 1 min before adding 1 µl of 6.25 mM sodium ascorbate.
- After 1 min, add 1 µl of 6.25 mM H_2O_2 to initiate the reaction and mix rapidly. The final concentrations are 125 µM for Fe^{2+} and 625 µM for both sodium ascor-

bate and H_2O_2 in the presence of 5 mM Mg^{2+}. The cleavage reaction is stopped after 1 min by the addition of thio-urea to a final concentration of 100 mM. The RNA is precipitated with 1 μl glycogen (10 μg/μl), 5 μl sodium acetate (3 M) and 60 μl 96% (v/v) ethanol.

Mg^{2+} competition of Fe^{2+} cleavage
To assure that Mg^{2+} and Fe^{2+} share the same or overlapping binding sites it is essential to carry out a Mg^{2+} competition experiment. The range of Fe^{2+}:Mg^{2+} ratios necessary for cleavage competition depends on the respective affinities of Fe^{2+} and Mg^{2+} to the appropriate binding site. In case of tetracycline, Fe^{2+} binds 100-fold more tightly than Mg^{2+} [41].

Mg^{2+} competition of Fe^{2+}-mediated cleavage

- Mix equal volumes of a 2.5 mM $FeCl_2$ solution and a Mg^{2+} stock solution. Notice that to achieve the desired Mg^{2+} concentration for competition, you have to take into account the Mg^{2+} already present in the reaction when calculating the Mg^{2+} concentration of the stock solution. In this case, for a final $MgCl_2$ concentration of 10 mM, e.g. mix 50 μl of 2.5 mM $FeCl_2$ and 50 μl of 140 mM $MgCl_2$. Then 1 μl of the Fe^{2+}/Mg^{2+} mixture is pipetted into the reaction tube and the cleavage reaction continued as above. The total reaction volume is 10 μl and a combined Mg^{2+} concentration of 10 mM is obtained by 3 mM originating from the native cleavage buffer and 7 mM resulting from the added Fe^{2+}/Mg^{2+} mixture.

Extraction of rRNA
The rRNA has to be extracted for the following primer extension analysis.

- Resuspend the pellet obtained after the ethanol precipitation in 200 μl ribosomal extraction buffer [REB: 0.3 M sodium acetate; 0.5% (w/v) SDS; 5 mM EDTA] at room temperature.
- In order to remove ribosomal proteins, carry out a phenolization followed by an isoamylalcohol/chloroform (1:24; v/v) treatment. Repeat this procedure twice.
- After an ethanol precipitation, resuspend the RNA in Millipore water and remove residual phenol by diethylether treatment. After a final ethanol precipitation, the RNA is resuspended in 10 μl Millipore water.

Primer extension and gel electrophoresis
Primer extension reaction and gel electrophoresis can be carried out as described in Chapter 14.

Mapping tetracycline binding to ribosomes by drug-directed Fenton cleavage of 16S rRNA
We identified three prominent Fe^{2+}-mediated cleavage sites in the 16S rRNA in the presence of tetracycline. All cleavage sites are in good agreement with published data for tetracycline from genetics [51, 52], biochemistry [53–55] and crystallogra-

phy [56, 57]. They are located in helices 29 (A1339–U1341) and 34 (C1195–A1197), and in the internal loop of helix 31 (A964–A969) (helical numbering according to [58]). Figure 15.3(B and C) shows sections of denaturing polyacrylamide gels with cleavage sites mapped to tetracycline binding site-1 which is formed by h31 and h34. According to crystal structures of 30S subunits complexed with tetracycline [56, 57], the affected bases overlap with bases within 10 Å distance of tetracycline bound to site-1 (Fig. 15.3A) and tetracycline binding site-4 which is formed by h29 (data not shown).

15.3
Comments and Troubleshooting

- Free Fe^{2+} is present in a large molar excess over the target RNA. Thus, much of the Fe^{2+} will not be bound to RNA, but will be free in solution where it can also generate hydroxyl radicals. Like Fe^{2+}/EDTA-generated hydroxyl radicals, they will cleave the target RNA non-specifically at all surface-exposed positions. This bulk cleavage will reduce the signal to noise ratio, but can be compensated by increasing the amount of target RNA in the reaction mixture. We typically used 500 ng of the group I intron RNA for cleavage with 250 µM Fe^{2+}. For cleavage of yeast tRNA^{Phe}, the final cleavage assay contained 10 µg RNA, 10 mM Mg^{2+} and 1 mM Fe^{2+} (C. Berens and R. Schroeder, unpublished). We have not performed experiments with higher RNA amounts or Fe^{2+} concentrations.
- To statistically ensure only a single cleavage event per RNA molecule, about 70% of the population should remain uncleaved after the reaction [59]. We achieve this with Fe^{2+}:Mg^{2+} ratios of 1:10 to 1:20 for the native group I introns or with very low Fe^{2+} concentrations (10 µM) for the non-folded ribozymes in the absence of divalent metal ions. Higher ratios lead to increased unspecific degradation of the RNA, lower ratios to reduced cleavage intensity as a result of the competition between Fe^{2+} and Mg^{2+} for the metal-ion binding sites. Due to the limitations on total RNA and Fe^{2+} that can be added to a reaction assay, this reduces the Mg^{2+} concentrations that can be used for Fenton cleavage to a maximum of 10–20 mM. For RNA molecules that require higher Mg^{2+} concentrations for correct folding, it should be attempted to reduce the divalent cation requirement by substituting spermine and spermidine for Mg^{2+}. These polyamines do not affect the cleavage reaction (C. Berens and R. Schroeder, unpublished).
- Within the limits described above, the native cleavage buffer and the denaturation/renaturation conditions should be adjusted to the requirements of the respective RNA molecule to be probed. For selection of buffer conditions, it is necessary to keep in mind that some buffer additives (EDTA) or reducing reagents (DTT, 2-mercaptoethanol) might scavenge radicals generated by the cleavage reaction.
- Differences in the reaction conditions for cleavage with Fe^{2+} or with Fe^{2+}/tetracycline indicate that the concentrations of the three chemical species (Fe^{2+},

A

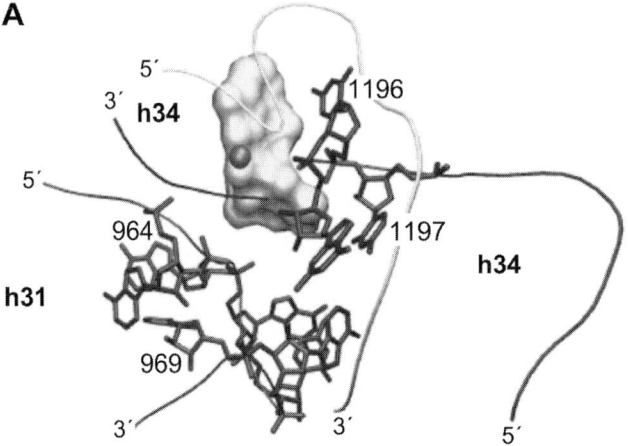

B

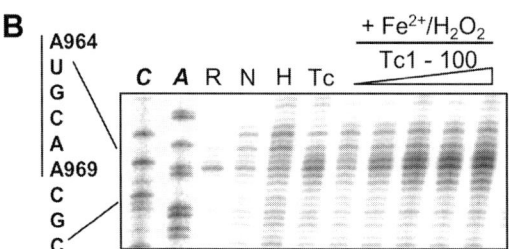

C

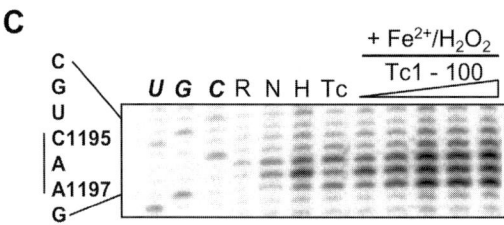

Fig. 15.3. Fe^{2+}-mediated hydroxyl radical cleavage of the 16S rRNA. (A) Surface structure of tetracycline complexed with Mg^{2+} (shown as a grey sphere) bound to the ribosomal binding site-1 which is formed by helix 34 (h34) and the internal loop of helix 31 (h31 [57]). The phosphodiester backbones of RNA strands containing bases that are attacked by hydroxyl radicals are shown as closed black ribbons, the unaffected backbone strand is shown as a light grey ribbon. Residues that are cleaved by Fe^{2+} are shown as black sticks. Bases 964, 969, 1196 and 1197 are marked for orientation. Coordinates were taken from the RCSB entry 1I97 [57]. (B) Autoradiograph of a polyacrylamide gel showing cleavage sites in the internal loop of h31. (C) Cleavage sites in h34. Lanes A, C, G, U: dideoxy sequencing lanes; R: unmodified RNA; N: control in which Fe^{2+} was omitted; H: Fe^{2+}/H_2O_2 cleaved RNA in the absence of antibiotics; Tc: unmodified RNA in the presence of 100 µM Tc and H_2O_2; Tc1–100: Fe^{2+}/H_2O_2 cleavage in the presence of 1, 3, 10, 30 and 100 µM Tc. Lines left of the sequence indicate regions of Fe^{2+}-mediated hydroxyl radical cleavage. *E. coli* 70S ribosomes were incubated with different amounts of tetracycline and treated with Fe^{2+}/H_2O_2 as described in Section 15.2. Cleavage sites were detected by primer extension and analyzed by electrophoresis on a denaturating 10% polyacrylamide gel.

sodium ascorbate and H_2O_2) might have to be varied to optimize the generation of hydroxyl radicals. We generally recommend using equimolar amounts of sodium ascorbate and H_2O_2. For naked RNA, it is not necessary to precipitate the 5′- or 3′-labeled RNA for initial evaluation and optimization of the experimental protocol. Addition of an equal volume of denaturing loading buffer and immediate electrophoresis gives data of sufficient quality. However, a purification step is still recommended for final mapping of the cleavage sites and for quantification of cleavage intensity.

- An incubation time of 1 min before the addition of sodium ascorbate and H_2O_2 was the shortest period of time in which the six reaction tubes that fit into the tabletop microcentrifuge could be manipulated easily. Extending the incubation time for Fe^{2+} and sodium ascorbate to 10 min did not lead to different cleavage patterns. The extension of the reaction time after the addition of H_2O_2 might increase cleavage intensities. However, we recommend changing the $FeCl_2$ concentration if the extent of cleavage intensity should be altered.
- Do not pre-mix the sodium ascorbate and $FeCl_2$ solutions before adding them to the RNA, as they will form a chelate-complex that effectively titrates the Fe^{2+} in the reaction mixture.
- It is necessary to perform control experiments with Fe^{2+} and H_2O_2 in the absence of tetracycline, in order to distinguish between tetracycline-mediated cleavage and cleavage caused by metal ions (either bound to specific binding pockets or diffusely associated with the surface of the RNA). Tetracycline is known to bind RNA unspecifically. This may lead to additional, unspecific cleavage sites at high concentrations. It is therefore advisable to titrate the reaction with tetracycline. In addition, one control should contain the highest amount of tetracycline used in the titration in the absence of Fe^{2+} and H_2O_2 since binding of tetracycline to RNA could cause a stop of reverse transcription. For all controls, the compound omitted is substituted by Millipore or double-distilled water.

References

1 J. A. Doudna, T. R. Cech, *Nature* **2002**, *418*, 222–228.
2 J. A. Piccirilli, J. S. Vyle, M. H. Caruthers, T. R. Cech, *Nature* **1993**, *361*, 85–88.
3 S. P. Ryder, S. A. Strobel, *J. Mol. Biol.* **1999**, *291*, 295–311.
4 C. Brunel, P. Romby, *Methods Enzymol.* **2000**, *318*, 3–21.
5 T. D. Tullius, B. A. Dombroski, M. E. Churchill, L. Kam, *Methods Enzymol.* **1987**, *155*, 537–558.
6 B. Balasubramanian, W. K. Pogozelski, T. D. Tullius, *Proc. Natl. Acad. Sci. USA* **1998**, *95*, 9738–9743.
7 D. W. Celander, T. R. Cech, *Biochemistry* **1990**, *29*, 1355–1361.
8 S. Udenfried, C. T. Clark, J. Axelrod, R. B. Brodie, *J. Biol. Chem.* **1954**, *208*, 731–739.
9 F. A. Cotton, G. Wilkinson, C. A. Murillo, M. Bochmann, R. N. Grimes, *Advanced Inorganic Chemistry*, 6th edn, Wiley, New York, **1999**.
10 P. Ansel-McKinney, L. Gehrke, in: *Analysis of mRNA Formation and Function*, J. D. Richter (ed.), Academic Press, New York, **1997**, pp. 285–303.
11 W. J. Dixon, J. J. Hayes, J. R. Levin,

M. F. Weidner, B. A. Dombroski, T. D. Tullius, *Methods Enzymol.* **1991**, *208*, 380–413.

12 C. Merryman, D. Moazed, G. Daubresse, H. F. Noller, *J. Mol. Biol.* **1999**, *285*, 107–113.

13 T. Nishiyama, H. Yamamoto, N. Shibuya, Y. Hatakeyama, A. Hachimori, T. Uchiumi, N. Nakashima, *Nucleic Acids Res.* **2003**, *31*, 2434–2442.

14 T. M. Rana, C. F. Meares, *J. Am. Chem. Soc.* **1990**, *112*, 2457–2458.

15 D. P. Greiner, R. Miyake, J. K. Moran, A. D. Jones, T. Negishi, A. Ishihama, C. F. Meares, *Bioconjug. Chem* **1997**, *8*, 44–48.

16 G. M. Heilek, R. Marusak, C. F. Meares, H. F. Noller, *Proc. Natl. Acad. Sci. USA* **1995**, *92*, 1113–1116.

17 T. Powers, H. F. Noller, *RNA* **1995**, *1*, 194–209.

18 G. M. Culver, J. H. Cate, G. Z. Yusupova, M. M. Yusupov, H. F. Noller, *Science* **1999**, *285*, 2133–2136.

19 V. Ramakrishnan, *Cell* **2002**, *108*, 557–572.

20 G. M. Culver, H. F. Noller, *Methods Enzymol.* **2000**, *318*, 461–475.

21 S. Joseph, H. F. Noller, *Methods Enzymol.* **2000**, *318*, 175–190.

22 L. F. Newcomb, H. F. Noller, *Biochemistry* **1999**, *38*, 945–951.

23 I. Huq, N. Tamilarasu, T. M. Rana, *Nucleic Acids Res.* **1999**, *27*, 1084–1093.

24 J. A. Latham, T. R. Cech, *Science* **1989**, *245*, 276–282.

25 A. Hüttenhofer, H. F. Noller, *Proc. Natl. Acad. Sci. USA* **1992**, *89*, 7851–7855.

26 J. Swisher, C. M. Duarte, L. J. Su, A. M. Pyle, *EMBO J.* **2001**, *20*, 2051–2061.

27 T. S. Heuer, P. S. Chandry, M. Belfort, D. W. Celander, T. R. Cech, *Proc. Natl. Acad. Sci. USA* **1991**, *88*, 11105–11109.

28 T. Pan, *Biochemistry* **1995**, *34*, 902–909.

29 X. Fang, T. Pan, T. R. Sosnick, *Biochemistry* **1999**, *38*, 16840–16846.

30 C. Y. Ralston, Q. He, M. Brenowitz, M. R. Chance, *Nat. Struct. Biol.* **2000**, *7*, 371–374.

31 B. Sclavi, M. Sullivan, M. R. Chance, M. Brenowitz, S. A. Woodson, *Science* **1998**, *279*, 1940–1943.

32 C. Y. Ralston, B. Sclavi, M. Sullivan, M. L. Deras, S. A. Woodson, M. R. Chance, M. Brenowitz, *Methods Enzymol.* **2000**, *317*, 353–368.

33 I. B. Brown, *Acta Crystallogr. B* **1988**, *44*, 545–553.

34 J. M. Farber, R. L. Levine, *J. Biol. Chem.* **1986**, *261*, 4574–4578.

35 C. H. Wei, W. Y. Chou, S. M. Huang, C. C. Lin, G. G. Chang, *Biochemistry* **1994**, *33*, 7931–7936.

36 J. J. Hlavaty, J. S. Benner, L. J. Hornstra, I. Schildkraut, *Biochemistry* **2000**, *39*, 3097–3105.

37 A. L. Feig, O. C. Uhlenbeck, in: *The RNA World*, 2nd edn, R. F. Gesteland, T. R. Cech, J. F. Atkins (eds), Cold Spring Harbor Laboratory Press, Cold Spring Harbor, NY, **1999**, pp. 287–319.

38 R. L. Gonzalez, I. Tinoco, *Methods Enzymol.* **2001**, *338*, 421–443.

39 C. Berens, B. Streicher, R. Schroeder, W. Hillen, *Chem. Biol.* **1998**, *5*, 163–175.

40 L. M. McMurry, M. L. Aldema-Ramos, S. B. Levy, *J. Bacteriol.* **2002**, *184*, 5113–5120.

41 N. Ettner, J. W. Metzger, T. Lederer, J. D. Hulmes, C. Kisker, W. Hinrichs, G. A. Ellestad, W. Hillen, *Biochemistry* **1995**, *34*, 22–31.

42 G. Bauer, C. Berens, S. J. Projan, W. Hillen, *J. Antimicrob. Chemother.* **2004**, *53*, 592–599.

43 H. Donis-Keller, A. M. Maxam, W. Gilbert, *Nucleic Acids Res.* **1977**, *4*, 2527–2538.

44 J. H. Cate, R. L. Hanna, J. A. Doudna, *Nat. Struct. Biol.* **1997**, *4*, 553–558.

45 J. H. Cate, J. A. Doudna, *Structure* **1996**, *4*, 1221–1229.

46 E. L. Christian, M. Yarus, *J. Mol. Biol.* **1992**, *228*, 743–758.

47 E. L. Christian, M. Yarus, *Biochemistry* **1993**, *32*, 4475–4480.

48 M. Lindqvist, K. Sandström, V. Liepins, R. Strömberg, A. Gräslund, *RNA* **2001**, *7*, 1115–1125.

49 B. Streicher, U. von Ahsen, R. Schroeder, *Nucleic Acids Res.* **1993**, *21*, 311–317.
50 D. R. Banatao, R. B. Altman, T. E. Klein, *Nucleic Acids Res.* **2003**, *31*, 4450–4460.
51 J. I. Ross, E. A. Eady, J. H. Cove, W. J. Cunliffe, *Antimicrob. Agents Chemother.* **1998**, *42*, 1702–1705.
52 C. A. Trieber, D. E. Taylor, *J. Bacteriol.* **2002**, *184*, 2131–2140.
53 D. Moazed, H. F. Noller, *Nature* **1987**, *327*, 389–394.
54 R. Oehler, N. Polacek, G. Steiner, A. Barta, *Nucleic Acids Res.* **1997**, *25*, 1219–1224.
55 J. W. Noah, M. A. Dolan, P. Babin, P. Wollenzien, *J. Biol. Chem.* **1999**, *274*, 16576–16581.
56 D. E. Brodersen, W. M. Clemons, A. P. Carter, R. J. Morgan-Warren, B. T. Wimberly, V. Ramakrishnan, *Cell* **2000**, *103*, 1143–1154.
57 M. Pioletti, F. Schlünzen, J. Harms, R. Zarivach, M. Glühmann, H. Avila, A. Bashan, H. Bartels, T. Auerbach, C. Jacobi, T. Hartsch, A. Yonath, F. Franceschi, *EMBO J.* **2001**, *20*, 1829–1839.
58 F. Mueller, R. Brimacombe, *J. Mol. Biol.* **1997**, *271*, 524–544.
59 M. Brenowitz, D. F. Senear, M. A. Shea, G. K. Ackers, *Methods Enzymol.* **1986**, *130*, 132–181.
60 K. Juneau, E. Podell, D. J. Harrington, T. R. Cech, *Structure* **2001**, *9*, 221–231.

16
Measuring the Stoichiometry of Magnesium Ions Bound to RNA

A. J. Andrews and Carol Fierke

16.1
Introduction

RNAs are large polyanions containing negative charges on the many backbone phosphodiester groups that interact with positively charged ions in the solution by charge–charge, or Coulombic, interactions [1]. This Coulombic field of negative charge attracts large numbers of positively charged counter-ions. These counterions typically consist of both monovalent ions, such as potassium or sodium, and divalent ions, such as magnesium. These ions can both loosely associate with the phosphodiester backbone of RNA mainly by electrostatic interactions, forming an "ionic cloud", or form specific interactions with the RNA backbone and bases to bind in a unique position [2]. Specific binding sites can include direct coordination of the cation by the RNA (inner-sphere interaction) or a hydrogen bond contact via a coordinated water molecule (outer-sphere interaction).

In this chapter we present a method to measure the total number of divalent magnesium ions, including ions that are either specifically or electrostatically bound, that interact with an RNA molecule under a given set of conditions. This method can be useful for determining how the composition of the electrostatic cloud is affected by solution conditions, and whether changes in the RNA structure or the addition of protein cofactors affect the number and composition of interacting ions [3]. For instance, alteration of a nucleotide in RNA or addition of a protein component may reduce the total ionic charge of the RNA and, therefore, the number of associated magnesium ions. Quantification of the number of ions that bind or interact with polyanions, such as RNA or DNA, becomes more difficult as the size of the polyanion, and therefore the number of bound ions, increases. In some cases, ions bound to the RNA can be distinguished from free ions using spectroscopic analysis, e.g. electron paramagnetic resonance spectroscopy has been used to measure the binding of manganese ions to RNA [4]. However, a more general approach is to physically separate the bound and free metal ions and then determine the concentration of ions in both fractions.

The main methods for separating ions bound to large RNAs are gel-filtration chromatography, equilibrium dialysis and forced equilibrium dialysis [5]. Forced

Handbook of RNA Biochemistry. Edited by R. K. Hartmann, A. Bindereif, A. Schön, E. Westhof
Copyright © 2005 WILEY-VCH Verlag GmbH & Co. KGaA, Weinheim
ISBN: 3-527-30826-1

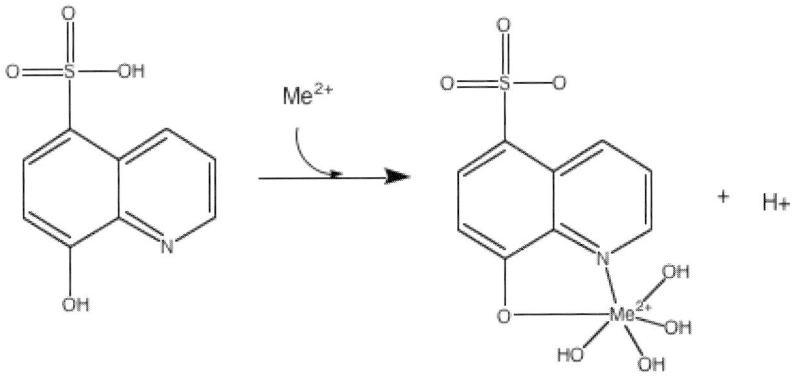

Fig. 16.1. The chemical structure of HQS binding to metal as seen in the nickel crystal structure [10]. The magnesium–HQS complex is assumed to have a similar structure.

dialysis has given the most consistent results and is the easiest to perform; therefore this method will be the main focus of this chapter. Once the bound and free ions are separated, the total concentration of ions in each sample can be measured using a number of techniques [inductively coupled plasma emission mass spectrometry (ICP-MS), atomic absorption or fluorescent dye binding], but most require expensive equipment and in-depth training, except for the fluorescent methods. A number of fluorophores have been used to detect metal ions including 8-hydroxyquinoline-5-sulfonic acid (HQS). HQS at neutral pH is minimally fluorescent, but becomes highly fluorescent upon binding to magnesium and forming a magnesium-8-quinolinol complex (Fig. 16.1). Here, we describe in detail the use of the fluorophore HQS for determining magnesium concentrations following the separation of bound and free magnesium ions.

16.2
Separation of Free Magnesium from RNA-bound Magnesium

Equilibrium dialysis is still the best way to separate small molecules from ones associated with larger molecules under equilibrium conditions. While numerous methods of dialysis exist, from the traditionnal simple dialysis tubing to fully automated machines, the deciding factors are the concentration and volume of RNA required, and the amount of time needed for the dialysis experiment to reach equilibrium. Long equilibration times and large volumes often limit traditional equilibrium dialysis experiments. The ideal method would use a small amount of sample, have a high degree of precision and be rapid. Forced dialysis is an equilibrium method with shorter equilibration times and smaller volumes than traditional dialysis methods. While new and more advanced methods for the dialysis of small samples are available, the use of ultrafiltration devices provides a quick and cheap alternative.

16.3
Forced Dialysis is the Preferred Method for Separating Bound and Free Magnesium Ions

The main advantages of forced dialysis compared to conventional dialysis are: (1) equilibration occurs in the absence of a dividing membrane, which significantly decreases the amount of time needed to reach equilibrium, (2) small volumes (approximately 100 µl) of sample are required, and (3) separation of the free and bound molecules is rapid. The forced dialysis method is accomplished by first allowing the sample to equilibrate at a given set of experimental conditions in the absence of a separating membrane. Separation is then carried out after equilibrium is reached. A simple method for carrying out this separation is to use a Microcon centrifugal filter device (Millipore, Billerica, MA; www.millipore.com) or other similar products. Microcon filter devices are manufactured with membranes that limit the size of the nucleic acid that can pass through the pores, with limitation sizes of 10–300 single-stranded nucleotides. Liquid is forced through the membrane by centrifugation (up to 14 000 g) which separates unbound small molecules from those associated with the larger nucleic acid (Fig. 16.2). Furthermore, equilibrium is maintained throughout the experiment as the concentration of unbound ligand remains constant. Two important notes are that Microcon devices, as manufactured, contain a small amount of glycerin in the filter and the filter has roughly 10 µl retention volume. Before beginning any dialysis experiment, it is necessary to confirm that your ligand of interest can pass through the membrane easily and that neither component will preferentially bind to the dialysis membrane.

The success of this experiment also depends significantly on the purity of the

Forced dialysis method

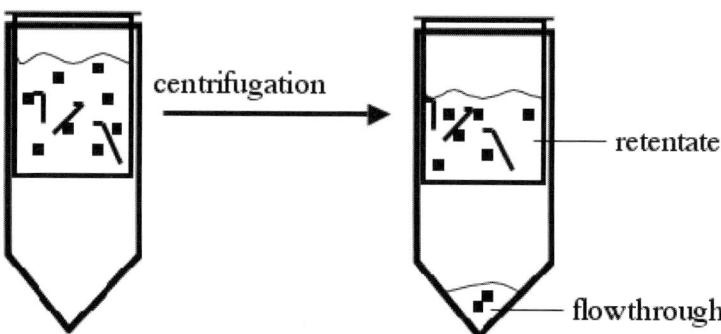

Fig. 16.2. A cartoon illustrating the forced dialysis method. Squares represent magnesium ions and larger lines are RNA molecules. After enough time has elapsed to reach equilibrium, the device is centrifuged so that a small amount of liquid is forced through the dialysis membrane. The retentate contains the bound magnesium plus the free magnesium and the flow-through contains only the free magnesium ions.

RNA and the reagents. RNA samples should be as clean as possible, with both contaminating metals (i.e. magnesium) and chelators (i.e. EDTA) removed. All solutions should be prepared using metal-free tubes and pipette tips. Buffers should be as close to physiological or assay conditions as possible, with close to neutral pH, low monovalent salt concentrations and varying concentrations of magnesium ions. All small molecules with high affinity for magnesium, such as EDTA, should be removed. Excess metal ions can be removed from microfuge tubes by soaking in 100 mM EDTA overnight followed by extensive washing with metal-free water. To remove the glycerin found in the membranes, the device should be washed by centrifugation of at least 500 μl of metal-free water followed by the same amount of buffer. The RNA should be prepared and refolded as usual, although the magnesium concentration should be kept as low as possible.

The minimum concentration of RNA needed for this experiment is dependent on both the binding affinity for magnesium and the stoichiometry for magnesium. There must be sufficient RNA to bind enough magnesium ions such that the total magnesium concentration ($[Mg^{2+}]_{tot} = [Mg^{2+}]_{bound} + [Mg^{2+}]_{free}$) is greater than the free magnesium concentration alone. The following equation (1) demonstrates that for a stoichiometry (n) of 1 and $K_{1/2}$ of 1 mM, the concentration of RNA would need to be 1 mM to see a 2-fold difference between the bound and free fractions:

$$\frac{[RNA \cdot Mg^{2+}]}{[RNA]_{total}} = \frac{n[Mg^{2+}]}{([Mg^{2+}] + K_{1/2})} \tag{1}$$

However, most RNAs have a stoichiometry for bound Mg^{2+} that is much higher than 1, which lowers the required concentration of RNA. For instance, if $n = 100$ and $K_{1/2} = 1$ mM, then only 20 μM RNA is required to achieve equal concentrations of free and bound magnesium. Therefore, as the size of the RNA and the resulting Coulombic field increases, the concentration of RNA required to see a measurable difference in fluorescence decreases. The number of metal ions binding to RNA can be estimated to be between 0.3 and 0.7 M^{2+}/nt [3, 6].

To initiate the experiment, the RNA is diluted into buffer containing magnesium. A recommended buffer volume for this experiment is 100 μl, but this depends on the size of the filter apparatus. This allows for the removal of sufficient volume for the analysis of free magnesium without changing the RNA concentration by a large amount. The half-time for equilibration can be estimated from the K_D and a reasonable guess as to the association rate constant (k_a), assuming a simple two-step binding reaction, as shown in Eq. (2):

$$t_{1/2} = \frac{\ln 2}{k_{obs}} = \frac{0.693}{k_a([Mg^{2+}] + K_D)} \tag{2}$$

Even assuming a value of K_D of 1 μM and a slow second-order association rate constant of 1×10^5 M^{-1} s^{-1}, the calculated $t_{1/2}$ is 3 s suggesting that a 30-s equilibrium time should be sufficient. In practice, the best way to test whether the incubation time is adequate to achieve equilibrium is to demonstrate that doubling (or halving) the equilibration time does not alter the final result.

Once equilibrium is reached, the sample can be added to the pre-washed Microcon device. After adding the RNA to the device, centrifuge the Microcon at 7000 g until a small volume has passed through (around 50 µl). The flow-through can then either be added back to the Microcon or replaced with an equal volume of the original magnesium buffer. Allow the reaction mixture to re-equilibrate (i.e. 15 min at 25 °C) then spin again at 7000 g to collect the flow-through containing only the free magnesium. This sample should be a small percentage (20% or less) of the total volume. This $[Mg^{2+}]_{Free}$ sample is ready for analysis and the $[Mg^{2+}]_{total}$ sample can be obtained directly from the solution that is retained in the top of the Microcon.

16.4
Alternative Methods for Separating Free and Bound Magnesium Ions

Size exclusion chromatography and equilibrium dialysis are alternative methods that can be used to separate free from bound magnesium ions. In all cases, care needs to be taken to make sure all solutions have no other divalent metals beside magnesium and no magnesium chelators.

Gel filtration columns using size-exclusion matrices such as Sephadex G-50 and others can be used to separate small molecules from large RNAs. However, even small columns take on the order of 5 min to run which allows ample time for re-equilibration of the metal ions during the separation. An alternative rapid separation method is gel-filtration spin columns [7, 8]. Using the gel-filtration spin columns, separation can be accomplished on the order of seconds, which greatly reduces the likelihood of re-equilibration during separation. However, for micromolar binding constants, the dissociation rate constant for metal ions from RNA could be on the order of $1-10 \text{ s}^{-1}$ ($t_{1/2} = 0.1-1 \text{ s}$) indicating that re-equilibration can occur on the same time scale as the separation by spin column method. Therefore, this technique is only applicable for very tight or slowly equilibrating magnesium-binding sites.

Equilibrium dialysis is probably the most well-known and common method of determining the stoichiometry and affinity of ligand binding sites. Equilibrium dialysis is often quite slow, taking many hours to complete since movement through the membrane can be slow [7]. Therefore, it is important to test both the magnesium equilibration time and the RNA stability. As before, great care should be taken to make sure that all solutions and tubes are free from metals or chelators that could interfere with the experiments as well as contaminating RNase. Therefore, the equipment used should be autoclaved if possible and then soaked in 70% ethanol followed by a 100% ethanol wash. After choosing the correct dialysis membrane, the next important step is to determine the equilibration time. This can be estimated from measuring the time required for a magnesium solution to reach equal concentrations in both chambers of the equilibrium dialysis apparatus in the absence of RNA. After the dialysis experiment is complete, the RNA should be analyzed by gel electrophoresis to determine the extent of degradation that has occurred.

16.5 Determining the Concentration of Free Magnesium in the Flow-through

Magnesium standards should be made in the same buffer concentrations as the RNA being tested. The magnesium standards will be used to make a standard curve to determine the concentration of magnesium in the sample. Three very important factors need to be addressed in the use of HQS to measure bound magnesium. (1) The RNA needs to be denatured to prevent high affinity sites from competing with HQS for binding Mg^{2+}. Therefore, guanidine–HCl is included in the assay buffer (5 M guanidine). (2) The pH should be near neutral since, at high pH, HQS ionizes which increases the background fluorescence and decreases the sensitivity of the metal analysis. (3) HQS should be in high enough concentration to completely bind the available magnesium ions. Therefore, the HQS concentration should be much larger than both the magnesium dissociation constant (K_D) and the [Mg^{2+}]. In summary, the recommended assay conditions are 5 M guanidine–HCl, 0.1 M Tris, pH 8 and 2 mM HQS. A small volume of the sample to be tested (20 µl or less) is mixed with 150 µl of the HQS solution and the fluorescence is then measured in a 120-µl cuvette at 25 °C. The excitation wavelength is set at 397 nm and the emission wavelength is 502 nm. At this point, the experimentally determined fluorescence can be compared to the standard curve of fluorescence as a function of magnesium concentration under the exact same conditions to calculate the concentrations of magnesium ions in the experimental samples. See Fig 16.3.

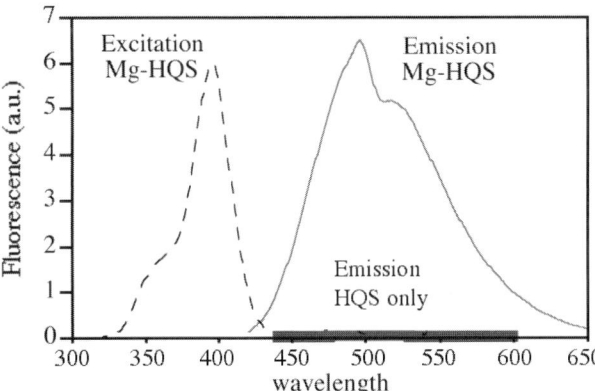

Fig. 16.3. The excitation and emission spectra of Mg^{2+}–HQS with the emission spectra of HQS shown as a reference. The spectra were collected on an Amico-Bowman Series 2 spectrometer (ThermoSpectronic, Rochester, NY) by exciting the flourophore at 397 nm. The slit width for both excitation and emission spectra was kept at 8 nm. The sample was placed in 0.1 M Tris, pH 8, 2 mM HQS and with or without 1 mM magnesium at 25 °C.

16.6
How to Determine the Concentration of Magnesium Bound to the RNA and the Number of Binding Sites on the RNA

In the forced dialysis method, the concentration of bound magnesium ions $[Mg^{2+}]_{bound}$ can be calculated by subtracting the concentration of free magnesium $[Mg^{2+}]_{free}$ determined from the flow-through from the total magnesium concentration $[Mg^{2+}]_{total}$ determined from the retentate. Similarly, in a standard equilibrium dialysis experiment the concentration of bound magnesium can be determined by measuring the difference between the magnesium concentration on the two sides of the dialysis membrane. The $[Mg^{2+}]_{free}$ is measured on the side without the RNA, while the concentration on the side with RNA equals the combination of RNA–magnesium and free magnesium ($[Mg^{2+}]_{bound} + [Mg^{2+}]_{free}$). Therefore the concentration of bound magnesium ions can be determined by subtracting the magnesium concentration on the "free" side from the concentration on the side with the RNA:

$$[Mg^{2+}]_{bound} = ([Mg^{2+}]_{bound} + [Mg^{2+}]_{free}) - [Mg^{2+}]_{free} \qquad (3)$$

As stated before, important control experiments include demonstrating that the membrane does not bind significant magnesium ions and that the RNA of interest does not affect the determination of the magnesium concentration. If the RNA competes with HSQ for binding magnesium, you can either: (1) decrease the magnesium affinity of the RNA by adding RNase, (2) measure the standard curve for magnesium in the presence of a known concentration of the competing RNA or (3) estimate the bound magnesium solely from the free magnesium by subtracting two times the free magnesium $[Mg^{2+}]_{free}$ from the total magnesium added:

$$[Mg^{2+}]_{bound} = [Mg^{2+}]_{total} - 2[Mg^{2+}]_{free} \qquad (4)$$

Once the concentrations of bound and free magnesium are determined at different magnesium concentrations, the $K_{1/2}$ for magnesium binding and the stoichiometry, n, can be determined by fitting Eq. (5) to these data (Fig. 16.4) [3, 7]. This analysis assumes that the magnesium is binding to the RNA via multiple, independent binding sites [5], where v equals magnesium bound divided by total RNA added. This assumption can be tested by making a Scatchard plot where $v/[Mg^{2+}]_{free}$ is plotted versus v; the slope of a linear fit equals $1/K_{1/2}$ (Eq. 6). If only one type of site is observed, the Scatchard plot will be linear. If multiple types of binding sites are observed, the Scatchard plots will not be linear [5]. In the case of most large RNAs, the ratio of specifically bound ions to loosely bound or interacting magnesium ions is so small that only one class of ions will be seen.

$$v = \frac{\left(\dfrac{n[Mg^{2+}]_{free}}{K_{1/2}}\right)}{\left(\dfrac{1 + [Mg^{2+}]_{free}}{K_{1/2}}\right)} \qquad (5)$$

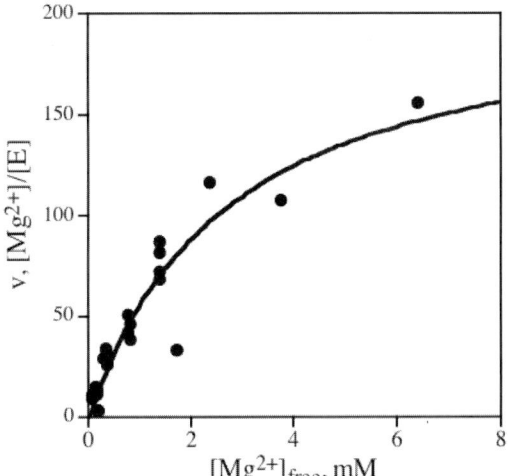

Fig. 16.4. A plot showing the amount of magnesium bound to the RNase P holoenzyme at various concentrations of magnesium ions determined using the forced dialysis method [3]. The stoichiometry of magnesium bound to the holoenzyme is 160 Mg^{2+} per enzyme and the $K_{1/2}$ is 1.5 mM [3].

The data can also be plotted as a Scatchard plot which results in a linear correlation:

$$\frac{v}{[Mg^{2+}]_{free}} = \frac{n}{K_{1/2}} - \frac{v}{K_{1/2}} \tag{6}$$

In this analysis we have made the assumption that magnesium ions will bind to RNA in a non-cooperative fashion (Eq. 1). However, if the Mg^{2-} ions bind cooperatively, the fraction of magnesium bound to RNA will be described by Eq. (7), so that the binding can be measured and analyzed in a manner similar to what has been described for non-cooperative binding.

$$\frac{[RNA \cdot Mg^{2+}]}{[RNA]_{total}} = \frac{[Mg^{2+}]}{([Mg^{2+}]^n + K_{1/2})} \tag{7}$$

16.7
Conclusion

This method is a straightforward way to quantitate the number of metal ions directly interacting with RNA by using common laboratory equipment. The important step in this method that is missing in other systems is the separation of free and bound magnesium ions. Other systems add HQS directly to the RNA and measure competition of the two ligands for the magnesium ions [9]. While these methods may work for one or two binding sites, the analysis of this type of competition experiment becomes extremely complicated as the number of binding sites increases. The measurement of the number of metal ions interacting with an RNA molecule under various conditions will advance our knowledge of RNA–metal interactions and will be useful for testing the validity of RNA modeling techniques.

16.8
Troubleshooting

(1) If you are using Microcons, make sure that you have removed the glycerol from the filter and you have not exceeded the maximum g force.
(2) Confirm that RNA, but not magnesium, is retained by the membrane and that neither RNA nor magnesium sticks to the membrane.
(3) Make sure that all of the solutions, tips and tubes are free from contaminating metals, chelators and RNases.
(4) Determine that the RNA is folded and stable throughout the experiment.
(5) Demonstrate that the reaction was incubated for sufficient time to reach equilibrium. Make sure that the magnesium in the experimental samples is determined under exactly the same conditions as the magnesium standard curve (buffer, temperature, wavelengths, voltage, etc.).

References

1 G. S. MANNING, J. RAY, *J. Biomol. Struct. Dyn.* **1998**, *16*, 461–476.
2 V. K. MISRA, D. E. DRAPER, *Biopolymers* **1998**, *48*, 113–135.
3 J. C. KURZ, C. A. FIERKE, *Biochemistry* **2002**, *41*, 9545–9558.
4 V. J. DeROSE, *Curr. Opin. Struct. Biol.* **2003**, *13*, 317–324.
5 C. R. CANTOR, P. R. SCHIMMEL, *Biophysical Chemistry*, Freeman, New York, **1980**.
6 A. L. FEIG, O. C. UHLENBECK, The role of metal ions in RNA biochemistry, in *The RNA World*, 2nd edn, R. F. GESTELAND, T. R. CECH, J. F. ATKINS, (eds), Cold Spring Harbor Laboratory Press, Cold Spring Harbor, NY, **1999**, pp. 287–319.
7 J. A. BEEBE, J. C. KURZ, C. A. FIERKE, *Biochemistry* **1996**, *35*, 10493–10505.
8 J. J. GARCIA, A. GOMEZ-PUYOU, E. MALDONADO, M. TUENA DE GOMEZ-PUYOU, *Eur. J. Biochem.* **1997**, *249*, 622–629.
9 V. SEREBROV, R. J. CLARKE, H. J. GROSS, L. KISSELEV, *Biochemistry* **2001**, *40*, 6688–6698.
10 S. B. RAJ, P. T. MUTHIAH, G. BOCELLI, L. RIGHI, *Acta Crystallogr.* **2001**, *E57*, m591–m594.

17
Nucleotide Analog Interference Mapping and Suppression: Specific Applications in Studies of RNA Tertiary Structure, Dynamic Helicase Mechanism and RNA–Protein Interactions

Olga Fedorova, Marc Boudvillain, Jane Kawaoka and Anna Marie Pyle

17.1
Background

17.1.1
The Role of Biochemical Methods in Structural Studies

Recent advances in structure determination of RNA and RNA–protein complexes by diffraction and NMR methods have radically expanded our understanding of RNA architecture [1–5]. Due to the complexity and resolution of these structures, the role of accompanying biochemical studies for construct design, testing of function and, ultimately, for interpretation of high-resolution structural data has never been greater. Classical methods such as photo-crosslinking, footprinting and chemical modification interference remain powerful tools [6]. However, a new set of biochemical tools for high-resolution structure determination and testing is now available in the form of Nucleotide Analog Interference Mapping (NAIM) and Nucleotide Analog Interference Suppression (NAIS) [7–10]. These methods are based on the selection for functionally active RNA molecules. Structural constraints provided by NAIM/NAIS experiments are uniquely powerful because they reflect the active RNA conformation. Therefore data from NAIM/NAIS experiments are particularly informative for meaningful structure–function analysis. NAIM and NAIS also provide biochemical data at an unprecedented level of resolution, as they interrogate individual atoms, and predict specific hydrogen bonds and RNA interaction motifs [9–12].

Even if it were possible to solve the crystal structure of every interesting RNA, the classical and NAIM/NAIS methods would remain essential, complementary approaches. The biochemical approaches can help distinguish functionally relevant structural information from crystal packing artifacts. NAIM and NAIS can probe whether hydrogen bonds predicted from crystallographic studies are actually important for molecular function or activity [5]. When a crystal structure provides important clues about catalytic mechanism, chemical details and models can be probed through biochemical methods. Finally, the very success of crystallographic

Handbook of RNA Biochemistry. Edited by R. K. Hartmann, A. Bindereif, A. Schön, E. Westhof
Copyright © 2005 WILEY-VCH Verlag GmbH & Co. KGaA, Weinheim
ISBN: 3-527-30826-1

studies often rests on the ability to design a functional construct that is capable of crystallizing. Biochemical methods guide in the identification of stable RNA structural domains and provide tests for activity of the resultant RNA. Thus, at all stages of investigation, high-resolution RNA structure determination is aided by a diverse arsenal of biochemical methods.

There remain many important RNA molecules that have eluded high-resolution analysis by diffraction and NMR. Even if one form of a structure is known, there is often no information about alternative molecular conformations that are important for function. High-resolution information is often found for small, stable subdomains of larger molecules which, in their intact form, have not been possible to crystallize [13]. Thus, there remains ample room for the application of biochemical methods in structure determination and modeling. A diversity of methods have historically been used to provide distance constraints and to elucidate long-range tertiary interactions [14–16]. This type of information has facilitated the construction of three-dimensional (3-D) models for RNA molecules. While methods such as UV crosslinking and footprinting have been valuable tools in modeling efforts, they are unable to differentiate active from inactive conformations of a molecule and they are not applicable for testing the role of specific RNA functional groups. Thus, there has been a need for biochemical approaches that provide functional information at high resolution.

Mutational analysis has traditionally been a powerful tool for elucidating the contribution of functional groups in an RNA [17]. In these studies, the identity of an entire nucleobase is changed or, in a more precise adaptation, a single functional group is altered. The latter modifications are incorporated through chemical synthesis of RNA (if the RNA is no longer than around 50 nt) or by a combination of chemical and enzymatic syntheses (for longer RNAs). The modified species are then analyzed for function in parallel with unmodified RNA (i.e. to evaluate binding or a reaction of interest). The relative importance of the functional group is then determined from the difference in activity between modified and unmodified molecules. This approach necessitates the synthesis of many different modified RNA molecules, in which important atoms are changed one at a time. The systematic screening of functionalities by this method requires an enormous synthetic effort. The approach has been successfully applied in many cases, including the determination of tertiary interactions between the group I intron core and its substrate helix [18], and the identification of catalytically important functional groups on domain 5 of a group II intron [17]. However, the method is simply inapplicable for screening the importance of functional groups throughout large RNA molecules such as group II introns, which are often around 1000 nt in length.

The problem of identifying specific RNA atoms that are essential to function is solved by NAIM. In a single experiment, this method screens a combinatorial library of modified RNA molecules for atoms that are important for function. The RNA of interest is prepared by *in vitro* transcription in the presence of a nucleoside analog thiotriphosphate that contains desired sugar or base modifications. Because it is susceptible to cleavage by iodine, the phosphorothioate linkage serves as a tag that reflects the location of an important modified residue. As the first application

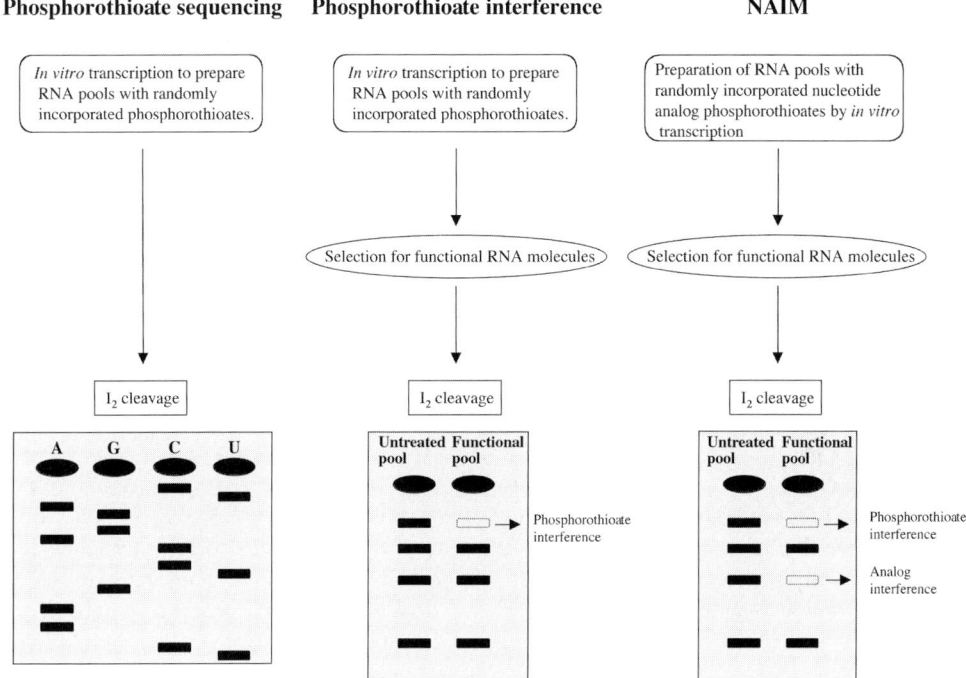

Fig. 17.1. Evolution of NAIM methods: from sequencing to the analog interference approach.

of phosphorothioate tagging, Gish and Eckstein reported a technique for RNA and DNA sequencing that was based on selective susceptibility of a phosphorothioate linkage to iodine cleavage (Fig. 17.1) [19]. The method originally employed iodoethane as an activating agent, which alkylated the sulfur, thus making the phosphorus more susceptible to a nucleophilic attack by an adjacent 2′-OH group followed by strand scission [19]. Iodoethane has since been replaced by ethanolic iodine solution. The exact mechanism of iodine interaction with the phosphorothioate linkage has never been studied in detail, contrary to iodoethane cleavage where all the possible pathways and products have been thoroughly investigated [19]. The putative mechanism of iodine cleavage, assuming that it is similar to iodoethane cleavage, is shown in Fig. 17.2.

Although not discussed further here, it is notable that these investigators also proposed a similar method for chemical probing of DNA. In that case, iodoethanol replaces iodine as the activating electrophile. While the nucleophile for strand scission of RNA is the adjacent 2′-OH group (Fig. 17.2), in DNA it is the hydroxyl group of the ethanol moiety [20].

The phosphorothioate tagging method quickly evolved when several research groups realized that it could be coupled with a selection to identify functionally important phosphate groups in an RNA (Fig. 17.1) [21–24]. The approach was taken

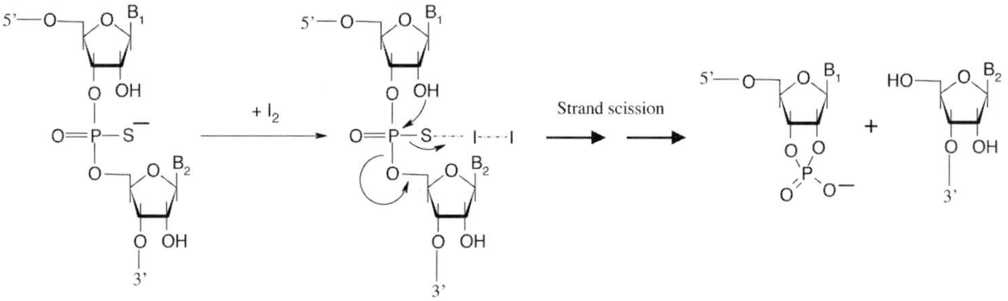

Fig. 17.2. The putative mechanism of cleavage of phosphorothioate-modified RNA by iodine. Iodine electrophilically attacks the sulfur, thus making phosphorus more susceptible to the nucleophilic attack by the adjacent 2′-hydroxyl. Sulfur (path a) or oxygen (path b) can be the leaving group. Products of interest, which make this reaction applicable for NAIM/NAIS, are formed as a result of strand scission. The actual reaction mechanism is still under discussion [19, 24].

yet to another level when it was expanded to include other nucleotide analogs that contained sugar and base modifications [11, 25–30]. In this approach, i.e. NAIM, the phosphorothioate linkage was still used as a tag, but it was incorporated along with a second modification of interest on either the base or the sugar. After a selection step, the phosphorothioate linkage reported not only the positions of important phosphate atoms, but also of other nucleotide atoms that influence function (Fig. 17.1). Recent studies have applied NAIM to the analysis of RNA structural motifs [11, 31], metal binding sites [32, 33] and mechanisms of ribozyme catalysis [34–36]. NAIM subsequently evolved into NAIS, which permits the precise identification of tertiary interaction partners [9, 12, 28]. In combination, the two methods now represent the most powerful biochemical method for RNA structure determination and analysis.

17.1.2
NAIM: A Combinatorial Approach for RNA Structure–Function Analysis

17.1.2.1 Description of the Method
NAIM involves the following steps [7].

(1) Preparation, by *in vitro* transcription, of an RNA library that is doped with a small population of modified phosphorothioate nucleotide analog.
(2) Selection and separation – a process of interest (i.e. binding, conformational change or catalytic reaction) is used to separate active from inactive pools of RNA.
(3) Visualization of interferences – iodine cleavage is used to detect the locations of modified atoms that affect function (Fig. 17.3A). In order to detect cleaved molecules, the RNA pool must be labeled at either the 3′ or 5′ terminus. In some cases, precursor molecules are labeled after transcription [7, 8], while in

17.1 Background | 263

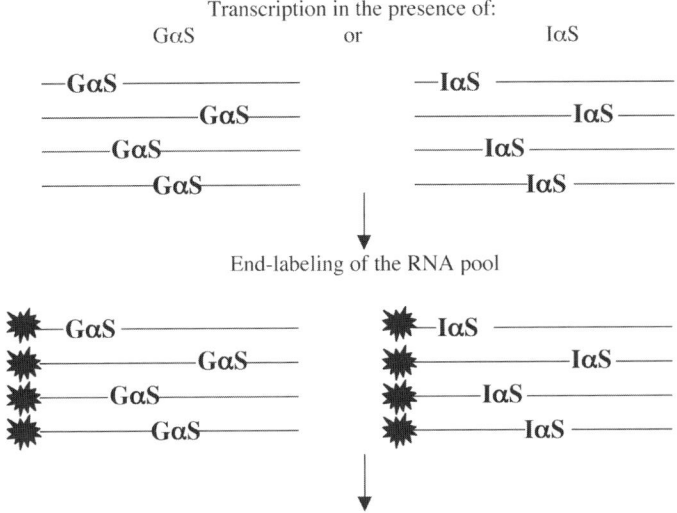

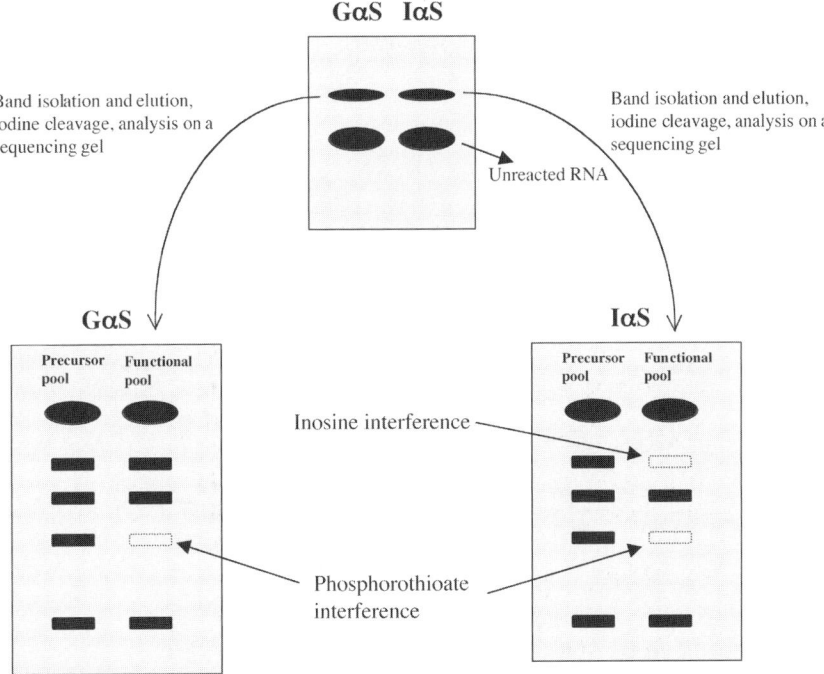

Fig. 17.3. A schematic representation of NAIM (A) and NAIS (B) assays. Note that in (B) the term "mutant RNA 1" is used not to describe a classical base mutation, but a single-atom substitution (2'-OH is changed to 2'-H). In NAIS literature this is frequently done for the purpose of simplicity.

264 | 17 Nucleotide Analog Interference Mapping and Suppression

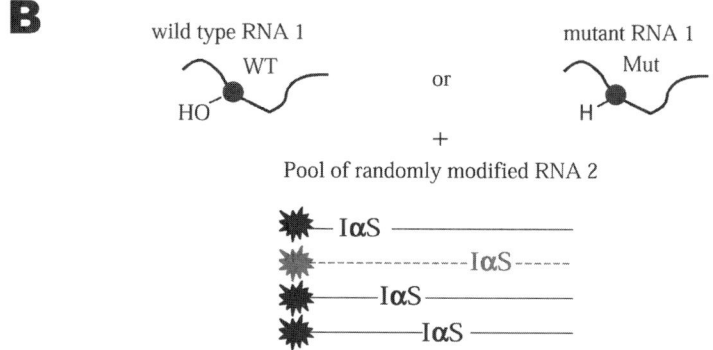

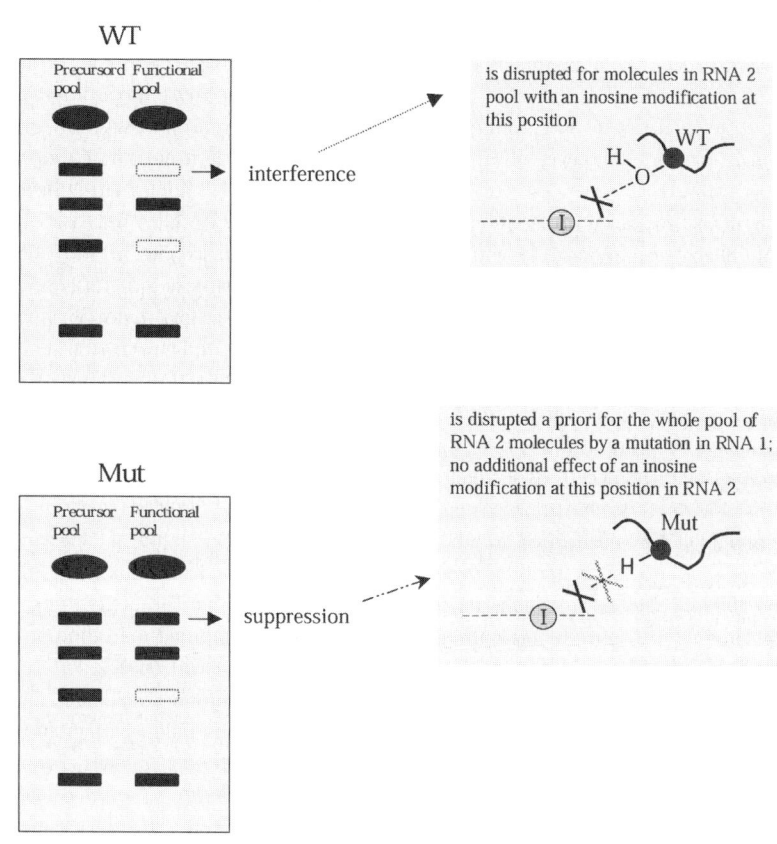

Fig. 17.3.B

other studies, functional molecules are radiolabeled during the selection step [28].

In preparing the pool of randomly modified RNAs, the ratio of a thiotriphosphate analog to a corresponding unmodified nucleoside triphosphate in the transcription mixture is adjusted so that the analog is incorporated with an efficiency of one or two modifications per molecule. Incorporation levels of up to 5% are considered satisfactory [8]. For optimal efficiency of the polymerase and for ease of analysis, only one modified analog is usually included at a time in a given transcription reaction [7, 8].

A typical NAIM experiment is exemplified as follows. To determine the functional importance of guanosine exocyclic amino groups in an RNA, it is transcribed in the presence of inosine phosphorothioate nucleotide triphosphates. The resulting pool of RNA molecules has a statistical probability of including a single inosine substitution at each guanosine position in the RNA (Fig. 17.3A). The RNA pool is then bound to another molecule or it undergoes a catalytic reaction of interest. If the inosine modification is located at a position where it interferes with binding or reaction, this RNA molecule will be depleted from the pool of functional molecules. The fraction of reacted molecules is then separated from unreacted molecules using electrophoresis or an affinity column, although separation methods vary widely depending on the application of interest. Then the reacted fraction (functional pool) as well as the precursor RNA pool and/or the unreacted fraction are treated with iodine and analyzed on a denaturing polyacrylamide gel (Fig. 17.3A). If the modification at a certain position interferes with function, the corresponding band will be missing or underrepresented in the lane that represents the pool of active RNAs (Fig. 17.3A). Since the phosphorothioate linkage itself can interfere with activity, all NAIM assays contain a control experiment where only phosphorothioate is incorporated and its effects on function are analyzed in parallel.

17.1.2.2 Applications

Identifying Atoms and Functional Groups Critical for Ribozyme Activity
NAIM has been widely used to map chemical groups that are important for the catalytic function of ribozymes such as the group I intron [11, 26, 27, 37], group II intron ribozymes [28] and the hairpin ribozyme [29]. Applications to the study of RNase P are discussed in Chapter 18.

Generally, the first insight into structural organization and 3-D architecture of a large ribozyme is through phylogenetic analysis, which allows identification of highly conserved nucleotides and, in some cases, tertiary interactions [38–40]. NAIM takes this type of approach to the next level and obtains information with atomic resolution. Since most tertiary interactions involve a network of contacts among sugar-phosphate backbone residues, traditional phylogenetic approaches often fail to identify critical tertiary contacts. However, an alternative form of phylogenetic information can be obtained by comparing the NAIM profiles of ribozymes that belong to different organisms. This approach has been referred to as "chemical phylogeny", and it has been successfully applied to the elucidation of universally conserved and alternative tertiary interactions in group I ribozymes [37].

NAIM "Signatures" for Elucidation of RNA Structural Motifs

Although NAIM provides a map of chemical groups that are important for RNA function, it remains challenging to translate this into concrete information on RNA tertiary structure. One can distill the code of interference patterns into probable types of structural motifs through judicious choice of modified nucleotides and careful comparison with known structures. For example, if an adenosine residue exhibits interference with 7-deazaAαS, m^{6}AαS, PurαS and 2APαS, it is likely to form hydrogen bonds from its Hoogsteen face [11, 41]. Similarly, if N^2-methyl guanosine or inosine substitutions interfere with a specific guanosine, its minor groove is likely to be involved in a tertiary interaction [27]. Further examples of analogs are provided in Fig. 17.4 and in [8].

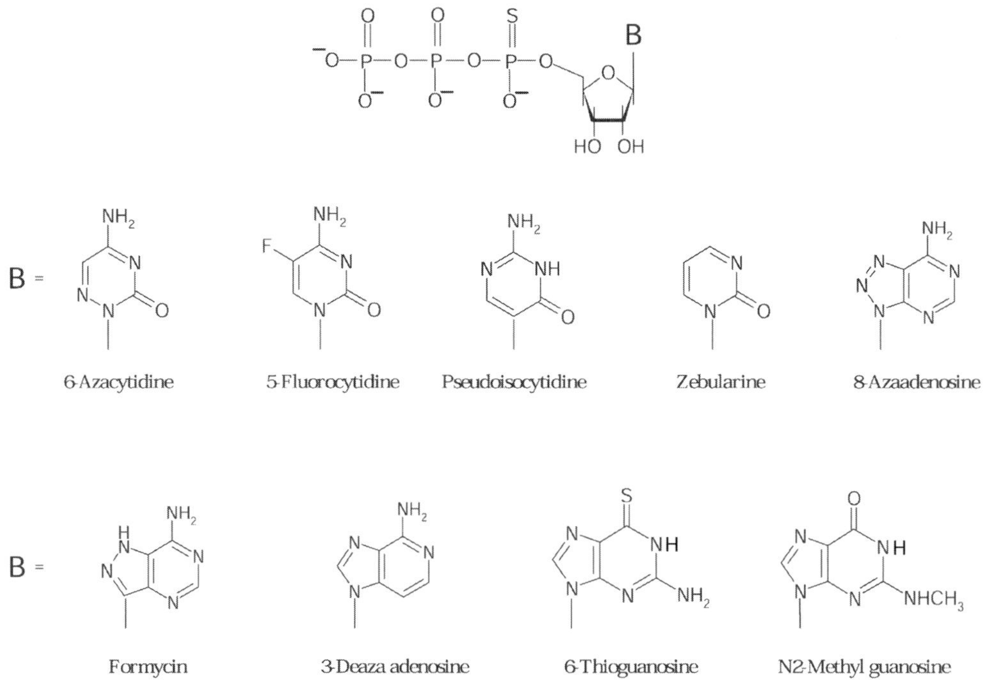

Fig. 17.4. Nucleotide analog α-thiotriphosphates synthesized by the Strobel group. These analogs are not commercially available, but can be synthesized using published protocols [8, 64, 66]. Commercially available analogs can be found in the Glen Research catalog (www.glenres.com).

The growing number of RNA crystal and NMR structures provide an invaluable library of known RNA structural motifs. From the hydrogen bonding patterns in these motifs, one can predict a pattern of interferences with certain nucleotide analogs and then screen an RNA of interest for such patterns. For example, using the crystal structures of two different ribozymes [2, 42], one can predict that a GAAA tetraloop involved in an interaction with a cognate tetraloop receptor will exhibit 2′-deoxy and 7-deaza adenosine interferences at the second and third adenosine positions of the loop. Such a pattern has been observed for a GAAA tetraloop on domain 5 of a group II intron, confirming its interaction with a cognate receptor in domain 1 [28]. Another very common RNA packing motif is the A-minor motif, which involves the interaction of an adenosine's N^1, N^3 and/or 2′-OH functional groups with 2′-OH groups and the minor groove edges of a GC or a GU base pair [43]. Using a combination of 2′-deoxy AαS and 3-deazaAαS analogs, Strobel et al. have identified A-minor motifs in the active site of the group I ribozyme [10, 31].

Identifying Nucleobases Involved in Ribozyme Catalysis

Proton transfer represents an important catalytic strategy of many protein enzymes. General acid or base catalysis usually requires ionization of specific functionalities, such as a histidine side chain. In some cases, the function of an important side chain requires perturbation of its normal pK_a [44]. An increasing number of ribozymes have now been shown to effect catalysis via proton transfer [45–47] and, in certain cases, this activity appears to be divalent metal-ion independent [48–50]. This suggests that nucleobases themselves may be directly involved in proton transfer during catalysis, particularly if they undergo a shift in pK_a value. The adenosine N^1/N^3 and cytosine N^3 functionalities have most typically been implicated, as they are most easily protonated. Strobel et al. have established specialized NAIM assays and analogs for identifying nucleotides with shifted pK_a values [34–36]. For this purpose, a series of adenosine and cytosine analogs have been designed that probe pK_a values at N^3 (Fig. 17.4). The resulting interference patterns have then been studied as a function of pH [34–36]. Nucleotide positions that show pH-dependent interferences with the new analogs are good candidates for participation in catalysis via proton transfer. However, unambiguous conclusions about direct involvement in catalysis cannot be made on the basis of NAIM alone. There is always the possibility that an ionized nucleobase participates in a pH-dependent tertiary interaction or folding event that is not directly involved in catalytic activity.

Locating Metal-binding Sites in RNA

The incorporation of single phosphorothioates (Rp or Sp isomers) at certain positions, followed by kinetic analysis in the presence of magnesium versus different thiophilic metal ions, has long been instrumental in establishing and characterizing catalytically important metal-binding sites in RNA [51–55]. When phosphorothioate substitution results in a drastic decrease in catalytic activity that is rescued by the addition of a thiophilic metal ion, such as Mn^{2+} or Cd^{2+}, then the corresponding Rp or Sp oxygen is implicated in direct coordination with Mg^{2+}. Random incorporation of phosphorothioate analogs through NAIM can be used to screen for potential metal-binding sites. However, the experiment is limited to Rp-phosphorothioates due to stereospecific incorporation of Sp-α phosphorothioate NTPs by T7 RNA polymerase. This NAIM approach has been explored in a variety of ribozyme systems [56–58], including the *Tetrahymena* group I intron [59]. Basu and Strobel revisited this system and found that metal ion rescue of 2′-deoxy or phosphorothioate interference effects may not always have been due to Mg^{2+} coordination [33], as metal binding at these sites was not confirmed by X-ray crystallography. However, they found that reduction of Mn^{2+} concentration to 0.3 mM eliminates all non-specific rescue artifacts and produces data that were in a good agreement with crystallographic results. Other thiophilic metal ions are sometimes used in rescue experiments [22, 52, 53]. However, in some cases they can inhibit ribozyme activity [60] or produce incomplete subsets of metal-binding sites [33]. The possibility of artifacts in metal rescue experiments underscores the need for proper controls and cautious interpretation of metal rescue data.

NAIM is not limited to the detection of divalent metal ion binding sites. Strongly associated monovalent ions can also be detected if they bind to a site through interaction with guanosine and uracil ketone groups. To probe for the presence of site-bound potassium ion, RNA is transcribed in the presence of 6-thioguanosine thiotriphosphate (Fig. 17.4) [32, 61]. If interference is observed, a metal ion rescue experiment is attempted using thiophilic thallium ion, which has properties similar to those of potassium. This approach resulted in the identification of a unique potassium-binding site located below the AA platform of a GNRA tetraloop receptor [32].

Determining the Free Energy Contributions of Individual Functional Groups on RNA

To identify RNA functional groups that are important for interacting with other molecules, one can set up an NAIM experiment in which binding is used for selection. A free energy profile can be constructed for each site of interference by measuring the magnitude of NAIM effects as a function of ligand concentration [62]. The efficacy of this approach has been successfully demonstrated on the signal recognition particle (SRP) system where the results obtained from NAIM experiments were in good agreement with mutational analysis [62].

17.1.3
NAIS: A Chemogenetic Tool for Identifying RNA Tertiary Contacts and Interaction Interfaces

17.1.3.1 **General Concepts**

Although NAIM is useful for identifying the locations of important atoms in an RNA molecule, it does not provide information on how these atoms interact with each other and it does not reveal tertiary interaction partners. However, by comparing the NAIM profiles of a wild-type molecule with that of a mutant, one can discern patterns of energetic communication and tertiary contact in a system. This variation of the NAIM experiment, which is called NAIS, is generally carried out after all interference (NAIM) data has been collected. The original NAIM patterns are used as a guide for the design of mutants with a critical nucleobase or a highly important atom altered, which are then produced by transcription or chemical synthesis. The resulting mutants are usually much less reactive than wild-type RNA, as important tertiary interactions are disrupted by introduced mutations.

In a two-part system for probing tertiary contacts along an interaction interface between RNA molecules (Fig. 17.3B), one of the RNA molecules (RNA 1) is typically mutated. The other RNA molecule (RNA 2) is transcribed in the presence of nucleotide analogs. The latter RNA is then screened for changes in its interference pattern upon reaction with either the mutant or the wild-type versions of RNA 1. Assume, for example, that the 2'-OH group of a nucleotide A in RNA 1 interacts with the exo-cyclic amine of the guanosine residue at position B in RNA 2 (Fig. 17.3B) and that this interaction is critical for a selectable function. Upon reaction with wild-type RNA 1, RNA 2 will exhibit strong interference with an inosine and/or 2-methyl guanosine analog at position B. However, when RNA 2 reacts with a mutant RNA 1 (under conditions altered to result in the same extent of reaction as with wild-type RNA 1) that contains a single 2'-deoxynucleotide at position A, the interaction between nucleotides A and B will have already been disrupted and the penalty for this disruption will have already been paid in the form of lesser reactivity by mutant RNA 1. Therefore, when RNA 2 interacts with the specified mutant RNA 1, there will be no interference observed at position B (Fig. 17.3B). This effect is generally referred to as "interference suppression". Note that this is not a "rescue", because the disappearance of the interference effect is not attributed to a restored interaction, but rather to elimination of an interaction partner.

17.1.3.2 Applications: Elucidating Tertiary Contacts in Group I and Group II Ribozymes

NAIS as a method to identify long-range tertiary contacts in large ribozymes was first demonstrated during studies on group I intron ribozymes. There it was used to predict an extended minor groove triple helix between the P1 helix and the single-stranded J8/7 region, in which each triple appeared to be mediated by at least one 2'-hydroxyl group [9]. In the group II intron ribozymes, for which little structural information exists, NAIS identified two tertiary interactions, $\kappa-\kappa'$ and $\lambda-\lambda'$, that are essential for visualizing active-site architecture in the group II intron core (Fig. 17.5) [12, 28]. Hydrogen bonding contacts elucidated by NAIS have been extensively used as constraints for building models of group I and II ribozyme active sites [9, 63].

In subsequent Sections 17.2 and 17.3, we will illustrate applications of NAIM and NAIS, respectively, by using three diverse systems as examples: (1) ribozyme studies on a group II intron, (2) studies of dynamic interactions between protein and RNA during the NPH-II helicase reaction, and (3) studies of RNA–protein interactions that are involved in transcription termination. Detailed protocols will be provided for each NAIM/NAIS step as it applies to these systems, although they can be tailored to meet the specific needs of the reader. To avoid repetition, Section

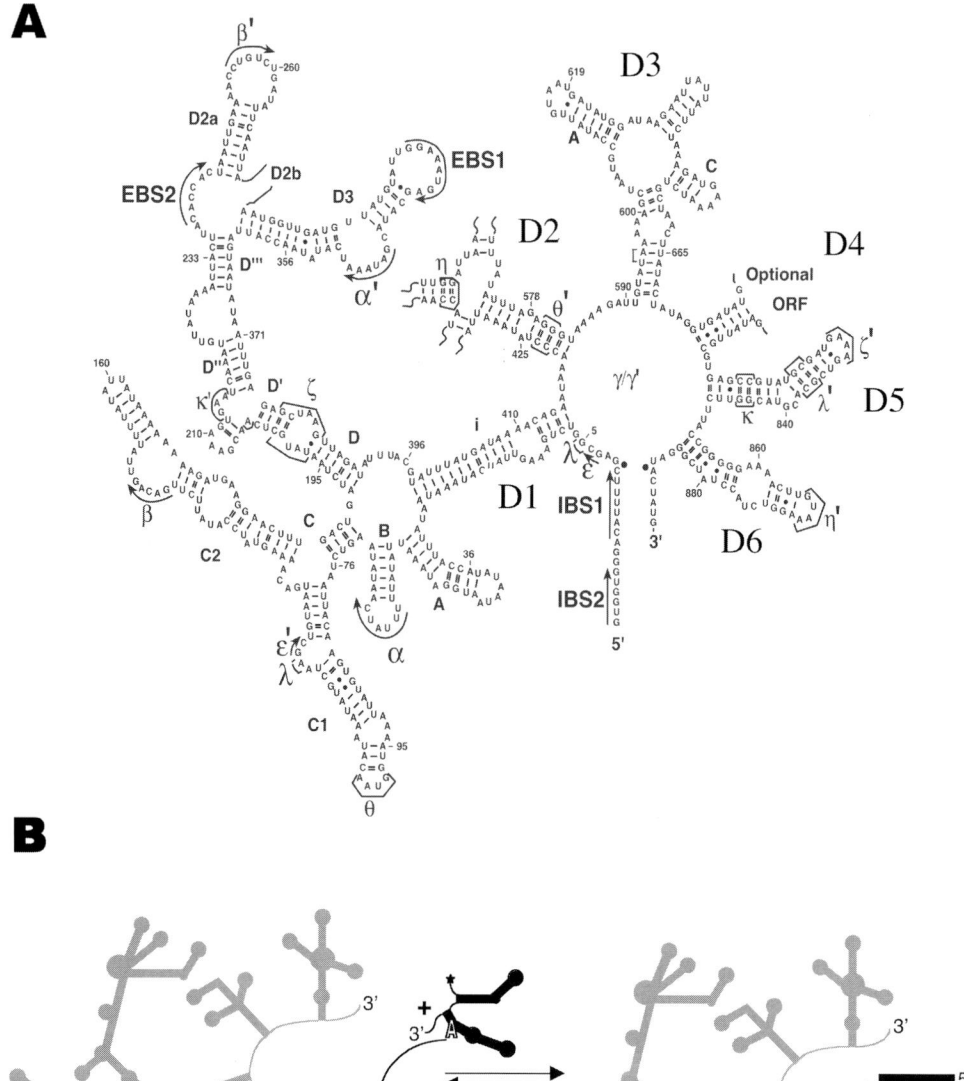

Fig. 17.5. (A) Secondary structure of the ai5γ group II intron. (B) Schematic representation of the *trans*-branching reaction used as a selection step in NAIM and NAIS studies of the ai5γ group II ribozyme.

17.2 begins with the basic protocols and methods that are used for all applications of NAIM. These are followed by specific examples, which differ primarily in the selection method used for NAIM. NAIM protocols will then be followed by NAIS methodologies that were developed to study group II intron tertiary structure (Section 17.3).

17.2
Experimental Protocols for NAIM

17.2.1
Nucleoside Analog Thiotriphosphates

Many phosphorothioate nucleotide analogs are commercially available from Glen Research (for European suppliers, see Chapter 18). Strobel et al. have expanded the existing collection by developing strategies and protocols for the efficient synthesis of other analogs (Fig. 17.4) [8]. Most analogs can be synthesized from unprotected nucleosides [generally available from Sigma, R.I. Chemical (Orange, CA) or ChemGenes (Wilmington, MA)] in two steps by adapting the procedure first described by Arabshahi and Frey [64]. The first step involves reaction of a nucleoside with thiophosphoryl chloride, resulting in the formation of 1,1-dichloro-1-thionucleoside phosphate, which is then converted to a nucleotide thiotriphosphate by reaction with tributylammonium pyrophosphate [8]. The synthetic procedures described above are not suitable for some cytidine analogs, which require the use of a salicyl phosphoramidite approach [65]. The synthesis of these analogs is described in detail by Oyelere and Strobel [66].

17.2.2
Preparation of Transcripts Containing Phosphorothioate Analogs

RNA molecules that contain randomly incorporated phosphorothioate analogs are prepared by *in vitro* transcription from a double-stranded DNA template (usually either a PCR product or a restriction digest of plasmid DNA). Transcription efficiency and fidelity tend to be low, however, unless the DNA template contains blunt or 5′-overhanging ends. Thus, restriction enzymes must be chosen accordingly [67].

To optimize transcription efficiency, only one phosphorothioate NTP analog at a time is usually added to the transcription reaction. Since in many cases nucleotide analog α-thiotriphosphates are not incorporated into the transcript with the same efficiency as unmodified triphosphates, the ratio of NTPαS to NTP needs to be optimized in order to achieve the desired level of analog incorporation (generally 5%). For commercially available analogs, such an optimization has already been performed and they are sold as 10× solutions at a concentration that is adjusted for 5% incorporation. For example, while conventional NTPs are added to the transcription at 1 mM final concentration, the modified analogs IαS and m^{6}AαS

Tab. 17.1. Concentrations (mM) of most common NTPαS analogs (and parental rNTP) required for around 5% NαS incorporations per transcript[1,2].

Incorporation	[NTPαS]	[rNTP]	RNAP
AαS, CαS, GαS, UαS	0.05	1	WT
dAαS, dCαS	1.5	1	Y639F
dGαS, dUαS	0.5	1	Y639F
IαS, m^{6}AαS	0.4	1	WT
C^7-AαS, C^7-GαS	0.1	1	WT
PurαS, 2APαS	2	0.5	WT
DAPαS	0.025	1	WT
m^{2}GαS	1.5	0.5	Y639F

[1] Further details on *in vitro* transcription with the Y639F mutant or wild-type (WT) T7 RNAP are described in the text.
[2] Adapted from [8].

are often added at a concentration of 0.4 mM. An optimal NTP/NTPαS ratio for many nucleotide analogs has been optimized and reported elsewhere (Table 17.1) [8]. Wild-type T7 RNA polymerase is used for the incorporation of most analogs, except for those containing minor groove substitutions. The N^2-methyl guanosine, 3-deaza adenosine and sugar modifications (2′-deoxy, 2′-fluoro, 2′-O-methyl) are usually incorporated using mutant Y639F polymerase [68], which has an enhanced tolerance for minor groove modifications [7].

For most constructs employed in ribozyme and helicase studies, the optimal yield of RNA can be obtained using the following transcription protocol.

(1) Prepare the reaction mixture containing 40 mM Tris–HCl, pH 7.5, 15 mM MgCl$_2$, 2 mM Spermidine, 5 mM DTT, 0.02 μg/μl DNA template and 0.08 μg/μl of wild-type T7 RNA polymerase or its mutant Y639F. In our hands, 1 ml transcription of exD123 or D56 was sufficient to carry out several NAIM/NAIS experiments.
(2) Incubate the transcription mixture at 37 °C for 3–4 h.
(3) Precipitate the transcript by the addition of 0.25 M NaCl, 0.02 M EDTA and 3 volumes of ethanol followed by incubation at −80 °C for 1 h and centrifugation.
(4) Re-suspend the pellet in 10 mM MOPS (pH 6.0), 1 mM EDTA (M$_{10}$E$_1$ buffer) that is supplemented with an equal volume of denaturing loading buffer (8 M urea, 0.05% each of xylene cyanol and bromophenol blue, 17% sucrose, 83 mM Tris, pH 7.5, and 1.7 mM EDTA) and purify on a denaturing polyacrylamide gel. Visualize the band corresponding to the desired transcript by UV shadowing (for a detailed description, see Chapter 3), excise from the gel and elute in M$_{10}$E$_1$ buffer at 4 °C for 1–3 h.
(5) Precipitate the RNA as above; re-suspend the pellet in M$_{10}$E$_1$ buffer and store at −80 °C.

Tips and troubleshooting
For the best yield of RNA, the transcription conditions always need to be optimized for each particular construct. One can vary, for example, the reaction time, the concentrations of DNA template, T7 RNA polymerase, $MgCl_2$ in the reaction buffer and triphosphates. The addition of 0.1% Triton X-100 (final concentration), extra DTT or RNase inhibitor can also help increase the transcription efficiency (see Chapter 1). If it is necessary to optimize the incorporation level for a nucleotide analog, the transcription is performed using different NTP:NTPαS ratios.

17.2.3
Radioactive Labeling of the RNA Pool

In order to employ the NAIM assay, one component of the reaction must be radiolabeled. This can be achieved by either labeling the entire RNA pool after transcription and purification or by reaction of the unlabeled RNA pool with a radioactively labeled molecule during the selection step. For example, prior to the selection that is based on helicase unwinding of duplex RNA, one strand of the duplex in the RNA pool is end-labeled with ^{32}P. The selection step for reaction by a group II intron ribozyme is performed by reacting a labeled pool of one RNA with an unlabeled pool of a second RNA or *vice versa*.

Radiolabeling of an NAIM transcript can be performed either at the 5′ end (kinase reaction) or at the 3′ end (via templated addition of a single α-^{32}P-labeled dAMP residue by Klenow polymerase) [69]. In order to 5′-end label an RNA transcript, one has to follow the protocol below.

(1) Prior to 5′-end-labeling, dephosphorylate transcribed RNA (around 60 pmol) with calf intestinal phosphatase (CIP, 40 U) in a final volume of 50 μl using the reaction buffer supplied by the manufacturer (Roche). Incubate the reaction mixture at 37 °C for 30 min, then phenol-extract and ethanol-precipitate the RNA.
(2) For 5′-end-labeling, incubate 10 pmol of dephosphorylated RNA with 1 μl of [γ-^{32}P]ATP (6000 Ci/mmol) and 10 U of T4 polynucleotide kinase (PNK) in a final volume of 10 μl at 37 °C for 1 h.
(3) Purify the labeled RNA on a denaturing polyacrylamide gel.

17.2.4
The Selection Step of NAIM: Three Applications for Studies of RNA Function

17.2.4.1 Group II Intron Ribozyme Activity: Selection through Transesterification

General aspects
Group II introns are complex ribozymes with a diverse catalytic repertoire [40, 70]. Self-splicing of these ribozymes occurs in two steps, with the first step occurring via competing parallel pathways: hydrolysis and branching [71, 72]. Despite the lack of primary sequence conservation, group II intron secondary structure is very

conserved. It consists of six domains, each of which has a distinct catalytic function (Fig. 17.5A) [38, 40, 70]. The largest, domain 1 (D1), folds independently [73], serves as a scaffold for binding other intronic domains as well as the 5′-exonic substrate, and contributes key elements of the intron active site. Domain 5 (D5) contains many active site residues and is absolutely critical for any catalytic reaction performed by the intron. Domain 3 (D3) is a catalytic effector that accelerates every reaction catalyzed by the intron. Domain 6 (D6) contains the bulged branch-point adenosine, which is essential for the transesterification pathway of splicing. Domain 2 (D2) mediates conformational rearrangements along the splicing pathway and Domain 4 (D4) often contains an open reading frame that encodes a protein cofactor.

One of the most remarkable features of group II intron ribozymes is their modularity, which allows reconstitution of an active ribozyme from two or more separately transcribed components. This modularity provides an opportunity to dissect the entire splicing pathway into separate reactions, which can be studied independently. One of the most informative constructs for studying group II intron catalysis is the *trans*-branching system, which divides the intron into two critical parts: exD123 (comprised of the 5′ exon and intronic domains 1–3) and D56 (containing domains 5 and 6) (Fig. 17.5B). When these two RNA components are combined in the presence of Mg^{2+}, the 2′-OH of the D6 branchpoint (in the D56 construct) attacks the 5′ splice site in exD123, resulting in the formation of a covalent linkage. The resulting 2′–3′–5′ branched D56/D123 molecule is highly stable and can be isolated. A pool of these branched molecules represents the population of species that were capable of undergoing the first step of group II intron self-splicing. Thus, the *trans*-branching assay represents a useful selection for NAIM and NAIS studies on the first step of splicing. NAIM assays that employ this *trans*-branching system have provided information on functionally important residues in both RNAs (exD123 and D56) (Fig. 17.6) [28] and NAIS applications have identified a number of critical tertiary interactions (Fig. 17.7) [12, 28].

Experimental procedures
The following protocol is used for identification of functional groups in D56 RNA that are important for branching.

(1) Transcribe D56 in the presence of nucleotide analogs and 5′-end-label the transcript as described above.
(2) Denature trace amounts of modified, labeled D56 (1–10 nM) and unlabeled, unmodified exD123 (1.5 μM) separately at 95 °C for 1 min in 40 mM MOPS pH 6.0. Cool down to about 42 °C by leaving the tubes at room temperature for around 1 min and then mix the two RNA fractions with the simultaneous addition of salts to final concentrations of 100 mM $MgCl_2$, 2 mM $Mn(OAc)_2$ and 0.5 M $(NH_4)_2SO_4$. Incubate the mixture at 42 °C until the fraction of branched product reaches around 20%.
(3) Separate unreacted D56 and branched product on a 5% polyacrylamide denaturing gel, recover both species from the gel and subject to iodine cleavage as described below.

A

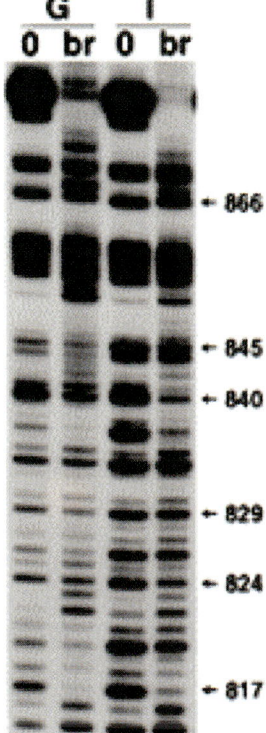

B

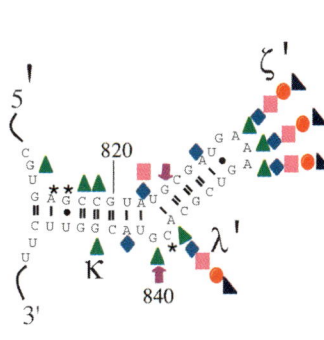

- ■ -7-deaza A
- ● -2,6-Diaminopurine
- ◆ -N6-Me A
- ▲ -2-Aminopurine
- ➡ -inosine
- * -phosphorothioate
- ▲ -2'-modifications

Fig. 17.6. Interference effects in the ai5γ group II intron. (A) A high-resolution sequencing gel summarizes the effects of GTPαS and inosine substitutions (G and I lanes, respectively) in D56 after iodine cleavage of the unreacted (0) and branched (br) fractions. A weak phosphorothioate effect at G840 could not be verified in subsequent experiments. Reprinted with permission from *The EMBO Journal*. (B) Summary of NAIM effects in D5 of the group II intron.

The experimental setup is essentially the same for mapping the 5'-half of D123 in the exD123 construct. In this case, exD123 is transcribed with nucleotide analogs and D56 is unmodified. The reaction is carried out according to the above protocol, and the branched product is purified in a 5% denaturing gel and subjected to sequencing by iodine. The cleavage pattern is compared to that of precursor exD123.

In order to map the 3'-half of exD123, the procedure is modified as follows.

(1) Transcribe exD123 in the presence of the nucleotide analogs of interest and label at the 3' end [69].

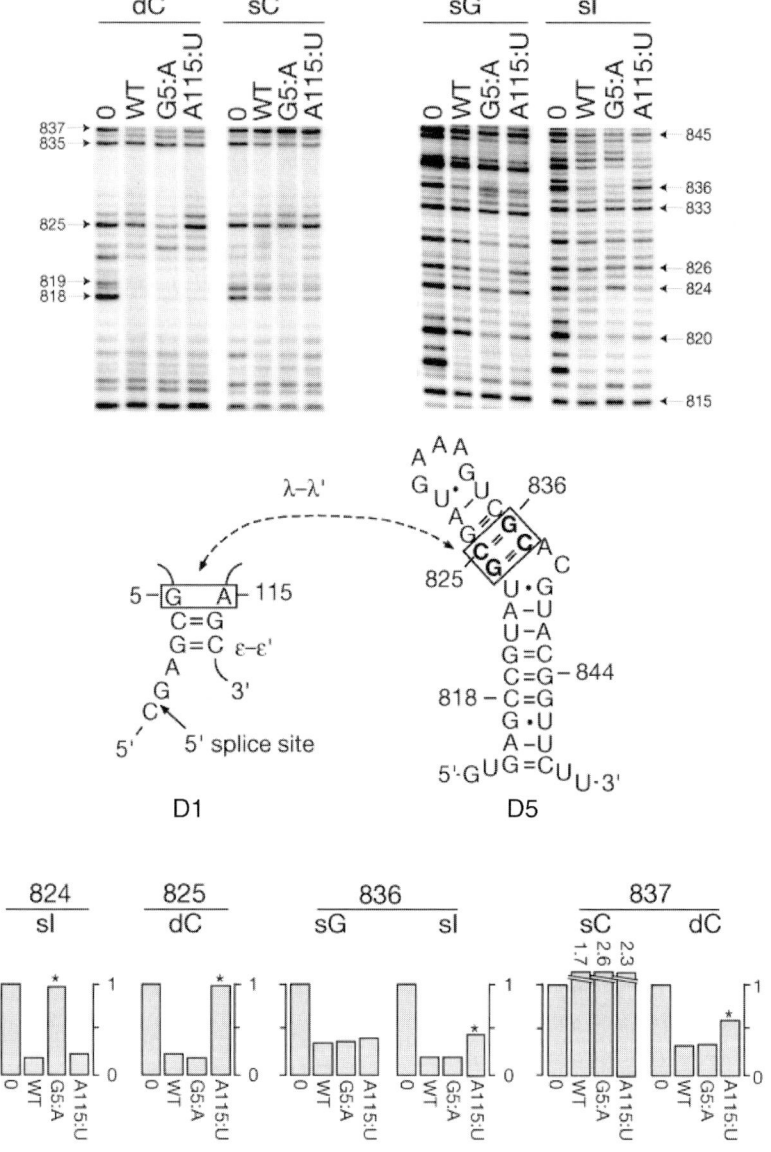

Fig. 17.7. Nucleotide analogue interference suppressions within D5 and D1 of the ai5γ group II intron. Autoradiographs correspond to iodine-cleaved transcripts modified with deoxy-CαS (dC), cytosineαS (sC), guanosineαS (sG) or inosineαS (sI). Comparative data are shown for NαS-containing D56 that was unreacted (0) or branched to wild-type exD123 (WT), exD123(G5:A) and exD123(A115:U) mutants. Band intensities were quantitated and corrected for background phosphorothioate effects. The normalized intensities shown in the bar diagram at the bottom represent mean values, each having a maximum variance of 20%, from two to four independent experiments. Asterisks indicate positions where NAIS effects were observed (loss or alleviation of interference compared with the wild-type).

(2) React trace amounts of radiolabeled exD123 (1–10 nM) with 1.5 µM unlabeled, unmodified D56 according to the same protocol as described above.
(3) Separate the branched product from the unreacted exD123 on a 5% denaturing gel; subject both RNAs to iodine cleavage.

Since the 3′ terminus of exD123 is labeled in this approach, one can also observe the formation of both branched and hydrolysis products, both of which can be analyzed for interference effects.

17.2.4.2 Reactivity of RNA Helicases: Selection by RNA Unwinding

General aspects
DExH/D proteins are involved in nearly every aspect of RNA metabolism and represent the largest group of RNA helicases [74, 75]. NPH-II is a prototypical member of the DExH/D family [76], and it was the first one shown to unwind RNA [77]. NPH-II has robust ATPase activity that is partially stimulated by nucleic acids [77]. NPH-II readily unwinds RNA substrates that contain a 3′ overhang [78] and it has a high degree of processivity during unwinding of long RNA duplexes [79]. The overall unwinding reaction is limited by the rate constant for unwinding initiation, which is 1–3 min^{-1}. NPH-II is both a robust motor for RNA strand displacement [79] and for the removal of proteins that are bound to RNA sites [80].

Currently there are about one dozen solved helicase structures. While these provide invaluable mechanistic clues, they are limited to providing a static picture, or "freeze-frames", of unwinding events. Mechanistic models for unwinding by an RNA helicase of the DExH/D family have not been well developed. NAIM has been used to probe the dynamic interactions that occur between NPH-II and its substrate during translocation (J. Kawaoka and A. M. Pyle, manuscript in preparation). NPH-II has been shown to specifically recognize a cluster of 2′-OH groups and a critical Rp-phosphate at the initiation site of unwinding (located in the first 4 bp of the duplex substrate). After initiation, the enzyme recognizes phased, periodic clusters of 2′-OH groups every 10 nt along the bottom strand of its substrate (J. Kawaoka and A. M. Pyle, manuscript in preparation). This periodic phasing suggests a defined translocation step size of discrete magnitude and is consistent with an inchworm-like model, as suggested for various helicases [81] including RecBC [82].

For suppressions at positions 824 and 825, significant phosphorothioate effects were not observed, so corresponding bars are not shown. The band intensity observed for unreacted material was arbitrarily set to one. For branched products, the size of the bars is related to 1/(interference effect), i.e. the smaller the bar, the stronger the interference. The values of band intensities over 1 are shown above the corresponding bars. In the middle of the figure, the intra-domain 1 ε–ε' interaction (base pairs G3–C117 and C4–G116) as well as the tertiary interaction (λ–λ') with D5 are illustrated; nucleotides G5 and A115, which were mutated, are shown boxed on the left. Reprinted with permission from Nature.

Substrate design

Substrates used in unwinding assays normally consist of two strands that are annealed through base-pairing. For most SF2 helicases like NPH-II, the typical substrate for *in vitro* studies is a duplex RNA that is flanked by a 3′-overhang. The overhanging strand (on which NPH-II tracks; J. Kawaoka and A. M. Pyle, manuscript in preparation) is called the "bottom or loading strand". The "top strand" is stripped away during the unwinding reaction. To evaluate unwinding, the two strands are synthesized independently, one of them is end-labeled with ^{32}P, then the two strands are annealed and the resultant duplex is purified on a native gel. For NAIM analysis of RNA functional groups that contribute to unwinding, the substrate strands were designed using mfold [91] to minimize intra-strand secondary structure. A typical top strand RNA used in these studies is 5′-CUG UGG CAU GUC CUA GCG UCG UAU CGA UCU GGU CGU CUCC-3′, which anneals to the complementary bottom strand with the following sequence: 5′-GGA GAC GAC CAG AUC GAU ACG ACG CUA GGA CAU GCC ACA GAC GUA CUA ACA GCA UCA AUG ACA UCA AUGA-3′ (nucleotides overlapping with top strand underlined).

Previous studies demonstrated that functional groups on the top strand do not contribute to NPH-II recognition during unwinding. Therefore, NAIM has only been used to determine the location of important functional groups on the bottom strand. A pool of modified bottom strands is created as follows.

(1) Transcribe bottom strands with phosphorothioate ribonucleotides (NTPαS) or phosphorothioate deoxyribonucleotides (dNTPαS) statistically incorporated to a level of 5% as described in Section 17.2.2. Add trace amounts of [α-^{32}P]UTP to the transcription mixture to label the transcripts internally, so that reaction products could be easily identified and cut from the gel.

(2) Gel-purify, dephosphorylate and 5′-end label transcripts as described in Sections 17.2.2 and 17.2.3.

(3) Test incorporation of analogs by treating around 30 000 c.p.m. of each modified transcript with 3 mM iodine (prepared freshly in EtOH) at 25 °C for 5 min. Analyze samples on a denaturing polyacrylamide sequencing gel and visualize by radioanalytic imaging.

NAIM selection – defining optimal unwinding conditions

Previous optimization experiments served to define the reaction conditions with the strongest influence on NPH-II unwinding [79]. Because it was not known how single deoxy (or phosphorothioate) modifications would affect NPH-II helicase activity, the NAIM assay was tested under a variety of conditions. Variables that were tested included extent of unwinding (5–50%), concentration of NaCl in the reaction (10–120 mM) and enzyme turnover (single versus multiple cycle kinetics). Nearly all interference effects were found to disappear under low salt conditions (NaCl < 30 mM), even when total unwinding proceeded to only 10%. In addition, interference was not seen under multiple cycle conditions (i.e. when helicase is allowed to rebind in the absence of a trap molecule), regardless of the extent of total unwinding (as low as 10% of the total population).

Under single cycle conditions, when the [NaCl] is around 70 mM, interference effects are strong. Also smaller degrees of total unwinding (below 20%) tend to amplify the weaker interference effects (around $I = 1.5$, as defined below). Therefore, these unwinding conditions [single cycle, low extent of unwinding (below 20%) and 70 mM NaCl] were selected for NAIM because they provided good selectivity while maintaining a high level of unwinding activity.

An NAIM helicase assay

(1) Prior to initiation of unwinding, pre-incubate NPH-II protein (15 nM) with duplex substrate (1–2 nM, labeled with ^{32}P at the 5′ end of the bottom strand) in 40 mM Tris, pH 8.0, 4 mM Mg(OAc)$_2$, 70 mM NaCl for 3 min at room temperature.
(2) Initiate reactions by the simultaneous addition of 3.5 mM ATP and 400 nM duplex RNA trap (which prevents helicase rebinding and causes the reaction to be under "single cycle" conditions). Total reaction volume can be 35–100 μl.
(3) Quench the unwinding reaction upon attaining a reaction extent of around 15%.
(4) Separate duplex and unwound fractions by native gel electrophoresis.
(5) Visualize the duplex and unwound species and isolate them from the gel.
(6) In order to make sure that the difference in the iodine cleavage pattern of the duplex and unwound fractions is due to the analog effect and not to the difference in accessibility to the iodine, isolate the bottom strand from the duplex fraction by denaturing PAGE before the cleavage procedure.
(7) Re-suspended all samples in a solution of 10 mM MOPS (pH 5.0), 1 mM EDTA, and subject to iodine cleavage (see below) and analysis on a denaturing polyacrylamide gel.
(8) Quantify and analyze interference effects using the same methods as for other NAIM applications (see below).

17.2.4.3 RNA–Protein Interactions: A One-pot Reaction for Studying Transcription Termination

General aspects
During the processive phase of transcription in bacteria, the fast-moving transcription elongation complex (TEC) is held together by an intricate network of cooperative interactions between the DNA template, RNA polymerase and the RNA product ([83] and references therein). To induce TEC dissociation, many transcription termination signals (termed intrinsic or rho-independent terminators) rely on the formation of a specific stem–loop structure within the nascent transcript that is directly upstream from a short U-rich 3′ end. These RNA functional elements largely contribute to the disruption of TEC-stabilizing interactions, albeit by a mechanism

that is not totally understood. To investigate the role of terminator components in the highly dynamic context of transcription elongation, NAIM experiments were applied [30]. At present, these experiments have been restricted to the study of RNA polymerases (RNAPs) that efficiently utilize NTPαS analogs, which include certain mutants of bacteriophage RNAPs (prototypes are the Y639F and Y639F/H784A mutants of T7 RNAP [68, 84]; M. Boudvillain, unpublished results). However, efforts to evolve suitable mutants of multi-subunit RNAPs are also underway, and these may broaden the scope of NAIM/NAIS-based transcription experiments in the near future (M. Boudvillain and R. Rahmouni, unpublished results).

The implementation of an NAIM strategy for studying transcription termination has resulted in a simple "one-pot" reaction that combines the first two steps of NAIM (preparation of the RNA pool *and* transcript selection) [30]. In this approach, linear DNA templates containing a terminator sequence downstream from a phage promoter are transcribed with an appropriate RNAP variant (here we will only consider the case of the Y639F mutant of T7 RNAP) and in the presence of a NTPαS analog. Due to the incomplete efficiency of most termination signals, only a fraction of the TECs are released at the termination points, whereas other TECs continue transcription to the end of the template (Fig. 17.8A). Purification of both types of transcript products (i.e. the NαS-modified transcripts that are released at the terminator, and those that are released at the template end) allows one to compare the functional groups that are important to the formation of each. By ^{32}P-labeling the two types of products and then treating them with iodine, one reveals the RNA atoms and functional groups that are important for transcription termination (Fig. 17.8C). Because high-resolution crystal structures of the TEC now exist for various RNAPs ([83] and references therein), it is possible to link some of the NαS effects with specific known interactions between the RNAP sidechains and the transcript [30].

Experimental procedure

(1) In a 0.5 ml tube, mix 1.1 pmol of DNA template (either linearized plasmid or purified PCR fragment) with 8 μl of 5 × transcription buffer (30 mM MgCl$_2$,

Fig. 17.8. NAIM analysis of transcription termination. (A) *In vitro* transcription of a linear DNA template (schematically depicted on the left) containing the sequence of the *rrnB* T1 terminator from *E. coli* (boxed) downstream from a T7 promoter (arrow). As shown on the gel (right), termination (T) and runoff (RO) transcripts are formed in comparable amounts during transcriptions with the Y639F and wild-type (WT) T7 RNAPs. (B) Selection of NAIM interference effects on *rrnB* T1 transcription termination [30] using either κ (or 1/κ) values (top) or λ discrimination factors (bottom). Broken lines correspond to standard interference thresholds (see text). (C) A representative gel showing Rp-phosphorothioate (GαS) and 2′-deoxy (dGαS) interference effects at G positions during intrinsic termination of transcription at the major site (T) of the *rrnB* T1 terminator; R = runoff transcripts. Positions of Rp-phosphorothioate (triangles) and 2′-deoxy (circles) modifications that favor (open symbols) or are detrimental (filled symbols) to transcript release at the T site are illustrated in the context of the terminator secondary structure.

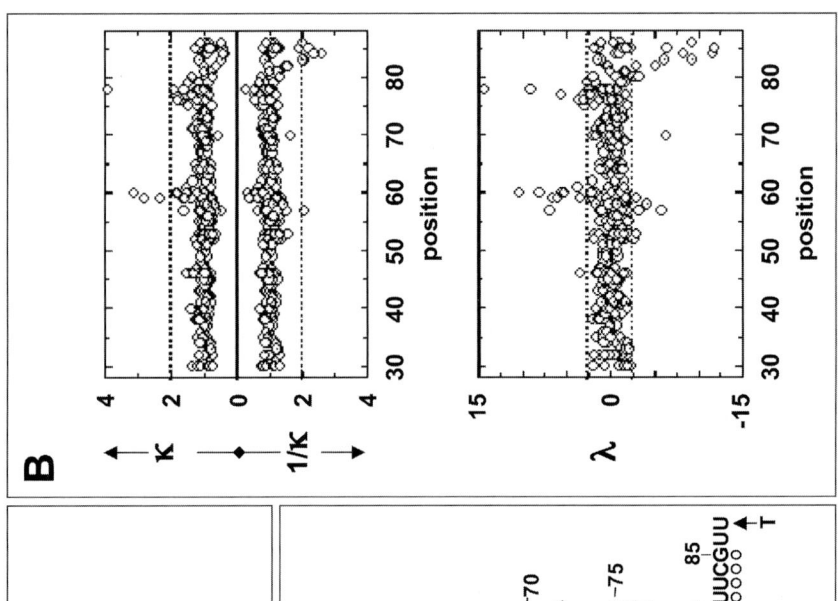
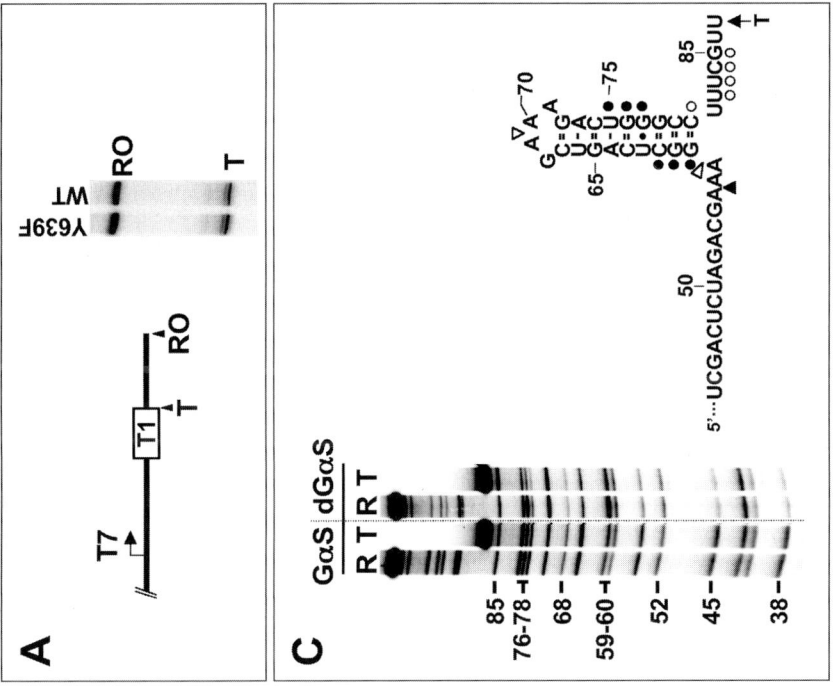

Fig. 17.8

50 mM NaCl, 100 mM HEPES, pH 7.5, 30 mM DTT, 0.05% Triton X-100, 10 mM spermidine), appropriate amounts of NTPαS analog and parental rNTP (Table 17.1), 20 nmol of each of the other rNTPs and 3 pmol of the Y639F RNAP. The final reaction volume should be 40 µl, which usually yields sufficient quantity of transcripts for subsequent NAIM analysis.

(2) Incubate transcription mixtures for 15–30 min at 37 °C before addition of single-stranded M13 DNA (1 µg) and KCl (250 mM) to prevent non-specific association between free RNAPs and transcripts [85].

(3) Quickly load the mixtures onto Microcon columns (100-kDa cut-off for around 100-nt transcripts) that had been saturated with BSA as described by the manufacturer (Millipore). After centrifugation of the columns for 2 min at 10 000 r.p.m. in a desktop centrifuge, collect the filtrates, which should then be free of any unwanted high-molecular-weight species [86]. Unwanted salts and nucleotides are also eliminated by gel filtration on a 1–2 ml Sephadex G-50 (Sigma) column.

(4) To remove triphosphate moieties at the 5' ends of transcripts, mix the column eluate with 2 µl of calf intestinal alkaline phosphatase (CIAP) (Roche Diagnostics) and incubate for 30 min at 37 °C. The enzyme efficiently dephosphorylates transcripts under these conditions, no additional buffer or salts are required.

(5) Add sodium acetate to 0.3 M final concentration and extract twice with one volume of a phenol:chloroform:isoamyl alcohol mix (25:24:1; Amresco) and twice with 1 volume of chloroform buffered with 10 mM MOPS, pH 7. Then, add 3 volumes of ethanol and precipitate the RNA for 2 h at −20 °C.

(6) Centrifuge for 30 min at 5000 g and 4 °C, discard the supernatant, briefly dry the pellet in a Speed Vac apparatus and re-suspend in 14 µl of $M_{10}E_1$.

(7) Add 3 µl of [γ-^{32}P]ATP (3000 Ci/mmol; Amersham Biosciences), 2 µl of 10 × kinase buffer (100 mM $MgCl_2$, 50 mM DTT, 700 mM Tris–Cl, pH 7.5) and 1 µl of T4 polynucleotide kinase (New England Biolabs). Incubate for 30 min at 37 °C, then phenol-extract the sample and precipitate it with ethanol as described above.

(8) Re-suspend the pellet in 10 µl of denaturing buffer (95% formamide, 5 mM EDTA, 0.1% xylene cyanol) and incubate for 1 min at 95 °C before loading on a denaturing polyacrylamide gel (8–10% gels are adequate for around 100-nt long transcripts and should also contain 7 M urea and 30% formamide to ensure strong denaturing conditions) that had been pre-heated to 60 °C. After gel migration (about 2 h at 40 W for adequate separation of around 100-nt transcripts on an 8% acrylamide, 20 × 40-cm gel), recover termination and runoff transcripts from gel slices (see Section 17.2.2) and then analyze through iodine sequencing as described below.

Tips and troubleshooting

Terminator stem–loop structures usually contain many GC pairs. In some instances, complete unfolding of the structure does not occur within the polyacrylamide gel, resulting in anomalous electrophoretic migration (e.g. [85]) even under the harsh denaturing conditions described above. This may result in improper partitioning of the transcription products and in an inextricable mixing of NαS effects on termination and unrelated RNA structural features (A. Schwartz, R. Rahmouni and M. Boudvillain, unpublished results). For similar reasons, results implicating NαS modifications that strengthen RNA base pairs (such as DAPαS) should be analyzed with great caution. In general, it is best to use terminators with no more than three or four consecutive GC pairs in the hairpin stem such as in the *rrnB* T1 terminator of *Escherichia coli* [30]. The DNA template should also be designed to yield transcripts as short as possible (70–100 nt should leave enough positions of neutral NαS incorporation [for calibration of NAIM effects; see Section 17.2.6] between the initiation and termination regions); the terminator sequence should also be introduced sufficiently upstream from the template end to yield termination and runoff transcripts with significantly different lengths. Appropriate DNA templates can be easily prepared by sub-cloning oligonucleotides within commercial vectors that bear a multi-cloning site surrounded by phage promoters such as the pSP73 plasmid (Promega).

In order to calibrate the assay, it is useful to compare NAIM signals obtained with both the Y639F and wild-type T7 RNAPs and analogs that are good substrates for the two enzymes (such as ITPαS or C7-RTPαS analogs). With the *rrnB* T1 terminator, we did not observe significant differences (M. Boudvillain, unpublished observations).

Transcripts of various lengths are usually released in the termination window albeit with different efficiencies (Fig. 17.8A). It is usually best to isolate and analyze those termination species separately as significant differences in NAIM patterns are likely to be observed [30].

The efficiency of transcript release by a given terminator may be affected by modifications of the experimental conditions [87]. This lack of control on the termination reaction precludes an easy adjustment of the detection threshold of NAIM effects (see also Section 17.2.6). For this reason, the use of λ discrimination factors in place of κ interference values (see Section 17.2.6) usually facilitates the identification of weak and moderate NAIM effects (Fig. 17.8B; [30]; M. Boudvillain, unpublished results). Although recent results suggest that, at least in some cases, the assay sensitivity may be controlled through variations of the RNA polymerase/DNA template ratio [88], we still prefer to rely on statistical discrimination (λ values) for detection of NAIM effects on transcription termination.

17.2.5
Iodine Cleavage of RNA Pools

The preceding selection methods result in three different pools of RNA, all of which are examined by iodine cleavage in order to assay function: (A) the selected pool of functional RNA molecules, (B) the selected pool of nonfunctional RNA molecules and (C) the unreacted starting pool of RNA molecules. By comparing

Pool A to Pool B or Pool C (or both B and C), one can deduce the role of specific atoms on function.

Experimental procedure

(1) Re-suspend RNAs in 10 µl of a solution containing 5 µl $M_{10}E_1$ buffer (10 mM MOPS, pH 6.0, 1 mM EDTA) and 5 µl of formamide, and denature at 95 °C for 1 min, followed by chilling on ice.
(2) Add a freshly prepared solution of iodine [1 µl of a 10 mM iodine (Sigma) solution dissolved in ethanol] and incubate the reaction mixture at 37 °C for 3 min.
(3) Precipitate the RNA by adding 240 µl of 0.3 M sodium acetate (pH 5.0), 1 µg tRNA carrier and 750 µl of ethanol.
(4) Analyze samples on a denaturing polyacrylamide gel. Depending on the application and the length of the RNA transcript, the percentage of acrylamide varies from 4 to 20%. It is essential to load samples that have *not* been treated with iodine on the same gel, to provide controls for non-specific RNA degradation. In many cases NAIM gels in published articles do not show lanes of iodine-untreated samples in order to preserve space; however, it is always assumed that the researcher has performed this important control. In order to facilitate the comparison of iodine cleavage patterns in lanes corresponding to precursor RNA and selection product, it is advisable to determine the amount of radioactive material in each sample and load equal amounts of radiolabeled RNA onto the gel. In this case, the 2-fold or greater interference effects will be easily detectable after radioanalytic imaging, even prior to quantitation.

Tips and troubleshooting
The stated iodine cleavage conditions may require optimization, depending on the RNA sample. The final concentration of iodine can vary from 0.1 to 1 mM, and often iodine cleavage reactions are carried out at room temperature. Incubation times can vary from 1 to 5 min. It is also possible to load the samples onto the gel directly without precipitation, thereby sparing additional sample handling. Frequently, however, this results in a salt front on the gel, which can adversely affect migration of the samples.

17.2.6
Analysis and Interpretation of NAIM Results

17.2.6.1 Quantification of Interference Effects
Dried polyacrylamide gels are subjected to radioanalytic imaging [using, for example, a PhosphorImager (Molecular Dynamics)]. For most applications, the PhosphorImager Imagequant program (Molecular Dynamics) is used to quantify the radioactive intensity of each band in precursor (pre) and product (pro) lanes for both unmodified (NαS) and modified (N'αS) nucleoside phosphorothioate-containing

RNAs. Then the interference at each position is calculated using the following equation [8, 11]:

$$\text{Int} = [\text{N}\alpha\text{S}_{\text{pro}}/\text{N}'\alpha\text{S}_{\text{pro}}]/[\text{N}\alpha\text{S}_{\text{pre}}/\text{N}'\alpha\text{S}_{\text{pre}}]$$

This allows one to normalize the interference value for the effect of the phosphorothioate itself as well as for the incorporation differences between unmodified and modified nucleotides. After interferences have been calculated for each nucleotide position, their values are adjusted to correct for variable sample loading on the gel, as previously described [8] (see Table 17.2): an average interference value is calculated for all positions that are within two standard deviations of the mean. Each interference value is then divided by this average. Adjusted interference values (usually referred to as κ values) that are higher than 2 or less than 0.5 are interpreted as interference or enhancement effects, respectively.

Utilization of λ discrimination factors
When the stringency of the NAIM assay cannot be easily modulated, a statistical filter, such as the λ discrimination factor [30], may help to discriminate weak and moderate interference effects (Table 17.2 and Fig. 17.8B). The λ discrimination factors are derived from κ values [8] as follows: κ values deviating by more than two standard deviations from the mean of a homogeneous data set (inosine effects, for instance) are not included in a new calculation of the standard deviation (SD'). Then for every κ value of the data set, the discrimination factor is defined as:

$$\lambda = (\kappa - 1)/\text{SD}' \quad \text{if } \kappa > 1$$
$$\lambda = (1 - 1/\kappa)/\text{SD}' \quad \text{if } \kappa < 1$$

Note that the sign of λ is completely arbitrary so that the above formulas can be formatted to suit an inverted reference system of 'negative' and 'positive' effects (as in [30]).

The use of λ factors presents several advantages. It normalizes NAIM signals for varying experimental quality and population extent among the different data sets and provides identical intensity scales for favorable ($\lambda < 0$) and detrimental ($\lambda > 0$) NAIM effects. Moreover, if one assumes that SD' adequately reflects the standard deviation of random (i.e. no interference) NAIM signals, then z tables, which are found in most statistics textbooks, may be used to select a confidence interval. For instance, an interference threshold of $|\lambda| \sim z = 2.5$ would correspond to a confidence interval for random NAIM signals above 98% (the probability, p, that a value of $|\lambda| > 2.5$ is indicative of a random signal rather than interference would be less than 0.0124). Of course, this would also depend on a number of assumptions such as signals being truly random, following a normal distribution, not being interdependent, etc. In any case, the sample size should be sufficiently large (at least over 20 random signals for every data set) and the κ values deter-

Tab. 17.2. Analysis of NAIM data obtained upon selection of reactive molecules (pro) from a pool of AαS-containing transcripts (pre).

Imager counts[1]		Int[2]	κ[3]	λ[4]
AαS$_{pre}$	AαS$_{pro}$			
53385	49431	1.08	0.96	−0.47
82981	66920	1.24	1.11	1.25
69918	97108	0.68	0.61	*−7.10*
62386	58855	1.06	0.95	−0.60
60916	59142	1.03	0.92	−0.99
51981	49982	1.04	0.93	−0.85
39672	20991	1.89	1.69	*7.71*
53039	44200	1.20	1.07	0.79
56126	41885	1.34	1.20	2.27
52541	53072	0.99	0.88	−1.55
16793	17313	0.97	0.87	−1.70
24755	23576	1.05	0.94	−0.72
28757	23571	1.22	1.09	1.02
18172	16225	1.12	1.00	0.00
35408	35057	1.01	0.90	−1.26
32444	29765	1.09	0.97	−0.35
39831	36210	1.10	0.98	−0.23
28660	26537	1.08	0.96	−0.47
27706	25654	1.08	0.96	−0.47
26325	21578	1.22	1.09	1.02
26404	21467	1.23	1.10	1.13
31698	26197	1.21	1.08	0.91
29349	24458	1.20	1.07	0.79
SD[5]		0.21	0.18	
Limits[6]				
$1 + 2 \times SD$		1.42	1.36	
$1/(1 + 2 \times SD)$		0.70	0.74	
Mean[7]		1.12	1.00	
SD′ [5,7]			0.09	

[1] Virtual imager counts were created to simulate a NAIM experiment. Each line corresponds to a single RNA position.
[2] For parental nucleotides such as AαS, Int values are determined with the formula: Int = NαS$_{pre}$/NαS$_{pro}$.
[3] The κ values are obtained by dividing the corresponding Int values by the mean (1.12), thereby correcting for differences in loading of the gel lanes.
[4] The λ discrimination factors were determined with the formulas given in the text. Statistically significant interference effects (threshold ± 2.5) are shown in italics. Such effects would not have been revealed by a standard analysis of the κ values ($0.5 < κ < 2$).
[5] Standard deviations were determined with Microsoft Excel; here, values marked by grey boxes were included.
[6] The formula $L_{low} = 1 - 2 \times SD$, which may also be used to calculate the lower limits, does not account for the inherently skewed distributions of Int and κ values.
[7] Data not included within limits (grey boxes) were excluded for mean and SD′ calculations.

mined with enough accuracy (in three to four independent experiments with an experimental error below 20%) for the λ values to reflect statistically significant interference effects ([30]; M. Boudvillain, unpublished results).

17.3
Experimental Protocols for NAIS

In this section we will describe experimental procedures specific for the NAIS assay on the group II intron system. Some of the protocols used in this method, such as *in vitro* transcription, end-labeling and iodine treatment are shared with a general NAIM assay (see Section 17.2).

17.3.1
Design and Creation of Mutant Constructs

General considerations
NAIS experiments are basically NAIM experiments on mutant and wild-type molecules that are performed in a side-by-side manner. It is generally thought that NAIS experiments are more meaningful when mutants contain single-atom changes. However, when studying molecules for which there is little structural information, it may be reasonable to initially perform NAIS using mutant constructs that contain full-base mutations. Based on the results in these coarse experiments, one can then refine the system by making single-atom mutants. This approach was essential for identifying critical tertiary interactions in the group II intron active site (such as κ–κ' and λ–λ') (Fig. 17.7) [12, 28].

When NAIM was first performed on the ai5γ group II intron, clusters of interference effects were observed in several regions of D1, including the very beginning of the intron (nt 1–5) and an asymmetric bulge in the C1 stem (Fig. 17.5A) [28]. When full-base mutations were introduced in these regions at positions 5 (G5:A) and 115 (A115:U), respectively, the resulting mutants exhibited suppression of specific interferences in D5 (Fig. 17.7), suggesting a complex tertiary interaction. This interaction (λ–λ') was then studied at higher resolution by NAIS that employed a series of D56 constructs containing single-atom changes [12]. Notably, when there is no appropriate analog available for studying certain atoms or functional groups by NAIM/NAIS (e.g. N^1 of adenosine), NAIM/NAIS can be easily complemented by chemical modification interference (e.g. using dimethylsulfate as a modifying reagent) [12].

Preparation of RNA molecules containing single-atom substitutions
RNAs that contain single-atom substitutions can be either purchased commercially or synthesized on an automated DNA–RNA synthesizer via standard solid-phase phosphoramidite method (Chapter 7). Although the variety of modified RNA oligonucleotides offered commercially is growing, one still has more options when synthesizing them in-house on an automated synthesizer (such as Applied

Biosystems, Pharmacia, etc). A variety of modified phosphoramidites with single-atom or functional group modifications can be purchased commercially (from Glen Research, ChemGenes, etc). When choosing a modified phosphoramidite for incorporation into an RNA oligonucleotide, it is important to ensure that base and/or 2'-OH protecting groups on the modified monomer are compatible with the synthetic cycle and can be easily and quantitatively removed by using standard RNA-deprotecting protocols. For example, if the desired modification is 2'-deoxy cytidine and the RNA oligo is to be deprotected by the Wincott procedure [89], it is advisable to use the phosphoramidite with an acetyl protecting group on the base, which is most compatible with the RNA deprotection protocol.

In our hands, the Wincott deprotection protocol is the most effective of all existing RNA deprotection procedures. Our protocol is similar but somewhat simpler than the procedure described in Chapter 7. Nevertheless, it allows one to synthesize the catalytic domain D5 of the group II intron which is as active as the transcribed D5 RNA.

(1) Our base-deprotection protocol is essentially the same as described in Chapter 7 with the following exceptions.
   (a) We use 40% methylamine in water instead of 8 M ethanolic methylamine in our base-deprotection mixture. Our mixture consists of concentrated (28–30%) ammonium hydroxide: methylamine (40% in water) (1:1 by volume).
   (b) We add 4 ml (not 2 ml) of base-deprotecting mixture to the polymer support and incubate it at 65 °C for 10 min (not 20–40 min) with occasional stirring.
(2) We generally synthesize trityl-off RNA and our 2'-OH deprotection protocol also differs from the one described in Chapter 7. After base deprotection, we use the following procedure.
   (a) Separate the supernatant from the support, aliquot into six to eight Eppendorf tubes and dry in the Speed Vac.
   (b) Re-suspended pellets in 250 μl of triethylamine:triethylamine *tris*(hydrofluoride):*N*-methylpyrrolidinone mix (0.75:1:1.5) and incubate 1.5 h at 65 °C.
   (c) Precipitate the deprotected RNA by adding 25 μl of 3 M KOAc and 1 ml of butanol to each tube, followed by incubation at −20 °C overnight.
   (d) Purify the RNA on a denaturing polyacrylamide gel. In order to facilitate gel purification and prevent gel overloading with salts, one may consider desalting the oligo on a C_{18} disposable cartridge (Sep-Pak, Waters; or OPC, Applied Biosystems).

For NAIS experiments on the group II intron system, D56 molecules containing single-atom substitutions were prepared by chemical synthesis. The D56 construct is 80 nucleotides long, so it was synthesized in two pieces using the procedure described above. The pieces were joined by splint-directed ligation using T4 DNA ligase (see Chapter 3), which has been particularly successful for preparing functionally active D56 molecules using the procedure described below.

(1) Combine 5′-phosphorylated downstream RNA fragment, the upstream fragment and a DNA splint oligonucleotide (60 nt) in equimolar ratio, denature in water at 95 °C for 1 min and slowly cool down to 30 °C.
(2) Supplement the mixture with 10 × buffer [500 mM Tris–HCl, 100 mM $MgCl_2$, 100 mM DTT, 10 mM ATP, 250 µg/ml BSA, pH 7.5 (25 °C); provided by the manufacturer (New England Biolabs)], RNase Inhibitor and T4 DNA ligase and incubate at 30 °C for 10–12 h.
(3) Ethanol-precipitate and gel-purify the ligated RNA.

In this method, the RNA concentration, reaction volume, and amounts of RNase inhibitor and ligase must be optimized and conditions are highly dependent on the system. For the preparative ligation we generally use 3′- and 5′-fragments of the D56 construct at 20 µM each in the total reaction volume of 100 µl containing 8 µl of RNase Inhibitor (40 U/µl) and 8 µl of T4 DNA ligase (400 U/µl).

17.3.2
Functional Analysis of Mutants for NAIS Experiments

Before using mutant molecules in an NAIS experiment, it is essential to evaluate activity using the same functional assay as for the NAIS selection step. If the mutation does not affect activity, then it is unlikely to disrupt important interactions and therefore useless for NAIS analysis. By contrast, if the mutant retains only traces of activity, it becomes very difficult to harvest sufficient material for subsequent iodine treatment. In our hands, the optimal reduction in mutant activity is between 3- and 20-fold relative to wild-type. The *trans*-branching reaction was used both for selection and for analysis of D56 mutant activity (Fig. 17.5B) [28].

17.3.3
The Selection Step for NAIS

The selection step for NAIS studies was carried out by following essentially the same protocol as described for the NAIM experiments on this system (see Section 17.2). However, the following considerations were incorporated into the experimental design.

(1) In order to compare the interference pattern of the wild-type and mutant RNAs, the corresponding reactions must be carried out under exactly the same conditions (buffer, pH, ionic conditions).
(2) The extent of the reaction for wild-type and mutant constructs must be the same (usually about 20% of product formation). This is generally achieved by varying the reaction time, which can be, for example, 20 min for the wild-type and 4 h for the mutant.
(3) If the reaction conditions must be varied to make the mutant more reactive (for example, the monovalent concentration is raised to 1 M instead of 0.5 M), also the wild-type has to be tested under these changed conditions to deter-

mine the reaction time at 20% product formation, and, most importantly, to make sure that the wild-type still shows the interferences of interest.

(4) Iodine treatment of the samples is performed according to the same protocol as described in Section 17.2. It is always necessary to have lanes of iodine-untreated reaction products for both wild-type and mutant RNAs on the gel next to iodine-treated samples, to make sure that the band attributed to the interference suppression in the presence of the mutant is not simply a degradation product due to the prolonged incubation time.

(5) In order to demonstrate that the tertiary interaction identified by NAIS is specific, it is helpful to carry out a reverse NAIS experiment. In this setup, the RNAs that contain the single-atom modification and the pool of analog-modified transcripts are switched. Ideally, a mutation or single-atom modification is introduced at the position where suppression was found in the previous experiment, and the other RNA is screened for interferences. If such an interference is observed, there is strong evidence that a specific interaction exists between the two RNAs. The validity of this approach has been successfully demonstrated on a group II intron system [12].

17.3.4
Data Analysis and Presentation

NAIS results are analyzed and quantified essentially in the same manner as NAIM data (see Section 17.2.6). The presentation of results usually includes a gel clearly showing analog interference with the wild-type RNA and suppression of this interference with the mutant RNA, and a bar graph with nucleotide analog effects for mutant molecules in comparison with the wild-type [9, 12] (Fig. 17.7). It is important to ensure that the mutation or a single-atom change results in a specific suppression of the analog interference at one or two positions with the rest of interference effects remaining unchanged, and not in five or more suppression effects at various positions throughout the molecule. The latter indicates a problem with the selection step, i.e. wild-type and mutant not treated under the same conditions or not reacted to the same extent. If one mutation causes suppression of interference at more than one position, reverse NAIS (see above) may allow one to distinguish between a complex multi-component tertiary interaction, e.g. $\lambda-\lambda'$ in the group II intron [12], and an artifact or an indirect effect.

Exacerbation of interferences caused by a mutation is more difficult to interpret. The presence of a mutation sometimes makes a system more susceptible to disturbance by modifications and can result in non-specific exacerbation of interferences with various analogs at different positions. At the same time, if a certain function is supported by a multi-component interaction involving some redundant elements, a mutation at one of these elements can make the system more sensitive to modifications at other components of this interaction. In this case the appearance of additional interferences compared to the wild-type is selectively caused by a specific mutation. This type of specific exacerbation has also been observed during studies of the ai5γ group II intron [90]. While wild-type RNA does not exhibit inosine interference at the θ' tetraloop receptor in D2 (Fig. 17.5A), it is observed upon incorporation of specific mutations in D3 (A627:G), suggesting a functional connection between this nucleotide and the $\theta-\theta'$ tertiary contact.

Acknowledgments

We thank S. Hamill for contributing data used in creating Fig. 17.6(B). A. M. P. and O. F. thank the Howard Hughes Medical Institute for the financial support of this work. Research on transcription termination (M. B.) has been supported in part by the Ligue Contre le Cancer (Comité du Loiret). This work is also supported by grants from the National Institutes of Health (NIH GM50313 and GM60620 to A.M.P.).

References

1. A. Ferre-D'Amare, J. A. Doudna, Annu. Rev. Biophys. Biomol. Struct. 1999, 28, 57–73.
2. J. H. Cate, A. R. Gooding, E. Podell, K. Zhou, B. L. Golden, C. E. Kundrot, T. R. Cech, J. A. Doudna, Science 1996, 273, 1678–1685.
3. N. Ban, P. Nissen, J. Hansen, P. B. Moore, T. A. Steitz, Science 2000, 289, 905–920.
4. P. Nissen, J. Hansen, N. Ban, P. B. Moore, T. A. Steitz, Science 2000, 289, 920–930.
5. R. Batey, R. P. Rambo, L. Lucast, B. Rha, J. A. Doudna, Science 2000, 287, 1232–1239.
6. J.-L. Chen, J. M. Nolan, M. E. Harris, N. R. Pace, EMBO J. 1998, 17, 1515–1525.
7. S. Ryder, L. Ortoleva-Donnelly, A. B. Kosek, S. A. Strobel, Methods Enzymol. 2000, 317, 92–109.
8. S. Ryder, S. A. Strobel, Methods 1999, 18, 38–50.
9. A. A. Szewczak, L. Ortoleva-Donnelly, S. P. Ryder, E. Moncoeur, S. A. Strobel, Nat. Struct. Biol. 1999, 5, 1037–1042.
10. J. Soukup, N. Minakawa, A. Matsuda, S. A. Strobel, Biochemistry 2002, 41, 10426–10438.
11. L. Ortoleva-Donnelly, A. A. Szewczak, R. R. Gutell, S. A. Strobel, RNA 1998, 4, 498–519.
12. M. Boudvillain, A. Delencastre, A. M. Pyle, Nature 2000, 406, 315–318.
13. L. Zhang, J. A. Doudna, Science 2002, 295, 2084–2088.
14. A. Burgin, N. R. Pace, EMBO J. 1990, 9, 4111–4118.
15. M. Harris, J. M. Nolan, A. Malhotra, J. W. Brown, S. C. Harvey, N. R. Pace, EMBO J. 1994, 13, 3953–3963.
16. S. Joseph, M. L. Whirl, D. Kondo, H. F. Noller, R. B. Altman, RNA 2000, 6, 220–232.
17. D. L. Abramovitz, R. A. Friedman, A. M. Pyle, Science 1996, 271, 1410–1413.
18. A. M. Pyle, F. L. Murphy, T. R. Cech, Nature 1992, 358, 123–128.
19. G. Gish, F. Eckstein, Science 1988, 240, 1520–1522.
20. K. L. Nakamaye, G. Gish, F. Eckstein, H. P. Vosberg, Nucleic Acids Res. 1988, 16, 9947–9959.
21. R. B. Waring, Nucleic Acids Res. 1989, 17, 10281–10293.
22. D. E. Ruffner, O. C. Uhlenbeck, Nucleic Acids Res. 1990, 18, 6025–6029.
23. K. L. Maschhoff, R. A. Padgett, Nucleic Acids Res. 1992, 20, 1949–1957.
24. D. Schatz, R. Leberman, F. Eckstein, Proc. Natl. Acad. Sci. USA 1991, 88, 6132–6136.

25 F. CONRAD, A. HANNE, R. K. GAUR, G. KRUPP, *Nucleic Acids Res.* **1995**, *23*, 1845–1853.
26 S. A. STROBEL, K. SHETTY, *Proc. Natl. Acad. Sci. USA* **1997**, *94*, 2903–2908.
27 L. ORTOLEVA-DONNELLY, M. KRONMAN, S. A. STROBEL, *Biochemistry* **1998**, *37*, 12933–12942.
28 M. BOUDVILLAIN, A. M. PYLE, *EMBO J.* **1998**, *17*, 7091–7104.
29 S. RYDER, S. A. STROBEL, *J. Mol. Biol.* **1999**, *291*, 295–311.
30 A. SCHWARTZ, A. R. RAHMOUNI, M. BOUDVILLAIN, *EMBO J.* **2003**, *22*, 3385–3394.
31 S. STROBEL, *Biochem. Soc. Trans.* **2002**, *30*, 1126–1131.
32 S. BASU, R. P. RAMBO, J. STRAUSS-SOUKUP, J. H. CATE, A. R. FERRE-D'AMARE, S. A. STROBEL, J. A. DOUDNA, *Nat. Struct. Biol.* **1998**, *5*, 986–992.
33 S. BASU, S. A. STROBEL, *RNA* **1999**, *5*, 1399–1407.
34 S. RYDER, A. K. OYELERE, J. L. PADILLA, D. KLOSTERMEIER, D. P. MILLAR, S. A. STROBEL, *RNA* **2001**, *7*, 1454–1463.
35 A. OYELERE, J. R. KARDON, S. A. STROBEL, *Biochemistry* **2002**, *41*, 3667–3675.
36 F. JONES, S. A. STROBEL, *Biochemistry* **2003**, *42*, 4265–4276.
37 J. K. STRAUSS-SOUKUP, S. A. STROBEL, *J. Mol. Biol.* **2000**, *302*, 339–358.
38 F. MICHEL, K. UMESONO, H. OZEKI, *Gene* **1989**, *82*, 5–30.
39 F. MICHEL, E. WESTHOF, *J. Mol. Biol.* **1990**, *216*, 585–610.
40 F. MICHEL, J.-L. FERAT, *Annu. Rev. Biochem.* **1995**, *64*, 435–461.
41 A. SZEWCZAK, L. ORTOLEVA-DONNELLY, M. ZIVARTS, A. K. OYELERE, A. V. KAZANTSEV, S. A. STROBEL, *Proc. Natl. Acad. Sci. USA* **1999**, *96*, 11183–11188.
42 H. M. PLEY, K. M. FLAHERTY, D. B. MCKAY, *Nature* **1994**, *372*, 111–113.
43 P. NISSEN, J. A. IPPOLITO, N. BAN, P. B. MOORE, T. A. STEITZ, *Proc. Natl. Acad. Sci. USA* **2001**, *98*, 4899–4903.
44 A. FERSHT, *Structure and Mechanism in Protein Science*, Freeman, New York, **1999**.
45 S. NAKANO, D. M. CHADALAVADA, P. C. BEVILACQUA, *Science* **2000**, *287*, 1493–1497.
46 A. FERRE-D'AMARE, K. ZHOU, J. A. DOUDNA, *Nature* **1998**, *395*, 567–574.
47 A. PERROTTA, I. SHIH, M. D. BEEN, *Science* **1999**, *286*, 123–126.
48 A. NESBITT, L. A. HEGG, M. J. FEDOR, *Chem. Biol.* **1997**, *4*, 619–630.
49 D. J. EARNSHAW, M. J. GAIT, *Nucleic Acids Res.* **1998**, *26*, 5551–5561.
50 J. B. MURRAY, A. A. SEYHAN, N. G. WALTER, J. M. BURKE, W. G. SCOTT, *Chem. Biol.* **1998**, *5*, 587–595.
51 P. GORDON, E. SONTHEIMER, J. PICCIRILLI, *Biochemistry* **2000**, *39*, 12939–12952.
52 P. GORDON, J. A. PICCIRILLI, *Nat. Struct. Biol.* **2001**, *8*, 893–898.
53 J. M. WARNECKE, J. P. FURSTE, W.-D. HARDT, V. A. ERDMANN, R. K. HARTMANN, *Proc. Natl. Acad. Sci. USA* **1996**, *93*, 8924–8928.
54 J. M. WARNECKE, R. HELD, S. BUSCH, R. K. HARTMANN, *J. Mol. Biol.* **1999**, *290*, 433–445.
55 A. PYLE, *J. Biol. Inorg. Chem.* **2002**, *7*, 679–690.
56 E. C. CHRISTIAN, M. YARUS, *Biochemistry* **1993**, *32*, 4475–4480.
57 G. CHANFREAU, A. JACQUIER, *Science* **1994**, *266*, 1383–1387.
58 V. SOOD, T. L. BEATTIE, R. A. COLLINS, *J. Mol. Biol.* **1998**, *282*, 741–750.
59 J. H. CATE, R. L. HANNA, J. A. DOUDNA, *Nat. Struct. Biol.* **1997**, *4*, 553–558.
60 T. PFEIFFER, A. TEKOS, J. M. WARNECKE, D. DRAINAS, D. R. ENGELKE, B. SERAPHIN, R. K. HARTMANN, *J. Mol. Biol.* **2000**, *298*, 559–565.
61 S. BASU, S. A. STROBEL, *Methods*, **2001**, *23*, 264–275.
62 J. COCHRANE, R. T. BATEY, S. A. STROBEL, *RNA* **2003**, *9*, 1282–1289.
63 J. SWISHER, C. M. DUARTE, L. J. SU, A. M. PYLE, *EMBO J.* **2001**, *20*, 2051–2061.
64 A. ARABSHAHI, P. A. FREY, *Biochem. Biophys. Res. Commun.* **1994**, *204*, 150–155.
65 J. LUDWIG, F. ECKSTEIN, *J. Org. Chem.* **1989**, *54*, 631–635.
66 A. OYELERE, S. A. STROBEL, *J. Am. Chem. Soc.* **2000**, *122*, 10259–10267.

67 E. Schenborn, R. C. Mierendorf, *Nucleic Acids Res.* **1985**, *13*, 6223–6236.
68 R. Sousa, R. Padilla, *EMBO J.* **1995**, *14*, 4609–4621.
69 Z. Huang, J. W. Szostak, *Nucleic Acids Res.* **1996**, *24*, 4360–4361.
70 A. M. Pyle, *Nucleic Acids and Molecular Biology*, F. Eckstein and D. M. J. Lilley (eds), Springer Verlag, New York, **1996**, pp. 75–107.
71 K. A. Jarrell, C. L. Peebles, R. C. Dietrich, S. L. Romiti, P. S. Perlman, *J. Biol. Chem.* **1988**, *263*, 3432–3439.
72 D. Daniels, W. J. Michels, A. M. Pyle, *J. Mol. Biol.* **1996**, *256*, 31–49.
73 P. Z. Qin, A. M. Pyle, *Biochemistry* **1997**, *36*, 4718–4730.
74 J. de la Cruz, D. Kressler, P. Linder, *Trends Biochem. Sci.* **1999**, *24*, 192–198.
75 E. Silverman, G. Edwalds-Gilbert, R. J. Lin, *Gene*, **2003**, *312*, 1–16.
76 A. Gorbalenya, E. V. Koonin, *Curr. Opin. Struct. Biol.* **1993**, *3*, 419–429.
77 S. Shuman, *Proc. Natl. Acad. Sci. USA* **1992**, *89*, 10935–10939.
78 C. H. Gross, S. Shuman, *J. Virol.* **1996**, *70*, 2615–2619.
79 E. Jankowsky, C. H. Gross, S. Shuman, A. M. Pyle, *Nature* **2000**, *403*, 447–451.
80 E. Jankowsky, C. H. Gross, S. Shuman, A. M. Pyle, *Science* **2001**, *291*, 121–125.
81 S. S. Velankar, P. Soultanas, M. S. Dillingham, H. S. Subramanya, D. B. Wigley, *Cell* **1999**, *97*, 75–84.
82 P. R. Bianco, S. C. Kowalczykowski, *Nature* **2000**, *405*, 368–372.
83 K. S. Murakami, S. A. Darst, *Curr. Opin. Struct. Biol.* **2003**, *13*, 31–39.
84 R. Padilla, R. Sousa, *Nucleic Acids Res.* **2002**, *30*, e138.
85 M. Kashlev, N. Komissarova, *J. Biol. Chem.* **2002**, *277*, 14501–14508.
86 V. Gopal, L. G. Brieba, R. Guajardo, W. T. McAllister, R. Sousa, *J. Mol. Biol.* **1999**, *290*, 411–431.
87 L. E. Macdonald, Y. Zhou, W. T. McAllister, *J. Mol. Biol.* **1993**, *232*, 1030–1047.
88 V. Epshtein, F. Toulme, A. R. Rahmouni, S. Borukhov, E. Nudler, *EMBO J.* **2003**, *22*, 4719–4727.
89 F. Wincott, A. DiRenzo, C. Shaffer, S. Grimm, D. Tracz, C. Workman, D. Sweedler, C. Gonzalez, S. Scaringe, N. Usman, *Nucleic Acids Res.* **1995**, *23*, 2677–2684.
90 O. Fedorova, T. Mitros, A. M. Pyle, *J. Mol. Biol.* **2003**, *330*, 197–209.
91 M. Zuker, *Nucleic Acids Res.* **2003**, *31*, 3406–3415.

18
Nucleotide Analog Interference Mapping: Application to the RNase P System

Simona Cuzic and Roland K. Hartmann

18.1
Introduction

18.1.1
Nucleotide Analog Interference Mapping (NAIM) – The Approach

In classical mutational studies, only the base moiety of a nucleotide can be replaced with one of the three natural alternatives. Even a simple C → U transition affects more than one functional group and exchange of the 4-amino for a keto group represents a rather radical chemical change that can have profound effects on RNA functionality. A more specific and versatile chemogenetic approach is NAIM, which allows us to probe the functional consequences of changes as minor as single-atom substitutions in the base, sugar or phosphate moiety. For example, in the case of a guanosine to inosine modification, the chemical alteration is restricted to deletion of the 2-amino group without additionally replacing the 6-keto with an amino group as in G → A mutations.

At the onset of NAIM studies, a pool of RNA molecules with limited numbers of randomly distributed nucleotide analogs is synthesized. Such a pool of RNAs is then subjected to a selection procedure to separate active variants from those with impaired function due to modification at a particular location. Subsequent comparative analysis of the distribution of modifications in the active RNA fraction and a reference fraction (e.g. the fraction of molecules with impaired function or the original unselected pool) reveals positions critical for function. The salient feature of the method is that all incorporated nucleotide analogs additionally carry a phosphorothioate modification (one non-bridging phosphate oxygen replaced with sulfur), which permits to specifically cleave the nucleic acid chain by iodine [1] exclusively at the sites of analog incorporation (for details, see Chapter 17). Iodine treatment thus results in A-, C-, G- or U-specific sequence ladders on denaturing polyacrylamide gels.

The partial modification of RNA is achieved by the presence of nucleoside α-thiotriphosphate analogs during *in vitro* transcription by T7 RNA polymerase, resulting in the aforementioned pool of RNA molecules, each carrying a low number

Handbook of RNA Biochemistry. Edited by R. K. Hartmann, A. Bindereif, A. Schön, E. Westhof
Copyright © 2005 WILEY-VCH Verlag GmbH & Co. KGaA, Weinheim
ISBN: 3-527-30826-1

of randomly distributed modifications. The elegance of the method lies in the capacity to simultaneously screen for the functional contribution of a particular chemical group at almost every, for example, A residue in an RNA chain. There are two major limitations: (1) residues that show a strong phosphorothioate interference effect *per se* will be insensitive to the effect of the additional modification, such as a 2′-deoxy substitution in case of 2′-deoxy NTPαS analogs and (2) phage RNA polymerases do not accept all kinds of nucleotide analogs as substrates (see Section 18.1.2 and Chapter 1). Chemical RNA synthesis expands the scope of possible modifications (e.g. introduction of Sp- in addition to Rp-phosphorothioates [2]), but such approaches are usually more tedious and require equipment for chemical RNA synthesis as well as special protocols in order to introduce a low level of randomly distributed modifications. Also, chemical RNA synthesis is practically limited to an RNA chain length of about 50 nt, thus excluding directly screening of larger RNAs, such as self-splicing introns or RNase P RNA.

The analogs available for NAIM studies can be divided into three categories [3], according to their main attributes: (1) if they primarily change the chemical properties of the substitutent, (2) delete a functional group or (3) introduce a bulky substituent. Depending on the type of modification introduced, NAIM experiments have the potential to reveal the following information:

- An Rp-phosphorothioate modification *per se* (AMPαS, GMPαS, CMPαS, UMPαS) may identify crucial coordination sites for Mg^{2+} ions. Substitution of sulfur for a non-bridging phosphate oxygen essentially abolishes inner-sphere coordination to Mg^{2+}, because Mg^{2+}, a "hard" Lewis acid, prefers to bind oxygen, a "hard" Lewis base, relative to the much more polarizable and thus "softer" sulfur [4]. However, addition of more thiophilic metal ions ("softer" Lewis acids) such as Mn^{2+} or Cd^{2+} may restore, to varying extent, metal ion binding to the thiophosphate, leading to a (partial) rescue of the functional defect [5].
- C^7-deaza purine analogs (c7-AMPαS, c7-GMPαS) are employed to reveal N^7 positions involved in hydrogen binding or metal ion coordination. The latter aspect may be particularly relevant if RNA structure and function is probed in the presence of transition metal ions, such as Mn^{2+} or Zn^{2+}, which form inner-sphere complexes with the N^7 of purines [6, 7].
- Ribose 2′-deoxy modifications allow to probe for 2′-hydroxyls involved in tertiary contacts.
- IMPαS, incorporated by T7 RNA polymerase instead of G nucleotides, is suited to probe the role of guanine exocyclic amino groups in hydrogen bonding. For *Escherichia coli* RNase P RNA, relatively few inosine interference effects were detected in regular helices, suggesting that helix destabilization by this modification is of minor importance for the function of the RNase P ribozyme [8]. However, destabilization of secondary structure can become important if inosines are part of a short intermolecular hybrid helix required to bind the substrate to the ribozyme, particularly under conditions where ribozyme molecules compete for a limited amount of substrate RNAs [9]. For such cases, combined analysis of IMPαS and N^2-methyl-GMPαS interference patterns was reported as a strategy

to differentiate between helix destabilization and loss of important tertiary interactions as the cause of interference [3, 9]. The N^2-methyl group can still form a hydrogen bond with the O^2 of cytosine in Watson–Crick base pairs, but has lost its capacity to participate in a bifurcated hydrogen bonding frequently observed in tertiary contacts that involve the 2-amino group of G residues [8–11].
- Analogs incorporated at A positions by T7 RNA polymerase, such as purine, N^6-methyl-adenosine, 2-aminopurine and 2,6-diaminopurine, all of which are commercially available, probe the N^6 position in terms of chemical properties and steric constraints and the tolerance for an additional 2-amino group on the minor groove edge of the base, respectively [12].

18.1.2
Critical Aspects of the Method

18.1.2.1 Analog Incorporation

T7 RNA polymerase incorporates Sp-NTPαS analogs, yielding Rp-phosphorothioate-modified RNAs due to inversion of configuration at the phosphorus atom during polymerization [13]. It has previously been documented that Sp-NTPαS analogs are incorporated with essentially the same efficiency as normal NTP substrates by T7 RNA polymerase [14]. However, for other phosphorothioate-tagged analogs, which carry an additional modification at the base or sugar moiety, incorporation efficiency is in most cases lower. One exception is ITPαS which is accurately incorporated in place of guanosine and with comparable efficiency [8, 15]. Many analogs (such as those with 2′-ribose modifications) are better accepted by the Y639F mutant T7 RNA polymerase which shows a greater tolerance toward changes of functional groups in the minor groove [16, 17]. The reader is referred here to Chapter 17 and the publication by Ryder and Strobel [16] for more details on individual analogs.

The extent of analog incorporation is adjusted to 2.5–10% (usually 5%) in NAIM studies, although incorporation efficiency may not exceed 1–2% in the case of some analogs [16]. A modification extent of 5% permits good detection and quantitation of iodine hydrolysis bands, but avoids two problems associated with higher modification extents: (1) RNA inactivation due to phosphorothioate-tagged analog incorporation at multiple sites per molecule, as seen with fully (100%) AMPαS-, CMPαS- or GMPαS-modified *E. coli* RNase P RNA [18] and (2) an increased probability that each RNA molecule carries a modification at a site of strong interference in addition to weakly interfering modifications; as a consequence, the sites of strong interference will, by themselves, fully determine the deficiency status of an RNA molecule irrespective of additional weaker interferences, thus masking the latter.

One should also be aware that analogs are not incorporated to the same extent at all transcript positions, the incorporation pattern being largely specific for the analog and the individual RNA under investigation. One observation is the lack or reduction of analog incorporation at homo-di- or homo-oligonucleotide stretches [16].

We usually follow a simple strategy to adjust the extent of analog incorporation, based on the roughly equal incorporation efficiency of NTP and Sp-NTPαS substrates. For example, to assess ITPαS incorporation efficiency, we transcribed RNase P RNA in the presence of 2.5, 5.0 and 10% ITPαS; in parallel, we performed transcriptions in the presence of 5% GTPαS and 95% GTP as the reference. Gel analysis of samples after iodine hydrolysis then revealed that transcripts with 2.5% IMPαS modification resulted in iodine hydrolysis band intensities comparable with those of transcripts carrying 5% GMPαS modifications. For other analogs, however, this relation is reversed due to the less efficient incorporation of analog versus its NTPαS reference substrate.

18.1.2.2 Functional Assays

The second crucial step in NAIM studies, aside from analog incorporation, is the selection assay that partitions functional and defective RNA molecules. We will mention here examples of such assays that have been used in NAIM analyses of RNase P RNA (examples of selection assays developed for other ribozyme systems are described in Chapter 17):

- Separation of modified RNase P RNA pools into precursor tRNA-binding and non-binding fractions via adsorption of biotinylated precursor tRNA to streptavidin–agarose beads [12].
- Separation of modified RNase P RNA pools into RNase P protein-binding and non-binding fractions via adsorption of the His-tagged protein to Ni-NTA agarose beads [19].
- Partitioning of self-cleaving RNase P RNA ribozyme–substrate conjugates into reacted and unreacted fractions and separation by denaturing PAGE [20–22].
- Separation of modified RNase P RNA pools into tRNA-binding and non-binding fractions by gel retardation [8, 18, 23, 24].

18.1.2.3 Factors Influencing the Outcome of NAIM Studies

Factors that influence the outcome of NAIM studies include:

- The functional aspect selected for. As an example, Boudvillain and Pyle [25] reported their *trans*-branching assay for group II intron self-splicing to be more sensitive to modification interference and thus perturbations of tertiary structure than *cis*-splicing assays. They argued that the *trans*-branching approach is so effective because the stabilization energy resulting from essential tertiary interactions has to counterbalance the entropic penalty inherent to the assembly of a two- or multi-component system.
- Reaction conditions of the selection assay (nature and concentration of mono- and divalent cations, pH and temperature). High salt conditions, for example, suppressed weaker interference effects in group II intron *cis*- and *trans*-splicing assays [26]. Variations in pH were shown to alleviate or exacerbate interference effects [20, 21], attributable to changes in rate-limiting steps of ribozyme-catalyzed reactions [24]. Several modifications that interfere with tRNA binding

to RNase P RNA were suppressed at higher RNA concentrations [18]. Likewise, all factors that alter the dissociation constant of complexes in *trans*-binding assays (such as the temperature) will, to some extent, change the pattern and strength of interference effects.
- Inefficient analog incorporation at certain positions in the RNA chain (e.g. at homonucleotide stretches) and gel artifacts, such as band compressions at a string of G residues [8], which can limit the information content of NAIM experiments. The reader should further be aware that fluctuations in the strength of interference effects between individual experiments are considerable. The extent of such fluctuations is expected to rise with (1) an increasing number of experimental steps involved throughout the entire procedure, (2) difficulties to kinetically control a reaction catalyzed by an enzyme or (3) difficulties to define and control the proportion of functional protein in the case of RNA–protein-binding studies (e.g. if a His-tagged protein is coupled to a Ni-NTA affinity resin; [19]).

18.1.3
Interpretation of Results

The *Rp*-phosphorothioate modification itself can cause interference effects that are due to disruption of hydrogen bonding or inner-sphere coordination of Mg^{2+} normally involving the *pro*-Rp oxygen at this location. At most sites, the sulfur substitution has little or no effect on RNA structure; in rare cases, however, it may substantially change local structure [27] and, as a consequence, could induce global conformational changes. When NAIM is used to probe, for example, RNA–ligand interactions, a phosphorothioate modification could thus affect ligand binding even if the site of modification is at considerable distance to the binding interface. The resulting interference effect will then be hard to interpret solely on the basis of NAIM data. Along the same lines, phosphorothioate interference effects are often found to be clustered in densely packed RNA core structures which play a key role in RNA architecture [18, 23]. Such RNA regions are apparently very sensitive even to minor structural and chemical perturbations caused by the sulfur substitution, and thus many interference effects are likely to reflect changes of global structure rather than direct contacts to an RNA ligand.

Partial suppression of phosphorothioate interference in the presence of Mn^{2+} is often considered to indicate direct metal ion coordination to the respective phosphate oxygen [28]. The ability of Mn^{2+}, but the inability of Mg^{2+}, to coordinate to the sulfur is made responsible for this effect. However, although Mg^{2+} and Mn^{2+} ions can occupy basically the same metal ion binding pockets, they may coordinate in a slightly different way, as originally observed in tRNA crystals [29]. Likewise, Mg^{2+} and Mn^{2+} interacted differently with the phosphate of A9 in two hammerhead ribozyme crystal structures [11, 30]. Nevertheless, Mn^{2+} rescue of phosphorothioate interferences in the central P4 helix of RNase P RNA [18, 20] provided the basis for detailed studies of metal ion coordination in this region using RNase P RNA constructs with single-site phosphorothioate modifications that were analyzed for kinetic defects as well as Mn^{2+} and Cd^{2+} rescue effects [31, 32].

Generally, interpretation of NAIM results remains ambiguous without further information from other approaches. For example, if NAIM is performed to identify functional groups that are involved in binding of a protein or another RNA, it is impossible to differentiate whether interference effects represent direct contact sites or indirect effects due to destabilization of the global RNA fold. However, additional information often permits to interpret interference data, as illustrated by two examples. For the *Tetrahymena* ribozyme, the crystal structure of the P4–P6 domain [10] revealed a hydrogen bonding interaction between the exocyclic amine of G212 in helix P4 and A184 in the A-rich bulge motif, which bridges the core helix P4 and the three-way junction of helices P5a, b and c. NAIM identified N^2-methyl-GMPαS, but not IMPαS, interference at G212 [9]. The authors concluded that deletion of the hydrogen bond due to inosine modification was not sufficient to significantly impair intron splicing, while introduction of the bulky N^2-methyl group in the P4 minor groove substantially destabilized this tightly packed region of the RNA. Another example comes from the RNase P RNA system. Comparative sequence analysis and results from biochemical studies in combination with computer-aided derivation of RNase P RNA architecture led to the proposal of the loop–helix tertiary interactions L18–P8 and L8–P4 [33, 34]. Further, Easterwood and Harvey [35] proposed a model of tRNA 3′-CCA end binding to the P15 loop of *E. coli* RNase P RNA, which involves formation of two consecutive base triples. A variety of nucleotides in the corresponding structural elements of RNase P RNA showed Rp-phosphorothioate, 2′-deoxy, inosine and/or c7-deaza interference effects in a tRNA binding assay [8, 18, 23, 24]. The interference data went surprisingly well with the above-mentioned interactions and allowed us to confirm and refine their atomic details [8, 24].

On the other hand, NAIM approaches can be an important tool in cases where it is unclear if a ribozyme crystal structure represents the catalytically competent conformer. A study combining point mutations and NAIM showed formation of a base triple in the core of the *Tetrahymena* ribozyme, which contributes to substrate helix docking and stabilization of active site conformation [36]. The formation of this base triple indicated that a crystallized form of the ribozyme [37] would require a substantial rearrangement to adopt an active conformation.

For *E. coli* RNase P RNA, results from a series of NAIM experiments permit to extract the following conclusions:

(1) Although largely different functional assays and assay conditions were employed (self-cleavage of ribozyme–substrate conjugates at 3 M Na$^+$ or NH$_4^+$ and 1–25 mM Mg^{2+}, gel-resolvable binding of mature tRNA to RNase P RNA in the presence of 0.1 M NH$_4^+$ and 0.1 M Mg^{2+}, RNase P RNA binding to immobilized precursor tRNA in the presence of 1 M NH$_4^+$ and 25 mM Ca^{2+}), a substantial number of identical interference effects has been observed ([24] and references therein; [22]). The conclusion is that modifications at these vulnerable sites destabilize the tertiary fold, thus affecting substrate and product binding as well as catalysis. In fact, the vast majority of interference effects likely reflects perturbation of RNA tertiary structure rather than representing

functional groups directly involved in catalysis or the binding of mature or precursor tRNA.

(2) Interference effects cluster in regions which (i) organize the conserved catalytic core of RNase P RNA (such as P4, J2/4, J3/4, J18/2 and J5/15), which (ii) are tightly packed with the help of multiple metal ions (the P2–P3/P1–P4 four helix junction [32]) and which (iii) organize the metal ion-rich core of the specificity domain (J11/12, P11 [8, 12, 18, 23, 24, 38]).

18.1.4
Nucleotide Analog Interference Suppression (NAIS)

NAIS is an extension of NAIM that permits to overcome the uncertainty of NAIM experiments which leave open if an interference effect reflects a direct contact between two interacting macromolecules. In NAIS, the partially analog-modified RNA pool (RNA 1) is analyzed for interference effects in two parallel setups: one employing the interacting RNA (RNA 2) in its wild-type form, as in NAIM, and the second using RNA 2 with a point mutation or a single chemical group altered at a location suspected to interact directly with RNA 1. If a functional group in RNA 1 normally interacts with the functional group in RNA 2 that has been changed by the aforementioned mutation or modification, then the interference effect observed in RNA 1 for disrupting this contact should disappear, because the contact has already been abolished by the mutation or modification in RNA 2. A critical point of such NAIS experiments is to adapt the functional assay with the mutant or modified RNA 2 in such a way that the extent of reaction or complex formation is the same as in the reference assay with wild-type RNA 2. For a more detailed description and illustration of NAIS, see Chapter 17.

18.2
NAIM Analysis of *Cis*-cleaving RNase P RNA–tRNA Conjugates

18.2.1
Characterization of a *Cis*-cleaving *E. coli* RNase P RNA–tRNA Conjugate

Cis-cleaving RNase P RNA (P RNA)–tRNA conjugates of the type shown in Fig. 18.1(A) open up the perspective to identify functional groups that are crucial to cleavage chemistry, either applying NAIM or NAIS. In previous related approaches, the tRNA substrate was tethered to internal positions of *E. coli* P RNA [20–22]. However, these constructs required 3 M monovalent salt for efficient self-cleavage and we were concerned that such conditions may suppress informative interference effects to an unwanted extent. We therefore pursued a somewhat different strategy, based on a transcript consisting of *E. coli* P RNA, a linker region and the 5′ half of a bacterial tRNA^{Gly} which is annealed to the tRNA 3′ half (added in excess over P RNA–tRNA 5′ half), creating a self-cleaving P RNA–tRNA conjugate (Fig. 18.1A). This bipartite system prevents self-cleavage at the tRNA 5′ end already

18.2 NAIM Analysis of Cis-cleaving RNase P RNA–tRNA Conjugates

during RNA preparation and permits to use tRNA 3' halves with single-site modifications for NAIS experiments. Before conducting NAIM experiments, it was essential to analyze the kinetic behavior of the conjugate in order to determine the conditions under which *cis*-cleavage only occurs. Moreover, we had to clarify what limits the reaction rate, for example cleavage chemistry or a refolding step. Our finding that the cleavage rate remained constant in the tested range of P RNA–tRNA 5' half concentrations (Fig. 18.1C) demonstrated that this type of ribozyme acts only *in cis* and not *in trans* under dilute conditions. Exclusion of *trans*-cleavage is essential for the separation of active ribozymes from less active and inactive RNA variants. The data points for the time course of the *cis*-cleavage reaction were best described by the sum of two first-order reactions (Fig. 18.1B), suggesting that there are at least two populations of ribozymes reacting with different velocities: the correctly folded population (around 20%, see Limit 1 in Fig. 18.1B, panel II) reacts fast, while the fraction of slower reacting conjugates either has to change its conformation before *cis*-cleavage can occur or uses an alternative, albeit slower reaction pathway. The linear relationship for log k_1 versus pH with a slope of about 1 in the range of pH 5.2–6.5 (Fig. 18.1D, panel I) indicates that the chemical step is rate-limiting in the initial phase of the reaction [39]. In contrast, the cleavage rate constant k_2 describing the slow turnover (Fig. 18.1D, panel II) was independent of pH, supporting the idea that k_2 reflects the rate of a slow refolding step. The rate constant k_1 was further shown to be independent of the linker length in the range of 45–130 nt (Fig. 18.1E); also, tRNA 3' half concentrations exceeding that of P RNA–tRNA 5' half (0.1 nM) by a factor of 800 did not further increase k_1 (Fig. 18.1F).

18.2.2
Application Example

In the following section we will describe, as an application example, IMPαS modification experiments using the *E. coli* P RNA–tRNA conjugate with a 53-nt long linker connecting the P RNA and tRNA moieties. Usually, one compares IMPαS with GMPαS interference patterns to be able to ascribe interference effects to the thioate and/or additional base modification (see Chapter 17). However, our initial intention was to determine to which extent IMPαS interference patterns obtained with the *cis*-cleaving conjugate overlap with those observed in a gel retardation assay selecting for high affinity tRNA binding to *E. coli* P RNA [8]. The protocol is tailored to NAIM analysis of the 3' half of P RNA. Usually, analysis from the 5' end is conducted in parallel in order to fully resolve interference effects for an RNA of the size of P RNA (around 400 nt). Corresponding analysis from the P RNA 5' end requires two major changes in the procedure outlined below and illustrated in Fig. 18.2: (i) in Step 1, the procedure is started exclusively with 5'-^{32}P-end-labeled pool RNA (instead of adding some 5'-end-labeled to predominantly unlabeled RNA), using about 15 times more radioactivity than specified in Protocol 5, and (ii) 3'-end-labeling (Step 6) is omitted. In addition, the simplest strategy for this setup is to directly excise the cleaved and uncleaved fractions from the same gel lane (no need to perform Step 4b), since 5'-end-labeled bands derived from

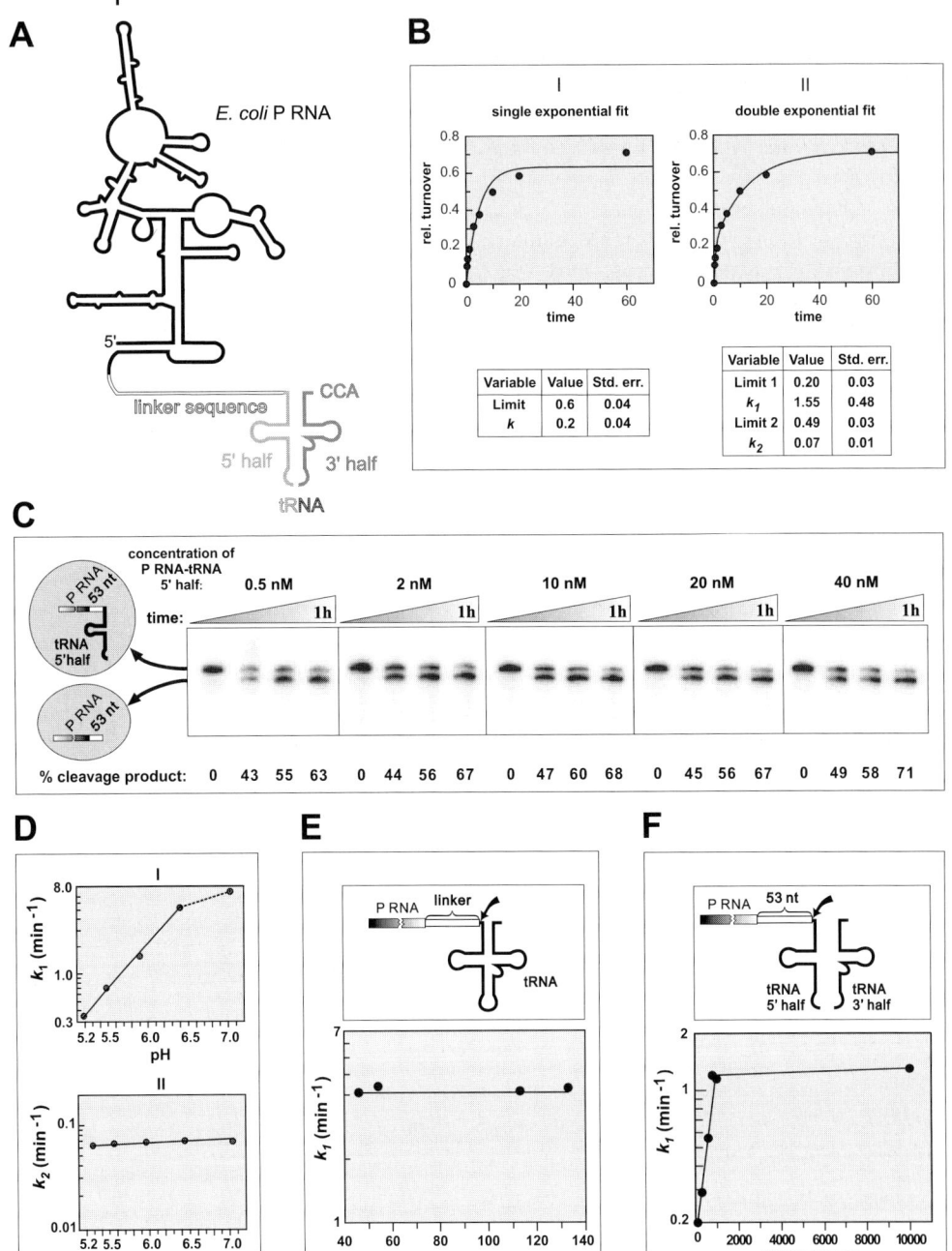

iodine hydrolysis within the P RNA moiety will co-migrate for the cleaved and uncleaved fraction in the final gel analysis (Step 7).

The experimental procedure involved the following steps, illustrated in Fig. 18.2:

(1) Transcription of P RNA–tRNA 5′ half carrying a low degree (2.5%) of randomly distributed IMPαS modifications, 5′-^{32}P-end-labeling of an aliquot of the RNA pool for the detection of cleaved and uncleaved molecules on denaturing PAA gels, and transcription of tRNA 3′ half (unmodified for NAIM). The concept for NAIS experiments is illustrated as well. For example, with the goal to identify functional groups in P RNA that interact with 2-hydroxyls in the T arm, two tRNA 3′ halves will be utilized: an all-ribose 3′ half and a variant thereof with a single 2′-deoxyribose (indicated by a filled triangle); interference patterns obtained with the two tRNA 3′ halves will then be compared in order to identify interference suppression effects when using the tRNA 3′ half carrying a single 2′-deoxy modification in the T arm. Experimental steps for NAIS will be identical to those of NAIM, except that the reaction conditions have to be adapted for the 2′-deoxy-modified 3′ half to give the same extent of reaction as for the all-ribose tRNA 3′ half.

Fig. 18.1. Characterization of a model *E. coli* P RNA–tRNA conjugate. (A) Schematic representation of the construct consisting of the P RNA moiety (black), a linker sequence (thin double lines) and the tRNA 5′ half (light grey); the complete tRNA substrate is reconstituted by annealing the tRNA 3′ half (dark grey). (B) Self-cleavage of the conjugate shown in (A), equipped with a 53-nt long linker. Assays were performed as follows. The P RNA–tRNA^{Gly} complex was formed by annealing the tRNA 3′ half to 5′-end-labeled P RNA–tRNA 5′ half in the presence of 100 mM NH$_4$Cl and 5 mM CaCl$_2$ to avoid uncontrolled self-cleavage. The P RNA–tRNA 5′ half concentration was 0.3 nM and that of tRNA 3′ half 300 nM. The annealing mixture was heated to 95 °C for 2 min, then transferred to and cooled down in a heating block adjusted to 50 °C and pre-incubated for 30 min at 50 °C. After pre-incubation, the mixture was diluted to 0.1 nM P RNA–tRNA 5′ half and 100 nM tRNA 3′ half by adding NH$_4$Cl, urea and CaCl$_2$ to a final concentration of 100, 100 and 5 mM, respectively. The reaction was started by adding MgCl$_2$ to a final concentration of 36 mM. Samples were withdrawn at different time points and analyzed by 8% denaturing PAGE. The time course was fit to either a single first-order (panel I) or two consecutive first-order reactions (panel II). (C) This panel documents that no significant *trans*-cleavage occurred under the conditions tested, since the cleavage rate was constant at different concentrations of P RNA–tRNA 5′ half (0.5–40 nM final concentration in the *cis*-cleavage assay). (D) pH dependence of k_1 (measuring the rate of the fast initial phase of the reaction) and k_2 (measuring the rate of the second slower phase of the reaction; see panel B II); the linear relationship of log k_1 versus pH with a slope of about 1 in the range of pH 5.2–6.5 suggests that the chemical step is rate-limiting in the initial phase of the reaction. (E) Influence of the linker length on the cleavage rate constant k_1. The linker length is defined as the number of nucleotides that separate the P RNA 3′-end from the tRNA 5′-end. No significant differences were observed in the cleavage rate constant k_1 among complexes with linker lengths between 45 and 133 nt. (F) To find the saturation limit of the tRNA 3′ half, its concentration was varied up to a 10,000-fold excess over P RNA–tRNA 5′ half (53-nt linker) whose concentration was 0.1 nM; tRNA 3′ half concentrations exceeding that of P RNA–tRNA 5′ half by more than a factor of 800 did not further increase the cleavage rate constant k_1. For reaction conditions in (D)–(F), see legend to (B).

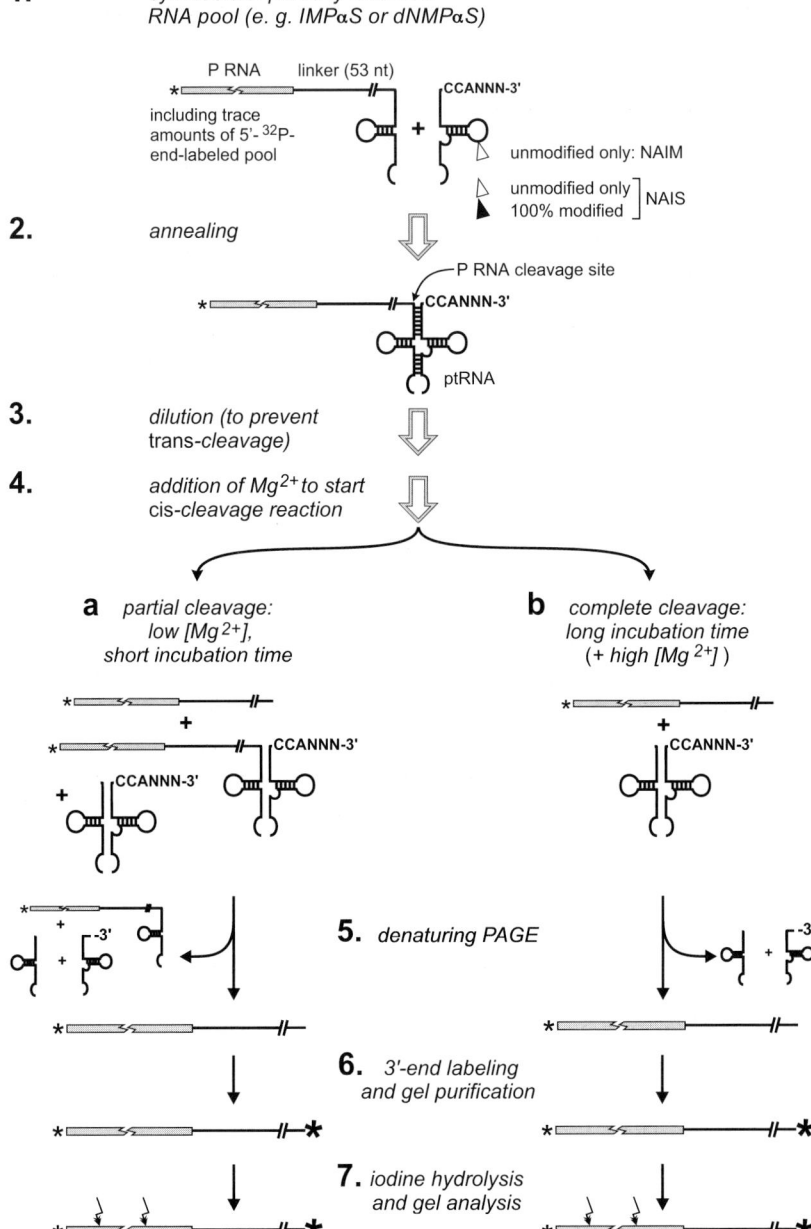

Fig. 18.2. Flow scheme for NAIM (NAIS) analysis of a *cis*-cleaving *E. coli* RNase P RNA–tRNA conjugate. For details, see Sections 18.2.2 and 18.2.4.

(2) Formation of the P RNA–tRNA^{Gly} complex by annealing the tRNA 3′ half (8.8 µM) to the P RNA–tRNA 5′ half (91 nM, including trace amounts of 5′-end-labeled material) in the presence of 100 mM NH$_4$Cl and 5 mM CaCl$_2$ at pH 5.9, conditions that prevent uncontrolled self-cleavage; the annealing mix is heated to 95 °C for 2 min, then transferred to and cooled down in a heating block adjusted to 50 °C and pre-incubated for 30 min at 50 °C.

(3) 3.8-fold dilution of the annealed mix to 24 nM P RNA–tRNA 5′ half and 2.3 µM tRNA 3′ half by addition of NH$_4$Cl, urea and CaCl$_2$ to final concentrations of 100, 100 and 5 mM, respectively.

(4) Starting the *cis*-cleavage reaction by addition of MgCl$_2$ to a final concentration of 36 mM.
   (a) Stopping the reaction after 2 min at 50 °C, resulting in 20–30% product formation corresponding to the fast phase of the reaction (Fig. 18.1B, panel II).
   (b) A parallel reaction was incubated for 2 h, resulting in essentially complete substrate conversion, serving as the reference RNA pool for NAIM analysis. The reason for taking this sample as the reference pool, and not the original, untreated RNA pool, is that the same length species as in the short incubation is generated (P RNA plus linker). One potential drawback is that modification at some positions may entirely block *cis*-cleavage, resulting in the absence of an iodine hydrolysis band for the 2-min fraction as well as the reference pool, with the effect that strong interference effects would escape notice. A strategy to circumvent this problem is to perform the 2-h incubation for the reference pool under *trans*-cleavage assay conditions by elevating the Mg^{2+} concentration. Yet another option is to first load the iodine-hydrolyzed starting pool (P RNA–tRNA 5′ half), or possibly the uncleaved fraction from Step 4a, onto the gel (Step 7 below), let it run for some time, and then load the RNA fraction that was cleaved within 2 min to compensate for its reduced length due to the absence of the tRNA 5′ half (Fig. 18.2). In any case, it is advisable to compare the iodine hydrolysis patterns of alternative reference pools to address the potential ambiguity mentioned above.

(5) Denaturing PAGE and elution of 5′-end-labeled cleavage products (P RNA plus linker).

(6) 3′-End-labeling of eluted RNAs, such that the radioactivity of the 3′-^{32}P-label exceeds that of the 5′-label by a factor of around 100.

(7) Iodine hydrolysis, denaturing PAGE analysis, phosphoimaging and quantification of band intensities.

18.2.3
Materials

Sp-ITPαS was custom-synthesized and purified by IBA (Göttingen, Germany; http://www.iba-go.com/). A variety of other nucleoside α-thiotriphosphates, mainly those with 2′-ribose or adenine base modifications in addition to ATPαS, CTPαS, GTPαS, ITPαS and UTPαS, are available from Glen Research and can be found in their catalogue under the keyword NAIM (http://www.glenres.com/index.html). The catalogue, which can be downloaded as a pdf, also lists their authorized distributors outside the US (e.g. Eurogentec in France). Likewise, IBA (see above) offers numerous nucleoside α-thiotriphosphates including several halogen-derivatized base analogs. For additional analogs, see Chapter 17.

18.2.4
Protocols

Protocol 1: Transcription of P RNA–tRNA 5′ half carrying randomly distributed IMPαS modifications

Transcripts can be initiated with the nucleoside guanosine that introduces a 5′-terminal hydroxyl group to permit direct 5′-end-labeling (Fig. 18.2, Step 1). Alternatively, one may perform a standard transcription initiating RNA chains with 5′-guanosine triphosphate, followed by a dephosphorylation step with alkaline phosphatase prior to 5′-end-labeling, as described in Chapters 3 and 17. With the protocol outlined below, the average yield of P RNA–tRNA 5′ half was in the range of 400–1000 μg (2.4–6 nmol) per 500 μl transcription mix.

(1) Before starting to prepare the reaction mix, incubate the guanosine stock solution at 75 °C in a thermoshaker until the solution becomes clear; then stop shaking, but leave the solution at 75 °C.
(2) Prepare the reaction mix – except for guanosine and T7 RNAP – at room temperature; add the components in the order they are presented in the table below, and afterwards pre-warm to 37 °C before addition of guanosine. Add the preheated guanosine solution rapidly to the reaction mix and vortex to avoid guanosine precipitation; start the reaction by addition of enzyme.

Transcription reaction, 500 μl:		Final concentration
RNase-free water	83.5 μl	
HEPES, pH 8.0, 1 M	40 μl	80 mM
DTT 100 mM	75 μl	15 mM
MgCl$_2$ 3 M	5.5 μl	33 mM
spermidine 100 mM	5 μl	1 mM
rNTP mix (25 mM each)	75 μl	3.75 mM (each)
ITPαS 4.33 mM	11 μl	0.095 mM
Template (linearized plasmid 3.2 kb) 0.5 μg/μl	40 μl	40 μg/ml
Pyrophosphatase 200 U/ml	5 μl	2 U/ml
• Pre-warm mixture to 37 °C, then add:		
Guanosine (30 mM, kept at 75 °C)	150 μl	9 mM
T7 RNAP 200 U/μl	10 μl	4000 U/ml

(3) Incubate for 4 h at 37 °C.
(4) Extract RNA once with an equal volume phenol (saturated with 10 mM Tris–HCl, 1 mM EDTA, pH 7.5–8.0) and twice with equal volumes of chloroform.
(5) Precipitate by adding 1/5 volumes 2 M NH$_4$Ac (pH 7.0) and 2.7 volumes

of ethanol. Mix and keep at −20 °C for at least 1 h; centrifuge at 4 °C and 16 000 g for 1 h in a desktop centrifuge.
(6) Dissolve the pellet in 50 µl RNase-free water and 50 µl gel loading buffer (see Protocol 2a).
(7) Purify the RNA by denaturing PAGE as described in Protocol 2a.

Protocol 2a: Purification of analog-modified P RNA–tRNA 5′ half pools by denaturing PAGE

- Gel loading buffer: 2.7 M urea; 1 × TBE; 67% (v/v) formamide; 0.01% (w/v) each bromophenol blue and xylene cyanol.
- RNA elution buffer 1: 200 mM Tris; 1 mM EDTA; 0.1% SDS; pH 7.0 at room temperature (this buffer may also be prepared without SDS, e.g. when the eluted RNA is afterwards subjected to 3′-end-labeling).
- RNA elution buffer 2: 1 M NH_4OAc; pH 7.0.

Prepare a 5% polyacrylamide (24:1 acrylamide:N,N'-methylene bisacrylamide) gel in 8 M urea and 1 × TBE, 15-cm wide, 35-cm long and 1- or 0.5-mm thick. The pocket size depends on the amount of RNA that has to be purified (about 10-cm pocket width for a 500-µl preparative transcription in case of 1 mm gel thickness).

(1) Load the RNA sample from Protocol 1, Step 6, onto the gel immediately after extensive rinsing of the pocket with a syringe to remove urea solution that has diffused from the gel matrix into the pocket; run the gel at 20–25 mA for about 3 h until the xylene cyanol has reached the bottom of the gel.
(2) Separate the glass plates and place the gel between two sheets of kitchen wrapping film.
(3) Visualize RNA band(s) by UV shadowing (for details, see Chapter 3). The exposure should be minimized to avoid UV-induced damage of the RNA. Mark the band of interest with a pen or marker and excise it with a sterile scalpel under normal light. Check correct excision by UV shadowing.
(4) Elute the RNA in the appropriate volume of RNA elution buffer 1 or 2 (3 ml for RNA purified from a 500-µl transcription assay) overnight at 4 °C with shaking.
(5) Collect the supernatant.
(6) For ethanol precipitation, add 1/5 volumes 2 M NH_4OAc (pH 7.0) and 2.7 volumes of ethanol (omit NH_4OAc when using elution buffer 2). Mix and keep at −20 °C for at least 1 h and centrifuge at 4 °C and 16 000 g for 1 h in a desktop centrifuge.
(7) Wash the pellet with 100–200 µl ice-cold 70% ethanol, centrifuge at 4 °C and 16 000 g for 5 min; air-dry the pellet and resuspend in 200 µl RNase-free water for RNA derived from a 500-µl transcription assay (the RNA concentration should be 2–5 µg/µl).
(8) Measure the RNA concentration by UV spectroscopy (see Chapter 4 and Appendix).

One may repeat the elution (Steps 4–8) to recover higher amounts of RNA.

Protocol 2b: Purification of aliquots of analog-modified pool RNA after ^{32}P-end-labeling
Follow Protocol 2a, with the following alterations:

- Load the radiolabeled RNA (5–20 pmol) into a 0.5-cm wide (1 mm gel thickness) or 1.3-cm wide (0.5 mm gel thickness) gel pocket.
- In Step 3, visualize the radiolabeled RNA with a phosphoimager (instead of UV shadowing) after an image plate has been exposed to the gel for 1–20 min, depending on the amount of radioactivity loaded on the gel.
- In Step 4, elute the RNA in 500–1000 μl elution buffer 1 or 2.
- Resuspend the RNA pellet after elution and ethanol precipitation in 10–20 μl RNase-free water.

Protocol 3: 5′-End-labeling of analog-modified pool RNA

(1) Prepare the reaction mix by adding the components in the order as they are presented in the table below; vortex, spin down and incubate at 37 °C for 60–120 min.
- 10 × T4 polynucleotide kinase (T4 PNK) buffer (forward reaction): 500 mM Tris–HCl, pH 7.6; 100 mM MgCl$_2$; 50 mM DTT; 1 mM spermidine; 1 mM EDTA

Labeling reaction, 15 μl:		Final concentration
10 × T4 PNK buffer (forward reaction)	1.5 μl	1×
25 mM DTT	1.5 μl	2.5 mM
RNase-free water	6 to 7 μl	
Pool RNA purified according to Protocol 2a	1–2 μl (10–20 pmol)	0.66–1.33 μM
[γ-^{32}P]ATP (3000 Ci/mmol, 10 μCi/μl, 3.3 μM)	3.0 μl	0.66 μM
10 U/μl T4 PNK	1 μl	0.66 U/μl

(2) After incubation, add 35 μl of RNase-free water and vortex; add 7 μl 2 M NH$_4$OAc and 94.5 μl ethanol for RNA precipitation; proceed as described in Protocol 2a, Step 6. Wash the pellet with 100 μl ice-cold 70% ethanol, centrifuge at 4 °C and 16 000 g for 5 min and air-dry the pellet.
(3) Resuspend the pellet in 10 μl gel loading buffer (Protocol 2a) and purify the radiolabeled RNA by 5% denaturing PAGE as described in Protocols 2a and b.
(4) After gel elution and ethanol precipitation resuspend the RNA pellet in 10–20 μl RNase-free water and determine the overall yield of labeled RNA by measuring 1 μl using a scintillation counter.

Protocol 4: 3′-End-labeling of analog-modified pool RNA
- 10 × T4 RNA ligase buffer: 500 mM HEPES–NaOH, pH 8.0; 100 mM MgCl$_2$; 100 mM DTT

18.2 NAIM Analysis of *Cis*-cleaving RNase P RNA–tRNA Conjugates

3'-end-labeling reaction, 6 μl:		Final concentration
Cleaved RNA after Step 5 of Fig. 18.2 (air-dried pellet)	5–10 pmol	0.83–1.66 μM
10 × T4 RNA ligase buffer	0.6 μl	1×
1.5 mM ATP	0.33 μl	82.5 μM
[5-^{32}P]pCp (3000 Ci/mmol, 10 μCi/μl)	4 μl	2.2 μM
10 U/μl T4 RNA ligase	1 μl	1.67 U/μl

(1) Prepare the reaction mix by adding the components in the order given in the table above; vortex, spin down and incubate at 4 °C overnight.
(2) Add 10 μl gel loading buffer and purify the radiolabeled RNA by 5% denaturing PAGE as described in Protocols 2a and b.
(3) After gel elution and ethanol precipitation according to Protocols 2a and b, resuspend the RNA pellet in 10–20 μl RNase-free water and determine the overall yield of labeled RNA by measuring 1 μl using a scintillation counter.

Protocol 5: Selection for *cis*-cleavage of P RNA–substrate conjugates
The analog-modified pool RNA (P RNA conjugated to tRNA 5' half, Fig. 18.1A) is first annealed to the tRNA 3' half in order to reconstitute the full-length substrate, followed by dilution and concomitant addition of Mg^{2+} to start the *cis*-cleavage reaction (Fig. 18.2).

- 4 × annealing buffer: 400 mM MES–NaOH, pH 5.9; 400 mM NH_4Cl.

(1) Prepare the annealing master mix:

Annealing mix 660 μl:		Final concentration
4 × annealing buffer	165 μl	1×
30 mM $CaCl_2$	110 μl	5 mM
24 μM IMPαS-modified P RNA–tRNA 5' half	2.5 μl	91 nM
26 000 c.p.m./μl radiolabeled IMPαS-modified P RNA–tRNA 5' half	5 μl	197 c.p.m./μl
166 μM tRNA 3' half	35 μl	8.8 μM
RNase-free water	342.5 μl	

(2) Distribute in 132-μl aliquots to five different tubes.
(3) Heat to 95 °C for 2 min, then transfer to and cool down in a heating block adjusted to 50 °C and pre-incubate for 30 min at 50 °C.
(4) Add to each aliquot of annealed mix the components listed in the table below; mix thoroughly and incubate for 2 min at 50 °C.

Self-cleavage reaction 506 μl:		Final concentration
Annealed mix	132 μl	0.26 × annealing buffer
300 mM $CaCl_2$	6.23 μl	5 mM
4 × annealing buffer	93.5 μl	0.74 × annealing buffer
5 M urea	10 μl	99 mM
RNase free water	258.2 μl	

(5) Then start the *cis*-cleavage reaction by adding 6.1 µl 3 M MgCl$_2$ to each of the five mixes (final concentration 36 mM).

(6) For four tubes, stop the reaction after 2 min (resulting in 20–30% product formation, representing the fraction of functional RNA) by placing them immediately on ice, followed by addition of 100 µl 2 M NH$_4$OAc, 2 µl of 20 µg/µl glycogen as carrier and 1.2 ml ethanol.

(7) Mix vigorously and store at −20 °C overnight.

(8) Keep the fifth tube at 50 °C for 2 h to allow the *cis*-cleavage reaction to proceed to quasi completion (represents the endpoint of the reaction, where essentially all RNA molecules of the original RNA pool have been cleaved); prepare ethanol precipitation as in Steps 6 and 7.

(9) Centrifuge all five samples for 30–60 min at 4 °C and 16 000 g, wash the pellets with 100 µl 70% ethanol, briefly centrifuge and air-dry the pellets; resuspend (combine) the pellets of tubes 1–4 in 15 µl gel loading buffer; resuspend the pellet of the fifth tube separately in 15 µl gel loading buffer; run the two samples on a 5% PAA/8 M urea gel (0.5-mm thick, pocket width 1.3 cm) as described in Protocol 2b.

(10) Expose an image plate to the gel for 10 min and visualize the bands using a phosphoimager. Excise the cleaved product band from the two lanes and elute each in 500–1000 µl elution buffer 2. Do not perform elution in buffer 1 because residual SDS may disturb the next step (3′-end-labeling). Ethanol-precipitate the eluted RNA and air-dry the pellets as in Protocol 2a.

(11) 3′-end-label the eluted RNA fractions according to Protocol 4.

Protocol 6: Iodine-induced hydrolysis of analog-modified RNA fractions after functional selection

(1) Prepare a fresh I$_2$ solution as in the table below:

I$_2$ solution, 50 µl:		Final concentration
10 mg/ml I$_2$ solution in ethanol	5 µl	1 mg/ml
Ethanol	5 µl	10%
RNase-free water	40 µl	

(2) Prepare the iodine hydrolysis reaction mix as follows:

Iodine-induced hydrolysis reaction, 50 µl:		Final concentration
3′ (or 5′)-end-labeled RNA	1–10 µl (50 000 c.p.m.)	1000 c.p.m./µl
100 mM HEPES pH 7.5	5 µl	10 mM
1 mg/ml I$_2$ solution from step 1	5 µl	0.1 mg/ml
RNase-free water	30–39 µl	

(3) Incubate the reaction mix for 10–20 min at 37 °C.

(4) Add 150 μl RNase-free water, and ethanol-precipitate by addition of 40 μl of 2 M NH$_4$OAc, 2 μl glycogen (20 mg/ml) and 540 μl ethanol.
(5) Resuspend the pellet in 10 μl gel loading buffer and apply to a 10% PAA gel containing 8 M urea (gel thickness: 0.5 mm; length: 35 cm; pocket width: 0.6–1.3 cm). Run the gel at 10 mA for 3–4 h until the distance of the xylene cyanol dye is 5 cm to the gel bottom. To also resolve longer iodine hydrolysis fragments, and thus a larger portion of the RNA molecule, the 10% denaturing PAA gel may be run for an extended period, or a lower gel percentage (e.g. 5%) may be used. However, according to our experience, longer runs or the use of 5% gels have the drawback of usually resulting in more diffuse bands.
(6) Remove the glass plates. Place the gel between one layer of gel drying (Whatman) paper and one layer of kitchen wrapping film.
(7) Dry the gel for 30 min at 70 °C in a gel dryer under vacuum. Switch off the heating and leave under vacuum for another 30 min.
(8) Expose an image plate to the dried gel overnight.
(9) Scan the image plate with a phosphoimager; encircle each band, either by a rectangle, an ellipsoid or by following the individual contours of the band; quantify the image quants therein and evaluate interference data as outlined in Section 18.2.5.

18.2.5
Data Evaluation

Iodine hydrolysis bands were visualized and quantified using a phosphoimager, in our case a Bio-Imaging Analyzer BAS-1000 or FLA 3000-2R (Fuji Film) and the analysis software PCBAS/AIDA (Raytest). An application example is illustrated in Fig. 18.3, representing two experiments run in parallel (lanes 1–4 and 5–8). The quantification boxes positioned with the program AIDA are drawn with thin black or white lines (white when black lines are masked by high band intensities). In the case of insufficient separation of individual bands (e.g. boxes 1 and 2), two or more bands are enclosed in a single box. Some bands, such as those of boxes 31, are hardly visible in Fig. 18.3 due to loss of resolution, but can be clearly differentiated from the background within the analysis program AIDA. The next step is to determine the normalization factor κ to compensate for differences in total radioactivity in lanes A versus B, where A represents the cleaved conjugate at the endpoint (here after 2 h) and B the fraction of conjugates *cis*-cleaved in the initial fast phase of the reaction (within 2 min). A' and B' represent data from another individual experiment; data from a third experiment (A" and B") were also included for the calculation of the average R value, but are not shown in Fig. 18.3 and Table 18.1. The normalization factor κ is then calculated from the ratio of the sum of all band intensities in lane A versus B ($\Sigma I_A / \Sigma I_B$). Interference and enhancement effects for individual bands (or two or more bands if quantified as one due to low gel resolution) are then determined by calculating the ratio $R = (\kappa \times I_B)/I_A$. Interference effects are associated with R values below 1.0, whereas enhancement effects will result in R values above 1.0. In a previous study [19], only R values below 0.82

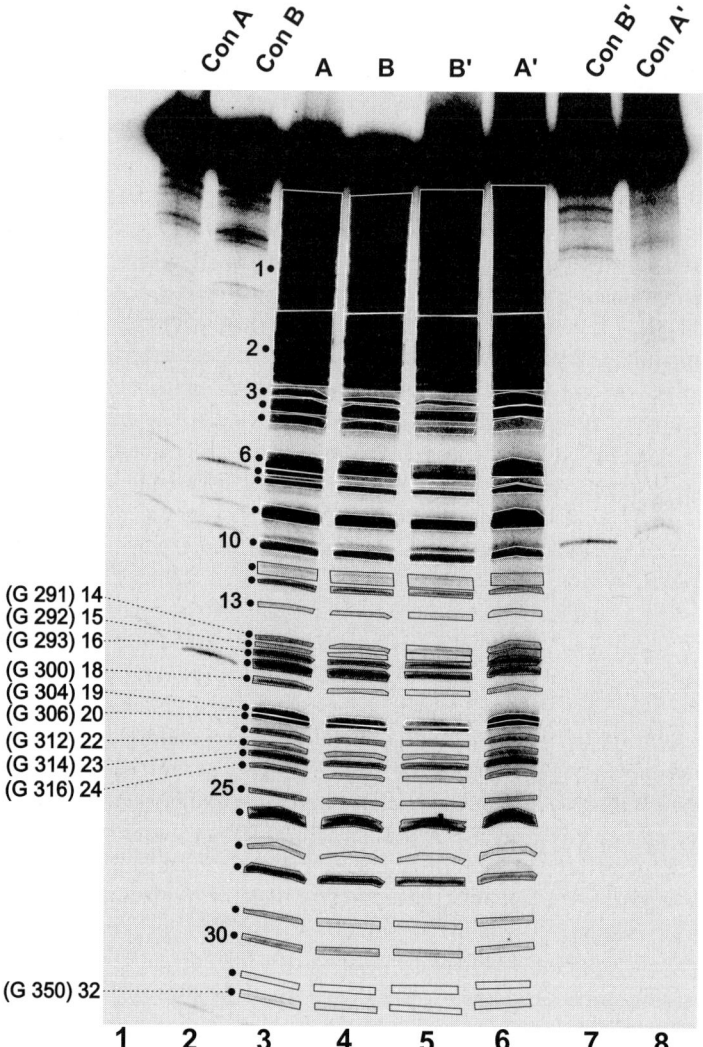

Fig. 18.3. NAIM experiment to identify IMPαS modifications that interfere with *cis*-cleavage of an *E. coli* P RNA–tRNA conjugate. 3′-end-labeled RNA samples, treated with iodine (lanes A, A′, B, B′) or not (lanes Con A, A′, B, B′) were loaded on a 10% PAA/8 M urea gel and separated by electrophoresis until xylene cyanol reached the bottom of the gel. Radioactive bands were visualized as described in Section 18.2.5. Lane A: pool of RNA molecules after 2 h of incubation, representing the endpoint of the reaction (Fig. 18.2, Step 4b); lane B: fraction of conjugates *cis*-cleaved in the initial fast phase of the reaction (within 2 min; Fig. 18.2, Step 4a). Lanes A′ and B′: same as lanes A and B, respectively, but representing a second individual experiment.

and above 1.2 were considered significant, but these cutoff values are arbitrary and depend on the quality of the data, such as the number of and fluctuations between individual experiments. In the example of Table 18.1, the data are based on three individual experiments, which we consider to be the absolute minimum for such studies. Figure 18.4(A) shows a graphical representation of the mean R values (including errors). We also evaluated the data following the calculation procedure described in Chapter 17 (Fig. 18.4B). Both evaluation procedures revealed G300 as a site of IMPαS interference, while several other weaker interferences are only identified according to the evaluation procedure presented in Fig. 18.4(A).

The gain of knowledge derived from NAIM experiments can be largely extended when comparing NAIM results obtained for the same system but with different functional assays. For example, comparison of the interference results shown in Fig. 18.4(A) with those observed in a gel retardation assay selecting for high affinity tRNA binding to *E. coli* P RNA [8] reveals substantial overlap. In the region of nucleotides 291–350, IMPαS modifications at G291–293, G300, G304, G306 and G314 caused interference in both functional assays, although with different amplitudes. It should be noted that, in the tRNA binding assay [8], interference at G300 was predominantly a phosphorothioate effect, already observable with the GMPαS modification alone, which, however, was not analyzed in the *cis*-cleavage assay (Fig. 18.3). Modifications at G291–293 directly weaken the interaction with the 3′-CCA terminus of tRNA [8, 40], explaining why interferences at these positions are detected in both functional assays. Modifications at G300, G304, G306 and G314 apparently also destabilize substrate binding to the ribozyme, either directly or by inducing conformational changes of ribozyme structure. The tRNA binding assay revealed additional IMPαS interferences at G329 and G356, but no effect at G312, G316 and G350 [8]. G312 and G316 are borderline cases due to the weakness of interference effects (Fig. 18.4A). However, G350 may represent a position whose interference is specific to the catalytic step, as evidence was provided that G350 contributes to the binding of catalytically important Mg^{2+} near the active site of RNase P RNA [41].

18.3
Troubleshooting

RNA transcription reaction did not work

- pH too low: check if all the reaction components were added in the required quantities, particularly the HEPES buffer; check the pH of the reaction mixture (should be in the range of 7.5–8.0).
- For further troubleshooting, see Chapter 1.

RNA degradation

- RNase contamination: prepare all solutions freshly using RNase-free water.

Tab. 18.1. Analysis of NAIM data obtained upon selection of fast-cleaving (within 2 min) E. coli P RNA–tRNA conjugates (fractions B and B′) compared with the IMPαS-modified pool of RNA molecules after 2 h of incubation, representing the endpoint of the reaction (fractions A and A′).

No.	Position	Fraction A I_A	Fraction B I_B	$I_B \times \kappa$	Fraction B′ $I_{A'}$	Fraction A′ $I_{B'}$	$I_{B'} \times \kappa'$	R	R′	R″	ØR	SD
1		267325.7	268508.1	285370.41	205632.4	247931.9	229510.56	1.07	1.16	1.02	1.07	0.03
2		99128.7	97563	103689.96	93011.7	96683.2	89499.64	1.06	0.96	0.98	0.10	0.03
3		5103.5	4019.6	4272.03	8152.5	3743.8	3465.64	0.84	0.43	1.08	0.78	0.19
4		11505.4	11559.2	12285.12	8920.8	9722.3	8999.93	1.07	1.01	1.04	1.04	0.02
5		6894.3	6143.4	6529.21	6028.7	5262.1	4871.13	0.95	0.81	0.99	0.92	0.06
6		16869.3	15114.1	16063.27	11072	13779.9	12756.05	0.96	1.15	1.11	1.07	0.06
7		5823.9	3511.6	3732.13	3464.9	2963.2	2743.03	0.65	0.79	0.96	0.8	0.09
8		6918.6	5624.5	5977.72	5825.2	5363.9	4965.36	0.86	0.85	0.96	0.89	0.03
9		20596.9	16283.7	17306.32	17225.6	14217.7	13161.32	0.85	0.76	1.15	0.92	0.12
10	271 + 270	9981.1	8461.5	8992.88	6623.5	8808	8153.57	0.90	1.23	1.21	1.11	0.11
11	275 + 276	4298.9	4396.8	4672.92	4542.9	4306.1	3986.26	1.09	0.88	1.22	1.06	0.10
12	280	3472.3	3499.8	3719.59	2768	3338	3089.99	1.07	1.12	0.93	1.04	0.06
13	285	1660.9	1402.6	1490.68	1699.1	1788.3	1655.43	0.90	0.97	1.38	1.09	0.15
14	291	2395.1	1871.8	1989.35	3471.2	2678.9	2479.84	0.83	0.71	0.86	0.80	0.04
15	292	2225.7	1535	1631.40	3451.9	2099.5	1943.51	0.73	0.56	0.84	0.71	0.08
16	293	5352.9	3555.6	3778.89	5085.5	4225.7	3911.73	0.71	0.77	0.80	0.76	0.027
17	296 + 297	8472.7	7253.3	7708.81	5296.3	5313.3	4918.52	0.91	0.93	1.36	1.07	0.15
18	300	3822.9	1686.3	1792.20	2301.7	1650.6	1527.96	0.47	0.66	0.63	0.59	0.06
19	304	6238.6	4291.2	4560.69	4714.8	3548.7	3285.03	0.73	0.70	0.88	0.77	0.06
20	306	7388.6	5131.3	5453.55	5342.3	4166.4	3856.84	0.74	0.72	0.81	0.76	0.03
21	310	3425.2	2531	2689.95	3154.1	2732.1	2529.10	0.79	0.80	0.95	0.85	0.05
22	312	2973	2091.6	2222.95	3304	2428.8	2248.34	0.75	0.68	0.97	0.80	0.09
23	314	5925.1	3661	3890.91	4617.2	4307.2	3987.17	0.66	0.86	0.81	0.78	0.06
24	316	2856.3	2112.4	2245.06	2818	1905.8	1764.20	0.79	0.63	1.02	0.81	0.12

Position	I_A	I_B	$I_{A'}$	$I_{B'}$	$I_{A''}$	$I_{B''}$	R	R'	R''	$\emptyset R$	SD
25	2667.7	2265.8	2408.09	2014.7	2228.7	2063.11	0.90	1.02	0.81	0.91	0.06
26	323 + 324	10050.2	9544.4	10143.79	10304.4	10058.5	9311.15	1.01	0.90	0.93	0.95	0.03
27	329	1942.9	1449.4	1540.42	1867.6	1696.1	1570.08	0.79	0.84	0.90	0.85	0.03
28	332	6697.3	6058.8	6439.29	3723.5	5509.8	5100.42	0.96	1.37	0.80	1.04	0.17
29	336	2334.2	2167.8	2303.94	2261.5	1788.6	1655.71	0.99	0.73	0.83	0.85	0.07
30	340	2966	2978.4	3165.44	2224.3	2556.2	2366.27	1.07	1.06	0.71	0.95	0.12
31	346	1264.6	804.2	854.70	702.1	632.2	585.23	0.68	0.83	1.14	0.88	0.14
32	350	1944.3	1486.7	1580.06	1461.1	1236.6	1144.72	0.81	0.78	0.68	0.76	0.04
Sum		540522.8	508563.9		443085.5	478672.1						

$\kappa = 1.06284146$; $\kappa' = 0.92565558$.
I = intensity; κ = normalization factor (see text).
Position: according to *E. coli* P RNA numbering system (see secondary structure in Fig. 13.2B of Chapter 13).
$R = I_B \times \kappa / I_A$; $R' = I_{B'} \times \kappa' / I_{A'}$; $R'' = I_{B''} \times \kappa'' / I_{A''}$; primary data ($I_{A''}$, $I_{B''}$; $I_{B''} \times \kappa$) not included in the table; $\emptyset R$ = mean of R, R' and R'';
SD = standard deviation of the mean.

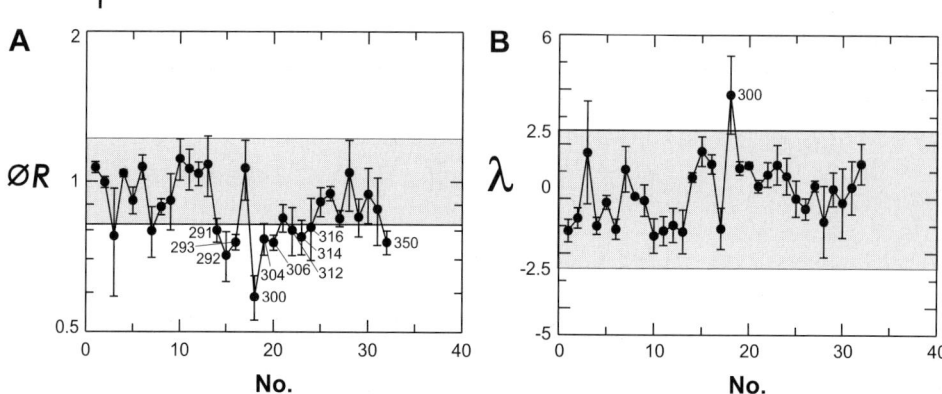

Fig. 18.4. Quantification and statistical analysis of IMPαS modifications that interfere with cis-cleavage of an E. coli P RNA–tRNA conjugate, based on three independent experiments. (A) Mean R values (∅R; error bars: standard deviation of the mean) were plotted against the numbers (No.) of quantification units according to Table 18.1. Effects with R values 0.82–1.2 were arbitrarily considered insignificant (grey-shaded area). At positions where data points correspond to individual G residues for which interference effects are suggested, the position according to the E. coli P RNA numbering system (see secondary structure in Fig. 2B of Chapter 13) is given next to the corresponding circle. (B) Data evaluation based on λ values as detailed in Chapter 17, starting from the primary intensity values (I_A and I_B; $I_{A'}$ and $I_{B'}$; $I_{A''}$ and $I_{B''}$) according to Table 18.1.

Inefficient RNA elution from denaturing PAA gels

- Use elution buffers described in Protocol 2a; do not use buffers containing HEPES instead of Tris.
- Lower the PAA concentration to an extent compatible with satisfactory gel resolution.

RNA is degraded after elution

- RNase contamination: prepare new elution buffer using RNase-free water; control pH (7.0).

Inefficient 3′- or 5′-end-labeling

- Check the RNA you want to label by denaturing PAGE to analyze if it was already degraded before the labeling reaction.
- RNA is degraded during the labeling reaction: use fresh solutions for labeling reaction.
- Be sure that enzymes from previous steps (e.g. alkaline phosphatase) have been completely removed.

- Be sure that the RNA sample is free of any residual chemicals that may inhibit the labeling reaction (e.g. SDS; use elution buffer 1 without SDS in Protocol 2a).
- Check enzyme activity: test old enzyme batch in parallel with a new one; also include a control substrate to rule out failure on the RNA level.

Iodine-induced hydrolysis failed or was inefficient

- Increase the volume (decrease RNA concentration) of your iodine hydrolysis mix; with a total of 50 000 c.p.m. of ^{32}P-labeled RNA, make sure not to exceed a radioactivity/volume ratio of 1000 c.p.m./µl in the iodine hydrolysis reaction.
- Always use freshly prepared iodine solutions.
- The incorporation of nucleotide analogs during *in vitro* transcription was insufficient:
  (1) Use freshly prepared stock solutions of analogs in RNase-free water, pH 7.0; use lithium salts if available.
  (2) Modification extent is too low: increase ratio of analog/natural nucleotide; the degree of modification has to be optimized for each RNA and type of modification (see Section 18.1.2.1).

Unsatisfactory gel performance after iodine cleavage (band smearing, curved bands, irregular shape of bands, unequal band migration in different lanes, insufficient band separation)

- Always freshly prepare acrylamide–urea gel solutions and filtrate them before use.
- Clean the glass plates thoroughly; make sure that the quality of glass plates is sufficient (plane surface, uniform thickness).
- Rinse the gel pockets extensively after taking out the comb and immediately before loading the samples.
- Adjust electrophoretic conditions such that the temperature of the glass plates does not exceed about 50 °C, e.g. let an 8% denaturing PAA gel (15-cm wide, 40-cm long, 0.5-mm thick) run at 10 mA.

References

1 G. GISH, F. ECKSTEIN, *Science* **1988**, 240, 1520–1522.
2 R. KNÖLL, R. BALD, J. P. FÜRSTE, *RNA* **1997**, 3, 132–140.
3 S. A. STROBEL, *Curr. Opin. Struct. Biol.* **1999**, 9, 346–352.
4 R. G. PEARSON, *J. Am. Chem. Soc.* **1963**, 85, 3533–3539.
5 J. M. WARNECKE, J. P. FÜRSTE, W.-D. HARDT, V. A. ERDMANN, R. K. HARTMANN, *Proc. Natl. Acad. Sci. USA* **1996**, 93, 8924–8928.
6 H. SIGEL, B. SONG, *Metal Ions in Biological Systems* 32, A. SIGEL, H. SIGEL (eds), Marcel Dekker, New York, **1996**.
7 J. R. RUBIN, J. WANG, M. SUNDARALINGAM, *Biochim. Biophys. Acta* **1983**, 756, 111–118.
8 C. HEIDE, T. PFEIFFER, J. M. NOLAN,

R. K. Hartmann, *RNA* **1999**, *5*, 102–116.

9 L. Ortoleva-Donnelly, M. Kronman, S. A. Strobel, *Biochemistry* **1998**, *37*, 12933–12942.

10 J. H. Cate, A. R. Gooding, E. Podell, K. Zhou, B. L. Golden, C. E. Kundrot, T. R. Cech, J. A. Doudna, *Science* **1996**, *273*, 1678–1685.

11 H. W. Pley, K. M. Flaherty, D. B. McKay, *Nature* **1994**, *372*, 111–113.

12 D. Siew, N. H. Zahler, A. G. Cassano, S. A. Strobel, M. E. Harris, *Biochemistry* **1999**, *38*, 1873–1883.

13 A. D. Griffiths, B. V. Potter, I. C. Eperon, *Nucleic Acids Res.* **1987**, *15*, 4145–4162.

14 E. L. Christian, M. Yarus, *J. Mol. Biol.* **1992**, *228*, 743–758.

15 S. A. Strobel, K. Shetty, *Proc. Natl. Acad. Sci. USA* **1997**, *94*, 2903–2908.

16 S. P. Ryder, S. A. Strobel, *Methods* **1999**, *18*, 38–50.

17 Y. Huang, F. Eckstein, R. Padilla, R. Sousa, *Biochemistry* **1997**, *36*, 8231–8242.

18 W.-D. Hardt, J. M. Warnecke, V. A. Erdmann, R. K. Hartmann, *EMBO J.* **1995**, *14*, 2935–2944.

19 C. Rox, R. Feltens, T. Pfeiffer, R. K. Hartmann, *J. Mol. Biol.* **2002**, *315*, 551–560.

20 M. E. Harris, N. R. Pace, *RNA* **1995**, *1*, 210–218.

21 A. V. Kazantsev, N. R. Pace, *RNA* **1998**, *4*, 937–947.

22 N. M. Kaye, E. L. Christian, M. E. Harris, *Biochemistry* **2002**, *41*, 4533–4545.

23 W.-D. Hardt, V. A. Erdmann, R. K. Hartmann, *RNA* **1996**, *2*, 1189–1198.

24 C. Heide, R. Feltens, R. K. Hartmann, *RNA* **2001**, *7*, 958–968.

25 M. Boudvillain, A. M. Pyle, *EMBO J.* **1998**, *17*, 7091–7104.

26 G. Chanfreau, A. Jacquier, *Science* **1994**, *266*, 1383–1387.

27 J. S. Smith, E. P. Nikonowicz, *Biochemistry* **2000**, *39*, 5642–5652.

28 E. L. Christian, M. Yarus, *Biochemistry* **1993**, *32*, 4475–4480.

29 A. Jack, J. E. Ladner, D. Rhodes, R. S. Brown, A. Klug, *J. Mol. Biol.* **1977**, *111*, 315–328.

30 W. G. Scott, J. T. Finch, A. Klug, *Cell* **1995**, *81*, 991–1002.

31 E. L. Christian, N. M. Kaye, M. E. Harris, *RNA* **2000**, *6*, 511–519.

32 E. L. Christian, N. M. Kaye, M. E. Harris, *EMBO J.* **2002**, *21*, 2253–2262.

33 J. W. Brown, J. M. Nolan, E. S. Haas, M. A. Rubio, F. Major, N. R. Pace, *Proc. Natl. Acad. Sci. USA* **1996**, *93*, 3001–3006.

34 C. Massire, L. Jaeger, E. Westhof, *J. Mol. Biol.* **1998**, *279*, 773–793.

35 T. R. Easterwood, S. C. Harvey, *RNA* **1997**, *3*, 577–585.

36 A. A. Szewczak, L. Ortoleva-Donnelly, M. V. Zivarts, A. K. Oyelere, A. V. Kazantsev, S. A. Strobel, *Proc. Natl. Acad. Sci. USA* **1999**, *96*, 11183–11188.

37 B. L. Golden, A. R. Gooding, E. R. Podell, T. R. Cech, *Science* **1998**, *282*, 259–264.

38 A. S. Krasilnikov, X. Yang, T. Pan, A. Mondragon, *Nature* **2003**, *421*, 760–764.

39 T. Persson, S. Cuzic, R. K. Hartmann, *J. Biol. Chem.* **2003**, *278*, 43394–43401.

40 S. Busch, L. A. Kirsebom, H. Notbohm, R. K. Hartmann, *J. Mol. Biol.* **2000**, *299*, 941–951.

41 T. A. Rasmussen, J. M. Nolan, *Gene* **2002**, *294*, 177–185.

19
Identification and Characterization of Metal Ion Binding by Thiophilic Metal Ion Rescue

Eric L. Christian

19.1
Introduction

One of the most important characteristics of RNA is its ability to fold into complex three-dimensional structures that participate both directly and indirectly in a wide range of biochemical functions. These closely packed structures result in a high degree of unfavorable electrostatic repulsion that must be offset by interactions with positively charged monovalent and divalent ions [1, 2]. While RNA secondary structure and some tertiary structure can form in the presence of monovalent ions alone, divalent ions are required by large RNAs to adopt their active conformations [3–5]. The large majority of divalent metal ion interactions are thought to be weak, non-specific and to exchange rapidly between positions of elevated negative charge density, precluding their identification by high-resolution methods such as X-ray crystallography or nuclear magnetic resonance (NMR) [2, 6–10]. In contrast, a relatively small fraction of divalent metal ions form tight and specific interactions that can be resolved by high-resolution methods [2, 7, 8, 11–16].

Metal ions have been shown to both direct RNA folding and to stabilize specific RNA structures [17]. In addition, metal ions in some RNA catalysts (ribozymes) can interact directly with substrate phosphates to catalyze phosphodiester bond hydrolysis or transesterfication reactions [18–25]. While several catalytic RNAs have been shown to have bases positioned in the appropriate structural environment to play this role, the pK_as of RNA functional groups generally lie far from neutral pH, inhibiting their participation in acid–base catalysis under physiological conditions [17]. Metal ions thus provide a crucial addition to the RNA functional group repertoire, and their identification is often key to our understanding of folding and catalytic mechanism. Locating specific metal ions within structural or catalytic RNAs, however, is experimentally challenging and thus only a few metal ion interactions have been linked to a specific aspect of RNA function.

The central role metal ions play in RNA structure and catalysis creates at least two distinct challenges for defining their specific biological function. First, unlike the analysis of a single or small number of metal ions associated with proteins, there are often many metal ions associated with RNA, (hundreds in the case of

Handbook of RNA Biochemistry. Edited by R. K. Hartmann, A. Bindereif, A. Schön, E. Westhof
Copyright © 2005 WILEY-VCH Verlag GmbH & Co. KGaA, Weinheim
ISBN: 3-527-30826-1

the larger ribozymes). Thus there is significant difficulty in distinguishing an individual metal ion from what has often been described as the "sea" of metal ions associated with an RNA molecule [5, 6, 13–15, 26–34]. Second, individual metal ions may have multiple roles in structure or catalysis, making quantification of their contribution to specific functions difficult to deconvolute. In addition, traditional methods of structural analysis such as crystallography and NMR are unlikely to completely define the functional properties of divalent metal ion binding. Different numbers of divalent metal ions have been defined in different structural studies for the same enzyme [35, 36] and different catalytic roles have been proposed for closely spaced active site metal ion candidates [37, 38]. Moreover, metal ion interactions important to the reaction transition state may exhibit only weak or transitory binding in the ground state, precluding their detection by high-resolution structural methods.

Despite these complications significant progress has been made in our understanding of the influence of metal ions on RNA folding using a range of biochemical approaches including footprinting [39, 40], thermal denaturation [41, 42], fluorescence energy transfer [43], UV absorbance and circular dichroism [44], small-angle X-ray scattering [45, 46], and single-molecule florescence [47, 48]. Notably, the information these experimental approaches provide reflects the combined influence of metal ions and does not address the role of individual metal ions at a specific location. Significant progress has also been made in our understanding of the role of metal ions in RNA catalysis using well-studied RNA enzymes and traditional enzyme kinetics [32, 49–52]. These studies have examined enzymatic RNA cleavage as a function of metal type and concentration and have provided evidence for catalytic mechanisms involving one or more metal ions. The usefulness of this approach, however, is limited as it does not provide a distinction between direct and indirect metal ion effects on catalysis, and it does not examine the functional role of an individual metal ion. Sites of metal ion binding have been examined by metal-dependent RNA cleavage studies, but this approach provides only the approximate location of the subset of metal ions able to cleave the RNA backbone rather than specific functional groups involved in metal coordination and does not necessarily reflect the position of the native magnesium ions (Mg^{2+}) [53, 54]. Thus, despite this arsenal of experimental approaches it remains difficult to conclusively address many of the most pressing questions regarding the role of divalent metal ions in RNA. These structure and function questions include identifying the number of metal ions in a folding domain or an RNA active site, the ligands involved in metal ion coordination, the energetic contribution of an individual ion to folding or catalysis, the step in the folding or catalytic pathway where specific metal binding exerts its effect, and the fundamental mechanisms by which an individual metal ion can direct RNA folding and catalysis.

Many of these questions can be addressed using the method of thiophilic metal ion rescue (or metal ion specificity switch) experiment (Fig. 19.1). This approach involves the atomic substitution of an individual oxygen atom in an RNA molecule with sulfur or nitrogen at a position that produces significant changes in RNA folding or catalytic rate (Fig. 19.1A and B) [55]. The majority of these experiments have

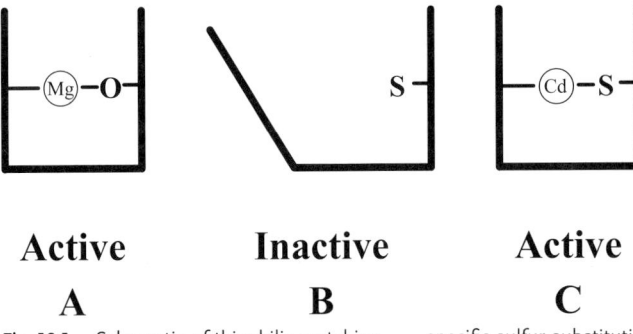

Fig. 19.1. Schematic of thiophilic metal ion rescue. (A) RNA activity dependent on specific Mg^{2+} coordination. (B) Loss of RNA activity due to the inability of Mg^{2+} to coordinate site-specific sulfur substitution. (C) Rescue of RNA activity by the addition of a thiophilic metal (e.g. Cd^{2+}) capable of direct coordination of sulfur as well as oxygen.

used phosphorothioate nucleotide analogs, which contain a sulfur substitution in place of a bridging or non-bridging phosphate oxygen. Such substitution at positions involved in metal ion binding, disrupts direct inner-sphere coordination of Mg^{2+}, as sulfur is a poor ligand for Mg^{2+} relative to oxygen (Fig. 19.1B, note: Mg^{2+} binding to oxygen is approximately 31 000-fold greater than that for sulfur [59]). Consistent with the direct disruption of metal ion coordination, and the central role that metal ions can play in RNA structure and function, phosphorothioate inhibition can produce substantial (greater the 10^4-fold) changes in the activity of ribozymes, rivaling the magnitude of effects observed in experiments that directly alter or delete essential active site elements in both RNA and protein enzymes [18, 20–22, 56, 57]. Moreover, these effects parallel the decrease in affinity of Mg^{2+} for sulfur relative to oxygen in model complexes [58, 59] and are consistent with the predicted energetic penalty for a change in ligand coordination of a Mg^{2+} ion [24]. In contrast to Mg^{2+}, "thiophilic" metal ions such as manganese (Mn^{2+}), cadmium (Cd^{2+}) or zinc (Zn^{2+}) can more readily accept both oxygen, nitrogen or sulfur in their inner coordination sphere and can thus alleviate deleterious effects of specific oxygen to sulfur or nitrogen substitutions in RNA that perturb the binding of Mg^{2+} (Fig. 19.1C) [55, 58–62]. The ability of low concentrations of such thiophilic metal ions to "rescue" deleterious effects of site-specific sulfur substitutions provides strong evidence for direct metal ion coordination [58].

Thiophilic metal ion rescue experiments have been used to identify likely metal contacts in the conserved structural domains and at the reactive phosphates of all three of the large ribozymes as well as metal coordination sites within small ribozymes and spliceosomal RNAs [18–22, 24, 25, 34, 52, 63–83], including interactions linked to the RNA active site [18–22, 24, 66, 67, 70, 73] and RNA folding [84, 85]. As will be discussed in detail in Section 19.2, the metal rescue approach has also been applied to determine the apparent affinity of a specific metal ion interaction by measuring the reaction rate over a range of metal ion concentrations [24, 25, 34, 72, 81, 86]. This quantitative characterization of individual metal ion interactions allows both the direct comparison of metal ion binding at different positions and the determination of the relative energetic contribution to the ribozyme reaction. These measurements are central to tests of specific structural and mechanistic proposals for the role of individual metal ions in enzyme function.

Metal binding sites identified by thiophilic metal ion rescue have generally been confirmed in the small (but growing) number of cases where alternative methodologies have been applied. These include independent confirmation of metal ion binding by X-ray crystallography [87–91], by ^{15}N- and ^{31}P-NMR [92–94], and by electron paramagnetic resonance [95], Q-band electron nuclear double resonance [95], and electron spin echo envelope modulation [96] spectroscopies. While some discrepancies have been observed in the detection of metal ion binding by the methods above [92, 97], the preponderance of independent analyses suggests that metal ion interactions detected by thiophilic metal ion rescue reflect genuine metal ion interactions within the native RNA structure.

It is important to underscore, however, that the method of thiophilic metal ion rescue is by no means definitive and has its own ambiguities and limitations that can lead to concluding something is a ligand when it is not or missing a functional group that is a ligand. Uncertainty in interpreting thiophilic metal ion rescue studies can stem from a wide range of potential indirect effects that may be induced by the phosphorothioate analog, the rescuing metal or the combination of both. Although the single atomic substitution of sulfur or nitrogen represents a minimal perturbation of enzymatic structure, changes in bond length, ionic radius, and local charge distribution can have significant effects beyond altering metal ion affinity [98–100]. For example, steric constraints imposed by the larger sulfur atom can alter the geometry of metal ion interactions to preclude metal ion rescue [101]. Thiophilic metal ions can also bind to positions other than the site of interest to indirectly affect the ribozyme's structure and reactivity [78, 102]. Moreover, different metals have different coordination properties that may be an issue in binding sites where precise geometry is important. In addition, while it is generally assumed that phosphorothioate modifications do not significantly perturb the native structure or induce *de novo* metal ion binding sites, it is difficult to exclude the possibility entirely.

The main limitations of thiophilic metal ion rescue are the small number of analogs available for this type of analysis and the inability to examine outer sphere interactions involved in metal ion coordination. This fact and the likelihood of geometric constraints can significantly restrict both the number of metal ions that can be identified as well as the number of contacts that can be identified in an individual coordination sphere. Moreover, even if the position of metal ion binding can be established, the determination of its biological role requires a detailed understanding of the reaction kinetics of the experimental system and the ability to analyze quantitatively an individual step in RNA function such as folding, substrate binding or enzymatic cleavage. Finally, while the general theory of thiophilic metal ion rescue is straightforward, well-controlled experimental design and interpretation of findings is not. Indeed, the absence of any one of a number of experimental considerations can substantially limit both the significance and the interpretability a great deal of experimental effort. Nevertheless, it is important to point out that many of the limitations and ambiguities noted above are not unique to thiophilic metal ion rescue and must be placed in the equally important context of the limitations and ambiguities of RNA (and protein) structural analysis in general, and in the significant difficulty in demonstrating that any given structure is functionally relevant.

Details for the proper application of thiophilic metal ion rescue will vary depending on the RNA being studied, the reaction conditions used, and the specific experimental question being addressed (e.g. whether the metal ions being examined are linked to folding, substrate binding, or catalysis). Thus, it is not possible in the scope of this short review chapter to cover the rationale or experimental detail for each of these circumstances. The study of ribozyme catalysis, however, provides an excellent illustration of the sensitivity of this probe for functional metal ions and for the measurement of energetic consequences of metal ion–RNA interactions. This review chapter highlights general experimental considerations with specific examples from the group I and group II introns, the hammerhead ribozyme, and bacterial ribonuclease P (RNase P) where the majority of thiophilic metal rescue studies have been performed. These experimental considerations, however, should also be central to other experimental systems and biological questions where thiophilic metal ion rescue can be applied.

19.2
General Considerations of Experimental Conditions

19.2.1
Metal Ion Stocks and Conditions

Highly purified preparations (above 99.99%) of $MgCl_2$, $MnCl_2$, $CdCl_2$, $ZnCl_2$ and other metal salts are available commercially (Aldrich), and should be stored tightly sealed to minimize absorption of water. Note that the relatively high pK_a of Mg^{2+} (11.4) makes stock solutions of this metal ion stable in water and most buffer conditions at or below pH 9. In contrast, the lower pK_as of thiophilic metal ions (10.6, 9.0 and 9.0 for $MnCl_2$, $CdCl_2$ and $ZnCl_2$, respectively) results in the formation of insoluble metal hydroxides at concentrations in the millimolar range above neutral pH [103, 104]. Thus solutions of thiophilic metal ions must be used immediately after preparation or made as concentrated, acidic stocks (pH 2) and diluted into buffer immediately prior to use.

Although optimal concentrations vary between experimental systems, the study of metal ions required for catalysis generally requires a baseline concentration Mg^{2+} to completely fold the RNA being studied and to minimize the effect of thiophilic metal ion binding at sites other than that associated with the phosphorothioate modification. Background levels of Mg^{2+} used to minimize non-specific effects in thiophilic metal ion rescue studies in group I, group II, bacterial RNase P and the hammerhead ribozyme, for example, are generally in the range of 5–

10 mM [34, 66, 72, 86]. The upper limit of total divalent metal ion concentration is also an important experimental consideration since elevated levels of divalent metal ion can both alter RNA structure and the rate-limiting step of the reaction, and produce significant non-specific cleavage of the RNA backbone. The range over which experiments can be conducted can restrict both the level and types of metal ions that can be used, and the ability to detect or completely rescue individual phosphorothioate positions. The concentration and types of metal ions that can be used in thiophilic metal ion rescue are those that can be shown by independent experiments to maintain the same rate limiting step of reaction (e.g. catalysis) as that of the native enzyme (see below).

Two additional factors can influence the concentration and type of divalent ion present in thiophilic metal ion rescue. First, elevated levels of RNA used to achieve single turnover conditions can have significant effects on the concentration of free metal ion in solution. At 10 μM an RNA of 400 nt such as RNase P will have 400 negative charges on its backbone that can bind several hundred Mg^{2+} ions or 2 mM of the free divalent metal ion. Changes in the concentration of free metal ion in the millimolar range may thus produce indirect effects if the total background magnesium concentration is not in sufficient excess to insure RNA folding. Second, while EDTA may be useful to prevent degradation due to metal-dependent hydrolysis, it preferentially binds thiophilic metal ions even in an excess of Mg^{2+}, which can lead to a significant overestimate of the amount of thiophilic metal ion required for rescue [105]. EDTA can generally be omitted from solutions without consequence, thus removing this potential complication. However, since EDTA chelates most thiophilic divalent ions five to eight orders-of-magnitude more tightly than Mg^{2+}, it is routinely added to control reactions done in the presence of Mg^{2+} to demonstrate that the observed effect is not due to the presence of contaminating thiophilic metal ions [105] (see below).

19.2.2
Consideration of Buffers and Monovalent Salt

Although a wide range of buffers can be used to examine metal ion rescue, it is important to avoid the use of buffers that chelate metal ions (e.g. citrate, phosphate or buffers containing acetate). Buffers used in thiophilic metal ion rescue studies include MES, MOPS, EPPS, HEPES, PIPES, Tris–HCl and BisTris–propane–HCl (Sigma, molecular biology grade) and are commonly used at a final concentration of 50 mM. Side by side comparisons of these buffers at a given pH generally do not produce large buffer-specific changes in the level of phosphorothioate inhibition or thiophilic metal ion rescue (usually less than 2-fold), allowing direct comparison of metal rescue studies under different buffer conditions. However, control experiments should be done to show that this is, in fact, the case.

In contrast, thiophilic metal ion rescue can be strongly influenced by both the concentration and type of monovalent salt. This observation is not surprising given that monovalent ions can compete directly with divalent ions for binding to the vast majority of charged positions, and thus are an important determinant of the final folded structure. The effect of monovalent salt becomes particularly important when the optimal concentration of monovalent salt is in excess of the concentration of divalent metal ion (e.g. 1 M M^+ versus 25 mM M^{2+} for *Escherichia coli* RNase P). While thiophilic metal ion rescue should be done under the same monovalent in which the kinetic pathway of the system of interest has been defined, changes in monovalent conditions can sometimes be advantageous under some experimental circumstances. Elevated levels of monovalent ion can act to both compensate for perturbations in diffuse metal ion binding and significantly dampen indirect effects due to changes in ionic strength upon the addition of thiophilic metal ions. Moreover, changes in monovalent ion species can significantly alter the level of metal ion rescue. Conditions that produce the best rescue, however, are not necessarily comparable to the behavior observed under a standard set of conditions. Comparisons of this type should be made with caution and with evidence that the rate-limiting step of the reaction has not been altered (see below). Careful consideration should also be given in using monovalent salts containing acetate, as elevated concentrations required for some systems may significantly alter the level of free divalent metal ion.

19.2.3
Incorporation of Phosphorothioate Analogs

The most commonly used phosphorothioate analogs contain sulfur substitutions at the non-bridging phosphate oxygens and are available commercially (Glen Research). These analogs are stored and handled in a manner analogous to that of other ribonucleotides (-80 °C in dH_2O). As noted above, sulfur substitutions have also been introduced at bridging phosphate oxygen positions [71, 73, 75]. However, these analogs are not available commercially and must therefore be chemically synthesized [106]. Phosphorothioate analogs are generally incorporated in one of two ways, by *in vitro* transcription or by solid phase chemical synthesis and ligation. These methods are covered in detail in Chapters 5 and 2.6.1, and thus only a brief overview and discussion of their relative merits with respect to thiophilic metal ion rescue will be presented here.

The incorporation of phosphorothioates by transcription generally involves the addition of low levels of nucleotide triphosphate analogs (ATPαS, CTPαS, GTPαS or UTPαS) to a standard transcription mixture to yield a population of randomly modified RNA that usually contain no more than one analog per molecule [107]. This mixture of randomly modified RNAs can be quickly analyzed for the effects of phosphorothioate substitution as well as for thiophilic metal ion rescue using the method of nucleotide analog interference modification (NAIM, see Chapter 17). This approach offers the advantage of being able to analyze all modified positions in a given molecule in a single experiment, a feature particularly important in the initial analysis of large RNAs. However, there are two important limitations to this approach that restrict both detection and thorough characterization of metal ion interactions. First, RNA polymerase is somewhat limited in the number of different analogs that can serve as substrates in transcription. For example, phosphorothioate analogs leading to the modification of the *pro*-Sp position cannot be incorporated by RNA polymerase, thus excluding half of the non-bridging phosphate oxygens that are likely to be involved in metal ion interactions. Second, the use of a mixed population of randomly modified RNAs make it particularly difficult to characterize the effect of an individual phosphorothioate modification and associated metal ion rescue beyond qualitative changes in folding, reaction rate or substrate binding. Thus, while random incorporation is useful in the initial survey of putative positions of metal ion binding, a uniform population of molecules with a site-specific substitution is ultimately necessary to utilize the power and quantitative potential of standard enzyme kinetics to confirm and characterize interactions involved in thiophilic metal ion rescue.

Site-specific phosphorothioate modifications are incorporated by chemical synthesis into short oligonucleotide fragments of approximately 10–20 nt (see Chapter 7). *Pro*-Sp and *pro*-Rp stereoisomers are separated by reverse-phase HPLC [108], with the *pro*-Rp peak usually emerging before *pro*-Sp. The identity of individual stereoisomers is verified by digestion with RNase T1 and snake venom phosphodiesterase, which cleave the *pro*-Sp isomer more slowly than *pro*-Rp, and P1 nuclease, which shows the opposite preference [109]. Note, however, that it can become increasingly difficult to separate the individual stereoisomers by HPLC for fragments greater than approximately 13 nt. The exact length will vary with sequence but separation can be enhanced by placing the phosphorothioate away from the center of the fragment toward the 5' or 3' end. Larger oligonucleotides containing racemic mixtures of the two stereoisomers can still be informative if the resulting reaction profile produces distinct kinetic phases (see below) and can be compared to unmodified and, optimally, purely *pro*-Rp isomers produced by transcription.

Isolated oligonucleotides are often 5'-end-labeled using $[\gamma\text{-}^{32}P]$ATP and T4 polynucleotide kinase and further purified by denaturing or non-denaturing PAGE prior to their use as substrates for joining to other RNA fragments. Purified oligonucleotides are generally ligated by the method of Moore and Sharp [110] (Chapter 3) to synthesized or transcribed RNA fragments containing the remaining sequence to yield a complete RNA. Oligonucleotide concentration and ligation efficiency can be assessed using UV absorbance or the specific activities of the radioactive RNAs.

Because ligation efficiency can be significantly reduced in reactions that contain more than a single ligation junction, oligonucleotide fragments are generally added to the 5' or 3' end of the larger RNA fragment rather than in the interior of the molecule as part of a reaction involving the joining of three or more fragments. Ligation efficiency is an important consideration in phosphorothioate studies since some ligation reactions are limited by structural constraints that reduce efficiency to only a few percent. Thus it may be difficult to produce enzyme or substrate in the quantities that are necessary for certain experimental conditions (e.g. single turnover conditions, see below). The insertion of functional groups in the interior of large RNAs such as the RNase P ribozymes can be accomplished by moving the native 5' and 3' ends of the RNA near the site of modification by circular permutation [111] to allow a simple two-part ligation. Note, however, that any modification of RNA structure must be kinetically characterized to show that it retains the functional properties of the native enzyme. Purified oligonucleotide fragments can also be positioned in the substrate or ribozyme by simple annealing using a protocol analogous to that described below for general folding of enzyme or substrate prior to enzymatic reactions. While this has been shown to be useful in the study of smaller RNAs, such as the pre-tRNA substrate for bacterial RNase P [112, 113], it is not clear whether this approach is well suited for studying the interiors of more complex RNAs where saturating levels of oligonucleotide binding may be difficult to obtain.

19.2.4
Enzyme–Substrate Concentration

The amount of RNA used in thiophilic metal ion rescue studies varies greatly depending on the phenomenon being studied (e.g. folding, binding or catalysis) and whether the experimental system is intra- or inter-molecular. However, under all circumstances, it is imperative that the relative RNA concentrations and experimental conditions isolate or uniquely reflect the phenomenon of interest. For the current discussion, enzyme reactions must be limited by the rate of catalysis rather than other kinetic events such as folding, substrate binding or product release. Because multiple turnover reactions for ribozyme systems in which thiophilic metal ion rescue has been applied are not generally limited by chemistry, single turnover conditions [e.g. the relative concentrations of enzyme and substrate (E:S) > 5:1] have been used to isolate the catalytic step. Control reactions at higher enzyme concentrations (e.g. E:S = 10:1 or 20:1) are generally required to demonstrate that the reaction is not dependent on enzyme concentration and that the substrate is completely bound in all cases. Note, however, that a wide range of other ratios of enzyme and substrate has been used to address different aspects of the kinetic profile [24].

Other methods to insure the reaction is dependent on the rate of cleavage involve the use of modifications at or adjacent to the scissile phosphate that inhibit the reaction [114, 115], as well as mutants that produce strong catalytic defects within the ribozyme itself [74]. In addition, enzymatic catalysis can be analyzed in the context of self-cleaving RNAs in which both ribozyme and substrate are contained in the same RNA fragment [74]. These enzymatic systems offer the advantage of dramatically increasing the effective local substrate concentration associated with the ribozyme to allow the analysis of cleavage of the enzyme substrate complex with radiochemical amounts ($\ll$ 1 pM) of ligated RNA as opposed to the significantly higher enzyme concentrations (e.g. 1–10 µM) generally required for intermolecular single turnover reactions. Constructs that tether ribozyme and substrate require different temperature, monovalent and divalent metal ion conditions than the native *trans* reaction, however, and can differ in both the position and magnitude of the metal ion effects observed, making direct comparison problematic in the absence of additional controls [74, 82].

19.2.5
General Kinetic Methods

As noted above, analysis of metal ions important for catalytic function is generally done under single turnover conditions with ribozyme in excess of a 5′-end-labeled substrate. Although protocols vary, ribozyme and substrate are usually denatured by heating (e.g. 95 °C for 2–3 min) in reaction buffer without divalent metal ions and cooled stepwise or gradually to the reaction temperature. Reactions in which ribozyme and substrates are renatured together to allow proper annealing (as in the case of the hammerhead and group I ribozymes) are typically initiated by the addition of divalent ions. In reactions where ribozyme and substrate are folded independently, divalent metal ions are added during or after the cooling phase (to minimize metal dependent hydrolysis) and reactions are initiated by mixing of enzyme and substrate. Renaturation and subsequent incubation of the cleavage reaction can be done in a heat block, water bath, or PCR machine, although the latter is often convenient to insure reproducibility of the reaction conditions.

Reaction volumes usually vary between 10 and 60 μl and aliquots (1–2 μl) taken at specified times are added to solutions (around 2–10 μl) containing formamide (above 80%) and an excess of EDTA (generally greater than 2-fold) to terminate the reaction. Termination solutions also often contain buffer (e.g. 1 mM Tris, pH 7.5) and 0.005% xylene cyanol and bromophenol blue when products will be separated by gel electrophoresis. Long time points (lasting several hours to several weeks) should be taken from reactions covered with mineral oil (50–100 μl) and kept in an incubator or PCR machine with a heated lid to minimize changes in reaction volume due to evaporation. Substrate and products are typically resolved by polyacrylamide/7 M urea gel electrophoresis and quantified on a PhosphoImager (Molecular Dynamics).

Data are fit to the appropriate kinetic equation using KaleidaGraph (Synergy Software) or SigmaPlot (Jandel Scientific), preferably with a sufficient number of time points (usually six to 10) to give non-linear least-squares fits with $R^2 > 0.98$. Reaction time courses are typically fit to a single exponential:

$$\frac{[P]}{[E \cdot S]_{total}} = A - Be^{-k_{obs}t} \tag{1}$$

where [P] is the amount of product formed at time t, $[E \cdot S]_{total}$ is the initial concentration of bound substrate, A represents the maximal extent of reaction and B is the amplitude of the exponential fit. Two independent exponentials are generally used if the reaction appears to have distinct (e.g. fast and slow) phases which may arise from the existence of multiple enzyme–substrate (E · S) complexes with different observed rate constants (see Eq. 2 where k_1 and B, and k_2 and C represent the observed rate and amplitude of independent reaction phases, respectively).

$$\frac{[P]}{[E \cdot S]_{total}} = A - \left(Be^{-k_1 t} + Ce^{-k_2 t}\right) \tag{2}$$

Very slow reactions (longer than 10 days) often only allow the measurement of the initial phase of the reaction and are typically fit to a linear equation or to an exponential equation if an endpoint can accurately be measured by another means [72]. Note that experimental error can be significant for large changes in the observed reaction rate. This is due to the inherent difficulty in obtaining accurate measurements of very slow rates ($\ll 10^{-4}$ min^{-1}) where the reactions cannot be followed to completion; caution should be exercised in comparing the magnitude of inhibitory effects due to site-specific substitution or inefficient thiophilic metal ion rescue. Independent experiments should also be conducted to show that the ribozyme remains fully active after extended reaction times.

The accurate measurement of the level of thiophilic metal ion rescue of a specific modification requires that it be distinguished from effects due to thiophilic metal ion binding at other positions. This requirement is based on the fact that changes in both the concentration and type of metal ion present can have significant effects on folding and activity of the unmodified ribozyme. Consequently, the observed rate of both the modified ribozyme or substrate (k_S) and the unmodified RNA control (k_O) must be measured in the presence and absence of added thiophilic metal ion [34]. For example, in the analysis of Cd^{2+} rescue of a specific phosphorothioate substitution, the observed rate of the modified RNA in the presence and absence of Cd^{2+} would be expressed by the terms k_S^{Cd} and k_S^{Mg}, while the observed reaction rate of the unmodified ribozyme under the same conditions would be expressed by the terms k_O^{Cd} and k_O^{Mg}. The relative rate of rescue (k_{rel}) of thiophilic metal ion binding at a specific site of phosphorothioate modification is determined by the fold rate enhancement of the modified ribozyme or substrate in the presence of the thiophilic metal ion over that observed in its absence normalized to the rate enhancement (or inhibition) observed in the absence of the phosphorothioate modification. For example, using the four rate constants in the example immediately above, $k_{rel} = (k_S^{Cd}/k_S^{Mg})/(k_O^{Cd}/k_O^{Mg})$ [34]. Rescue observed at thiophilic metal ion concentrations that show comparatively little or no effect on an RNA lacking a specific modification provides strong evidence for thiophilic metal ion binding to the specific modification itself. It must be emphasized, however, that the ability to interpret k_{rel} in this way requires independent evidence that the reactions being compared share the same rate-limiting step, that the effect of other metal ion sites is the same for both reactions and there is no direct effect of the rescuing metal ion on the "normal" reaction (see below).

19.2.6
Measurement of Apparent Metal Ion Affinity

As noted in the introduction, measurement of the dependence of reaction rate on the concentration of rescuing metal ion(s) allows the determination of the apparent metal ion affinity for binding to the sulfur substitution (Fig. 19.2) [34, 72].

The dependence of the relative rate (k_{obs}^{rel}) on the concentration of added thiophilic metal ion (M_S) is determined by the relative fraction of RNA with site-specifically bound thiophilic metal ion (Eq. 3) which is a function of the thiophilic

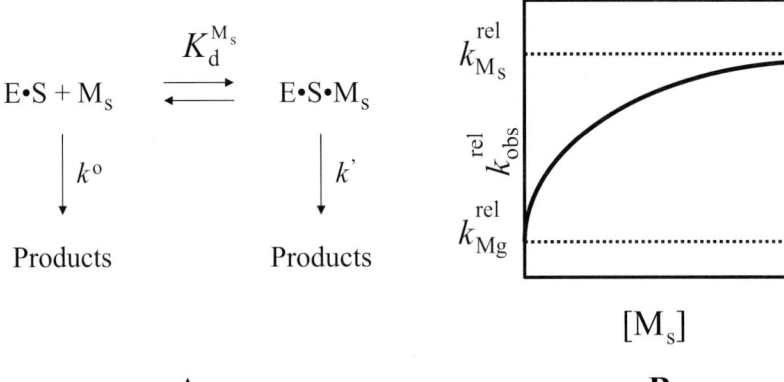

Fig. 19.2. Dependence of the reaction rate on thiophilic metal ion binding. (A) Kinetic scheme showing thiophilic metal ion binding ($K_d^{M_S}$) to an enzyme–substrate complex that produces an observed reaction rate (k') that is distinct from that observed in the absence of thiophilic metal ion binding (k^0). (B) Plot of the observed relative rate (k_{obs}^{rel}) as a function of thiophilic metal ion concentration ($[M_S]$). Note that the relative contribution of the individual rates (k_{Mg}^{rel} and $k_{M_S}^{rel}$) to k_{obs}^{rel} is directly proportional to the fraction of RNA bound to the thiophilic metal ion. Determination of apparent thiophilic metal ion affinity is derived from equations described in the text.

metal ion's affinity ($K_d^{M_S}$) (Eq. 4) [34]:

$$k_{obs}^{rel} = k_{Mg}^{rel} \times \{\text{fraction }^{Mg}E\} + k_{M_S}^{rel} \times \{\text{fraction }^{M_S}E\} \tag{3}$$

$$k_{obs}^{rel} = k_{Mg}^{rel} \times \frac{K_d^{M_S}}{[M_S] + K_d^{M_S}} + k_{M_S}^{rel} \times \frac{[M_S]}{[M_S] + K_d^{M_S}} \tag{4}$$

Note that under conditions where background (e.g. Mg^{2+}) metal ion competes with rescuing thiophilic metal ion (as is often seen) the observed dissociation constants, $K_d^{M_S}$, are "apparent" values. Under conditions where added thiophilic metal ions produce non-specific inhibition (e.g. inhibition of unmodified RNA), $K_d^{M_S}$ has been measured using a model analogous to that described by Equation 4 but includes the inhibitory binding of an additional thiophilic metal ion ($K_i^{M_S}$) [72]:

$$k_{obs}^{rel} = k_{Mg}^{rel} \times \frac{[M_S]}{\{[M_S] + K_d^{M_S}\}\left\{1 + \frac{[M_S]}{K_i^{M_S}}\right\}} \tag{5}$$

Observed reaction rates are often normalized to facilitate comparison of metal ion dependence between reactions that show large differences in the absolute change in reaction rate, such as those involving reactions where thiophilic metal ion rescue is incomplete [72]. Under conditions where saturating concentrations of thiophilic metal ion cannot be obtained, lower limits for $K_d^{M_S}$ can be estimated from curve fits if partial saturation is observed, or the lower limit set as the highest concentration of thiophilic metal ion tested when significant levels of saturation are not evident [72, 86]. Since the level of rescue is proportional to the fraction of ribozyme bound by thiophilic metal ion, comparisons of relative affinity can also be made in the absence of complete thiophilic metal ion saturation by comparing the concentration of added thiophilic metal ion required to produce the same level of rate enhancement [25]. Note that variation in the individual observed reaction rates from experiment to experiment can be significant, and that greater precision can be achieved by comparing values of relative rate or affinity obtained in side-by-side experiments.

Because of the large range in the observed reaction rate and thiophilic metal ion concentrations used in these experiments it is often advantageous for the interpretation of the data to plot k_{obs}^{rel} vs. $[M_S]$ on a log-log scale (Fig. 19.3 [72]):

$$\log k_{obs}^{rel} = \log \left\{ k_{Mg}^{rel} \times \frac{K_d^{M_S}}{[M_S] + K_d^{M_S}} + k_{M_S}^{rel} \times \frac{[M_S]}{[M_S] + K_d^{M_S}} \right\} \quad (6)$$

Single metal ion binding is consistent with plots in which k_{obs}^{rel} increases linearly with added thiophilic metal ion at concentrations below $K_d^{M_S}$ (Fig. 19.3A) [72]:

$$\log k_{obs}^{rel} = \log \left\{ k_{Mg}^{rel} + k_{M_S}^{rel} \times \frac{[M_S]}{[M_S] + K_d^{M_S}} \right\}; \quad [M_S] \ll K_d^{M_S} \quad (7)$$

Note that the apparent non-linearity of plots at thiophilic metal ion concentrations well below $K_d^{M_S}$ is due to the contribution of enzymatic activity in the absence of thiophilic metal ion (k_{Mg}^{rel}). Log–log plots that exhibit a steeper dependence of the observed reaction rate on added thiophilic metal ion concentration when compared to that for single metal ion binding indicate the binding of multiple metal ions. In this case experimental data are generally fit to a non-linear form of the Hill equation:

$$k_{obs}^{rel} = \frac{n K_d^{M_S} k_{max} [M_S]^n}{1 + K_d^{M_S} [M_S]^n} \quad (8)$$

where n is the Hill coefficient that gauges the cooperative dependence of thiophilic metal ion stoichiometry on ribozyme function ($n = 1$ reflects no cooperativity, $n = 2$ reflects the cooperative effect of *at least* two metal ions) and $K_d^{M_S}$ represents the thiophilic metal ion concentration required to attain half the maximal reaction rate (k_{max}) (Fig. 19.3A) [72]. Note, however, that the Hill equation does not directly imply cooperative *binding* when reactivity is used as the experimental signal since "cooperative" effects on function can also be observed if metal ions bind independently. In addition, it is important to remember that the Hill constant reflects the *minimal* number of metal ions. In the current example, $n_{Hill} = 2$ indicates the involvement of 2 *or more* metal ions in rescue.

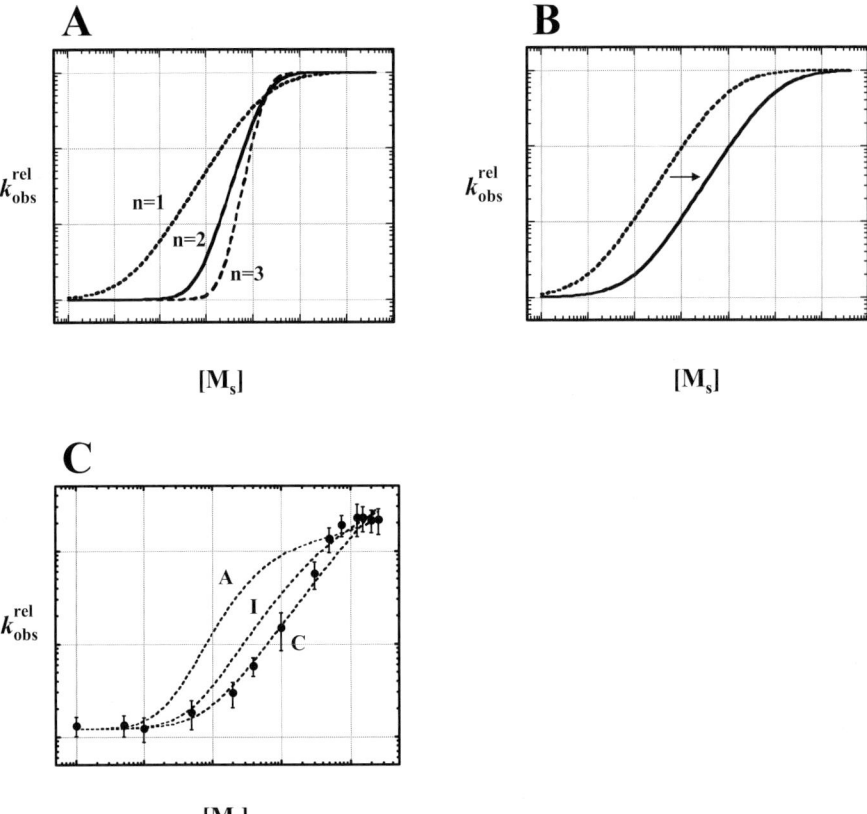

Fig. 19.3. Characterization of metal ion binding. (A) Log-log plot of the observed relative reaction rate (k_{obs}^{rel}) as a function of thiophilic metal ion concentration ([M_S]). Curves labeled $n = 1$, $n = 2$ and $n = 3$ describe the binding of one, two and three metal ions, respectively. The lower and upper limits of k_{obs}^{rel} reflect k_{Mg}^{rel} and $k_{M_S}^{rel}$, respectively. (B) Plot same as in (A) but showing a 10-fold shift in single metal ion binding affinity that may result from changes in RNA structure or altered levels of monovalent or divalent ions. (C) Experimental data plotted with respect to theoretical models for anti-cooperative (A), independent (I) and cooperative (C) binding.

It is important to appreciate that the accurate measure of $K_d^{M_S}$ and n_{Hill} generally requires thiophilic metal ion rescue of the reaction rate to be larger than 10-fold. In cases where an apparent rescue falls below this threshold, changes in reaction conditions can be used to augment the experimental signal. For example, significant increase in the size of a thio effect has been achieved using background levels of Ca^{2+} in place of Mg^{2+} to allow RNA folding but to reduce the rate of catalysis [72]. Any changes to the reaction conditions, however, will require demonstrating that the rate-limiting step of the reaction has not been altered (see below).

Further insight into the nature of cooperative binding has been made through the use of theoretical models such as that shown below in Scheme 19.1, which de-

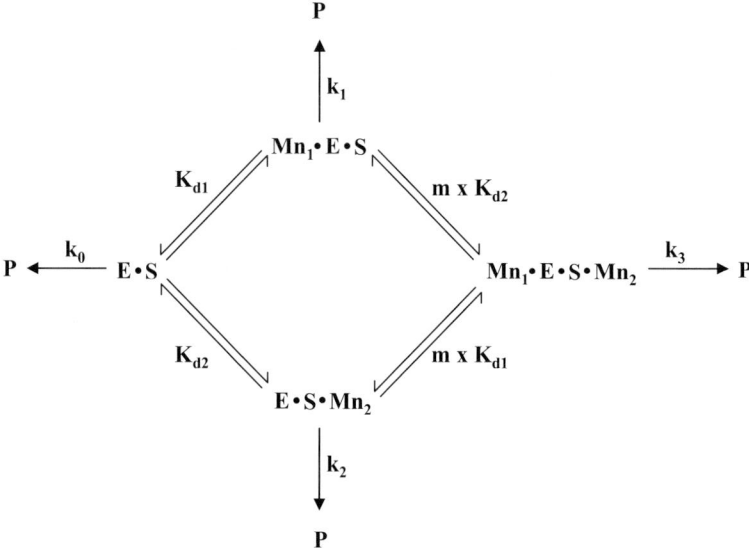

Scheme 19.1

scribes a general kinetic mechanism for the binding of two metal ions and Eq. (9), which describes the predicted dependence of the observed cleavage rate on the concentration of thiophilic metal ion [72]:

$$k_{obs}^{rel} = \frac{k_{Mg}^{rel} + k_1 \frac{[M_S]}{K_{d1}^{M_S}} + k_2 \frac{[M_S]}{K_{d2}^{M_S}} + k_3 \frac{[M_S]^2}{mK_{d1}^{M_S} K_{d2}^{M_S}}}{1 + \frac{[M_S]}{K_{d1}^{M_S}} + \frac{[M_S]}{K_{d2}^{M_S}} + \frac{[M_S]^2}{mK_{d1}^{M_S} K_{d2}^{M_S}}} \quad (9)$$

Using this framework, metal ion binding can be modeled as cooperative ($m > 1$), anti-cooperative ($m < 1$) or independent ($m = 1$). Given measured values for k_{obs}, thiophilic metal ion concentration, and the observed rate in the absence of thiophilic metal ion, simulations can be carried out over a range of assigned values of metal ion affinity ($K_d^{M_S}$) and degree of cooperativity (m) and plotted with the experimental data to assess fit to models for metal ion binding (Fig. 19.3C). Additional experiments to distinguish between theoretical models are described below.

19.2.7
Characterization of Metal Ion Binding

Quantification of metal ion binding provides the basis to directly compare affinities observed by modifications at different positions, to assess their relative energetic contribution to the ribozyme reaction and to determine whether metal ion rescue at distinct positions is consistent with the binding of the same or distinct metal ions. This latter point is of particular importance as it can provide both distance constraints to determine RNA (and in particular active site) structure and specific features of the catalytic mechanism itself [25, 72]. In general, the finding of similar affinities (e.g. within experimental error) is consistent with rescue by a common ion, while the observation of distinctly different affinities (above 10-fold) provides strong evidence for rescue by different metal ions (for additional considerations see below).

In order to *directly* compare metal ion affinities one must be able to measure the affinity of a rescuing thiophilic metal ion that is *not* perturbed by a given modification (for a detailed discussion, see [34]). That is, although a given modification is required for thiophilic metal ion rescue, reaction conditions must be set such that the affinity of the rescuing metal ion is not altered by the presence of sulfur or nitrogen substitution. Perturbation of metal ion affinity by the sulfur or nitrogen modification is highly dependent on local RNA structure and the specific combination of modification and thiophilic metal ion involved, and is thus not necessarily indicative of the native metal ion interaction or an accurate basis for *direct* comparison of kinetic and thermodynamic values. In the group I ribozyme from *Tetrahymena*, the measurement of metal ion affinity unperturbed by site-specific substitution has been met by examining thiophilic rescue by several distinct kinetic methods. One particularly robust approach that is unique to this system takes advantage of the observation that oligonucleotide substrates in this system bind in two steps, first as an "open complex" by simple base pairing with the ribozyme, and second by formation of a "closed complex" which involves docking of the substrate helix into the active site via tertiary interactions [34]. Introduction of 2′-deoxy or 2-methoxy modifications at or upstream of the cleavage site results in ground state binding predominantly in the open complex with no effect on subsequent reaction steps. Although rescue occurs through the interaction of the thiophilic metal ion in the transition state, the transition state itself is transient and does not affect the observed metal ion affinity. The measurement of thiophilic metal ion affinity (K_d^{Ms}) using substrate bound in the open complex thus reflects the apparent metal ion affinity to the unmodified enzyme.

K_d^{Ms} can also be measured under conditions where the concentration of enzyme is subsaturating with respect to modified substrate and well below K_d^{Ms} [34]. In this case the active site is left sufficiently unoccupied such that thiophilic metal ion affinity is not perturbed by atomic (e.g. sulfur) substitution. In addition, thiophilic metal ion affinities independent of modification can be calculated from (1) the measurement of K_d^{Ms} under conditions where substrate is bound and (2) the independent determination of the effect of thiophilic metal ion on the affinity of modified and unmodified substrates, using simple thermodynamic cycles involving alternative pathways of binding and catalysis [34]. Note, however, that the ability to measure thiophilic metal ion binding independent of the phosphorothioate modification is made possible because of the detailed kinetic understanding of the *Tetrahymena* group I ribozyme and that it may be difficult to meet this criteria in other systems where the kinetic mechanism is less defined.

Although thiophilic metal ion binding to sulfur or nitrogen does not necessarily reflect the metal ion interaction in the native context, quantitative measurement of such an affinity along with other characteristics can nevertheless provide experimental evidence for the number of metal ions involved in binding and for the coordination of individual ligands to the same or different metal ions. Experiments often involve comparative analysis of the effect of additional modifications on n_{Hill} or K_d^{Ms} of thiophilic metal ion binding. Such comparisons can be made in the context of a second phosphorothioate modification or additional modification(s) of the backbone or base. The most common form of alternative (non-phosphorothioate) modifications are changes in ring nitrogens (e.g. N^7-deaza) and 2′ hydroxyl positions (e.g. 2′-H, -NH$_2$) because of the propensity of metal ions to interact with these functional groups [34, 72, 78, 81, 86]. Additional mutations that result in a shift to weaker or stronger metal ion binding indicate that the modified functional groups are thermodynamically-linked. This result is consistent with a model in which the functional groups in question are common ligands to the same metal ion (e.g. Fig. 19.3B). In contrast, additional mutations that produce no change in affinity are more consistent with independent interactions [72, 81, 86]. Similarly, second-site modifications that alter affinity but do not perturb the single metal ion binding characteristic ($n_{Hill} = 1$) of the original sulfur or nitrogen modification are consistent with the binding of both ligands to the same metal ion, while combined mutations that alter the level of apparent cooperativity (e.g. from $n_{Hill} = 1$ to $n_{Hill} = 2$) are indicative of independent metal ion binding (Fig. 19.3A) [25, 34]. In addition, second-site modification can allow detection of metal ion interaction to the original sulfur or nitrogen substitution by providing a ligand or structural environment more favorable to binding, positioning, or specificity of a particular thiophilic metal ion [25, 72, 73, 86] (see below). Note that it is important to show that the observed change in signal is specific to the functional group(s) in question, with other sites having no significant effect [72, 86].

The binding of metal ions to individual functional groups can also be compared by measuring the competing effect of background metal ions (such as Ca^{2+} or Mg^{2+}) on the affinity of the rescuing thiophilic metal ion [25]. Such experiments are most easily interpreted using metal ions that do not directly contribute to the observed rescue of the reaction rate, but rather appear to weaken the apparent binding of thiophilic metal ion by direct competition. Increasing concentrations of competing metal ion should result in a decrease in the apparent affinity of thiophilic metal ion binding without changing the shape or apparent cooperativity of the binding curve (e.g. Fig. 19.3B). A shift in rescuing thiophilic metal ion binding should only be observed at concentrations of competing metal ion in the range of that observed for K_d^{Ms}. Therefore, the presence or absence of a shift in K_d^{Ms} at different concentrations of competing metal ion can be used to discern whether thiophilic metal ion affinity at one position is similar or distinct from that previously observed at other positions [25]. This approach is particularly useful under conditions where saturating concentrations of thiophilic metal ion cannot be obtained and apparent affinity must be measured indirectly from comparing the amount of thiophilic metal ion required to achieve the same enhancement of the observed rate for modified and control RNAs (see above). In addition, the observation of a shift in apparent K_d^{Ms} at increasing concentrations of background metal ion suggests (but does not prove) that the background metal ion occupies the binding site rescued by thiophilic metal ion and reflects a metal binding site in the native ribozyme [24]. As with many of the experiments above, it is advantageous to identify analogous shifts in K_d^{Ms} with different competing metal ions to demonstrate the generality of the observation.

19.2.8
Further Tests of Metal Ion Cooperativity

Although the observed metal ion dependence can appear to be consistent with a model for single metal ion binding, there are a number of circumstances where the binding of two metal ions can produce the same results. For example, thiophilic metal ion dependence consistent with single metal ion binding would also be observed if a modification allowed the first of two separate metal ions to bind tightly and did not produce a rate defect until a second metal ion was bound at a lower affinity site [72, 86]. The presence of a higher affinity site can be probed by examining the extent of rescue at different concentrations of ribozyme–substrate complex [72]. Because the model for tight metal binding requires the binding of a much lower affinity ion to observe rescue, significant rescue can only be observed if the concentration of thiophilic metal ion exceeds that of the ribozyme substrate complex. In contrast, the observed rescue of a single metal ion is not affected by enzyme–substrate concentration.

Thiophilic metal ion dependence consistent with single metal ion binding would also be observed for independent or cooperative binding of multiple metal ions with similar affinities, but with anti-cooperative effects on catalysis in which binding of one ion reduces the stimulatory effect of the other [72]. Moreover, affinities differing by less than 10-fold are difficult to resolve as independent elements of a binding curve. These models for multiple metal ion binding, as well as that above involving the tight binding of a second metal ion can be addressed using base deletions or functional group modifications that perturb the observed thiophilic metal ion dependence. These modifications are not expected to produce identical changes in affinity of two distinct metal ions and can reveal biphasic binding predicted for models involving two metal ions [72, 116]. Distinct metal binding sites also have the potential for distinct affinities for different thiophilic metal ion species and thus may produce biphasic dependence when different thiophilic metal ions are used. Note, however, that different ligands to the same metal ion may have distinct effects on binding activity due to changes in the steric or electrostatic environment engendered by the phosphorothioate modification, as has been observed in protein enzymes [101]. Thus, the observation of single metal ion dependence, even in the presence of modifications that perturb metal ion affinity or in the presence of different thiophilic metal ions that support rescue of the observed rate, can provide evidence that suggests, but does prove, the involvement of only a single metal ion [72].

19.3
Additional Considerations

19.3.1
Verification of k_{rel}

Because the interpretation of thiophilic metal ion rescue studies is dependent on the extent to which k_{rel} reflects the effect of an individual functional group on the native reaction, it is crucial to provide independent evidence that that this is in fact the case. As noted above the ability to interpret k_{rel} in this way requires: (1) the reactions being compared share the same rate-limiting step, (2) the effect of other metal ion sites is the same for both reactions and (3) there is no direct effect of the rescuing metal ion on the "normal" reaction. First, one cannot overstate the importance of isolating the step of interest and providing evidence that the step of the reaction studied under conditions of metal ion rescue is the same as that for the native ribozyme. The establishment of this condition is particularly vital since a shift in rate-limiting step (which is likely in cases where the perturbation of the observed rate is large or the rescue is incomplete) can result in apparent (i.e. false) rescue and thus an inaccurate and misleading interpretation [78]. However, this control is a difficult criterion to fulfill given the small number of tests that can be used to identify an individual step in a ribozyme reaction and the fact that even these tests are not definitive. While the kinetic conditions required to isolate an individual step such as catalysis are system dependent and beyond the scope of this chapter, it can generally be said that the insight that can be gained from thiophilic metal ion rescue studies is directly proportional to the degree to which the kinetic scheme of a ribozyme has been defined. Indeed the ability to examine the effect of metal ion binding on different steps of a given reaction lies at the heart of defining the metal ion's functional role.

In the kinetically defined ribozyme systems from the Hammerhead, the group I and group II ribozymes and bacterial RNase P, single turnover conditions have been defined in which the observed rate of cleavage shows a log-linear dependence on pH [49, 75, 104, 117, 118]. Although the mechanisms of cleavage differ, the log-linear dependence of observed rate on pH is nevertheless considered consistent with rate-limiting catalysis, and not reflective of upstream events such as substrate binding or a conformation change prior to catalysis. Thus changes in reaction conditions (e.g. changes introduced into the RNA structure or in the type or concentration of divalent metal ion) that provide evidence of a change in rate limiting step of the reaction preclude meaningful comparison with the wild-type enzyme and interpretation of thiophilic metal rescue studies. Conversely, modification-induced changes in RNA structure that do not alter the apparent pH dependence of the observed reaction rate provide evidence (but do not prove) that the rate-limiting step of the reaction has not been perturbed, and allow the effects of modification and thiophilic metal ion rescue to be more directly and quantitatively compared. Similarly, perturbations from base deletion, mutation, or additional functional group modification that produce analogous effects for all RNA species and conditions being compared in thiophilic rescue also provide evidence (but do not prove) that the ribozyme–substrate complexes of the RNAs being compared follow the same mechanism and involve the same molecular interactions [72, 75, 119, 120].

For the study of catalysis, additional care should also be taken to make sure that steps upstream of the cleavage step are in rapid equilibrium. Support for the presence of rapid equilibrium can be obtained (1) by showing that the rate of the reaction is not affected by either the time of incubation with thiophilic metal ion before the initiation of the reaction or by the order of addition of thiophilic metal ion and other reaction components, (2) by showing that reactions follow good first order reaction kinetics without a lag phase, or (3) by showing that pulse chase studies do not perturb the extent and apparent rate of reaction [24]. Although the methods for accurately measuring a given rate-limiting step will differ in different systems, determination of the thiophilic metal ion rescue over a range of conditions provides an important gauge of the robustness and generality of the observed effect and the extent to which the effect can be reliably measured.

The second criterion that the effect of other metal ion sites is the same for both reactions reflects the need for equivalent indirect effects of metal ion binding in order to justify normalizing or canceling-out of these effects when the observed rate of one RNA is divided by the other to produce k_{rel}. Differences in indirect effects of metal ion binding can produce significant under or over estimates of apparent metal ion affinity or cooperativity and should be suspected in plots of k_{rel} versus thiophilic metal ion concentration that cannot be fit to a standard binding isotherm. The likelihood of differences in indirect metal ion binding effects can be assessed by comparing the observed rate dependence on metal ion concentration for thiophilic metal ions that do not rescue, for the native metal ion Mg^{2+}, and for alternative background metal ions such as Ca^{2+}.

The third criterion that there be no direct effect of the rescuing metal ion on the "normal" reaction addresses the issue of whether the binding of thiophilic metal ion alters the chemical mechanism and is thus not representative of the native reaction. Changes in the reaction mechanism can result form changes in local geometry, alterative coordination, or the introduction of novel metal sites within the active site or elsewhere. The likelihood of such changes can be assessed by examining individual reaction characteristics (e.g. k_{obs}, n_{Hill}, pH dependence) over the full range of metal ion conditions tested with particular attention to changes in an individual parameter being greater reaction than expected for a small change in thiophilic metal ion concentration. In this respect, clear interpretation of thiophilic metal ion concentrations that produce a thousand fold enhancement in k_{obs} for unmodified ribozyme is problematic while changes of 10-fold or less are unlikely to reflect changes in reaction mechanism.

19.3.2
Contributions to Complexity of Reaction Kinetics

Reaction kinetics in thiophilic metal ion rescue studies are often complex (containing two or more phases). This complexity is typically the result of differences in the level or type of molecular substitution or heterogeneity in RNA folding, any of which can significantly affect the interpretation of the experiment results. Stereoisomers of individual phosphorothioates can have distinct effects on reaction rate; consequently incomplete separation during purification can lead to biphasic kinetics. Cross-contamination of the individual stereoisomers can be verified by testing whether the observed rate and amplitude of each kinetic fraction show a reciprocal correlation between the pro-R_p and pro-S_p purified fractions and by comparing the observed rates to that of unmodified RNA [66].

Reaction kinetics of RNAs containing inhibitory phosphorothioate substitutions can reveal a small burst phase of approximately 1–3% (but can be much higher), which reflects the level of contaminating unmodified nucleotide analog and must be excluded from calculations of rate defects due to the sulfur substitution itself. The presence of unmodified phosphate is likely the result of problems in synthesis or purification, but may also arise from desulfurization during the course of the reaction [121]. Short (10 or less) oligonucleotides containing either phosphate or phosphorothioate often migrate differently on higher percentage (20% or higher) denaturing polyacrylamide gels, allowing the relative level of oxygen contamination to be measured throughout purification and enzymatic analysis from a comparison of modified and unmodified oligonucleotides or fragments derived from RNase T1 nuclease cleavage. Note that contamination with unmodified RNA or a stereoisomer that is not inhibitory can result in a significant underestimation of the observed level of a specific phosphorothioate effect if substrate dissociation is fast relative to cleavage even under single turnover conditions [66].

Distinct kinetic phases can also result from disruption of structural elements important for RNA folding to produce alternative conformations that may be inactive and exchange slowly with the native structure [122, 123]. The presence of alternative conformations may be monitored by non-denaturing gel electrophoresis [82], and the fraction of correctly folded RNA can be determined from active site titration and burst kinetics [124, 125]. Although some additional insight into the nature of structural complexity may be gained through chemical modification, cross-linking or metal-dependent cleavage, the resolution and conclusions that can be drawn from these techniques is somewhat limited. Finally, it is possible that complex reaction kinetics is due to the formation of multimeric complexes. This possibility can be excluded by demonstrating that the rates and relative amplitudes of the individual kinetic phases are not changed upon dilution of the ribozyme–substrate complex by more than 100-fold [66].

19.3.3
Size and Significance of Observed Effects

Because effects on the observed rate from atomic substitution and thiophilic metal ion rescue vary over a broad range and can be significantly larger than 1000-fold, it is tempting to consider the size of the effect as a measure of its biological importance. However, as noted above, there are many factors that contribute to the observed effect that may mask the energetic contributions of specific metal ion interactions and the nature of metal ion binding in the native state. In particular, geometric constraints present in the native metal binding site or imposed by atomic substitution or binding of non-native metal species preclude prediction of the presence or magnitude of thiophilic rescue based simply on the affinity of a metal ion for a given functional group [98–100]. Both sulfur and thiophilic metal ions are significantly larger than their oxygen and Mg^{2+} counterparts, and significantly alter the bond to phosphate, which in turn can alter the position of the backbone as well as the position or occupancy by the metal ion itself [126–128]. It is also important to note that while Mn^{2+} has a greater thiophilicity than Mg^{2+}, it nevertheless retains a strong preference for oxygen that may mute the magnitude of rescue, particularly in cases where more than one metal ion is involved in coordinating a single sulfur substitution and contributes more than one unfavorable interaction [25, 58, 129]. The combination of sulfur and thiophilic metal ion may also lead to an interaction not present in the native structure such as a change from an outer-sphere to an inner-sphere interaction [5]. In addition, the ligands for Mg^{2+} and thiophilic ions may not be equivalent and may reflect distinct binding sites that are mutually exclusive, either because of electrostatic repulsion or conformational rearrangements [78]. Electrostatic or conformational rearrangements may also allow thiophilic metal ions to bind better than Mg^{2+} in the active conformation of the ribozyme or produce alternative structures capable of activity [5, 72, 78]. Finally, thiophilic metal ions may be bound or better positioned in the transition state compared to the ground state of the reaction [66, 72]. The significance of incomplete rescue should therefore be interpreted with caution and the absence of rescue interpreted simply as a negative result. Nevertheless, rescue effects less than 10-fold fall close to the uncertainty from experimental error. Thus, while this does not mean that smaller effects are not indicative of functionally important interactions, it does increase the burden of distinguishing direct from indirect effects. The finding of similar effects from phosphorothioate substitution or thiophilic metal ion rescue in a structurally distinct but evolutionarily related system, however, can help to provide evidence for the significance of more subtle effects [81, 130]. While phylogenetic comparison bares the burden of finding two or more systems that can meet the criteria stated above, it can provide a powerful control to help rule out experimental artifacts and to isolate structural features central to function rather than idiosyncratic to a particular RNA species.

19.4
Conclusion

As can be seen from the considerations outlined above, application of thiophilic metal ion rescue involves careful design of the experiment within a well-defined kinetic scheme to control for numerous sources of indirect effects on experimental signal. However, its proper application has been invaluable in defining the biochemical role of metal ions in RNA structure and enzymatic catalysis, and as an independent and essential approach to validating high-resolution studies.

Acknowledgments

Sincere thanks go to Drs. Jonatha Gott, Frank Campbell, Nathan Zahler, Mike Harris, Joe Piccirilli and Dan Herschlag for helpful discussions, comments and careful reading of the manuscript.

References

1 Misra, V. K., Draper, D. E., *Biopolymers* **1998**, *48*, 113–135.
2 Misra, V. K., Shiman, R., Draper, D. E., *Biopolymers* **2003**, *69*, 118–136.
3 Shelton, V. M., Sosnick, T. R., Pan, T., *Biochemistry* **2001**, *40*, 3629–3638.
4 Wadkins, T. S., Shih, I., Perrotta, A. T., Been, M. D., *J. Mol. Biol.* **2001**, *305*, 1045–1055.
5 DeRose, V. J., *Curr. Opin. Struct. Biol.* **2003**, *13*, 317–324.
6 Gluick, T. C., Wills, N. M., Gesteland, R. F., Draper, D. E., *Biochemistry* **1997**, *36*, 16173–16186.
7 Horton, T. E., Clardy, D. R., DeRose, V. J., *Biochemistry* **1998**, *37*, 18094–18101.
8 Schreier, A. A., Schimmel, P. R., *J. Mol. Biol.* **1974**, *86*, 601–620.
9 Yamada, A., Akasaka, K., Hatano, H., *Biopolymers* **1976**, *15*, 1315–1331.
10 Anderson, C. F., Record, M. T., Jr, *Annu. Rev. Phys. Chem.* **1995**, *46*, 657–700.
11 Ott, G., Arnold, L., Limmer, S., *Nucleic Acids Res.* **1993**, *21*, 5859–5864.
12 Cate, J. H., Gooding, A. R., Podell, E., Zhou, K., Golden, B. L., Kundrot, C. E., Cech, T. R., Doudna, J. A., *Science* **1996**, *273*, 1678–1685.
13 Cate, J. H., Doudna, J. A., *Structure* **1996**, *4*, 1221–1229.
14 Cate, J. H., Hanna, R. L., Doudna, J. A., *Nat. Struct. Biol.* **1997**, *4*, 553–558.
15 Walter, F., Murchie, A. I., Thomson, J. B., Lilley, D. M., *Biochemistry* **1998**, *37*, 14195–14203.
16 Pyle, A. M., *J. Biol. Inorg. Chem.* **2002**, *7*, 679–690.
17 Hanna, R., Doudna, J. A., *Curr. Opin. Chem. Biol.* **2000**, *4*, 166–170.
18 Piccirilli, J. A., Vyle, J. S., Caruthers, M. H., Cech, T. R., *Nature* **1993**, *361*, 85–88.
19 Sontheimer, E. J., Sun, S., Piccirilli, J. A., *Nature* **1997**, *388*, 801–805.
20 Warnecke, J. M., Furste, J. P., Hardt, W. D., Erdmann, V. A., Hartmann, R. K., *Proc. Natl. Acad. Sci. USA* **1996**, *93*, 8924–8928.
21 Weinstein, L. B., Jones, B. C., Cosstick, R., Cech, T. R., *Nature* **1997**, *388*, 805–808.
22 Chen, Y., Li, X., Gegenheimer, P., *Biochemistry* **1997**, *36*, 2425–2438.
23 Narlikar, G. J., Herschlag, D., *Annu. Rev. Biochem.* **1997**, *66*, 19–59.
24 Shan, S. O., Herschlag, D., *Biochemistry* **1999**, *38*, 10958–10975.
25 Shan, S., Kravchuk, A. V., Piccirilli, J. A., Herschlag, D., *Biochemistry* **2001**, *40*, 5161–5171.
26 Bujalowski, W., Graeser, E., McLaughlin, L. W., Proschke, D., *Biochemistry* **1986**, *25*, 6365–6371.
27 Quigley, G. J., Teeter, M. M., Rich, A., *Proc. Natl. Acad. Sci. USA* **1978**, *75*, 64–68.
28 Celander, D. W., Cech, T. R., *Science* **1991**, *251*, 401–407.
29 Pan, T., Long, D. M., Uhlenceck, O. C., *Divalent Metal Ions in RNA Folding and Catalysis*, Cold Spring Harbor Laboratory Press, Cold Spring Harbor, NY, **1993**.
30 Pan, T., *Biochemistry* **1995**, *34*, 902–909.
31 Bassi, G. S., Murchie, A. I., Lilley, D. M., *RNA* **1996**, *2*, 756–768.
32 Beebe, J. A., Kurz, J. C., Fierke, C. A., *Biochemistry* **1996**, *35*, 10493–10505.

33 Draper, D. E., *Trends Biochem. Sci.* **1996**, *21*, 145–149.
34 Shan, S., Yoshida, A., Sun, S., Piccirilli, J. A., Herschlag, D., *Proc. Natl. Acad. Sci. USA* **1999**, *96*, 12299–12304.
35 Zhang, Y., Liang, J. Y., Huang, S., Ke, H., Lipscomb, W. N., *Biochemistry* **1993**, *32*, 1844–1857.
36 Bone, R., Frank, L., Springer, J. P., Atack, J. R., *Biochemistry* **1994**, *33*, 9468–9476.
37 Wilcox, D. E., *Chem. Rev.* **1996**, *96*, 2435–2458.
38 Cowan, J. A., Ohyama, T., Howard, K., Rausch, J. W., Cowan, S. M., Le Grice, S. F., *J. Biol. Inorg. Chem.* **2000**, *5*, 67–74.
39 Rangan, P., Masquida, B., Westhof, E., Woodson, S. A., *Proc. Natl. Acad. Sci. USA* **2003**, *100*, 1574–1579.
40 Hampel, K. J., Burke, J. M., *Biochemistry* **2003**, *42*, 4421–4429.
41 Misra, V. K., Draper, D. E., *J. Mol. Biol.* **2002**, *317*, 507–521.
42 Nakano, S., Cerrone, A. L., Bevilacqua, P. C., *Biochemistry* **2003**, *42*, 2982–2994.
43 Hammann, C., Lilley, D. M., *ChemBioChem* **2002**, *3*, 690–700.
44 Pan, T., Sosnick, T. R., *Nat. Struct. Biol.* **1997**, *4*, 931–938.
45 Fang, X. W., Thiyagarajan, P., Sosnick, T. R., Pan, T., *Proc. Natl. Acad. Sci. USA* **2002**, *99*, 8518–8523.
46 Russell, R., Millett, I. S., Tate, M. W., Kwok, L. W., Nakatani, B., Gruner, S. M., Mochrie, S. G., Pande, V., Doniach, S., Herschlag, D., Pollack, L., *Proc. Natl. Acad. Sci. USA* **2002**, *99*, 4266–4271.
47 Zhuang, X., Rief, M., *Curr. Opin. Struct. Biol.* **2003**, *13*, 88–97.
48 Sosnick, T. R., Pan, T., *Curr. Opin. Struct. Biol.* **2003**, *13*, 309–316.
49 Smith, D., Pace, N. R., *Biochemistry* **1993**, *32*, 5273–5281.
50 McConnell, T. S., Herschlag, D., Cech, T. R., *Biochemistry* **1997**, *36*, 8293–8303.
51 Lott, W. B., Pontius, B. W., von Hippel, P. H., *Proc. Natl. Acad. Sci. USA* **1998**, *95*, 542–547.
52 Warnecke, J. M., Held, R., Busch, S., Hartmann, R. K., *J. Mol. Biol.* **1999**, *290*, 433–445.
53 Brannvall, M., Mikkelsen, N. E., Kirsebom, L. A., *Nucleic Acids Res.* **2001**, *29*, 1426–1432.
54 Sigel, R. K., Vaidya, A., Pyle, A. M., *Nat. Struct. Biol.* **2000**, *7*, 1111–1116.
55 Eckstein, F., Gish, G., *Trends Biochem. Sci.* **1989**, *14*, 97–100.
56 Peracchi, A., Beigelman, L., Usman, N., Herschlag, D., *Proc. Natl. Acad. Sci. USA* **1996**, *93*, 11522–11527.
57 McKay, D. B., *RNA* **1996**, *2*, 395–403.
58 Jaffe, E. K., Cohn, M., *J. Biol. Chem.* **1978**, *253*, 4823–4825.
59 Pecoraro, V. L., Hermes, J. D., Cleland, W. W., *Biochemistry* **1984**, *23*, 5262–5271.
60 Burgers, P. M., Eckstein, F., *J. Biol. Chem.* **1979**, *254*, 6889–6893.
61 Burgers, P. M., Eckstein, F., *J. Biol. Chem.* **1980**, *255*, 8229–8233.
62 Connolly, B. A., Eckstein, F., *J. Biol. Chem.* **1981**, *256*, 9450–9456.
63 Christian, E. L., Yarus, M., *Biochemistry* **1993**, *32*, 4475–4480.
64 Chanfreau, G., Jacquier, A., *Science* **1994**, *266*, 1383–1387.
65 Hardt, W. D., Warnecke, J. M., Erdmann, V. A., Hartmann, R. K., *EMBO J.* **1995**, *14*, 2935–2944.
66 Peracchi, A., Beigelman, L., Scott, E. C., Uhlenbeck, O. C., Herschlag, D., *J. Biol. Chem.* **1997**, *272*, 26822–26826.
67 Sjogren, A. S., Pettersson, E., Sjoberg, B. M., Stromberg, R., *Nucleic Acids Res.* **1997**, *25*, 648–653.
68 Ortoleva-Donnelly, L., Szewczak, A. A., Gutell, R. R., Strobel, S. A., *RNA* **1998**, *4*, 498–519.
69 Sood, V. D., Beattie, T. L., Collins, R. A., *J. Mol. Biol.* **1998**, *282*, 741–750.
70 Scott, E. C., Uhlenbeck, O. C., *Nucleic Acids Res.* **1999**, *27*, 479–484.
71 Sontheimer, E. J., Gordon, P. M., Piccirilli, J. A., *Genes Dev.* **1999**, *13*, 1729–1741.
72 Wang, S., Karbstein, K., Peracchi, A., Beigelman, L., Herschlag, D., *Biochemistry* **1999**, *38*, 14363–14378.
73 Yoshida, A., Sun, S., Piccirilli, J. A., *Nat. Struct. Biol.* **1999**, *6*, 318–321.

74 CHRISTIAN, E. L., KAYE, N. M., HARRIS, M. E., *RNA* **2000**, *6*, 511–519.
75 GORDON, P. M., SONTHEIMER, E. J., PICCIRILLI, J. A., *Biochemistry* **2000**, *39*, 12939–12952.
76 GORDON, P. M., SONTHEIMER, E. J., PICCIRILLI, J. A., *RNA* **2000**, *6*, 199–205.
77 PFEIFFER, T., TEKOS, A., WARNECKE, J. M., DRAINAS, D., ENGELKE, D. R., SERAPHIN, B., HARTMANN, R. K., *J. Mol. Biol.* **2000**, *298*, 559–565.
78 SHAN, S. O., HERSCHLAG, D., *RNA* **2000**, *6*, 795–813.
79 YEAN, S. L., WUENSCHELL, G., TERMINI, J., LIN, R. J., *Nature* **2000**, *408*, 881–884.
80 WARNECKE, J. M., SONTHEIMER, E. J., PICCIRILLI, J. A., HARTMANN, R. K., *Nucleic Acids Res.* **2000**, *28*, 720–727.
81 GORDON, P. M., PICCIRILLI, J. A., *Nat. Struct. Biol.* **2001**, *8*, 893–898.
82 CRARY, S. M., KURZ, J. C., FIERKE, C. A., *RNA* **2002**, *8*, 933–947.
83 SZEWCZAK, A. A., KOSEK, A. B., PICCIRILLI, J. A., STROBEL, S. A., *Biochemistry* **2002**, *41*, 2516–2525.
84 SHAN, S. O., HERSCHLAG, D., *RNA* **2002**, *8*, 861–872.
85 LINDQVIST, M., SANDSTROM, K., LIEPINS, V., STROMBERG, R., GRASLUND, A., *RNA* **2001**, *7*, 1115–1125.
86 CHRISTIAN, E. L., KAYE, N. M., HARRIS, M. E., *EMBO J.* **2002**, *21*, 2253–2262.
87 PLEY, H. W., FLAHERTY, K. M., MCKAY, D. B., *Nature* **1994**, *372*, 68–74.
88 SCOTT, W. G., MURRAY, J. B., ARNOLD, J. R., STODDARD, B. L., KLUG, A., *Science* **1996**, *274*, 2065–2069.
89 CURLEY, J. F., JOYCE, C. M., PICCIRILLI, J. A., *J. Am. Chem. Soc.* **1997**, *119*, 12691–12692.
90 BRAUTIGAM, C. A., SUN, S., PICCIRILLI, J. A., STEITZ, T. A., *Biochemistry* **1999**, *38*, 696–704.
91 MURRAY, J. B., DUNHAM, C. M., SCOTT, W. G., *J. Mol. Biol.* **2002**, *315*, 121–130.
92 MADERIA, M., HUNSICKER, L. M., DEROSE, V. J., *Biochemistry* **2000**, *39*, 12113–12120.
93 TANAKA, Y., KOJIMA, C., MORITA, E. H., KASAI, Y., YAMASAKI, K., ONO, A., KAINOSHO, M., TAIRA, K., *J. Am. Chem. Soc.* **2002**, *124*, 4595–4601.
94 HUPPLER, A., NIKSTAD, L. J., ALLMANN, A. M., BROW, D. A., BUTCHER, S. E., *Nat. Struct. Biol.* **2002**, *9*, 431–435.
95 MORRISSEY, S. R., HORTON, T. E., DEROSE, V. J., *J. Am. Chem. Soc.* **2000**, *122*, 3473–3481.
96 MORRISSEY, S. R., HORTON, T. E., GRANT, C. V., HOOGSTRATTEN, C. G., BRITT, R. D., DEROSE, V. J., *J. Am. Chem. Soc.* **1999**, *121*, 9215–9218.
97 SUZUMURA, K., YOSHINARI, K., TANAKA, Y., TAKAGI, Y., KASAI, Y., WARASHINA, M., KUWABARA, T., ORITA, M., TAIRA, K., *J. Am. Chem. Soc.* **2002**, *124*, 8230–8236.
98 SMITH, J. S., NIKONOWICZ, E. P., *Biochemistry* **2000**, *39*, 5642–5652.
99 MADERIA, M., HORTON, T. E., DEROSE, V. J., *Biochemistry* **2000**, *39*, 8193–8200.
100 HORTON, T. E., MADERIA, M., DEROSE, V. J., *Biochemistry* **2000**, *39*, 8201–8207.
101 BRAUTIGAM, C. A., SUN, S., PICCIRILLI, J. A., STEITZ, T. A., *Biochemistry* **1999**, *38*, 696–704.
102 BASU, S., STROBEL, S. A., *RNA* **1999**, *5*, 1399–1407.
103 KRAGETEN, J. *Atlas of Metal–Ligand Equilibria in Aqueous Solution*, Halsted Press, New York, **1978**.
104 DAHM, S. C., DERRICK, W. B., UHLENBECK, O. C., *Biochemistry* **1993**, *32*, 13040–13045.
105 ANDEREGG, G., in: *Comprehensive Coordination Chemistry: The Synthesis, Reactions, Properties, Applications of Coordination Compounds*. WILKINSON, G., GILLARD, R. D., MCCLEVERTY, J. A. (eds), Pergamon Press, Oxford, **1987**, pp. 777–792.
106 SUN, S., YOSHIDA, A., PICCIRILLI, J. A., *RNA* **1997**, *3*, 1352–1363.
107 CHRISTIAN, E. L., YARUS, M., *J. Mol. Biol.* **1992**, *228*, 743–758.
108 SLIM, G., GAIT, M. J., *Nucleic Acids Res.* **1991**, *19*, 1183–1188.
109 BURGERS, P. M., ECKSTEIN, F., *Proc. Natl. Acad. Sci. USA* **1978**, *75*, 4798–4800.

110 Moore, M. J., Sharp, P. A., *Science* **1992**, *256*, 992–997.
111 Harris, M. E., Christian, E. L., *Methods* **1999**, *18*, 51–59.
112 Hansen, A., Pfeiffer, T., Zuleeg, T., Limmer, S., Ciesiolka, J., Feltens, R., Hartmann, R. K., *Mol. Microbiol.* **2001**, *41*, 131–143.
113 Zahler, N. H., Christian, E. L., Harris, M. E., *RNA* **2003**, *9*, 734–745.
114 Herschlag, D., Eckstein, F., Cech, T. R., *Biochemistry* **1993**, *32*, 8312–8321.
115 Yoshida, A., Shan, S., Herschlag, D., Piccirilli, J. A., *Chem. Biol.* **2000**, *7*, 85–96.
116 Christian, E. L., Zahler, N. H., Kaye, N. M., Harris, M. E., *Methods* **2002**, *28*, 307–322.
117 Herschlag, D., Khosla, M., *Biochemistry* **1994**, *33*, 5291–52287.
118 Pyle, A. M., Green, J. B., *Biochemistry* **1994**, *33*, 2716–2725.
119 Herschlag, D., Piccirilli, J. A., Cech, T. R., *Biochemistry* **1991**, *30*, 4844–4854.
120 Michels, W. J., Jr, Pyle, A. M., *Biochemistry* **1995**, *34*, 2965–2977.
121 Hamm, M. L., Nikolic, D., van Breemen, R. B., Piccirilli, J. A., *J. Am. Chem. Soc.* **2000**, *122*, 12069–12078.
122 Clouet-D'Orval, B., Uhlenbeck, O. C., *RNA* **1996**, *2*, 483–491.
123 Ruffner, D. E., Dahm, S. C., Uhlenbeck, O. C., *Gene* **1989**, *82*, 31–41.
124 Beebe, J. A., Fierke, C. A., *Biochemistry* **1994**, *33*, 10294–10304.
125 Cornish-Bowden, A. J., *Fundamentals of Enzyme Kinetics*, Portland Press, London, **1995**.
126 Weast, R. C., *Handbook of Physics and Chemistry*, CRC Press, Boca Raton, FL, **1989**.
127 Shannon, R. D., *Acta Crystallogr. A* **1976**, *32*, 751–767.
128 Liang, C. X., Allen, L. C., *J. Am. Chem. Soc.* **1987**, *109*, 6449–6453.
129 Strobel, S. A., Ortoleva-Donnelly, L., *Chem. Biol.* **1999**, *6*, 153–165.
130 Kuo, L. Y., Piccirilli, J. A., *Biochim. Biophys. Acta* **2001**, *1522*, 158–166.

20
Identification of Divalent Metal Ion Binding Sites in RNA/DNA-metabolizing Enzymes by Fe(II)-mediated Hydroxyl Radical Cleavage

Yan-Guo Ren, Niklas Henriksson and Anders Virtanen

20.1
Introduction

The presence and requirement for divalent metal ions in the active sites of nucleic acid metabolizing enzymes which participate in phospho(di)ester formation and breakage has emerged as a common theme (reviewed in [1, 2]). One of the best-studied active sites involved in the cleavage of a phosphodiester bond is the 3′ exonucleolytic site of *Escherichia coli* DNA polymerase (Pol) I, which has been characterized by a combination of genetics, biochemistry and structural techniques (see [3–5] and references therein). Most importantly, crystallographic studies provided direct evidence that divalent metal ions are coordinated in this active site directly or via bridging water to oxygens in four acidic amino acid residues as well as to a non-bridging oxygen at the scissile phosphodiester. The divalent metal ions in the 3′-exonucleolytic site of DNA Pol I play a critical role during catalysis, and it has been proposed that the nucleophile (water or hydroxide ion) attacking the scissile phosphate during cleavage is oriented by one metal ion and two amino acid residues. After cleavage, a second divalent metal ion stabilizes the negative charge on the leaving group. A similar mechanism, where one divalent metal ion activates the initially attacking nucleophile while the other stabilizes the leaving group, is used by several other enzymes involved in breaking and forming phospho(di)ester bonds, such as endo- or exonucleases, kinases, phosphatases and polymerases (reviewed in [1, 2, 6]). Taken together, it has become apparent that a very fruitful strategy to study the active site of any enzyme participating in phospho(di)ester formation and breakage is to identify and characterize its divalent metal ion binding sites.

Here we will describe protocols that we have used to characterize and map divalent metal ion binding sites in the active sites of the Klenow fragment of *E. coli* DNA Pol I and human poly(A)-specific ribonuclease (PARN) [7, 8]. PARN is a 3′ exonuclease that efficiently degrades mRNA poly(A) tails [9–17] and belongs to the RNase D family of nucleases [8, 10, 18], of which the 3′ exonuclease domain of *E. coli* DNA Pol I is one of the best-studied examples. The method of Fe(II)-mediated hydroxyl radical cleavage [8] described here has been applied to map di-

Handbook of RNA Biochemistry. Edited by R. K. Hartmann, A. Bindereif, A. Schön, E. Westhof
Copyright © 2005 WILEY-VCH Verlag GmbH & Co. KGaA, Weinheim
ISBN: 3-527-30826-1

valent metal ion binding sites in a large variety of metalloenzymes (see, e.g. [19–25] and references therein). We want to emphasize, however, that this method needs to be combined with several other approaches before a complete picture of a divalent metal ion binding site can be drawn. In our case, the Fe(II)-mediated hydroxyl radical cleavage assays of PARN were preceded by two important steps: (1) bioinformatic identification of amino acids potentially located in the active site of the enzyme and (2) site-directed mutagenesis of amino acids expected to be important for catalysis. Several protocols and descriptions of these two steps are available elsewhere (see, e.g. [26, 27]) and thus will not be given here.

20.2
Probing Divalent Metal Ion Binding Sites

One of the most important prerequisites for the successful analysis of any enzyme is the availability of sufficiently large and pure preparations of the enzyme. It is therefore important to spend some time to determine a simple and efficient protocol for the expression and purification of a recombinant form of the enzyme of interest. A large variety of expression systems are commercially available and we have successfully used several of them (e.g. the pET system from Novagen or the pCAL system from Stratagene). When choosing your expression system it is important to investigate if the recombinant form of the enzyme has the same key properties as the non-recombinant one, which is often not the case. For example, we observed when studying human poly(A) polymerases that the position, N- or C-terminally, of the tag used for affinity purification significantly affected the K_m parameter of the enzyme (our unpublished observation).

20.2.1
Fe(II)-mediated Hydroxyl Radical Cleavage

The induction of hydroxyl radicals through the Fenton reaction [28] (Fig. 20.1) in the vicinity of Fe(II) ion binding sites has become a powerful tool to identify divalent metal ion binding sites in protein and RNA enzymes. In the presence of reductants, such as ascorbic acid or DTT, Fe(II) generates hydroxyl radicals which efficiently cleave the polypeptide or nucleic acid backbone in the vicinity of the Fe(II) binding site.

Before performing the Fe(II)-mediated hydroxyl radical cleavage assay, it is advisable to investigate if the enzyme under study is active in the presence of Fe(II) [29].

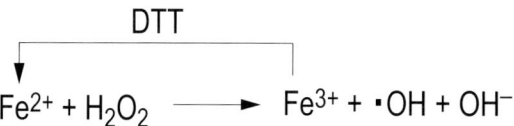

Fig. 20.1. The Fenton reaction.

For PARN we could readily detect enzymatic activity when we replaced the essential divalent metal ion Mg(II) with Fe(II) [8]. A positive result from this simple assay argues immediately that Fe(II) functions catalytically, which implies that some of the Fe(II) ion binding sites overlap with binding sites for Mg(II) ions. It is important to remember that the latter statement is one of the key assumptions of your analysis since you will argue that some of the binding sites for Fe(II) ions that you eventually will identify correspond to binding sites for the natural metal cofactors in the active site of the enzyme.

Technically, the Fe(II)-mediated hydroxyl radical cleavage assay is easy to perform. The enzyme (0.5–10 μM) is incubated in a buffer containing 50–100 mM HEPES, pH 7.0, 10 mM DTT and 0.2–20 μM $Fe(NH_4)_2SO_4$. The exact reaction volume, incubation time and temperature as well as the amount of enzyme incubated have to be determined empirically in order to resolve visible and distinct cleavage products by SDS–PAGE. As a rule, the reaction volume is about 10 μl, and incubation times and temperatures are between 2 and 30 min and 0 and 37 °C. Protocol 1 describes the conditions we used for PARN [8], while Protocol 2 describes our conditions for the Klenow Pol fragment [7]. Often, a small amount of H_2O_2 (approximately 0.1–0.2% v/v) has to be added, as well as a small amount of NaCl. In the case of Fe(II)-mediated cleavage of PARN and Klenow Pol we could omit H_2O_2, while we included 5 mM NaCl in reactions containing the Klenow Pol fragment. The presence of substrate can also influence the cleavage pattern or the efficiency of cleavage. For PARN, the inclusion of the substrate poly(A_{50}) improved the cleavage reaction significantly, while the addition of DNA to Klenow Pol had a minor effect only.

The reaction is terminated by the addition of one reaction volume of $2 \times$ SDS loading buffer and directly fractionated by SDS–PAGE. Subsequently, the cleavage products are visualized by silver staining if a non-radioactive polypeptide was reacted, or by autoradiography if the polypeptide was radioactively labeled. Figure 20.2 shows a typical result obtained when Klenow Pol is subjected to Fe(II)-mediated cleavage. Please note the importance of the control lanes 2, 3, 5 and 6. Lane 2 controls for the dependence on Fe(II), lane 3 demonstrates the essential role of the reducing agent DTT, lane 6, including the chelator EDTA, documents the requirement for Fe(II) and/or possibly traces of other divalent metal ions brought in with the enzyme preparation, and lane 5 in comparison with lane 4 suggests that Fe(II) and Mg(II) occupy overlapping binding sites since addition of Mg(II) suppresses the appearance of the main cleavage product. Note that the reactions in lanes 2, 3, 5 and 6 also control for the presence or appearance of non-specific cleavage products.

20.2.2
How to Map Divalent Metal Ion Binding Sites

The Fe(II)-induced cleavage site(s) will be in the vicinity, within a few Ångstroms, of the binding site(s) for Fe(II). Thus, a major effort is dedicated to localizing the cleavage site. For this purpose, radioactive labeling of the N or C terminus of the

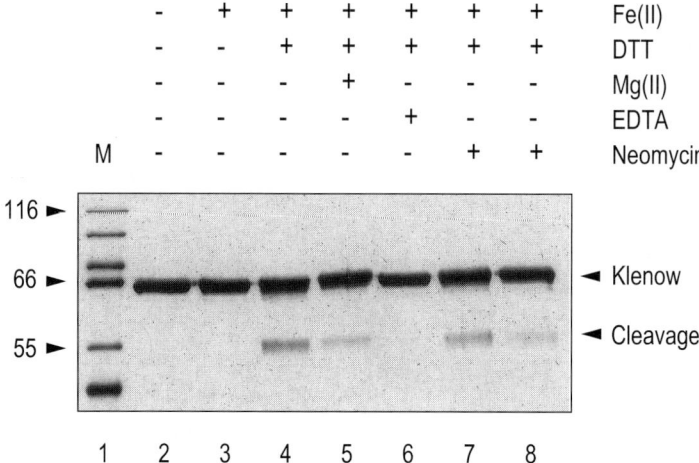

Fig. 20.2. Fe(II)-mediated cleavage of recombinant Klenow Pol fragment. A sample of 2 μg of Klenow Pol fragment was incubated as described in Protocol 2. The resulting cleavage products were analyzed by SDS–PAGE followed by silver staining. Presence (+) or absence (−) of indicated reagent; Fe(II) (lanes 3–8), 20 μM Fe(NH$_4$)$_2$(SO$_4$)$_2$; DTT (lanes 4–8), 10 mM DTT; Mg(II) (lane 5), 10 mM MgCl$_2$; EDTA (lane 6), 10 mM EDTA·Na$_2$; neomycin (lanes 7 and 8), 2 and 10 mM neomycin B. The molecular size marker was fractionated in lane 1. The arrowheads on the right mark the position of Klenow Pol fragment and its cleavage product; arrowheads on the left depict selected size markers with their indicated molecular weight (in kDa).

enzyme and size fractionation of cleavage fragments by SDS–PAGE is often used [30–34]. A good assignment of the cleavage position then depends on an accurate correlation of molecular size and migrational distance. Another advantage of using radioactively labeled polypeptides is the possibility to quantify the cleavage reaction. Protocol 3 describes our procedure for radioactive labeling of PARN at the N terminus [8]. In this particular case, we have made use of a protein kinase recognition motif that was present in the N-terminal tag encoded in the commercially available expression plasmid pET33 (Novagen).

However, due to the tertiary structure of proteins, induced cleavages and Fe(II) binding sites are not always close to each other in the primary sequence. As a matter of fact, accurate mapping of the cleavage sites is not always required. This is very well exemplified by our studies of the active site of PARN. Here we could apply a different strategy instead, since bioinformatic characterization followed by site-directed mutagenesis had already revealed amino acids presumably located in the active site of the enzyme. Thus, we simply investigated if any of these introduced mutations affected the appearance of the Fe(II)-induced cleavage products in comparison with the wild-type enzyme. An altered cleavage pattern for the mutant polypeptide then indicated that the mutated amino acid is required for Fe(II)-mediated cleavage. The effects caused by the mutations that we have observed ranged from minute decreases to complete disappearance of cleavage product. An excellent way to quantify such effects is to determine an apparent K_D (appK_D) for the Fe(II) ions causing the cleavages [22]. For this purpose, an increasing amount of Fe(II) is added to the individual reactions and the cleavage product at each concentration of Fe(II) is quantified. Finally, an appK_D is calculated, e.g. using Lineweaver–Burk formalism. It is worthwhile mentioning that a difference in the calculated appK_D for two or more cleavage sites in the same polypeptide implies that (1) the cleavages are induced by different Fe(II) ions and, thus, (2) multiple binding sites for Fe(II) ions have been identified.

20.2.3
How to Use Aminoglycosides as Functional and Structural Probes

Aminoglycosides bind frequently to negatively charged binding pockets present in both protein enzymes and RNA (see [7, 35–37] and references therein). Often these binding sites overlap with binding sites for divalent metal ions, and experimental evidence suggests that aminoglycosides displace functionally important divalent metal ions upon binding and thereby perturb the function of RNA and protein metalloenzymes (e.g. [7, 36, 37]). Aminoglycosides have therefore turned out to be convenient probes in studies of divalent metal ion binding and function. For application of the strategy we used in our studies of PARN, Klenow Pol and poly(A) polymerase ([7] and unpublished data), one may follow the experimental scheme outlined below (see [7] for a detailed description):

(1) Investigate if aminoglycosides inhibit the enzymatic activity by simply adding increasing amounts of aminoglycoside to the reaction. The chemical properties of the aminoglycoside will, of course, influence how efficient it inhibits enzyme activity and one should therefore investigate a repertoire of commercially available aminoglycosides. As a rule of thumb: the higher its number of positively charged amino groups, the more efficiently the aminoglycoside will inhibit the enzyme. The interaction is highly electrostatic; the pH of the reaction therefore plays a decisive role and should usually be below 7.0. The inhibition constants are often in the micromolar range.
(2) Once conditions for inhibition have been established, one should investigate if the aminoglycoside perturbs the Fe(II)-mediated cleavage reaction. For this purpose, include increasing amounts of aminoglycoside in the Fe(II)-mediated cleavage reaction, followed by SDS–PAGE. This is illustrated in Fig. 20.2 (lanes 7 and 8) for neomycin B and the Klenow Pol fragment.
(3) Finally, you should investigate if increasing amounts of a second divalent metal ion, such as Mg^{2+}, relieve the inhibition.

Provided certain aminoglycosides bind to the metalloenzyme of interest with reasonable affinity and specifically displace active site metal ion(s) as inferred from suppression of Fe(II)-mediated cleavage, one has established an elegant experimental platform to investigate the structural and functional role of divalent metal ions in much detail.

20.3
Protocols

Protocol 1: Fe(II)-mediated cleavage of PARN

(1) Label PARN with ^{32}P at the N-terminus using [γ-^{32}P]ATP and bovine heart protein kinase (see Protocol 3).
(2) Prepare the following stock solutions: 0.5 M HEPES–KOH, pH 7.0, 20 µM Fe(NH$_4$)$_2$(SO$_4$)$_2$ [Fe(NH$_4$)$_2$(SO$_4$)$_2$·6 H$_2$O; Sigma F 3754], 50 mM DTT, 0.3 µM poly(A$_{50}$) and 2 × SDS–PAGE loading buffer (0.25 mM Tris–HCl, pH 6.8 at room temperature, 20% glycerol, 2% SDS, 0.025% bromophenol blue) supplemented with 0.2 M DTT.
(3) For each reaction, mix on ice 5 pmol of ^{32}P-labeled PARN, 1 µl 0.5 M HEPES–KOH, pH 7.0, 1 µl 20 µM Fe(NH$_4$)$_2$(SO$_4$)$_2$ and adjust the volume to 8 µl with H$_2$O.
(4) Start the reaction by the addition of 2 µl 50 mM DTT. Mix by gently flicking the tube with your finger and transfer to 37 °C. Incubate for 15–30 min.
(5) Stop the reaction by the addition of 10 µl of 2 × SDS–PAGE loading buffer supplemented with 0.2 M DTT.
(6) Boil the sample for 3 min.
(7) Load on a 10% SDS–polyacrylamide gel with a 4% stacking gel.
(8) Run the gel until the dye reaches the bottom.
(9) Fix and dry the gel, and expose an X-ray film or phosphoimage screen.

Protocol 2: Fe(II)-mediated cleavage of Klenow Pol

(1) Prepare a pure preparation of Klenow Pol at approximately 1–2 mg/ml in 20 mM HEPES–KOH, pH 7.0, and 5 mM NaCl.
(2) Prepare the following stock solutions: 0.5 M HEPES–KOH, pH 7.0, 50 mM NaCl, 100 µM Fe(NH$_4$)$_2$(SO$_4$)$_2$, 50 mM DTT and 2 × SDS–PAGE loading buffer (see Protocol 1) supplemented with 0.2 M DTT.
(3) For each reaction, mix on ice 2–4 µg Klenow Pol, 2 µl 0.5 M HEPES–KOH, pH 7.0, 1 µl 50 mM NaCl, 2 µl 100 µM Fe(NH$_4$)$_2$(SO$_4$)$_2$ and adjust the volume to 8 µl with H$_2$O.
(4) Start the reaction by the addition of 2 µl 50 mM DTT. Mix by gently flicking the tube with your finger and transfer to 37 °C. Incubate for 15–30 min.
(5) Stop the reaction by the addition of 10 µl of 2 × SDS–PAGE loading buffer supplemented with 0.2 M DTT.
(6) Boil the sample for 3 min.
(7) Load on a 10% SDS–polyacrylamide gel with a 4% stacking gel.
(8) Run the gel until the dye reaches the bottom.

(9) Fix and silver stain the gel (Sigma ProteoSilver Silver Staining Kit or Amersham Biosciences PlusOne Silver Staining Kit).

Protocol 3: Radioactive labeling of recombinant polypeptide
The recombinant polypeptide should contain a protein kinase recognition motif, either at the N or C terminus. A number of recombinant protein expression systems (e.g. pET33; Novagen) will provide such a motif in-frame with the affinity tag and placed in the multiple cloning site. We have successfully labeled polypeptides expressed by the bacterial pET33 expression system.

(1) Apply 30 µl of the purified recombinant polypeptide at 10 nM onto a G-50 spin column (MicroSpin G-50 columns; Amersham Biosciences) pre-equilibrated with 20 mM Tris–HCl, pH 7.5, 0.1 M NaCl and 12 mM $MgCl_2$. Spin the column at 2000 g for 1 min.
(2) Mix the eluate with 1 µl [γ-^{32}P]ATP (10 mCi/ml, 3000 Ci/mmol), 1 µl 50 µM ATP and 3 µl (10 U/µl) of a freshly dissolved batch of bovine heart protein kinase A catalytic subunit (Sigma P 2645, supplied as lyophilized powder).
(3) Incubate for 30 min on ice.
(4) Apply the labeling mixture onto a G-50 spin column equilibrated in 25 mM HEPES–KOH, pH 7.0, and 100 mM NaCl. Spin the column at 2000 g for 1 min. Collect the eluate and check the efficiency of labeling by fractionating a small sample by SDS–PAGE.

20.4 Notes and Troubleshooting

No Fe(II)-mediated cleavage detected

(1) Use fresh Fe(II) and DTT solutions.
(2) Check if the pH of the reaction is altered. Usually, a higher pH (>7) facilitates Fe(II)-mediated cleavage while a low pH (<6) abolishes cleavage.
(3) Make sure that there is no metal ion-chelating compound such as EDTA present or check if contaminating divalent metal ions compete with Fe(II) in the reaction.
(4) Try titrating H_2O_2 into the reaction.

No distinct cleavage product(s) detected by SDS–PAGE after Fe(II)-mediated cleavage

(1) The concentration of Fe(II) is too high. Optimize the concentration of Fe(II) in the reaction.
(2) If H_2O_2 is used, optimize the concentration of H_2O_2 or omit it.
(3) Shorten the incubation time and/or lower the temperature.
(4) Try addition of enzyme substrate or product.

Low efficiency of ^{32}P-incorporation after kinase labeling reaction

(1) Be sure to use a freshly dissolved batch of protein kinase (see Protocol 3).
(2) Check if the protein kinase recognition sequence tag is intact and has not been removed by protein degradation.

References

1 T. A. STEITZ, J. A. STEITZ, *Proc. Natl. Acad. Sci. USA* **1993**, *90*, 6498–6502.
2 C. M. JOYCE, T. A. STEITZ, *J. Bacteriol.* **1995**, *177*, 6321–6329.
3 A. BERNAD, L. BLANCO, J. M. LAZARO, G. MARTIN, M. SALAS, *Cell* **1989**, *59*, 219–228.
4 C. A. BRAUTIGAM, T. A. STEITZ, *J. Mol. Biol.* **1998**, *277*, 363–377.
5 C. M. JOYCE, T. A. STEITZ, *Annu. Rev. Biochem.* **1994**, *63*, 777–822.
6 T. A. STEITZ, *Nature* **1998**, *391*, 231–232.
7 Y. G. REN, J. MARTINEZ, L. A. KIRSEBOM, A. VIRTANEN, *RNA* **2002**, *8*, 1393–1400.
8 Y. G. REN, J. MARTINEZ, A. VIRTANEN, *J. Biol. Chem.* **2002**, *277*, 5982–5987.
9 C. G. KÖRNER, E. WAHLE, *J. Biol. Chem.* **1997**, *272*, 10448–10456.
10 C. G. KÖRNER, M. WORMINGTON, M. MUCKENTHALER, S. SCHNEIDER, E. DEHLIN, E. WAHLE, *EMBO J.* **1998**, *17*, 5427–5437.
11 J. MARTINEZ, Y. G. REN, P. NILSSON, M. EHRENBERG, A. VIRTANEN, *J. Biol. Chem.* **2001**, *276*, 27923–27929.
12 J. MARTINEZ, Y. G. REN, A. C. THURESSON, U. HELLMAN, J. ASTROM, A. VIRTANEN, *J. Biol. Chem.* **2000**, *275*, 24222–24230.
13 J. ÅSTRÖM, A. ÅSTRÖM, A. VIRTANEN, *EMBO J.* **1991**, *10*, 3067–3071.
14 J. ÅSTRÖM, A. ÅSTRÖM, A. VIRTANEN, *J. Biol. Chem.* **1992**, *267*, 18154–18159.
15 E. DEHLIN, M. WORMINGTON, C. G. KÖRNER, E. WAHLE, *EMBO J.* **2000**, *19*, 1079–1086.
16 P. R. COPELAND, M. WORMINGTON, *RNA* **2001**, *7*, 875–886.
17 M. GAO, D. T. FRITZ, L. P. FORD, J. WILUSZ, *Mol. Cell* **2000**, *5*, 479–488.
18 I. S. MIAN, *Nucleic Acids Res.* **1997**, *25*, 3187–3195.
19 J. M. FARBER, R. L. LEVINE, *J. Biol. Chem.* **1986**, *261*, 4574–4578.
20 M. R. ERMACORA, J. M. DELFINO, B. CUENOUD, A. SCHEPARTZ, R. O. FOX, *Proc. Natl. Acad. Sci. USA* **1992**, *89*, 6383–6387.
21 N. B. GRODSKY, S. SOUNDAR, R. F. COLMAN, *Biochemistry* **2000**, *39*, 2193–2200.
22 A. MUSTAEV, M. KOZLOV, V. MARKOVTSOV, E. ZAYCHIKOV, L. DENISSOVA, A. GOLDFARB, *Proc. Natl. Acad. Sci. USA* **1997**, *94*, 6641–6645.
23 J. LYKKE-ANDERSEN, R. A. GARRETT, J. KJEMS, *EMBO J.* **1997**, *16*, 3272–3281.
24 G. N. GODSON, J. SCHOENICH, W. SUN, A. A. MUSTAEV, *Biochemistry* **2000**, *39*, 332–339.
25 J. J. HLAVATY, J. S. BENNER, L. J. HORNSTRA, I. SCHILDKRAUT, *Biochemistry* **2000**, *39*, 3097–3105.
26 F. M. AUSUBEL, R. BRENT, R. E. KINGSTON, D. D. MOORE, J. G. SEIDMAN, J. A. SMITH, K. STRUHL (eds), *Current Protocols in Molecular Biology*, Wiley, New York, **2004**.
27 J. SAMBROOK, D. RUSSEL, *Molecular Cloning: A Laboratory Manual*, 3rd edn, Cold Spring Harbor Laboratory Press, Cold Spring Harbor, NY, **2001**.
28 H. J. H. FENTON, *Proc. Chem. Soc.* **1893**, *9*, 113.
29 I. B. BROWN, *Acta Crystallogr. B* **1988**, *44*, 545–553.
30 T. H. JENSEN, A. JENSEN, J. KJEMS, *Gene* **1995**, *162*, 235–237.
31 T. H. JENSEN, A. JENSEN, A. M. SZILVAY, J. KJEMS, *FEBS Lett.* **1997**, *414*, 50–54.

32 T. H. JENSEN, H. LEFFERS, J. KJEMS, *J. Biol. Chem.* **1995**, *270*, 13777–13784.
33 R. HORI, S. PYO, M. CAREY, *Proc. Natl. Acad. Sci. USA* **1995**, *92*, 6047–6051.
34 M. ZHONG, L. LIN, N. R. KALLENBACH, *Proc. Natl. Acad. Sci. USA* **1995**, *92*, 2111–2115.
35 T. HERMANN, E. WESTHOF, *J. Mol. Biol.* **1998**, *276*, 903–912.
36 N. E. MIKKELSEN, M. BRÄNNVALL, A. VIRTANEN, L. KIRSEBOM, *Proc. Natl. Acad. Sci. USA* **1999**, *96*, 6155–6160.
37 N. E. MIKKELSEN, K. JOHANSSON, A. VIRTANEN, L. A. KIRSEBOM, *Nat. Struct. Biol.* **2001**, *8*, 510–514.

21
Protein–RNA Crosslinking in Native Ribonucleoprotein Particles

Henning Urlaub, Klaus Hartmuth and Reinhard Lührmann

21.1
Introduction

Protein–RNA interactions lie at the structural and functional heart of ribonucleoprotein (RNP) particles. They govern such fundamental cellular processes as pre-mRNA processing, rRNA maturation, post-transcriptional control (mRNA stability), RNA export, translation and translational control. In this chapter, we present a method we have developed in recent years that allows us to characterize sites of direct protein–RNA contact in native particles, after the contacts have been made permanent by UV crosslinking [1–4].

Our method is especially suitable in situations where the objects of investigation are native RNP particles for which the RNA and the protein compositions are known, while little or no information is available on which proteins are in contact with RNA or where such contacts take place. The method has further been proven to be of value for the identification of direct RNA–protein contact sites in RNP particles reconstituted *in vitro* in which several proteins interact with the RNA component.

In the protocols listed, we refer to isolated U snRNP particles from HeLa cells [5, 6] involved in pre-mRNA processing (for review, see [7]). Importantly, we would like to note that the entire approach can be regarded as a general one, so that the protocols can be easily adapted to investigations of other native RNP particles, or of RNP particles reconstituted *in vitro*.

21.2
Overall Strategy

The overall experimental strategy that we have used for the identification of protein–RNA contact sites in native RNP particles comprises crosslinking of RNP particles by UV irradiation at 254 nm, which fixes protein–RNA interactions covalently by generating a zero-length crosslink, followed by analytical procedures to identify the exact nucleotide(s) on the RNA where the crosslink occurred and to identify the crosslinked polypeptide.

Handbook of RNA Biochemistry. Edited by R. K. Hartmann, A. Bindereif, A. Schön, E. Westhof
Copyright © 2005 WILEY-VCH Verlag GmbH & Co. KGaA, Weinheim
ISBN: 3-527-30826-1

The approach is primarily a primer extension analysis of the crosslinked RNA derived from the UV-irradiated native particles. The correct assignment of putative protein–RNA crosslinking sites on the RNA requires parallel analysis of UV-irradiated "naked" (protein-free) RNA and of non-irradiated naked RNA. Comparison of the reverse transcriptase patterns obtained in these three experiments leads to the identification of the RNA bases at which proteins are crosslinked. This first set of experiments gives an excellent overview if a certain protein – or several proteins of multiprotein complexes – is/are in direct contact with the RNA, but it yields no information about which protein of the RNA is crosslinked.

The identification of the corresponding crosslinked protein is achieved by immunoprecipitation combined with primer extension analysis. Thereby, one can define which protein of the RNP is crosslinked to the bases of the RNA that have been identified in the first set of experiments. It is obvious that this type of identification depends on the availability of antibodies against the different proteins and upon the efficiency with which each antibody precipitates its corresponding protein, especially under mild denaturing conditions (for details, see below). The advantage of the method is that it can reveal multiple crosslinks between one protein and its cognate RNA in native particles. For example, we found in this manner that in U1 snRNP particles the U1 70K protein is in contact with two nucleotides in the loop of stem I of the U1 snRNA [1]. Another example is the U4/U6-specific protein 61K: this was found in contact with two distinct sites on the U4 snRNA in native tri-snRNP particles, i.e. the loop in the 5′ stem–loop of the U4 snRNA and nucleotides upstream of the 5′ stem–loop [4].

21.3
UV Crosslinking

UV crosslinking of RNP particles is a straightforward technique. UV crosslinking at 254 nm generates a covalent bond between an amino acid side chain of a protein and a base of the RNA, whenever the relative position of the two components is favorable. In earlier studies we found that UV irradiation of native complexes at 254 nm leads to crosslinking of the side chains of the following amino acids: methionine, tyrosine, histidine, leucine, phenylalanine, and cysteine ([1–4, 8, 9] and our unpublished observations). On the basis of work with halopyrimidine-substituted RNAs, Koch et al. [10] suggested two possible mechanisms for UV-induced protein–RNA crosslinking events: (1) UV-induced electron transfer from the amino acid residue to the halopyrimidine followed by a loss of halide and subsequent radical combination or (2) UV-induced homolysis of the carbon–halogen bond followed by radical addition to the aromatic ring of the amino acid residue. The fact that we have also found highly specific amino acid-RNA crosslinks in non-substituted RNAs of native complexes strongly supports the first mechanism.

Our approach is highly specific, but it has some limitations. The crosslinking yield is relatively low when compared with that of crosslinking in particles reconstituted *in vitro* that carry an RNA species site-specifically labeled with a crosslinking moiety [11–19]. Furthermore, not all proteins that are tightly bound by RNA can also be directly crosslinked by UV irradiation. Examples of this are the U1A protein bound to the U1 snRNA particle ([20] and our unpublished observations) and the human 15.5K protein bound to the human U4 snRNA ([21] and our unpublished observations). On the other hand, every direct UV-induced protein–RNA crosslink found, in particular in native RNP particles, must reflect a "real" interaction because of the short distance between the crosslinked entities. For that reason such crosslinks are referred to as "zero-length". Moreover, work with particles reconstituted *in vitro* that carry a site-specifically labeled RNA is dependent on the efficiency of reconstitution and may even produce false-positive results if heterogeneous populations are generated as a result of incomplete assembly.

Protocol 1 describes the UV irradiation procedure. Some critical points have to be considered when one performs UV crosslinking experiments with native RNP particles:

(1) Concentration of the RNP particles. For UV crosslinking, purified native RNP particles are typically adjusted to a concentration of not more than about 0.1 mg/ml. The final concentration of native particles in solution as such is not critical, as native particles are fully assembled, and inter-particle crosslinking events in native RNP particles are highly unlikely. This item becomes much more of a problem when particles reconstituted *in vitro* are studied. Because an excess of protein over RNA has to be used for the efficient *in vitro* reconstitution, non-specific crosslinks due to the excess of protein may pose a problem [see Troubleshooting (1) for details]. For RNP particles reconstituted *in vitro*, an RNA concentration of 0.1 pmol/µl is in general well sufficient.

(2) Choice of buffer. First of all, the buffer should not contain a high concentration of reagents that are known to scavenge radicals, for example glycerol. UV crosslinking is a UV-induced radical reaction that generates a new covalent bond between the side chain of an amino acid and a base of the RNA. Thus, radical scavengers drastically reduce the crosslinking yield and, for example, glycerol concentrations should be kept as low as possible. Further, since samples are irradiated in small droplets (see below) any detergents in the buffer must be avoided, as the droplets will start to spread out over the sample plate. Finally, if treatment with proteinase K is necessary (see Protocol 2) the buffer should not contain potassium ions.

(3) Crosslinking conditions. These include the choice of UV lamp, the distance between the lamp(s) and the sample and the irradiation time. Our laboratory uses a specially constructed device for UV irradiation at 254 nm (see Fig. 21.1 and Protocol 1). Alternatively, other commercially available devices can be used (e.g. a UV Stratalinker 2400, Stratagene, La Jolla, USA); however, the conditions of UV irradiation, in particular the irradiation time as a function of the power of the UV source and the distance of the lamp(s) to the sample, have to be adjusted accordingly (see below). The samples can be irradiated in different ways, in droplets on a glass slide or Parafilm (Pechiney Plastic Packing, Menasha, USA), or in open plastic tubes. In cases of high sample volumes, custom-made larger glass dishes (4–12 cm in diameter) with a planar surface can be used. Pre-cooling of the samples (4 °C) and the glassware is essential. In our

21.4 Identification of UV-induced Protein–RNA Crosslinking Sites by Primer Extension Analysis

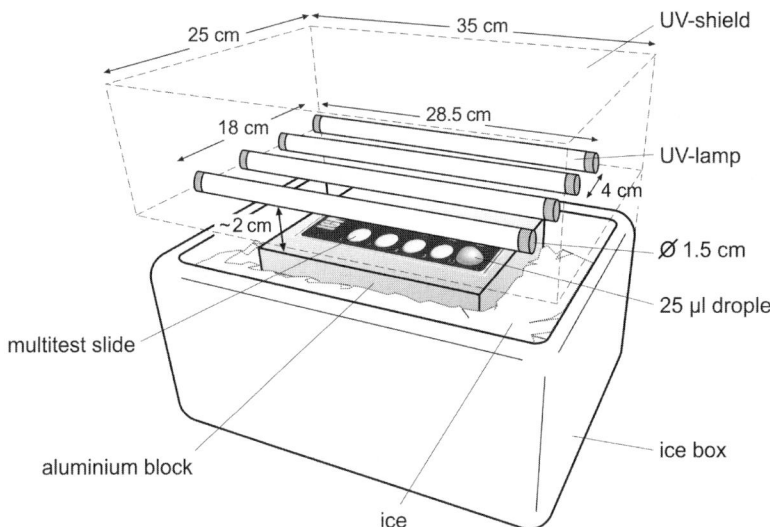

Fig. 21.1. Schematic drawing of the custom made UV crosslinking device.

hands, 25-μl droplets on a pre-cooled 10-well multitest slide (see Protocol 1) work best. For smaller volumes (e.g. 10 μl) we use Eppendorf tubes (Eppendorf AG, Germany) mounted directly under the UV source. We irradiate the samples at a distance of 2 cm from the source (corresponding to the height of a tube mounted directly under an 8-W lamp). The glass slide with the samples is put on top of an aluminium block placed in ice. In addition to these items, the most critical point is the duration of direct UV irradiation. We have observed that the maximum yield of crosslinks under our conditions is obtained after 2 min. Longer irradiation (3 min) does not increase the crosslinking yield significantly and further extended irradiation times lead to substantial loss of particles. On the other hand, when one is working with more rigid RNP particles such as ribosomes [8, 9], UV irradiation times may be prolonged. In any case, as a starting point we recommend performing Protocol 1 with different durations of UV irradiation.

21.4
Identification of UV-induced Protein–RNA Crosslinking Sites by Primer Extension Analysis

After exhaustive hydrolysis of the protein moiety of crosslinked RNP particles by proteinase K treatment, a few amino acids remain covalently attached to the RNA at the sites of crosslinking. In a primer extension reaction, RNA is primed with a 5'-^{32}P-labeled DNA oligonucleotide complementary to a chosen region on the RNA. The reverse transcriptase enzyme then adds dNTPs, which are complementary to the nucleotides of the RNA, to the 3' end of the labeled DNA primer and thus generates radioactively labeled DNA molecules that are complementary to the entire RNA sequence. At a nucleotide on the RNA that has been covalently modified (i.e. by crosslinking), no complementary DNA nucleotide can be added by the reverse transcriptase, owing to either incomplete Watson–Crick base pairing or to steric hindrance due to the presence of crosslinked amino acids. Thus, this nucleotide will cause a stop, or at least a "stuttering", of the reverse transcriptase. It should be noted that the reverse transcriptase stops one nucleotide before the actual crosslinking site. The complementary DNA generated has a certain length and the stop sites (i.e. the length of the generated DNA) can be deduced from a sequencing gel when analyzed next to a marker of complementary DNA that has been generated with the help of dideoxynucleotides [22].

Crosslinking induced by UV irradiation at short wavelengths can also cause intra-RNA crosslinks or induce strand breaks in the RNA; both of these also lead to stops or stuttering of the reverse transcriptase. Therefore, the reverse transcriptase patterns from three RNAs must be compared: (1) RNA from UV-irradiated RNP particles, (2) UV-irradiated naked RNA and (3) non-irradiated naked RNA.

Comparison of the primer extension reaction from these sets of experiments on a high-resolution sequencing gel leads to the identification of putative protein–RNA crosslinking sites. Figure 21.2(A) illustrates the principle of the three experiments necessary for the identification of protein–RNA crosslinking sites and Fig. 21.2(B) gives an example of identified protein–RNA crosslinking sites in native UV-irradiated U1 snRNPs [1, 2].

Protocol 2 describes the purification of RNA derived from UV-irradiated RNP particles. Protocol 3 describes the experimental steps that are required in order to generate naked UV-irradiated and non-irradiated RNAs for controls. Protocol 4 gives a detailed description of the primer extension reaction (including DNA primer purification and labeling) and the subsequent gel electrophoresis that are needed for the reproducible visualization and identification of the crosslinking sites on the RNAs.

For a correct assignment of protein–RNA crosslinking sites in native particles by this method we would like to emphasize several important points:

(1) The first set of experiments probes putative crosslinking sites in samples that contain an excess of non-crosslinked (or unmodified) RNA and in which only a small percentage of the RNA is modified by the UV irradiation (depending on the crosslinking yield and on the degree of UV-induced damage). Therefore, signals corresponding to the full-length RNA will be strongest, as seen by autoradiography of a sequencing gel. To systematically compare crosslinks on RNA from UV-irradiated particles with those on UV-irradiated naked RNA, the amount of RNA probed by primer extension must be the same. In order to ensure that similar amounts of crosslinked material were loaded, the signals corresponding to full-length RNA should be of comparable intensity in both sample preparations. In those cases where UV-induced crosslinks significantly reduce the signal intensity of the full-length transcript, the intensity of the naturally occurring stops on the RNA before the crosslinks should be of comparable

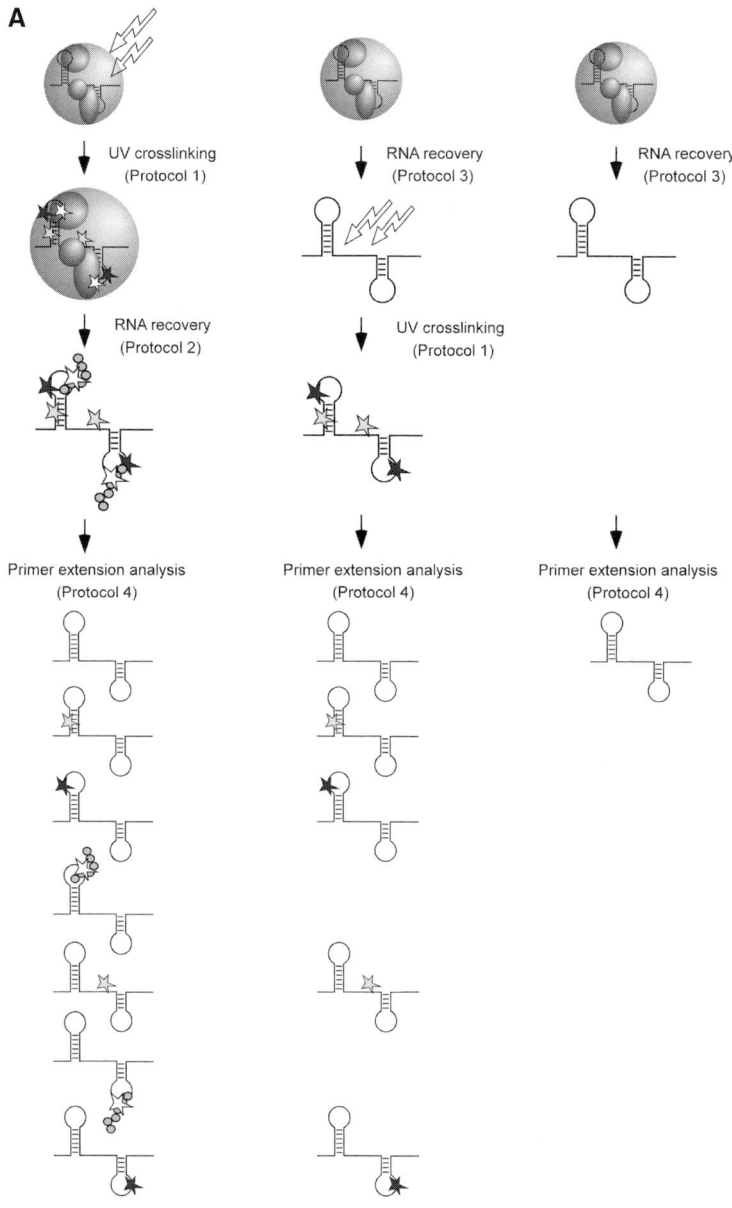

Fig. 21.2. (A) Schematic representation of the initial experiments necessary for the identification of putative protein–RNA crosslinking sites in native particles. RNA derived either from UV-irradiated particles, or RNA that was stripped of proteins before UV irradiation, or non-irradiated "naked" RNA is analyzed by primer extension analysis. White stars indicate nucleotides on the RNA that are covalently modified by crosslinked proteins, grey stars indicated intra-RNA crosslinks and black stars indicate UV-induced strand breaks on the RNA. Grey balls indicate those amino acids that remain covalently attached to the sites of crosslinking after digestion of the RNP particles with proteinase K.

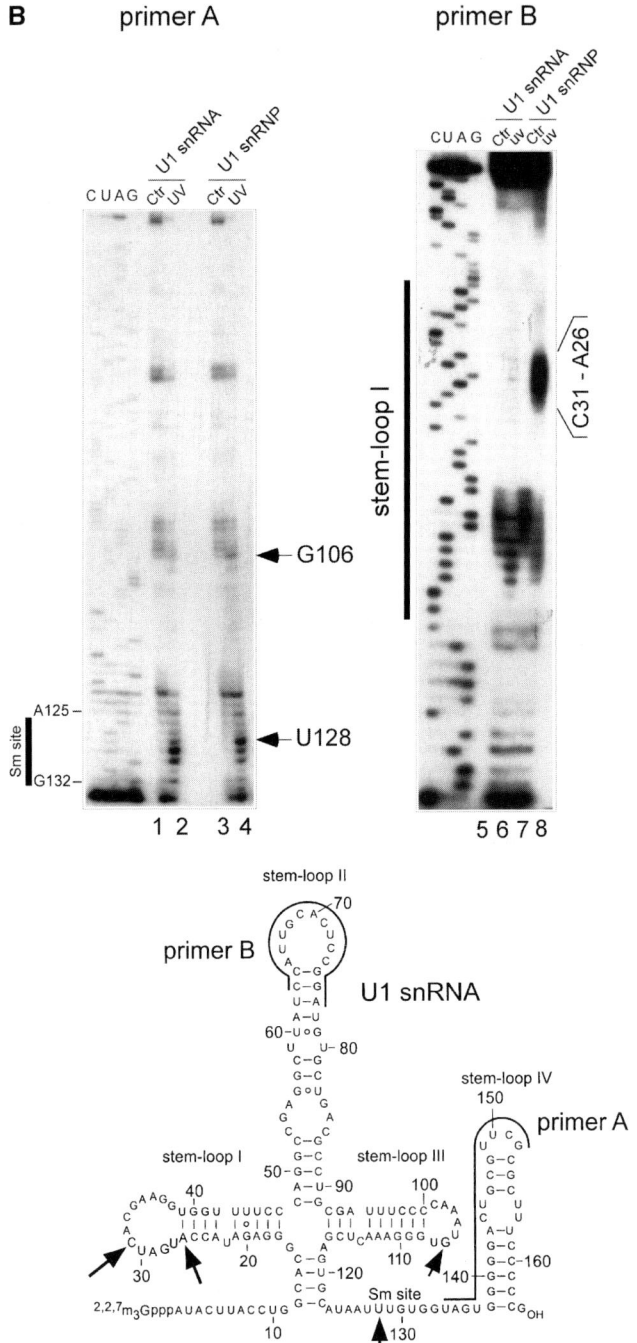

Fig. 21.2B

intensity. Alternatively, the sum of the intensities of all protein-independent stops occurring on the RNA must be the same in both experiments.

(2) If possible, the RNA should be probed with different primers for a comprehensive analysis of all crosslinking sites. When crosslinking sites are located far upstream (5′) of the primer binding site, the signals from the reverse transcriptase tend to be too weak. The detection of a putative crosslinking site is then no longer possible. As a rule of thumb, we use two primers (one matching at the extreme 3′ end and one in the middle) for probing an RNA molecule with 120–150 nt (see also Fig. 21.2B). It is furthermore obvious that putative crosslinking sites at the extreme 3′ end of the RNA cannot be detected. The detection of crosslinking sites close to the primer-binding region is difficult. In principle it is possible to detect a reverse transcriptase stop one nucleotide 5′ to the binding site of the primer. In practice, the signal of the radioactively labeled primer is very strong and will likely mask such a putative reverse transcriptase stop. This might be circumvented by empirically adjusting the conditions of electrophoresis such as to run the gel until the primer is about to migrate out of it and by extensive pre-running of the sequencing gel (Protocol 4.4).

(3) For reproducible clean primer extension reactions, the commercially obtained primer should be gel-purified prior to use (see Protocol 4.1). Furthermore, we recommend X-ray films for visualization of crosslinking sites (see Protocol 4.4); on autoradiographs the bands appear much sharper and less fuzzy when compared with phosphoimager scans. This thus facilitates the correct assignment of putative protein–RNA crosslinking sites.

21.5
Identification of Crosslinked Proteins

Once the overall protein–RNA crosslinking pattern in UV-irradiated RNP particles has been determined, the major challenge is to identify the corresponding crosslinked protein. This is achieved by performing immunoprecipitation of UV-irradiated RNP particles under conditions where protein–protein interactions within the particles are disrupted and only a single protein is precipitated (Fig. 21.3A).

Fig. 21.2. (B) Example of the analysis of protein–RNA crosslinking sites on the U1 snRNA in the native U1 snRNP. The primers A and B used in this experiment are complementary to nucleotides 134–152 and 63–77 of U1 snRNA, respectively. Lanes 1, 3, 5 and 7: controls with naked U1 snRNA isolated from non-irradiated U1 snRNP particles; lanes 2 and 6: UV-irradiated naked U1 snRNA (Protocol 3); lanes 4 and 8 are U1 snRNA derived from UV-irradiated U1 snRNP particles (Protocols 1 and 2). C, U, A and G: dideoxy sequence markers. Comparison of the reverse transcriptase pattern of the single lanes reveals protein–RNA crosslinking sites at nucleotides U128, G106 and A26 to C31. Black bars on the left indicate the Sm site and the stem–loop I sequence on U1 snRNA, respectively. A schematic diagram of U1 snRNA is shown below the panels. Arrows indicate the crosslinking sites, while the positions of the primers A and B are indicated by black lines.

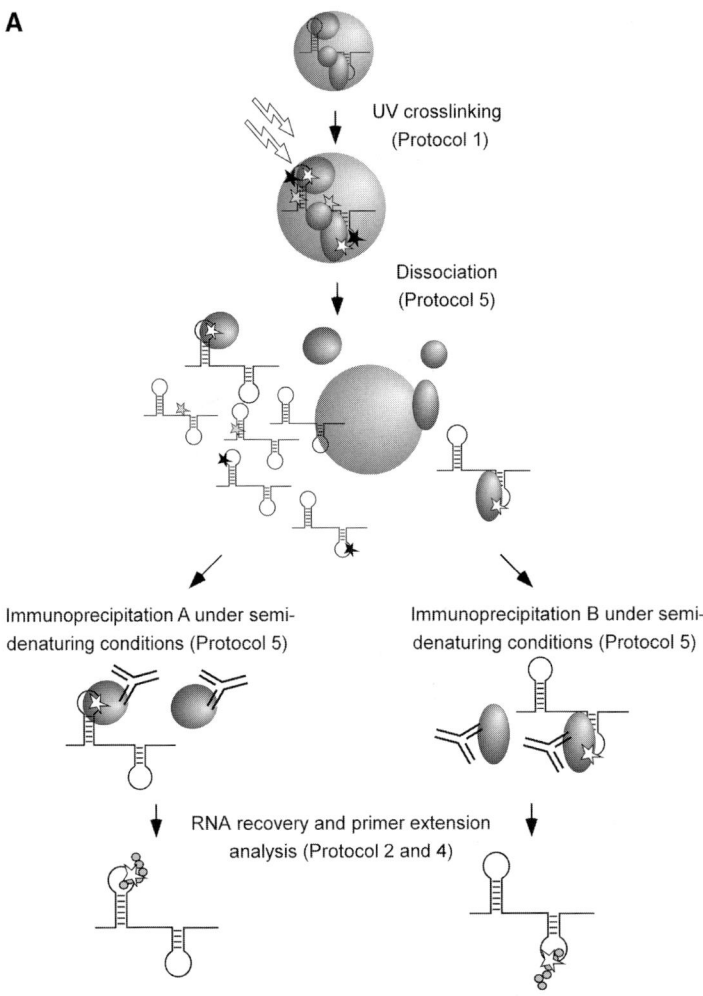

Fig. 21.3. (A) Schematic representation of the identification of crosslinked proteins in native RNPs by immunoprecipitation with two different antibodies combined with primer extension analysis. See legend to Fig. 21.2(A) and Protocol 5 for details. (B) Application of the procedure to the identification of protein–RNA crosslinks in U1 snRNPs. Immunoprecipitation was performed with antibodies against the SmF (α-SmF), SmG (α-SmG) and U1 70K (α-70K) proteins (Protocol 5). Primer extension analysis of the co-precipitated RNA was performed with the two primers A and B from Fig. 21.2(B) (Protocol 4). Lanes 1, 3 and 5: non-irradiated U1 snRNPs subjected to the immunoprecipitation and primer extension analysis (controls). Lanes 2, 4 and 6: primer extension analysis of the co-precipitated U1 snRNAs after immunoprecipitation of UV-irradiated U1 snRNP particles. The immunoprecipitations with anti-SmG and anti-70K antibodies showed stops at U129 and U30/A32, respectively. Therefore, the SmG protein crosslinks to U128 and U1 70K protein crosslinks to A29 and C31. The weak signal observed for the SmF protein can be explained by strong protein–protein interactions between the Sm proteins SmG and SmF which cannot be completely disrupted in 2% SDS. See legend of Fig. 21.2 and text, and [2] for further details.

21.5 Identification of Crosslinked Proteins | 363

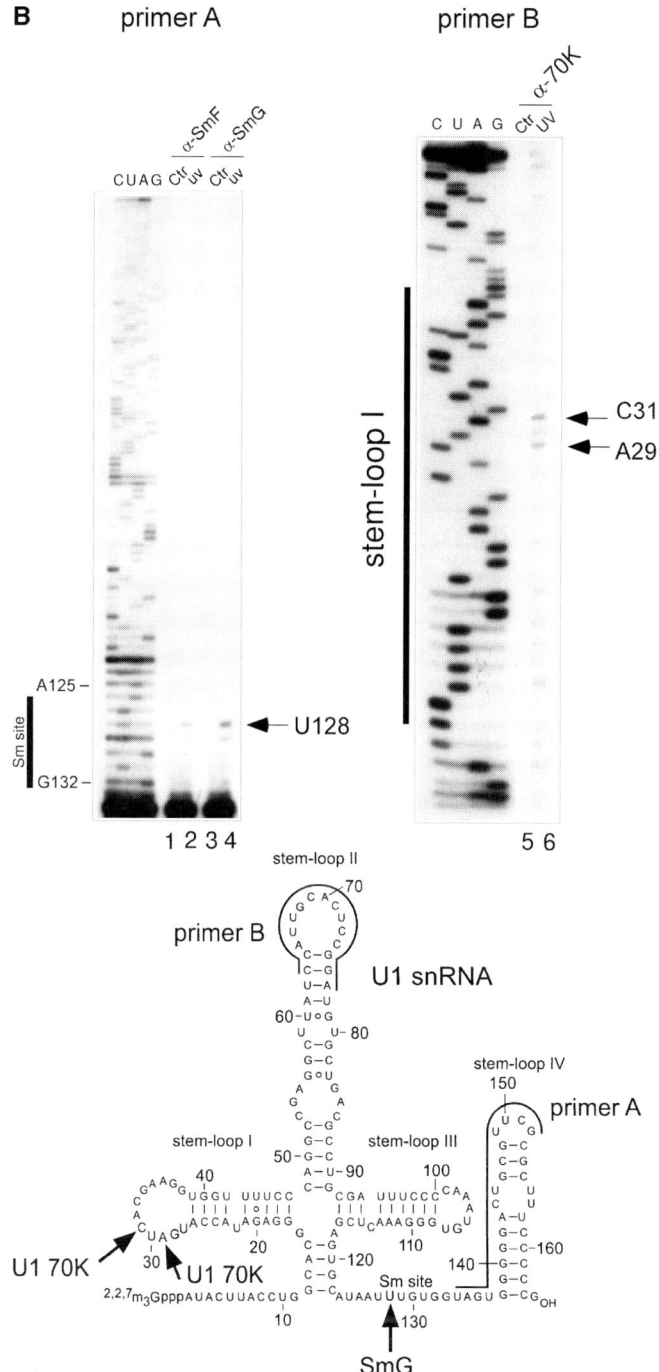

Fig. 21.3B

This approach results in the precipitation of most of the non-crosslinked protein together with the portion of the same protein that is crosslinked to RNA. After digestion of the entire protein moiety with proteinase K and extraction of the RNA, the only RNA molecules isolated are those crosslinked to the precipitated protein. Importantly, non-crosslinked RNA molecules are not precipitated. Consequently, in a subsequent primer extension analysis, the only transcripts detected are those with the reverse transcriptase stop at the sites of protein–RNA crosslinking (Fig. 21.3A). Full-length transcripts should not be visible. In practice, however, additional background stops or stops due to UV-induced RNA damage or intra-RNA crosslinking events (Fig. 21.3B) are usually visible. However, the bands due to the stops at the crosslinking sites of the precipitated protein have significantly greater intensities. Figure 21.3(A) illustrates the principle of the immunoprecipitation combined with primer extension analysis of the co-precipitated RNA and Fig. 21.3(B) summarizes results that we have obtained from our crosslinking experiments with native U1 snRNPs [1–3].

This type of analysis is dependent on the number of antibodies available for the different proteins of the particle and on how efficiently each antibody precipitates its corresponding protein, in particular under semi-denaturing conditions that disrupt protein–protein interaction but still preserve the reactivity of the antibodies during immunoprecipitation. As a control, the performance of similar experiments with the pre-immune sera of the corresponding antibodies is highly recommended.

Protocol 5 describes in detail all the steps necessary for this analysis. The most critical steps during the analysis are the procedures for dissociation of the RNP particles before immunoprecipitation and for washing of the beads after immunoprecipitation to remove non-specifically bound material. In our hands, the dissociation of the particles before immunoprecipitation usually works best in the presence of 1% SDS and is improved by subsequent heating of the samples to 70 °C. However, in several cases we observed that higher concentrations of SDS are required to dissociate the particles completely [2, 3]. Washing of the Protein A–Sepharose beads after precipitation should include an additional washing step in a new tube. Furthermore, we observed that washing the samples with buffer containing detergent (e.g. Nonidet P-40) leads to a dramatic increase of non-specific background signals in the subsequent primer extension. We have no explanation for this.

21.6
Troubleshooting

(1) The above procedures can be applied to the analysis of RNP particles assembled *in vitro* from a known number of defined components. The RNA is most easily synthesized *in vitro* by phage RNA polymerases (see Chapter 1) or by chemical synthesis (see Chapter 7). The protein(s) can be produced in either

Escherichia coli or insect cells, or they can be purified from a readily available biological source [22]. When embarking on such a project, a number of considerations should be borne in mind at the outset. The major problem is directly related to the efficiency of RNP assembly *in vitro*. It would be difficult to discern artificial crosslinking events due to incomplete or non-specific assembly from genuine crosslinks. Similar problems will arise if the protein preparation is contaminated by interfering proteins or by eubacterial RNA. This is in particular an obstacle when bacterially expressed RNA-binding proteins are being studied.

(2) Technically speaking, the most demanding aspect of the procedures outlined is that the interpretability of the final result is heavily dependent on the recovery of RNP and RNA in a large number of consecutive experimental manipulations. Great care has to be taken that the ethanol precipitations are quantitative. Similarly, recovery of RNA or RNP from the glass plate after UV irradiation may pose a problem. Also, care has to be taken that all steps requiring the resuspension of a dry RNA pellet in buffer are performed with the necessary patience.

(3) The numerous manipulations required to achieve the aims of the experiments are also possible entry points for contaminations by RNases. Standard precautions have to be taken at the outset. The most important ones are: gloves must be worn at all times; when preparing solutions, only double-distilled or Millipore Q water should be used; all solutions should be sterilized by filtering through 0.2-µm nitrocellulose filters.

(4) The primer extension itself should be performed at least 3 times with RNA obtained from independent experiments and it may be necessary to use the RNA from one particular experiment with two different primers.

(5) The most critical point of the entire analysis is the immunoprecipitation combined with primer extension analysis. False-positive results are obtained if immunoprecipitation is performed under conditions where the RNP complexes are not fully disrupted. For example, the U5 snRNP-specific proteins 40K, 116K, 200K and 220K form a remarkably stable heteromeric protein complex [23]. Using immunoprecipitation combined with primer extension analysis, we demonstrated that the U5 220K protein crosslinks to loop 1 of U5 snRNA [1]. However, immunoprecipitation under less stringent conditions compared to Protocol 5 [e.g. dissociation in 0.05% (v/v) SDS; see Fig. 21.4] resulted in co-precipitation of crosslinked RNA with antibodies against each of the proteins and thus a comparable reverse transcriptase pattern for the four proteins (data not shown). A similar situation was observed when crosslinking sites in the highly conserved Sm site of U1 snRNPs were analyzed. The seven U snRNP-specific Sm proteins form a highly stable heteromeric ring-like structure that interacts with the Sm site [24]. Immunoprecipitation combined with primer extension analysis under our standard conditions [1% SDS (v/v), 5% (v/v) Triton X-100] revealed a similar crosslinking pattern of the SmF and SmG protein to U128. To demonstrate that only SmG is in contact with U128 in the

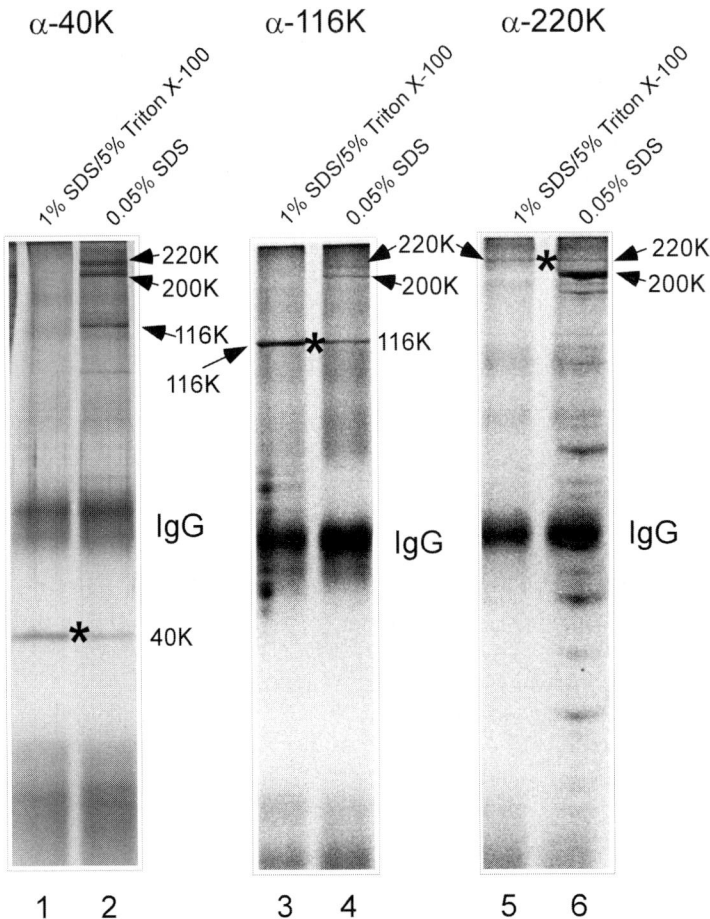

Fig. 21.4. Immunoprecipitation of single U5 snRNP-specific proteins from the U5 snRNP through dissociation of the U5 snRNPs by SDS/Triton X-100. U5 snRNPs were dissociated in 1% SDS and 5% Triton X-100 (lanes 1, 3 and 5, see Protocol 5.3) or in the presence of 0.05% (v/v) SDS only (lanes 2, 4 and 6). The immunoprecipitation was performed with covalently coupled antibodies (see Protocol 5.2) against the U5 snRNP specific proteins 40K (α-40K), 116K (α-116K) and 220K (α-220K). Proteins were visualized by silver staining. IgG: residual antibody released from the beads. Proteins precipitated under stringent conditions are marked with an asterisk.

U1 snRNA, we had to raise the SDS concentration to 2% (v/v) to allow complete dissociation of SmF and SmG proteins [2, 3]. The high specificity of our immunoprecipitation procedure is demonstrated in Fig. 21.4 for the precipitation of the above-mentioned U5 snRNP-specific proteins 40K, 116K and 220K. The silver-stained SDS–polyacrylamide gels of the precipitated proteins shows that a single protein can only be precipitated upon denaturation in the presence of 1% SDS with subsequent addition of Triton X-100 to 5%.

(6) Another important aspect is the reliability of the reactivity of the antibodies under these harsh conditions. To exclude the possibility that negative results (i.e. failure to detect protein-RNA crosslinks) are due to a poor reactivity of the antibodies, each antibody can be tested for its capability to precipitate a single protein from the RNPs (see above). Protocol 5 describes all steps necessary to visualize the precipitated proteins on a silver-stained SDS–polyacrylamide gel. Here, it is essential to couple the antibodies covalently to the beads in order to minimize the IgG background (Protocol 5.2).

21.7 Protocols

Protocol 1: UV irradiation of RNPs

(1) Starting materials are purified snRNP particles [5, 6], spliceosomal complexes [25–29] or reconstituted RNP particles [4]. The sample concentration is adjusted to approximately 0.1 mg/ml. Any buffer is suitable, provided it conforms to the criteria as stated above [Section 21.3 (2)]. The particles that we analyze are in 20 mM Tris–HCl (pH 7.0), 370 mM NaCl, 1.5 mM $MgCl_2$, 0.5 mM DTT (in case of U1 snRNPs) or in 20 mM HEPES–KOH (pH 7.9), 1.5 mM $MgCl_2$, 250 mM NaCl, 0.5 mM DTT, 0.2 mM EDTA (in case of 25S[U4/U6.U5] tri-snRNPs).
(2) Divide the sample into droplets of 25 µl and place the droplets carefully onto pre-cooled 10-well multitest slides (ICN Biomedical, USA). Ensure that the droplets stay intact and do not spread over the slide.
(3) Irradiate for 2 min at 254 nm at a distance of 4 cm from the UV source. We use a custom-made holder with four 8-W germicidal lamps (G8T5, Herolab, Germany) mounted in parallel, 4 cm apart (Fig. 21.1).
(4) Carefully pipette the droplets from the glass slide back into a new tube. It is essential that recovery of the sample from the slide is as complete as possible.
(5) Recover the RNA from crosslinked complexes as outlined in Protocol 2 or perform the immunoprecipitation (Protocol 5) before.

Protocol 2: RNA recovery from UV-irradiated RNPs

(1) To 50 µl of the pooled irradiated samples obtained in Protocol 1, add 40 µl of the buffer in which the native particles were initially purified.
(2) Add SDS to a final concentration of 1% (v/v), by adding 10 µl 10% SDS (v/v) to the above volume.
(3) Incubate for 10 min at 70 °C with gentle agitation and then allow the sample to cool down to room temperature over a period of 5 min.
(4) First add EDTA to 7.5 mM (1.5 µl of 0.5 M EDTA to above volume), and then add proteinase K to around 1 mg/ml (w/v) [10 µl proteinase K (10 mg/ml; Roche, Germany) to above volume]. Incubate the samples for a minimum of 30 min at 37 °C.

(5) Extract the RNA by adding 100 µl phenol/chloroform/isoamyl alcohol (PCI; Roth, Germany) and subsequent vigorous shaking for 2 min. Centrifuge (5 min, 13 000 r.p.m., 10 000 g) and transfer the aqueous phase containing the RNA to a new tube.
(6) Add 20 µg glycogen (Roche, Germany) and 1/10 of the sample volume 3 M sodium acetate, pH 5.3, and precipitate in 3 sample volumes of ethanol (p.a. grade; Merck, Germany) for a minimum of 2 h at −20 °C.
(7) Collect the RNA by centrifugation (20 min, 13 000 r.p.m. at 4 °C) and discard the supernatant. RNA recovery is monitored by inspection of the glycogen pellet, which must be clearly visible.
(8) Dissolve the RNA in 100 µl 0.3 M NaOAc, pH 5.3, and precipitate once more in 3 volumes ethanol for a minimum of 2 h at −20 °C. Collect the RNA by centrifugation (see Step 7) and dry the sample for 3 min in a Speed Vac.
(9) Dissolve the RNA in 6.5 µl CE buffer (10 mM cacodylic acid–KOH, pH 7.0, 0.2 mM EDTA, pH 8.0) with shaking for 10 min. The RNA is stored at −20 °C.

Protocol 3: UV irradiation of naked RNA

(1) As starting material, use twice as much naked RNA as that contained in the corresponding RNP particle employed in Protocol 1 (Step 1) to compensate for loss of RNA on the glass slides during UV irradiation.
(2) Perform the proteinase K digestion and RNA extraction essentially as described in Protocol 2, Steps 2–8, except that glycogen is omitted.
(3) Dissolve the RNA in 50 µl of the buffer used for the RNP (see Protocol 1, step 1).
(4) Perform UV irradiation and sample recovery exactly as described in Protocol 1, Steps 2–5.
(5) Further processing of the samples is as described in Protocol 2, Steps 6–9. In those cases where the starting material was not doubled, the RNA is resuspended in 3.5 µl instead of 6.5 µl CE buffer.
(6) Primer extension is performed as outlined in Protocol 4.

Protocol 4: Primer extension analysis

Protocol 4.1: Purification of the primer
The primer is obtained from any commercial source. It must be gel-purified for reproducibly clean primer extension reactions. Approximately 5 nmol of the primer is first dissolved in 100 µl of 80% formamide, 0.5 × TBE, 0.001% xylene cyanol and 0.001% bromophenol blue, and denatured at 96 °C for 3 min. After cooling to room temperature, it is loaded onto a 25-mm wide slot of a 1 mm thick, 15–20-cm long, denaturing 20% polyacrylamide/8.3 M urea gel and electrophoresed at approximately 1.5 W/cm. Electrophoresis time depends on the primer length (2 h for a 24mer). The region of the gel, which contains the primer, is identified by UV shadowing and excised. The gel slice is wetted with elution buffer (20 mM Tris–HCl, pH 7.5, 0.2 mM EDTA, pH 8, 0.15 M NaCl, 0.5% SDS) and cut into small cubes, which are subsequently transferred to a tube and overlaid with 300–500 µl

elution buffer. Elution is performed by diffusion (16 h at 30 °C). The eluate is recovered, extracted with phenol/chloroform, and precipitated twice essentially as described for the RNA extraction in Protocol 2, Steps 5–9, except that 300 μl PCI is used (Step 5), and that the primer is dissolved at 5 pmol/μl CE (Step 9).

Protocol 4.2: $5'$-^{32}P-labeling of the primer

(1) For one 10-μl reaction, the following components are mixed: 2 μl (10 pmol) of purified DNA oligonucleotide, 3 μl CE buffer, 1 μl 10 × T4 polynucleotide kinase (PNK) buffer (0.7 M Tris–HCl, pH 7.6, 0.1 M $MgCl_2$, 0.05 M DTT), 1 μl T4 PNK (New England Biolabs, USA), and 6 μl [γ-^{32}P]ATP (6000 Ci/mmol; Amersham Biosciences, UK). Incubate for 40 min at 37 °C.
(2) 50 μl CE buffer are added to the reaction and unincorporated nucleotides are removed by G-50 or G-25 Sephadex spin column chromatography (Amersham Biosciences, UK) according to the manufacturer's instructions. The volume is adjusted to 100 μl with CE and the extent of incorporation is determined (usually about 0.8 to 1.0×10^6 c.p.m. per pmol of primer).
(3) Residual protein and other impurities are removed essentially as described in protocol 2, Steps 5–9 with the following changes: (i) 10 μg glycogen is used in Step 6; (ii) the labeled primer is resuspended in 40 μl CE buffer (Step 9).

Protocol 4.3: Primer extension reaction

The following different samples are probed by primer extension analysis: (i) UV-irradiated RNAs from Protocols 2 and 3; (ii) non-irradiated RNA; (iii) crosslinked RNA isolated after immunoprecipitation; (iv–vii) template RNAs (either native RNA isolated according to Protocol 2 or RNA transcribed *in vitro* with bacteriophage RNA polymerases from an appropriate plasmid template) for the sequencing reactions used as markers. Template RNAs for marker synthesis should have a concentration of 0.2 pmol/μl. The experimental procedure for the primer extension closely follows that described in [30].

(1) For each RNA sample to be analyzed, 1.5 μl of a hybridization mix (HY) is required. It is composed of 0.25 μl 10 × hybridization buffer (0.5 M Tris–HCl, pH 8.4, 0.6 M NaCl, 0.1 M DTT), 0.5 μl ^{32}P-labeled DNA oligonucleotide, and 0.75 μl H_2O. Enough HY mix for the number of samples to be processed must be prepared.
(2) To anneal the primer, 1 μl of the RNA is first mixed with 1.5 μl of the HY mix, then heated for 60 s at 96 °C, and allowed to cool at room temperature for 5 min. Samples are briefly centrifuged.
(3) 1 μl of ddNTP is added to each of the four marker RNA samples (0.5 mM ddGTP, ddATP, ddTTP or ddCTP; Amersham Biosciences, UK).
(4) For each RNA sample, 2.5 μl of a reverse transcriptase mix (RT) is now required. It is composed of 0.25 μl 10 × reverse transcriptase buffer (0.5 M Tris–HCl, pH 8.4, 0.1 M $MgCl_2$, 0.6 M NaCl, 0.1 M DTT), 0.1 μl dNTPs (5 mM

each dGTP, dATP, dTTP and dCTP; Amersham Biosciences, UK), 0.08 μl (about 2 U) reverse transcriptase (30 U/μl, Seikagaku, Japan) and 2.07 μl H_2O. Enough RT mix for the number of samples to be processed must be prepared. 2.5 μl RT mix is added per sample, mixed and incubated for 45 min at 42.5 °C. A hybridization oven is recommended to avoid condensation at the lid of the tube.

(5) 6.5 μl loading buffer [8.3 M urea, 0.5 × TBE, 0.001% (w/v) bromophenol blue, 0.001% (w/v) xylene cyanol] is added to all samples, except for the markers, which receive 10 μl. Samples can be stored at −20 °C for at least 1 week.

Protocol 4.4: Gel electrophoresis
The transcribed cDNA products are analyzed on a 9.6% polyacrylamide (acrylamide:bisacrylamide, 19:1)/8.3 M urea gel in 1 × TBE in a Gibco/BRL Model S2 apparatus (0.5-mm thick gel) with 1 × TBE as electrophoresis buffer. Pre-electrophoresis is for 30 min at 65 W. Electrophoresis is at 65 W for a time depending on the length of the primer (approximately 2 h for a 24mer). For autoradiography, the sequencing gels are first transferred to a used X-ray film for support and covered with kitchen wrapping film. Alternatively, sequencing gels can be fixed in 40% methanol/10% acetic acid, transferred to Whatman 3MM paper and dried under vacuum (Bio-Rad model 583 gel dryer). A BioMax film (Kodak) is exposed to the gel at −70 °C for 1–10 days in the presence of intensifying screens. The long exposure times are required when performing the immunoprecipitation experiments combined with primer extension analysis, because of the inherently low yields of immunoprecipitation.

Protocol 5: Immunoprecipitation of the RNA-protein crosslinks

Protocol 5.1: Non-covalent coupling of antibodies to Protein A–Sepharose
Immunoprecipitation was found to be optimal with per assay 15 μl packed matrix volume of Protein A–Sepharose beads (Amersham Biosciences, UK) coupled with antiserum. Depending on the number of samples that are assayed, proportionally more bead slurry can be coupled with correspondingly increased amounts of antiserum. The coupled beads can be distributed afterward between the different tubes.

(1) For coupling of the antibody to beads an amount of slurry (30 μl) corresponding to 15 μl of beads (packed volume) is taken and washed 3 times with 500 μl aliquots of PBS (20 mM Na_2HPO_4, pH 8.0, 130 mM NaCl).
(2) The antiserum is diluted with PBS to 500 μl and added to the washed beads. Normally, 50 μl of antiserum is sufficient for one immunoprecipitation, but this volume may have to be adjusted, depending on the titer of the antiserum. Coupling is performed overnight by head-over-tail rotation at 4 °C.
(3) After coupling, beads are washed 3 times with 500 μl PBS. Tubes are changed by transferring the beads with the last washing aliquot to a new tube using a plastic pipette tip with a cut-off end. The washed beads with the coupled antibody are then overlaid with 15 μl PBS and kept on ice until use.

Protocol 5.2: Covalent coupling of antibodies to Protein A–Sepharose
Covalent coupling of antibodies is recommended when the capability of the antibodies to selectively precipitate a single protein from an RNP under semi-denaturing conditions (see Protocol 5.3) is tested, in order to exclude the possibility that negative results are due to the poor reactivity of the antibodies under these conditions [see also Troubleshooting (6)].

For covalent coupling it is recommended to increase the total amount of Protein A–Sepharose beads and the amount of antiserum is usually twice the volume of the beads.

(1) 30 µl Protein A–Sepharose beads are washed with PBS (Protocol 5.1) and then incubated with 60 µl of antiserum in a final volume of 500 µl PBS, 0.05% (v/v) Nonidet P-40 (NP-40) with head-over-tail rotation overnight at 4 °C.
(2) The beads are washed 5 times with 500 µl PBS, 0.05% NP-40 at 4 °C.
(3) The antibody-coupled beads are equilibrated 2 times with 300 µl 200 mM Na borate (pH 9.0) at room temperature.
(4) Crosslinking of the antibodies to the beads is achieved by incubation with 500 µl DMP (dimethyl pimelinidate dihydrochloride; Sigma, USA) at a final concentration of 5.2 mg/ml in 200 mM Na borate for 1 h at room temperature with head-over-tail rotation. Note that the pH of the solution must be above 8.3.
(5) The supernatant is removed as completely as possible and the reaction is stopped by addition of 300 µl 0.2 M ethanolamine–HCl, pH 8.0, to the beads and further incubation with head-over-tail rotation for 1 h.
(6) The beads are then washed with PBS (Protocol 5.1) and residual non-crosslinked antibodies are removed by three additional washes with 500 µl 0.1 M glycine–HCl, pH 2.7.
(7) After a final wash with PBS containing 0.02% NaN_3, the slurry can be stored for at least 6 months at 4 °C.
(8) The covalently coupled antibody beads are now used in the immunoprecipitation exactly as described in Protocol 5.3. After the final wash beads are incubated with an appropriate volume of SDS sample buffer [125 mM Tris–HCl, pH 6.8, 1% (v/v) SDS, 5% (v/v) glycerol, 10 mM DTT, 0.005% (w/v) bromophenol blue] and heated for 5 min at 70 °C. Beads are spun down with maximum speed (10 000 g) and the supernatant is loaded onto an SDS–PAA gel [31].
(9) Proteins are visualized by silver staining according to [32].

Protocol 5.3: Dissociation of RNP particles and immunoprecipitation
For the immunoprecipitation experiments native or reconstituted RNP particles in a volume of 50 µl of appropriate buffer (for buffer conditions see Protocol 1, Step 1) are used. For the reliable assignment of crosslinks it is essential to include a sample that was not UV-irradiated, but otherwise treated in an identical manner.

(1) Add SDS to a final concentration of 1% (v/v) to the samples from Protocol 1 and incubate for 10 min at 70 °C on a shaker. Use 2% SDS (v/v) in those cases

where the protein–protein interactions are known or were found to be extremely strong (see above).

(2) Allow the samples to cool at room temperature for 5 min. Then, add Triton X-100 (density 1.06 g/l, molecular biology grade; Sigma, USA) to a final concentration of 5% (v/v). Use of the concentrated Triton X-100 stock solution is necessary to keep the final volume as low as possible. Gently mixing is necessary to completely dissolve the added Triton X-100, which initially forms a separate phase at the bottom of the tube.

(3) Adjust the sample volume to 350 µl with PBS and add the mixture to the prepared antibody-coupled beads (Protocol 5.1). Incubate with head-over-tail rotation for 1–1.5 h at 4 °C.

(4) Wash the samples 4 times with 500 µl aliquots of PBS and transfer the slurry into a new tube at the fourth washing step. Wash the beads once more with 500 µl PBS. Carefully check recovery of the beads during the washing procedure by inspecting the amount of beads visible in the tube after each step; any loss of material must be avoided.

(5) Remove the supernatant as completely as possible, then add 90 µl of buffer (see Protocol 1, Step 1) and proceed with proteinase K digestion and RNA recovery essentially as described in Protocol 2, Steps 2–9, except that shaking is for 5 min (Step 5) and that the RNA is dissolved in 3.5 µl (Step 9).

(6) Proceed with the primer extension as outlined in Protocol 4.

Acknowledgments

We thank our colleagues for their excellent technical assistance at the time when we established our method, in particular Peter Kempkes for HeLa cell fermentation, Axel Baduin and Winfried Lorenz for the snRNP preparations, and Irene Öchsner for her expertise with the snRNP-specific antibodies. This work is supported by a grant from the BMBF (031U251B), Fonds der Chemischen Industrie and DFG Forschergruppe (LU 294/12-1) to R. L.

References

1 H. URLAUB, K. HARTMUTH, S. KOSTKA, G. GRELLE, R. LÜHRMANN, *J. Biol. Chem.* **2000**, *275*, 41458–41468.
2 H. URLAUB, V. RAKER, S. KOSTKA, R. LÜHRMANN, *EMBO J.* **2001**, *20*, 187–196.
3 H. URLAUB, K. HARTMUTH, R, LÜHRMANN, *Methods* **2002**, *26*, 170–181.
4 S. NOTTROTT, H. URLAUB, R. LÜHRMANN, *EMBO J.* **2002**, *21*, 5527–5538.
5 C. L. WILL, B. KASTNER, R. LÜHRMANN, Analysis of ribonucleoprotein interactions in: *RNA Processing, Vol. I, A Practical Approach*, S. J. HIGGINS, B. D. HAMES (eds), IRL Press, Oxford, **1994**.
6 B. KASTNER, Purification and electron microscopy of spliceosomal snRNPs, in: *RNP Particles, Splicing and Autoimmune Diseases*, J. SCHENKEL (ed.), Springer, Berlin, **1998**.
7 C. B. BURGE, T. TUSCHL, P. A. SHARP,

Splicing of precursors to mRNA by the spliceosome, in: *The RNA World*, R. F. GESTELAND, T. R. CECH, J. F. ATKINS (eds), Cold Spring Harbor Laboratory Press, Cold Spring Harbor, NY, **1999**.

8 H. URLAUB, V. KRUFT, O. BISCHOF, E. C. MÜLLER, B. WITTMANN-LIEBOLD, *EMBO J*. **1995**, *14*, 4578–4588.

9 H. URLAUB, B. THIEDE, E. C. MÜLLER, R. BRIMACOMBE, B. WITTMANN-LIEBOLD, *J. Biol. Chem.* **1997**, *272*, 14547–14555.

10 K. M. MEISENHEIMER, P. L. MEISENHEIMER, T. H. KOCH, *Methods Enzymol.* **2000**, *318*, 88–104.

11 R. BRIMACOMBE, W. STIEGE, A. KYRIATSOULIS, P. MALY, *Methods Enzymol.* **1988**, *164*, 287–309.

12 B. S. COOPERMAN, *Methods Enzymol.* **1988**, *164*, 341–361.

13 M. J. MOORE, P. A. SHARP, *Science* **1992**, *256*, 992–997.

14 J. R. WYATT, E. J. SONTHEIMER, J. A. STEITZ, *Genes Dev.* **1992**, *6*, 2542–2553.

15 I. DIX, C. S. RUSSELL, R. T. O'KEEFE, A. J. NEWMAN, J. D. BEGGS, *RNA* **1998**, *4*, 1675–1686.

16 R. REED, M. D. CHIARA, *Methods* **1999**, *18*, 3–12.

17 M. M. KONARSKA, *Methods* **1999**, *18*, 22–28.

18 Y. T. YU, *Methods Enzymol.* **2000**, *318*, 71–88.

19 B. RHODE, K. HARTMUTH, H. URLAUB, R. LÜHRMANN, *RNA* **2003**, *9*, 1542–1551.

20 C. OUBRIDGE, N. ITO, P. R. EVANS, C. H. TEO, K. NAGAI, *Nature* **1994**, *372*, 432–438.

21 S. NOTTROTT, K. HARTMUTH, P. FABRIZIO, H. URLAUB, I. VIDOVIC, R. FICNER, R. LÜHRMANN, *EMBO J*, **1999**, *18*, 6119–6123.

22 J. SAMBROOK, E. F. FRITSCH, T. MANIATIS, *Molecular Cloning: A laboratory Manual*, 2nd edn, Cold Spring Harbor Laboratory Press, Cold Spring Harbor, NY, **1989**.

23 T. ACHSEL, K. AHRENS, H. BRAHMS, S. TEIGELKAMP, R. LÜHRMANN, *Mol. Cell. Biol.* **1998**, *18*, 6756–6766.

24 C. KAMBACH, S. WALKE, R. YOUNG, J. M. AVIS, E. DE LA FORTELLE, V. A. RAKER, R. LÜHRMANN, J. LI, K. NAGAI, *Cell* **1999**, *96*, 375–387.

25 R. REED, L. PALANDIJAN, Spliceosome assembly, in: *Eukaryotic mRNA Processing*, A. R. KRAINER (ed.), IRL Press, Oxford, **1997**.

26 M. S. JURICA, M. J. MOORE, *Methods* **2002**, *28*, 336–345.

27 E. M. MAKAROV, O. V. MAKAROVA, H. URLAUB, M. GENTZEL, C. L. WILL, M. WILM, R. LÜHRMANN, *Science* **2002**, *298*, 2205–2208.

28 K. HARTMUTH, H. URLAUB, H.-P. VORNLOCHER, C. L. WILL, M. GENTZEL, M. WILM, R. LÜHRMANN, *Proc. Natl. Acad. Sci. USA* **2002**, *99*, 16719–16724.

29 K. HARTMUTH, H.-P. VORNLOCHER, R. LÜHRMANN, *Methods Mol. Biol.* **2004**, *257*, 47–64.

30 A. J. ZAUG, T. R. CECH, *RNA* **1995**, *1*, 363–374.

31 U. K. LAEMMLI, *Nature* **1979**, *227*, 680–685.

32 H. BLUM, H. BEIER, H. J. GROSS, *Electrophoresis* **1987**, *8*, 93–99.

22
Probing RNA Structure by Photoaffinity Crosslinking with 4-Thiouridine and 6-Thioguanosine

Michael E. Harris and Eric L. Christian

22.1
Introduction

Chemical crosslinking, including photoaffinity crosslinking, has been widely used to gain insight into structures associated with the biological function of large, structurally complex RNAs and ribonucleoproteins (RNPs). Examples include analysis of catalytic RNAs and the major cellular RNPs, the ribosome [1–4] and the spliceosome [5–9]. Combined with continuing improvements in the ability to generate RNAs with site-specific modifications, crosslinking continues to be a key analytical method for investigating structure–function relationships. If carried out with due care, crosslinking experiments can establish that specific residues are (or were) proximal when the crosslinking reaction occurred. Thus, when applied in a targeted way this information together with kinetic and thermodynamic studies of structure variants can be used to reveal residues involved in catalysis and molecular recognition. If sufficient information is available from other biochemical and comparative analyses, it can be possible to use the information gained from crosslinking as constraints for molecular modeling [1, 10–14]. Although the resolution of structures obtained this way is necessarily low (generally of the order of ± 10 Å), they present an explicit context for designing new structure–function experiments and for interpreting structural information.

Although a wide variety of chemical and photo-crosslinking reagents are available, 4-thiouridine and 6-thioguanosine are excellent choices due to their simple molecular structure, relative stability and high reactivity (Fig. 22.1) [15–20]. s^4U and s^6G introduce only minimal perturbations of the native structure since they differ from their corresponding "parent" nucleoside by a single atomic substitution, the replacement of a nucleobase oxygen by sulfur. This substitution renders the reagent sensitive to UV light and exposure yields reactive sulfur radical that can react efficiently with functional groups that are in proximity. Crosslinking reactions involving these reagents can be very efficient, making it an easier task to isolate sufficient quantities of crosslinked species for mapping of crosslinked nucleotides and assessment of retention of biological activity. Additionally, these reagents are advantageous in that they are relatively short range (around 3 Å), and thus in

Handbook of RNA Biochemistry. Edited by R. K. Hartmann, A. Bindereif, A. Schön, E. Westhof
Copyright © 2005 WILEY-VCH Verlag GmbH & Co. KGaA, Weinheim
ISBN: 3-527-30826-1

6-thioguanidine (s⁶G)   4-thiouracil (s⁴U)

Fig. 22.1. Structures of 6-thioguanidine and 4-thiouracil.

principle provide spatial information that is higher resolution than, for example, azido derivatives that generally introduce a linker between the RNA and photoagent that can be 10 Å or greater.

There are several excellent and up to date reviews available that describe methods for generation and incorporation of these and other photoaffinity reagents including chapters in this work [2, 21–25]. Because the choice of crosslinking reagent and method of incorporation will depend on the specific experimental application, the reader is referred to these important resources. Here, we will focus on simple procedures and considerations for generating and isolating crosslinked RNAs, and for primer extension mapping of crosslinked nucleotides. In the examples given below, crosslinking is applied to identify active site components within the RNase P ribozyme–substrate complex [16, 26]. The description is designed to be sufficiently general in order to be of maximum use as a guideline for an experimenter at least at the graduate level who is considering the application of photocrosslinking of RNA in their research. However, a basic understanding of techniques for handling nucleic acids is assumed.

It is important to note that the descriptions included here are by necessity brief and only a starting point because of the significant condition dependence of the crosslinking reactions and variability in the physical behavior of different RNAs. It cannot be overemphasized that achieving efficient crosslinking and obtaining clean and convincing primer extension mapping data will require significant effort toward optimization of different experimental parameters. In this section we attempt to describe the logic behind the choice of the basic experimental parameters we have used, and to illustrate the experimental constraints and controls necessary for interpretation of crosslinking data in terms of biological function.

As with any experimental approach it is important to first consider the difficulties inherent to its application and limitations to interpretation of the data. Despite its conceptual simplicity, successfully applying any crosslinking approach can be sometimes difficult and time consuming. Despite one's best efforts at optimization, the crosslinking reaction itself can be inefficient due to inherently unfavorable geometry or the chemical environment at the site of photoagent incorporation. Although methods using radioactive labeling that have good sensitivity are used to map crosslink sites, the clearest and best results are obtained when nanogram quantities of the crosslinked species can be obtained. Similarly, it can sometimes be hard to generate high-quality primary data since several manipulations of RNA are required (i.e. photoagent modification, crosslinking, gel purification, etc.). As described in more detail below, interpretation of primer extension–termination mapping of crosslinked sites can be difficult and great care must be taken to insure that the data truly reflect the formation of novel crosslinks. Much of the ambiguity can be resolved with the appropriate controls, and the most important considerations in this regard are outlined below.

Because of these issues, it is critical to test in the most direct way possible that the crosslinking data reflects the functional form or native folded structure of the RNA and to consider what experimental evidence can be brought to bear to establish the functional relevance of the data set. Optimally, this goal can be achieved by assaying directly whether the crosslinked RNA retains biological activity. However, this can be a problem when probing the functional core of an RNA since the crosslink itself can alter chemical groups important for biological function. Alternatively, the proximity data from crosslinking can be considered in light of other structural constraints from, for example, phylogenetic comparative studies, chemical and enzymatic probing and high-resolution structures of homologous molecules.

Important new insights into the validity as well as the limitations of crosslinking as an approach for exploring RNA structure comes from the comparison of the recent three-dimensional structure of the ribosome and the extensive collection of biochemical structure probing data [1]. Overall a large percentage of the crosslinking data were consistent with the structure from X-ray crystallography; however, the resolution of the structural information was less than expected given the chemical structure and size of the different crosslinking reagents used. Furthermore, no individual crosslinking reagent appeared to be superior with respect to validity of the data; however, the method of detection did have an important impact since most of the lower quality data was obtained by primer extension mapping. Most likely this limited accuracy is due to misidentification of non-specific terminations as crosslink sites. The highest quality data was obtained by direct physical mapping of the crosslinked nucleotides, underscoring the importance of optimization of the crosslinking procedure. Despite this track record, primer extension mapping is still a convenient method due to its sensitivity and flexibility; however, obtaining more direct data such as gel mobility, RNase H mapping or optimally by fingerprinting is obviously desirable.

Despite these limitations and considerations, crosslinking approaches can provide important structural information in those numerous instances when it is impossible to obtain material in adequate amounts or in sufficient purity for high-resolution structural analysis. Often it is desirable to probe structure in a context such as within cell extracts where high-resolution studies are impractical. In principle, crosslinking reports on the structure or structures as they occur in solution and in instances when conditions can be found to favor one conformation over another, it can be possible to use crosslinking to define the characteristic structural features of these different states. Perhaps the greatest advantage is sensitivity, since relatively small amounts of crosslinked material are needed for mapping. Once the sites of crosslinking are defined, the formation of a specific crosslink can be used analytically, again with high sensitivity using radiolabeled RNA.

22.2 Description

22.2.1 General Considerations: Reaction Conditions and Concentrations of Interacting Species

It is important to take into consideration that RNA structure, and thus its biological activity, can be highly dependent on reaction conditions. Individual RNAs often adopt multiple conformations and obviously it is necessary that the crosslinking experiment be performed under conditions that favor the correct structural form or the structure of interest [27–29]. Therefore, it must be considered how to optimally fold the RNA sample prior to initiating the crosslinking reaction. Thus, it is important to have as detailed an understanding as possible about the influence of mono- and divalent ion concentrations and identity as well as pH on the biological activity and RNA structure. It is also useful to examine the effect of these parameters on the crosslinking reaction as well, since gaining the highest efficiency possible is important for subsequent identification of crosslinked nucleotides and analysis of the retention of biological activity of the purified crosslinked species.

Crosslinking is very useful for initial analysis of intermolecular interactions, and can be used to define the potential interface between two RNAs or between RNA and a specific protein. Because of the aforementioned penitent for misfolding and condition dependence, RNAs can self-associate or bind in non-productive ways. Similarly, even specific RNA-binding proteins can interact weakly with RNA in a non-specific fashion. Thus, it is important to consider the relative concentrations of the interacting species in the reaction to minimize the potential for formation of non-specific complexes. Examining the effect of macromolecular concentration on the crosslinking reaction can provide insight into whether the information gained accurately reflects formation of high affinity or biologically active complexes.

In the following example the interaction between the RNase P ribozyme and its substrate were examined using intermolecular crosslinking with s^4U- and s^6G-modified tRNA precursors (pre-tRNA) (Fig. 22.2) [16, 26]. RNase P is a widespread and essential ribonucleoprotein enzyme that generates the 5′ end of mature tRNAs via a site-specific phosphodiester bond hydrolysis reaction [30–32]. In bacteria, RNase P enzymes are heterodimers composed of a small, but essential, protein subunit and a larger RNA component that is the catalytic subunit of the enzyme. Whereas most ribozymes catalyze self-cleavage or self-splicing reactions and have to be engineered to work in *trans*, for RNase P RNA catalysis of a multiple turnover reaction is intrinsic to its biological function. Although RNase P, like many other RNA processing RNPs, can recognize a broad spectrum of substrates, the mechanistic basis for its multiple substrate recognition properties is not clearly defined.

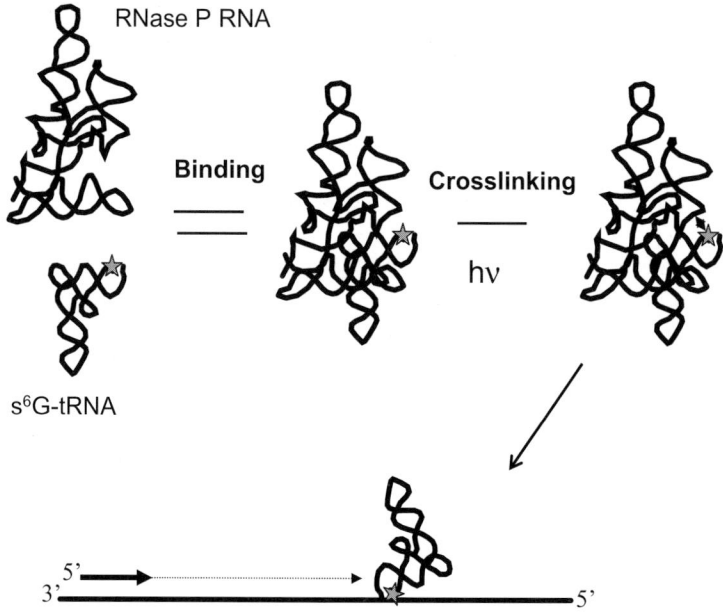

Fig. 22.2. Overview of photoaffinity crosslinking and primer extension mapping. The RNase P RNA is represented in this example as a black ribbon diagram. The photoagent modified pre-tRNA substrate is shown in grey. The position of the photoagent is indicated by a star. As described in the text, the two RNAs are allowed to bind (Binding) and the photoagent is activated by exposure to the appropriate wavelength of UV light (Crosslinking). Subsequently, the appropriate crosslinked RNA species are isolated, generally by gel purification and the sites of crosslinking determined by primer extension mapping. The radiolabeled primer used in the reaction is indicated by an arrow.

Thus, the interactions between RNase P RNA and pre-tRNA substrates, in particular those that underlie specificity, continue to be the subject of considerable interest [33].

To identify residues in the RNase P ribozyme that are proximal to the substrate cleavage site we positioned s^4U and s^6G on either side of the reactive phosphodiester bond in a model tRNA precursor. Kinetic and thermodynamic studies demonstrated that the inclusion of the photoagent at the substrate cleavage site did not interfere with high affinity binding and that the modified substrate was processed at a rate that was essentially identical to the unmodified substrate. To insure proper folding of the two RNAs and efficient formation of the enzyme–substrate complex the following procedure was used. The RNAs are resuspended separately in reaction buffer, in this case (2 M ammonium acetate; 50 mM Tris–HCl, pH 8.0) for refolding. The RNA-containing solutions are heated to 90 °C for 1 min in a programmable heating block (MJ Research) and then cooled to room temperature using a standard water bath over a period of approximately 20 min. Divalent metal ions, in this example 25 mM CaCl$_2$, are added and the RNAs incubated at 37 °C for 15–30 min to insure as much of the RNA as possible has attained the native, folded form. Equal volumes of substrate and enzyme RNA are mixed and incubated for 2 min. In this instance Ca^{2+} is used to replace the optimal metal ion for the reaction, Mg^{2+}, in order to slow the rate of catalysis and permit the assessment of the binding affinity of the substrate [34].

Preparative intermolecular crosslinking reactions generally contained 100 nM photoagent-containing pre-tRNA and 1 µM RNase P ribozyme in order to insure that the majority of the photoagent-modified substrate was bound to the ribozyme. Importantly, it could be demonstrated that formation of crosslinks was dependent on the presence of the ribozyme and occurred in a concentration-dependent manner over a broad range of concentrations (Fig. 22.3). Additionally, the same

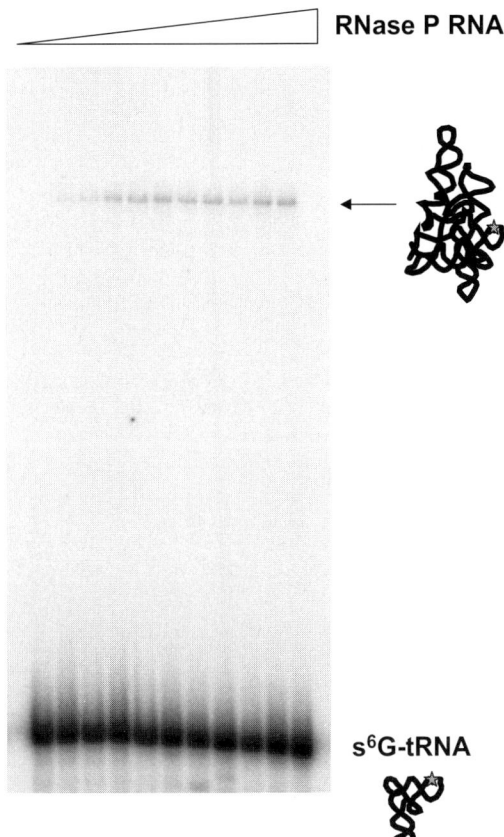

Fig. 22.3. Analysis of the formation of crosslinked species by gel electrophoresis. In this example, radiolabeled and photoagent modified pre-tRNA (s^{6}G-tRNA) was incubated with increasing concentrations of RNase P ribozyme and the reactions exposed to UV light. The formation of a single crosslinked species that requires the presence of the ribozyme and is dependent on its concentration is indicated by the arrow.

crosslinked species were detected at both high and low concentrations of the ribozyme. The concentration dependence clearly demonstrates that the crosslinks are intermolecular in nature and reflect the structure of high-affinity complexes between the two RNAs.

22.2.2
Generation and Isolation of Crosslinked RNAs

Once the conditions and concentrations of the reaction are set or optimized crosslinking is easily initiated by irradiation with the appropriate wavelength of light. Subsequently, the reactions are analyzed for the formation of new crosslinked species. Identification and isolation is almost always accomplished by taking advantage of the altered mobility of the crosslinked RNAs relative to uncrosslinked RNA on denaturing polyacrylamide gels. The crosslinked RNAs are subsequently eluted from the gel and recovered by ethanol precipitation using standard methods.

For analytical reactions in which the photoagent-modified pre-tRNA substrate was also radioactively labeled, aliquots of 12 μl were transferred to a parafilm covered aluminum block. A convenient source is the block from a standard dry-bath incubator, pre-cooled in ice for at least 1 h prior to the experiment. We found that crosslinking occurred optimally at 4 °C. Parafilm and samples were placed on the block just before irradiation to minimize dilution or contamination by condensation. The samples were irradiated for 5–15 min at 366 nm at a distance of 3 cm using a model UVGL-58 ultraviolet lamp from UVP, Upland, CA. A standard (3–4 mm) thick glass plate was placed between the lamp and the sample to help filter out shorter wavelengths of UV light that can damage the RNA sample. Aliquots were recovered from the block, diluted to 200 μl with 10 mM Tris–HCl, pH 8.0, 0.5 mM EDTA, 0.3 M sodium acetate, then extracted twice with 50/50% phenol/chloroform and once with chloroform alone and precipitated by addition of 3 volumes of ethanol.

Because both inter- and intramolecular crosslinking alter the linear topology of the targeted RNA, one can generally identify and isolate crosslinked species based on their slower mobility in denaturing acrylamide gels (Fig. 22.3). Appropriate controls should be run in parallel in which the photoagent is omitted from the reaction in order to demonstrate that the formation of the more slowly migrating species depends on presence of the crosslinking reagent and not from adventitious crosslinking due to ambient UV light. Similarly, control samples that are not irradiated must also be compared since crosslinking can occur during sample workup that may not necessarily reflect the functional structure. Additionally, for intermolecular crosslinking it is essential to demonstrate that formation of the crosslinked species requires the presence of the interacting partner RNA or protein and that its formation is concentration dependent.

Once the specificity of the crosslinking reaction is established, the next step is to isolate sufficient quantities of the individual crosslinked species to map the crosslink sites. Keeping in mind that picomole amounts of material will be optimal for primer extension mapping it is necessary to scale the crosslinking reaction up accordingly. We have had good success in simply "spiking" preparative reactions with a small quantity of radiolabeled RNA to use a marker for gel purification. For electrophoresis, the sample can be loaded in a continuous well across the top of the gel and electrophoresis conditions optimized using analytic reactions to achieve the best degree of separation between crosslinked and uncrosslinked RNA. Standard methods are acceptable for location of bands by autoradiography, excision of the appropriate gel slices and elution and recovery of the RNA. We have found that addition of 0.01 μg/μl glycogen as a carrier greatly improves recovery from larger volumes of gel elution buffer and does not interfere in subsequent primer extension mapping experiments.

22.2.3
Primer Extension Mapping of Crosslinked Nucleotides

The general principle behind primer extension mapping of crosslinked nucleotides is that polymerase will continue to synthesize a DNA stand up to, but not beyond the site of crosslinking due to chemical disruption of the template strand [15]. These specific terminations observed in reactions containing crosslinked RNA as a template and not observed in control, uncrosslinked RNA samples are interpreted as sites whether the photoagent has formed a new covalent bond (Fig. 22.2). Terminations are interpreted as occurring one nucleotide 5' to the site of crosslinking. A key advantage is that only relatively small amounts of template RNA are required (1–0.1 pmol); however, best results are obtained when at least picomolar amounts are available. Additionally, the technique is relatively easy, rapid and requires reagents and equipment that are widely available.

However, the biggest problem here is that reverse transcriptase will pause at specific sites on virtually any RNA template. Therefore, it is essential to distinguish between a termination that is due to a crosslink and one that is due to RNA structure, non-specific radiation damage or degradation. To control for these phenomena it is important to perform a parallel analysis of RNA taken through the protocol but not subjected to irradiation, as well as a control RNA sample irradiated in the absence of ligand, or with a ligand population that has not been modified with the photoagent. Additionally, corroborative information from RNase H mapping [22] or the mobility of the crosslinked species in denaturing gels [10] can be used to gain information on where crosslinks are located and this information can be important for resolving any potential ambiguities in the primer extension termination results.

For primer extension analysis of intermolecular crosslinks between photoagent-modified pre-tRNA and the RNase P ribozyme 0.2 pmol of 5'-^{32}P-end-labeled primer are annealed to 0.05–0.2 pmol of gel-purified crosslinked RNA in a total volume of 5 μl. The annealing solution is composed of 50 mM Tris–HCl, pH 8.3, 15 mM NaCl and 10 mM dithiothreitol. Individual samples are heated to 65 °C for 3 min and then set immediately on dry ice. The annealed samples are thawed on ice and 1 μl of 30 mM MgCl$_2$ is added followed by addition of the four deoxynucleotide triphosphates to a final concentration of 400 μM. These reactions (8 μl) are initiated by the addition of 2 U (in 2 μl) of AMV reverse transcriptase (Boehringer Mannheim) and then incubated at 47 °C for 5 min. Reactions are then quenched by the addition of an equal volume of 0.5 M NaCl, 20 mM EDTA and 0.5 μg glycogen, and the extension products recovered by ethanol precipitation.

Primer extension products are then resolved next to dideoxy sequencing standards on denaturing polyacrylamide gels. The concentration of dideoxynucleotide added can be varied from 5 to 100 μM to detect nucleotides from less than 10 nt to more than several hundred away from the primer binding site. After recovery by ethanol precipitation and washing with 80% ethanol to remove excess salt the radiolabeled products are resuspended in a small volume (2–5 μl) of formamide loading buffer (95% formamide, 150 mM Tris–HCl, pH 8.0, 15 mM EDTA, and trace amounts of bromophenol blue and xylene cyanol FF). Only a fraction of the reaction (e.g. 2 μl) is loaded in an individual lane such that the sample just covers the bottom of an individual well in order to generate the sharpest banding pattern possible.

Obtaining clean primer extension results necessarily requires significant attention to optimization of the reaction parameters. We find that there can be large differences in the quality of data and the pattern of non-specific terminations with different primers, and some attempt to compare two or more sites of primer binding was necessary in some cases. Additionally, the optimal primer concentration is often also idiosyncratic to individual oligonucleotide sequences. An additional area in which the procedure can be optimized is in the annealing procedure where it is useful to compare slow cooling to rapid cooling. Additionally, we have found that increasing the reaction temperature can result in fewer non-specific transcription terminations, but temperatures in excess of 50 °C result in enzyme denaturation or inhibition. Despite the number of parameters that can be varied, generally a few days spent optimizing these few aspects of the procedure using control, uncrosslinked RNA will be time very well spent, since the payoff will be in obtaining clearer and therefore more convincing primary data. See Fig. 22.4.

22.3
Troubleshooting

The key problems associated with these particular methods are inefficient formation of crosslinked species and ambiguous or unclear primer extension mapping results. In the case of the former, locating the photoagent to a nearby region of the molecule may overcome unfavorable geometric constraints. Additionally, as mentioned above, it is usually important to optimize the folding of the RNAs of interest in order to insure that the maximum fraction of the sample is in the correctly folded and biologically active form. If the particular site of photoagent attachment is sufficiently interesting then trying additional, longer-range crosslinking agents such as phenylazides [23] can be used to increase crosslinking efficiency.

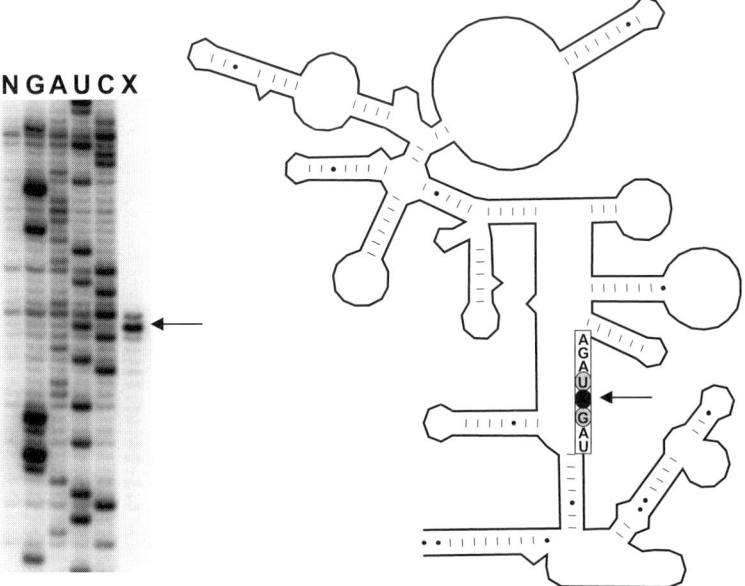

Fig. 22.4. Primer extension mapping. Products from reactions containing gel-purified crosslinked RNA (X) as well as control RNA from a control reaction that was not irradiated were resolved by gel electrophoresis (N). Lanes G, A, U and C contain products from sequencing reactions containing the appropriate dideoxynucleotide. An arrow indicates the position of the termination due to crosslink formation. A diagram of the secondary structure of *Bacillus subtilis* RNase P is shown on the right with the position of the crosslink indicated by an arrow.

One of the key difficulties that we have encountered is in protecting the photo-agent modified RNA from ambient UV light. Although it can be somewhat awkward, working up the photoagent modified RNA and performing the crosslinking reactions in a darkened laboratory environment is important for obtaining the cleanest possible results.

In our experience, by far the most challenging aspect is to generate appropriately clean primer extension mapping data. As with all procedures involving RNA, making sure that the sample is not exposed to heating in the presence of metal ions is essential. Degradation of the RNA sample is apparent as an intense background of terminations in the control RNA primer extensions. Beyond repetition of the experiment to confirm the reproducibility of the results, good experimental technique and careful handling of the RNA is implicit. We have found that longer incubation times for the reverse transcriptase reaction can increase the background and thus it is important to consider assessing the effects of varying the reaction time. Also, some pilot experiments addressing what conditions are most appropriate for annealing of the radiolabeled primer will also pay off in the long run.

References

1 M. WHIRL-CARRILLO, I. S. GABASHVILI, M. BADA, D. R. BANATAO, R. B. ALTMAN, *RNA* **2002**, *8*, 279.
2 D. I. JUZUMIENE, P. WOLLENZIEN, *RNA* **2001**, *7*, 71.
3 P. V. BARANOV, S. S. DOKUDOVSKAYA, T. S. ORETSKAYA, O. A. DONTSOVA, A. A. BOGDANOV, R. BRIMACOMBE, *Nucleic Acids Res.* **1997**, *25*, 2266.
4 T. KAO, D. L. MILLER, M. ABO, J. OFENGAND, *J. Mol. Biol.* **1983**, *166*, 383.
5 P. A. MARONEY, Y. T. YU, M. JANKOWSKA, T. W. NILSEN, *RNA* **1996**, *2*, 735.
6 A. J. NEWMAN, S. TEIGELKAMP, J. D. BEGGS, *RNA* **1995**, *1*, 968.
7 D. A. WASSARMAN, J. A. STEITZ, *Science* **1992**, *257*, 1918.
8 C. H. KIM, J. ABELSON, *RNA* **1996**, *2*, 995.
9 Y. T. YU, J. A. STEITZ, *Proc. Natl Acad. Sci. USA* **1997**, *94*, 6030.
10 M. E. HARRIS, A. V. KAZANTSEV, J. L. CHEN, N. R. PACE, *RNA* **1997**, *3*, 561.
11 J. L. CHEN, J. M. NOLAN, M. E. HARRIS, N. R. PACE, *EMBO J.* **1998**, *17*, 1515.
12 R. PINARD, D. LAMBERT, J. E. HECKMAN, J. A. ESTEBAN, C. W. t. GUNDLACH, K. J. HAMPEL, G. D. GLICK, N. G. WALTER, F. MAJOR, J. M. BURKE, *J. Mol. Biol.* **2001**, *307*, 51.
13 A. MALHOTRA, S. C. HARVEY, *J. Mol. Biol.* **1994**, *240*, 308.
14 F. MUELLER, H. STARK, M. VAN HEEL, J. RINKE-APPEL, R. BRIMACOMBE, *J. Mol. Biol.* **1997**, *271*, 566.
15 M. E. HARRIS, E. L. CHRISTIAN, *Methods* **1999**, *18*, 51.
16 E. L. CHRISTIAN, D. S. MCPHEETERS, M. E. HARRIS, *Biochemistry* **1998**, *37*, 17618.
17 P. V. SERGIEV, I. N. LAVRIK, V. A. WLASOFF, S. S. DOKUDOVSKAYA, O. A. DONTSOVA, A. A. BOGDANOV, R. BRIMACOMBE, *RNA* **1997**, *3*, 464.
18 E. J. SONTHEIMER, *Mol. Biol. Rep.* **1994**, *20*, 35.
19 Y. L. DUBREUIL, A. EXPERT-BEZANCON, A. FAVRE, *Nucleic Acids Res.* **1991**, *19*, 3653.
20 A. FAVRE, C. SAINTOME, J. L. FOURREY, P. CLIVIO, P. LAUGAA, *J. Photochem. Photobiol. B* **1998**, *42*, 109.
21 B. S. COOPERMAN, R. W. ALEXANDER, Y. BUKHTIYAROV, S. N. VLADIMIROV, Z. DRUZINA, R. WANG, N. ZUNO, *Methods Enzymol.* **2000**, *318*, 118.
22 Y. T. YU, *Methods Enzymol.* **2000**, *318*, 71.
23 A. B. BURGIN, N. R. PACE, *EMBO J.* **1990**, *9*, 4111.
24 M. M. KONARSKA, *Methods* **1999**, *18*, 22.
25 B. C. THOMAS, A. V. KAZANTSEV, J. L. CHEN, N. R. PACE, *Methods Enzymol.* **2000**, *318*, 136.
26 E. L. CHRISTIAN, M. E. HARRIS, *Biochemistry* **1999**, *38*, 12629.
27 O. C. UHLENBECK, *RNA* **1995**, *1*, 4.
28 D. K. TREIBER, J. R. WILLIAMSON, *Curr. Opin. Struct. Biol.* **2001**, *11*, 309.
29 D. K. TREIBER, J. R. WILLIAMSON, D. THIRUMALAI, N. LEE, S. A. WOODSON, D. KLIMOV, *Curr. Opin. Struct. Biol* **2001**, *11*, 309.
30 J. HSIEH, A. J. ANDREWS, C. A. FIERKE, *Biopolymers* **2004**, *73*, 79.
31 M. E. HARRIS, E. L. CHRISTIAN, *Curr. Opin. Struct. Biol.* **2003**, *13*, 325.
32 V. GOPALAN, A. VIOQUE, S. ALTMAN, *J. Biol. Chem.* **2002**, *277*, 6759.
33 E. L. CHRISTIAN, N. H. ZAHLER, N. M. KAYE, M. E. HARRIS, *Methods* **2002**, *28*, 307.
34 D. SMITH, A. B. BURGIN, E. S. HAAS, N. R. PACE, *J. Biol. Chem.* **1992**, *267*, 2429.

II.2
Biophysical Methods

23
Structural Analysis of RNA and RNA–Protein Complexes by Small-angle X-ray Scattering

Tao Pan and Tobin R. Sosnick

23.1
Introduction

Small-angle X-ray scattering (SAXS) is a solution technique that measures the size and shape of an individual or a complex of macromolecules. The method is well suited for the analysis of RNA structure, folding and association with proteins. For example, partially folded states that are not readily measured using other biophysical methods such as NMR or crystallography can be readily studied using SAXS.

In the typical SAXS experiment, X-rays are scattered by the sample and the scattering profile is measured at very low angles θ (Fig. 23.1A). The radius of gyration, R_g, of the particle can be determined from the width of the scattering profile. For globular objects, the profile can be approximated as a Gaussian $I(Q) = I_0 e^{-Q^2 R_g^2/3}$, where $Q = 2\pi \sin \theta / \lambda$ and λ is the X-ray wavelength. R_g is the root-mean-square of the distances of all regions to the center of mass of the particle weighted by their (excess) electron density. Typically, R_g is obtained from the slope of the Guinier plot of $\ln I(Q)$ versus Q^2.

The entire scattering profile can be used to obtain the R_g as well as other shape information such as the maximum distance in the particle, d_{max}, and the pair-distribution function $P(r)$ [1]. $P(r)$ is the probability distribution of distances between scattering atoms within the macromolecule (Fig. 23.1B). It has a maximum at the most probable distance in the object (e.g. slightly larger than the radius for a sphere) and goes to zero at d_{max} (e.g. the diameter). The R_g value can also be calculated from the second moment of the $P(r)$ distribution.

SAXS can be used to test the consistency of either high- or low-resolution structural models obtained experimentally or from modeling. The R_g value and the $P(r)$ function measured in a SAXS experiment can be compared with the predicted values [2]. As SAXS is a solution technique, this capability is particularly useful for determining whether a structure undergoes a conformational change upon crystallization or electron microscopy preparation. Multi-domain RNAs or RNA–protein complexes can be modeled using the crystal or NMR structures of the individual components. Generally, constructing a unique three-dimensional model from one-

Handbook of RNA Biochemistry. Edited by R. K. Hartmann, A. Bindereif, A. Schön, E. Westhof
Copyright © 2005 WILEY-VCH Verlag GmbH & Co. KGaA, Weinheim
ISBN: 3-527-30826-1

A

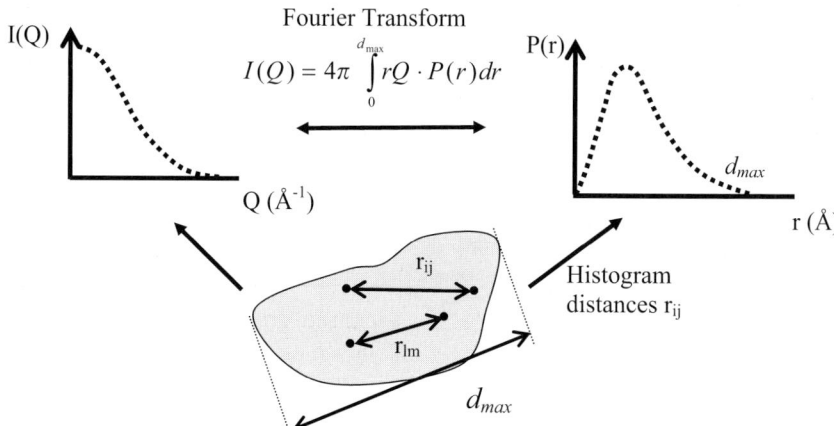

B

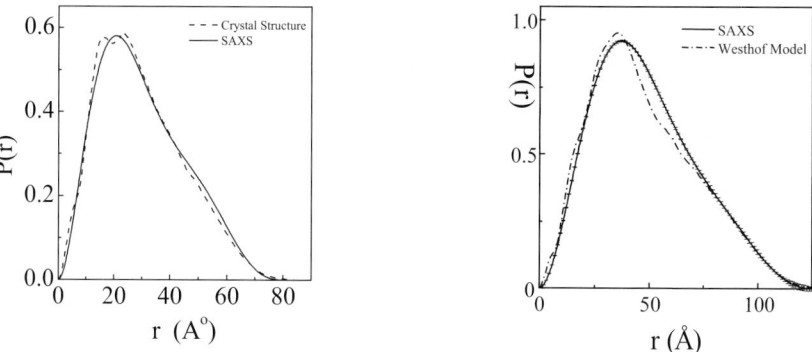

Fig. 23.1. (A) Relationship between $I(Q), P(r)$ and the macromolecule of interest. (B) Comparing the $P(r)$ function of the crystal structure of yeast tRNA^{Phe} (left) and the Westhof model of the catalytic domain of *B. subtilis* RNase P RNA (right) with the SAXS data.

dimensional scattering data is difficult in the absence of other information, although significant progress has been made in this area [3].

Another useful parameter from SAXS measurements is the absolute scattering intensity at zero angle, I_0. For a monodispersed system, I_0 is related to the molecular weight of the scattering species (MW), macromolecular weight concentration (C, in mg/ml) and the electron density (ρ) according to [4]:

$$I_0 \propto (\rho_{macromolecule} - \rho_{solvent})^2 \times MW \times C \qquad (1)$$

This proportionality allows the determination of the oligomerization state of the RNA or RNA–protein complex when a suitable RNA mass standard is used such as the yeast tRNA^{Phe}. Because the crystal structure is known for this tRNA, its monodispersity can be confirmed from the measured R_g and $P(r)$ values with those calculated from the crystal structure. Therefore, the I_0 value of the tRNA^{Phe} can be used as a molecular weight and size standard for the studies of RNA or RNA–protein complexes of interest [5].

An advantage of SAXS in studying RNA–protein complexes is that RNA scatters more strongly than protein due to its higher electron density. Generally, the relative scattering power of the RNA to the protein is $(\rho_{RNA} - \rho_{solvent})^2 / (\rho_{protein} - \rho_{solvent})^2 \sim$ 5-fold [6]. This property allows the size and the shape of the RNA to be analyzed in the presence of moderate amounts of protein – a feature particularly useful for identifying conformational changes in RNA upon protein binding.

The downside of the sensitivity of SAXS method to size and to oligomeric state is that even mild amounts of aggregation can lead to spurious $P(r)$ and R_g values. Before synchrotron facilities were available, a typical SAXS experiment at a home source required relatively high RNA concentrations, e.g. 10 mg/ml. A SAXS experiment at a synchrotron such as the Advanced Photon Source (APS) at the Argonne National Laboratory, however, only requires RNA samples in the range of 0.1–1 mg/ml (4–40 µM yeast tRNA^{Phe}) and proportionally lower for larger RNAs. Data acquisition takes just a few seconds, which further reduces the chances of aggregation and sample degradation. In addition, lower concentrations can be critical for samples that are difficult to obtain or permit measurements to be conducted over a variety of conditions.

23.2
Description of the Method

23.2.1
General Requirements

Biological SAXS experiments require dedicated instruments designed for low angle measurements and optimized for low background levels. As mentioned, synchrotron-based measurements have a significant number of advantages (Fig. 23.2). In addition, a programmable titrator in conjunction with a flow-through sample cell enables an entire titration series (e.g. varying the ion or denaturant concentration) to be expediently carried out with a single sample in an hour or less.

The following steps are recommended for an SAXS experiment:

(1) Find a synchrotron-based beam-line with personnel who have experience with biological samples. As biological samples scatter weakly, their signal can be orders of magnitude weaker than background levels. Hence, minimizing background levels at the outset of an experiment is highly advantageous in terms of

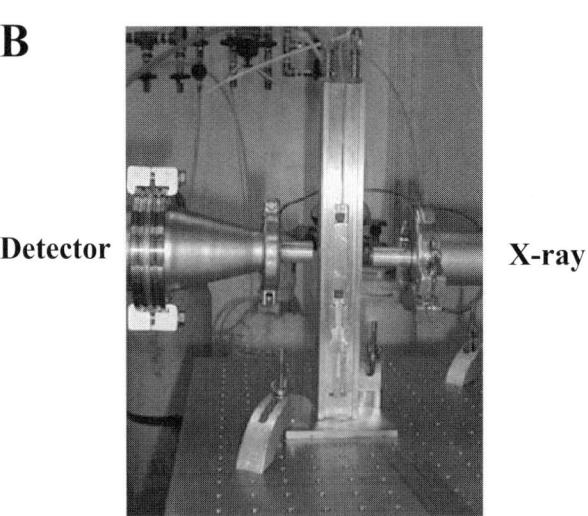

Fig. 23.2. (A) Experimental setup of SAXS at a synchrotron facility. (B) The sample holder used in our SAXS studies, sandwiched between the X-ray beam on the right and the CCD detector on the left.

sample requirements and reproducibility. Frequent interactions are necessary between the user and the scientists; fortunately, many facilities now have established user programs which provide expert support.

(2) Sample requirement. Approximately 200 μl of around 0.3 mg/ml RNA of about 50 kDa is needed for a single SAXS measurement. Large RNAs scatter more strongly than small RNAs ($I_0 \propto MW$), so the operational concentration can be

proportionally lower for large RNAs. For a titration series of 10–20 data points, at least 1 ml of sample is recommended.

(3) Time requirement. The most time-consuming component is the instrumental setup which is done by the staff at the synchrotron facility. A single SAXS measurement takes about 10 min, most of which is spent on sample manipulation. At the BioCat and BSSERC ID-12 beam-lines at the APS, the actual data acquisition takes just a few seconds or less. A titration series can take approximately 1 h, including the background (buffer) measurements.

(4) Data analysis. To confirm that measurements are generating useful information, preliminary data analysis should be conducted during the course of the experiment. This analysis generally includes background subtractions and Guinier analysis to obtain I_0 and R_g. These quantities enable the user to confirm that background levels are minimized, the signal is reproducible and the sample is monodispersed – three critical elements of a high-quality experiment.

23.2.2
An Example for the Application of SAXS

RNase P is an essential enzyme required for the 5′ maturation of all tRNAs and is conserved in all three kingdoms of life [7, 8]. In bacteria, RNase P is composed of one RNA subunit of 330–420 nt and a small protein subunit of 13–15 kDa. The bacterial RNase P RNA subunit (P RNA) alone is catalytically active, but the protein subunit is required for full activity under physiological conditions. We used SAXS to address the following questions for the RNase P ribozyme from *Bacillus subtilis* (for more details see [9, 10]):

(1) What is the oligomerization state of the P RNA alone and complexed with the RNase P protein (P protein) at varying monovalent ion concentration?
(2) What is the overall shape of the P RNA alone and complexed with the P protein?
(3) What is the size and shape of the ribozyme–substrate complexes?

23.3
General Information

RNase P RNA from various organisms was obtained by standard transcription using T7 RNA polymerase and purified using denaturing gels. The P protein was prepared from an overexpression clone. The concentration of the P RNA and the P protein was determined by UV absorbance using previously determined extinction coefficients.

The P RNA was renatured as follows: (1) heat the RNA in the buffer alone at

90 °C for 2 min, (2) incubate at room temperature for 3 min, (3) add MgCl$_2$ to 10 mM final concentration, (4) incubate at 50 °C for 5 min and (5) add NH$_4$Cl or KCl to the desired concentration.

To reconstitute the RNase P holoenzyme, an equal amount of P protein was added to the renatured P RNA, followed by the incubation at 37 °C for 5 min.

SAXS experiments were carried out at the SAXS instrument on the BESSRC ID-12 beam-line of the APS at the Argonne National Laboratory located in Illinois, USA [11]. Data were collected using a nine-element mosaic CCD area detector (15 cm × 15 cm) and exposure times were 1–6 s for each measurement. Sample–detector distance was 3 m; the energy of X-ray radiation was set to 13.5 keV. Computer-controlled Hamilton syringes injected sample into a thermostated flow cell made of 1.5-mm diameter cylindrical quartz capillary (Fig. 23.2). The background scattering was from a buffer solution in the identical configuration. Samples were measured under constant flow conditions in order to reduce the possibility of radiation damage.

23.4
Question 1: The Oligomerization State of P RNA and the RNase P Holoenzyme

I_0 and $P(r)$ function are used to deduce the oligomerization state of P RNA in the absence and presence of the P protein. Without the P protein, the I_0 ratio of P RNA and yeast tRNA^{Phe} at the same weight concentration (0.1–1 mg/ml) is proportional to their molecular weight ratio (5.6 ± 0.6 versus 5.4). Yeast tRNA^{Phe} has been shown conclusively to be a monomer under these conditions [5], and experimentally derived $P(r)$ and R_g for this tRNA agree well with the crystal structure. Hence, this result shows that P RNA is a monomer without the P protein.

The *B. subtilis* holoenzyme contains two P RNA molecules as indicated by SAXS (Fig. 23.3A). The I_0 of the P RNA–P protein complex reconstituted at 1:1 RNA–protein is twice the I_0 of P RNA alone. The higher scattering signal from RNA and the significantly larger size of the P RNA over P protein ensure that the observed I_0 is almost entirely derived from the scattering from P RNA. Consistent with the dimer formation, the $P(r)$ function of the holoenzyme has two times more distance pairs compared to those of the P RNA alone. As the stoichiometry of the P RNA and P protein is 1:1 in the holoenzyme, SAXS results show that the *B. subtilis* holoenzyme contains two P RNA and two P protein subunits.

The scattering profile of P RNA without the P protein shows considerable variation in the absence and the presence of 0.1 M NH$_4$Cl or 0.1 M KCl. The native structure of P RNA is composed of two independently folding domains. The variation in the scattering data of the P RNA monomer at different solution conditions may be explained by a difference in the relative orientation of the two domains.

The fraction of the holoenzyme dimer and monomer as a function of monovalent salt also is determined by SAXS (Fig. 23.3B). Changes in the I_0 and the R_g values are used to deduce the dimer fraction. Comparing 0.8 versus 0.1 M NH$_4$Cl, I_0 is 2-fold at 0.8 M NH$_4$Cl and the R_g value is also reduced from 57 to 50 Å at

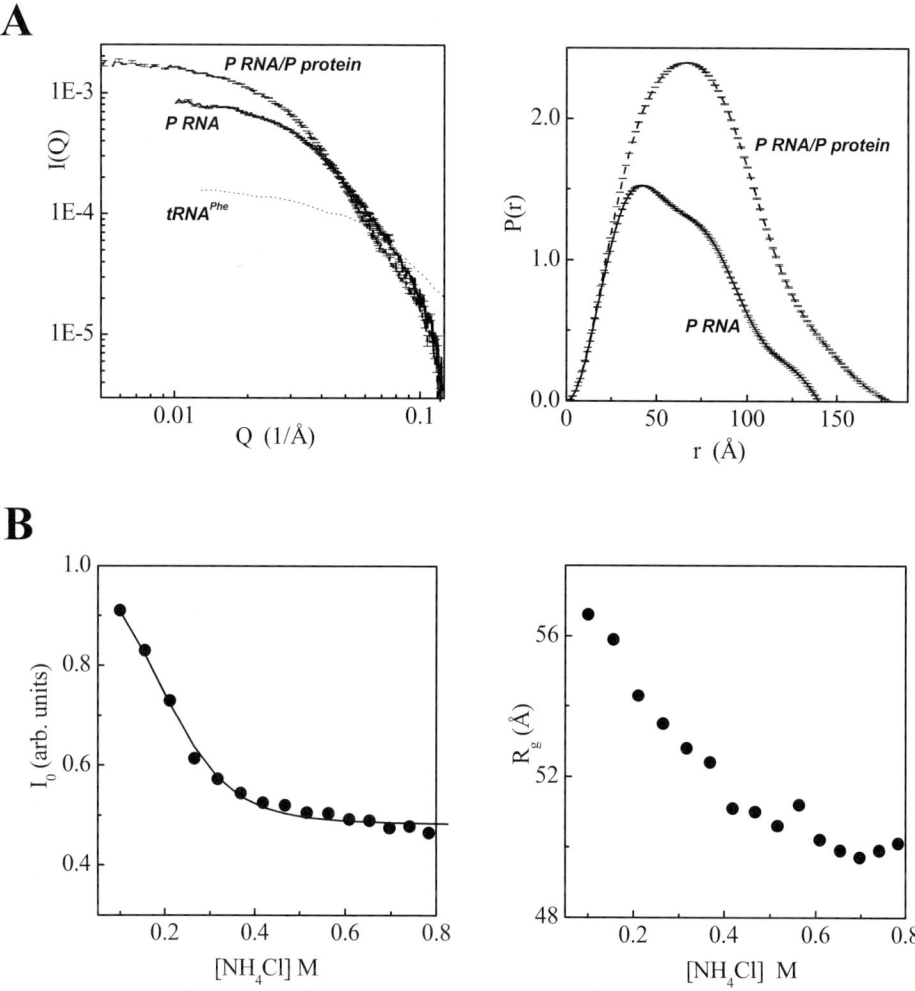

Fig. 23.3. (A) Scattering profile (left) and the $P(r)$ function (right) of the P RNA alone and complexed with the P protein. (B) The dependence of the holoenzyme dimer:monomer equilibrium on the concentration of NH_4Cl, monitored by I_0 (left) and the R_g value (right). Changes in the I_0 value can be fit (solid line) to obtain a fifth-power dependence of the dimerization constant on the NH_4Cl concentration.

0.8 M NH_4Cl. These results show that at the holoenzyme is predominantly a dimer at 0.1 M NH_4Cl, but is a monomer at 0.8 M NH_4Cl. The holoenzyme is in a dimer–monomer equilibrium at the intermediate NH_4Cl concentrations.

The dimer–monomer equilibrium has a fifth-power dependence on the NH_4Cl concentration according to the SAXS data, i.e. the dimerization constant, K_D, is reduced by 2^5-fold when the NH_4Cl concentration is increased by 2-fold. At 0.1 M NH_4Cl, the K_D value is around 50 nM.

23.5
Question 2: The Overall Shape

The shape of the P RNA monomer and the holoenzyme is modeled and compared to the R_g and $P(r)$ function from SAXS. The modeling is performed in two steps. First, the P RNA model from Westhof et al. [12] is modified to allow a better fit to the SAXS data. Second, two P RNAs are brought together and the dimer model compared to the SAXS results.

The R_g of the P RNA monomer derived from the Westhof model is smaller than those determined by SAXS (43 versus 47–55 Å). The $P(r)$ function of the Westhof model also has a narrower distribution of mass pairs than that derived from the SAXS data. The P RNA in the Westhof model, however, was constructed in the context of a bound tRNA substrate. As discussed in the next section, the presence of tRNA may affect the conformation of the P RNA.

The Westhof model was modified to determine whether a change in the domain orientation alone could explain the difference in the R_g and $P(r)$ (Fig. 23.4A). In our case, the structures of both domains are kept the same as those in the Westhof model. However, one of the two domain–domain connections is used as a hinge and the other connection is extended to allow one domain to be rotated away from the other domain. Rotation of the inter-domain angle from around 30° in the Westhof model to around 60° and 90° changes the R_g for the new P RNA models from 43 to 47–53 Å, respectively, much closer to the measured R_g with and without 0.1 M salt. Similarly, the $P(r)$ functions of the models with altered domain orientations agree better with the SAXS data.

Two P RNA molecules with modified inter-domain orientations are brought together with the catalytic domain of one P RNA proximal to the specificity domain of the other P RNA and vice versa to generate the holoenzyme model (Fig. 23.4B). To obtain a better fit to the SAXS data, the precise angle between the domains in the holoenzyme is similar, but not identical, to that in either P RNA monomer model. Changing the angle between the domains should be feasible because P protein binding could easily compensate for any potential energetic cost of altering the domain orientation. The model has similar $P(r)$ and R_g (56 Å) to the experimentally measured $P(r)$ and R_g (57 Å).

23.6
Question 3: The Holoenzyme–Substrate Complexes

Although the pre-tRNA substrates used in almost all biochemical studies contain a single tRNA, the cellular substrate is more diverse. *B. subtilis* has a total of 86 tRNA genes, only eight of which produce single tRNA transcripts [13]. The remaining 78 tRNAs are arranged in 13 operons that produce 2–21 tRNAs per transcript. At least in *B. subtilis*, the holoenzyme is likely to encounter tRNA transcripts containing two or more tRNAs. Therefore, two types of substrate are used in the

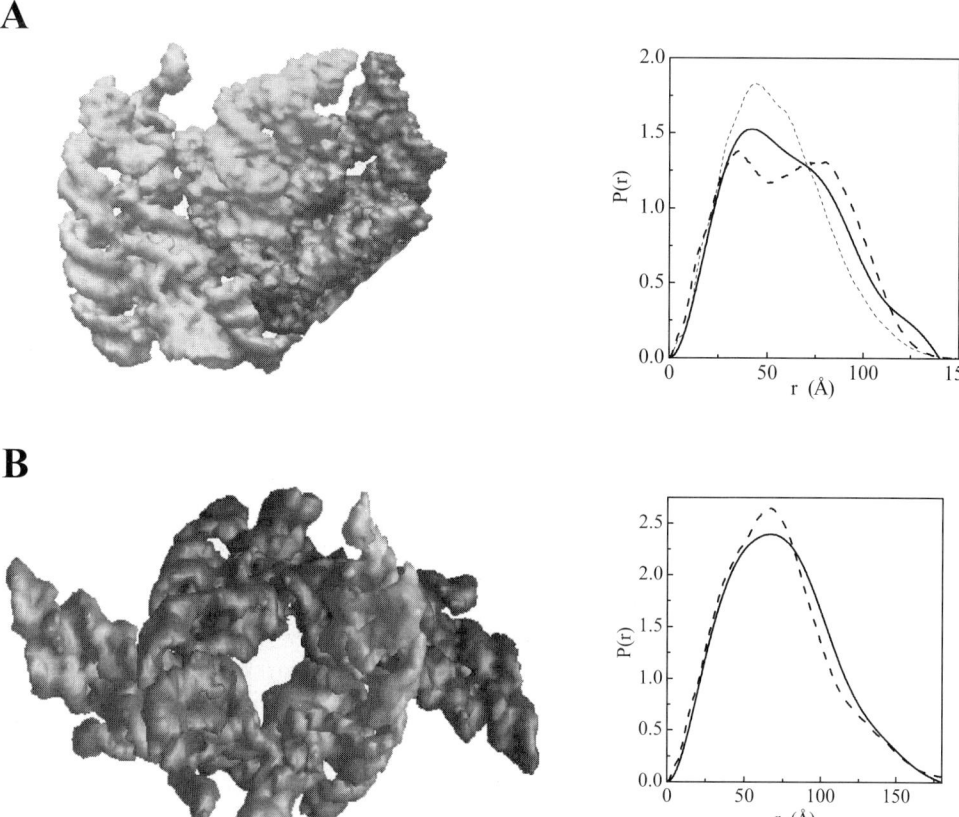

Fig. 23.4. (A) Models of the P RNA (left) and the P(r) functions calculated from these models compared to the SAXS data (right). The Westhof model of the *B. subtilis* P RNA is shown in light gray. The modified model has a different inter-domain angle and is shown in dark gray. The P(r) from the Westhof model is shown as a thin dashed line, from the modified model as a thick dashed line and from the SAXS data as a solid line. (B) A model of the holoenzyme dimer (left) and the P(r) functions (right) calculated from the model (dashed line) and derived from the SAXS data (solid line).

SAXS study – one containing a single tRNA precursor and the other containing two tRNA precursors.

Substrate binding of the holoenzyme is again analyzed by SAXS (Fig. 23.5). Upon substrate addition and complex formation, the I_0 value is proportional to the second power of the molecular weight ratio of the ES complex and the holoenzyme alone, i.e. $I_0^{ES}/I_0^E = (MW^{ES}/MW^E)^2$. Changes in the R_g value also provide information on the structure of the ribozyme–substrate complexes.

The formation of the holoenzyme–substrate complex is monitored at 0.1 and 0.8 M NH_4Cl upon varying the molar enzyme:substrate ratio from 0 to 2. The

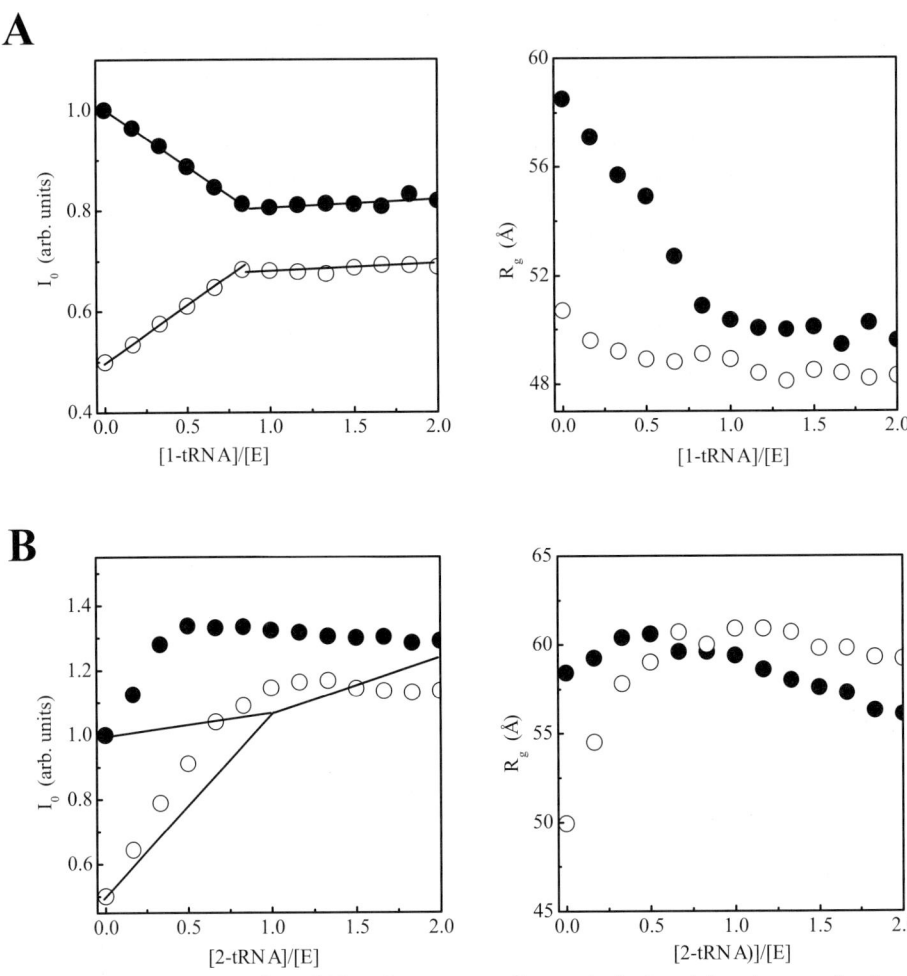

Fig. 23.5. SAXS studies of RNase P holoenzyme–substrate complexes. (A) Changes in the I_0 and R_g values as a function of the S:E ratio using a one-tRNA substrate at 0.1 M NH_4Cl (filled circles) and 0.8 M NH_4Cl (open circles). Only around 83% of this substrate is capable of binding to the enzyme under this condition, resulting in the non-identical I_0 and R_g value at high S:E ratio. (B) Changes in the I_0 and R_g values as a function of the S:E ratio using a two-tRNA substrate at 0.1 M NH_4Cl (filled circles) and 0.8 M NH_4Cl (open circles). Only around 60% of this substrate is capable of binding to the enzyme under this condition, resulting in the non-ideal I_0 and R_g value at higher S:E ratio. The solid lines show the predicted I_0 changes if all ES complexes are monomers.

holoenzyme concentration is kept constant at 2.4 µM to ensure that the initial holoenzyme is a dimer at 0.1 M NH_4Cl or a monomer at 0.8 M NH_4Cl. Under these conditions, the affinity of P RNA for a tRNA substrate is strong enough so that all properly folded substrates are bound to the P RNA when the stoichiometry of substrate to ribozyme is less than 1.

When the holoenzyme initially is a dimer at 0.1 M NH_4Cl, the I_0 value decreases upon the addition of the one-tRNA substrate until the molar ratio of the substrate to the total RNase P holoenzyme is approximately 1. Further addition of pre-tRNA^{Phe} produces only a very slight increase in the I_0 value, accounted for by the presence of the small, uncomplexed pre-tRNA substrate. When the holoenzyme is initially a monomer at 0.8 M NH_4Cl, the I_0 value increases upon the addition of pre-tRNA^{Phe} until the molar ratio of the substrate to the total holoenzyme is approximately one. These results show that the ES complex with a one-tRNA substrate is a monomer under all conditions.

Different ES complexes are observed when the two-tRNA substrate is bound to the holoenzyme. The ES complex is almost exclusively a dimer at 0.1 M NH_4Cl when the holoenzyme is in molar excess over the substrate, taking into account that only around 60% of this substrate is properly folded. Upon addition of more two-tRNA substrate, monomeric ES complex begins to form, presumably due to the holoenzyme binding to just one of the two tRNAs in this substrate. Now, both dimeric and monomeric substrate can exist at the same time. Both monomeric and dimeric ES complexes still form at 0.8 M NH_4Cl, but the fraction of the dimeric complex is significantly lower. These results show that the ES complex with a two-tRNA substrate is a mixture of dimer and monomer under very different conditions.

23.7
Troubleshooting

23.7.1
Problem 1: Radiation Damage and Aggregation

Due to the extremely high flux at a synchrotron source, radiation damage is a serious concern. The intense X-ray generates hydroxyl radicals that react with the RNA and proteins in the sample to result in aggregation. This problem can often be visualized upon comparing the scattering profiles of the same sample between multiple exposures.

There are two simple remedies to deal with the aggregation problem. First, the Tris and other organic buffers scavenge free radicals, so that their presence can significantly reduce radiation damage. Phosphate or other inorganic buffers are not recommended unless supplemented by a radical scavenger. Second, the measurement should be carried out under constant flow conditions, so that a fresh batch of molecules is exposed to X-ray at all times during measurement. Constant flow can be achieved using the sample-handling device depicted in this article.

23.7.2
Problem 2: High Scattering Background

Beyond optimizing beam-line performance, RNA samples purified by denaturing gel electrophoresis sometimes contain minute amounts of polyacrylamide particles that also scatter X-rays. Because these particles may be much bigger than the molecules of interest, the scattering profile may contain signals derived from these particles. Passing the RNA samples through a 0.22-µm filter often alleviates this problem.

23.7.3
Problem 3: Scattering Results cannot be Fit to Simple Models

Before jumping to a complicated conclusion, it is often advisable to determine the fractional folding or activity of the RNA and proteins present in the sample by another method. Even though a single RNA species is present, it may exist in two or more conformational populations at significant fractions. This subject is particularly important for interpreting titration experiments where the concentration of one or more components is varied.

Another trivial explanation for a complicated result is the loss of integrity of the RNA or the proteins during the experiment. Ideally, samples after exposure to X-ray should be examined by other methods after the SAXS measurements to ensure that the majority of the RNA is still intact.

The relatively high concentrations of RNA used in SAXS measurements can reduce the concentration of free cations. This reduction can result in an apparent concentration shift in the divalent cation requirement in the studies of RNA folding.

23.8
Conclusions/Outlook

The advance of synchrotron technologies has given this old biophysical method new life. In addition to the RNase P work described here, SAXS has also been applied recently to RNA folding studies [5, 14–19]. The two advantages of the SAXS method are the determination of the shape and size change in real-time and the accommodation of SAXS data to structural models. We anticipate more broad applications of SAXS to the understanding of other, even more complex biological systems in the near future.

Acknowledgments

All SAXS experiments were carried out at the Argonne National Laboratory in collaboration with Dr P. Thiyagarajan. Dr Xing-wang Fang designed the sample holder and carried out all SAXS experiments described here. We thank the NIH and the US Department of Energy for their financial support.

References

1. Moore, P. B., *J. Appl. Crystallogr.* **1980**, *13*, 168–175.
2. Svergun, D., C. Barberato, M. H. Koch, *J. Appl. Crystallgr.* **1995**, *28*, 768–773.
3. Svergun, D. I., V. V. Volkov, M. B. Kozin, H. B. Stuhrmann, *Acta Crystallogr. A* **1996**, *52*, 419–426.
4. Glatter, O., O. Kratky, *Small Angle X-ray Scattering*. Academic Press, London, **1982**.
5. Fang, X., K. Littrell, X. Yang, S. J. Henderson, S. Siefert, P. Thiyagarajan, T. Pan, T. R. Sosnick, *Biochemistry* **2000**, *39*, 11107–11113.
6. Cantor, C., P. Schimmel, *Biophysical Chemistry: Part II*. Freeman, New York, **1980**.
7. Altman, S., L. Kirsebom, in *The RNA World*, 2nd edn, Gesteland, R. F., Cech, T. R., Atkins, J. F. (eds), Cold Spring Harbor Laboratory Press, Cold Spring Harbor, NY, **1999**, pp. 351–380.
8. Frank, D. N., N. R. Pace, *Annu. Rev. Biochem.* **1998**, *67*, 153–180.
9. Fang, X. W., X. J. Yang, K. Littrell, S. Niranjanakumari, P. Thiyagarajan, C. A. Fierke, T. R. Sosnick, T. Pan, *RNA* **2001**, *7*, 233–241.
10. Barrera, A., X. Fang, J. Jacob, E. Casey, P. Thiyagarajan, T. Pan, *Biochemistry* **2002**, *41*, 12986–12994.
11. Seifert, S., R. E. Winans, D. M. Tiede, P. Thiyagarajan, *J. Appl. Crystallogr.* **2000**, *33*, 782–784.
12. Massire, C., L. Jaeger, E. Westhof, *J. Mol. Biol.* **1998**, *279*, 773–793.
13. Kunst, F., N. Ogasawara, I. Moszer, A. M. Albertini, G. Alloni, V. Azevedo, M. G. Bertero, P. Bessieres, et al. *Nature* **1997**, *390*, 249–256.
14. Russell, R., I. S. Millett, S. Doniach, D. Herschlag, *Nat. Struct. Biol.* **2000**, *7*, 367–370.
15. Fang, X. W., B. L. Golden, K. Littrell, V. Shelton, P. Thiyagarajan, T. Pan, T. R. Sosnick, *Proc. Natl. Acad. Sci. USA* **2001**, *98*, 4355–4360.
16. Fang, X. W., P. Thiyagarajan, T. R. Sosnick, T. Pan, *Proc. Natl. Acad. Sci. USA* **2002**, *99*, 8518–8523.
17. Russell, R., X. Zhuang, H. P. Babcock, I. S. Millett, S. Doniach, S. Chu, D. Herschlag, *Proc. Natl. Acad. Sci. USA* **2002**, *99*, 155–160.
18. Russell, R., I. S. Millett, M. W. Tate, L. W. Kwok, B. Nakatani, S. M. Gruner, S. G. Mochrie, V. Pande, S. Doniach, D. Herschlag, L. Pollack, *Proc. Natl. Acad. Sci. USA* **2002**, *99*, 4266–4271.
19. Das, R., L. W. Kwok, I. S. Millett, Y. Bai, T. T. Mills, J. Jacob, G. S. Maskel, S. Seifert, S. G. J. Mochrie, P. Thiyagarajan, S. Doniach, L. Pollack, D. Herschlag, *J. Mol. Biol.* **2003**, *332*, 311–319.

24
Temperature-Gradient Gel Electrophoresis of RNA

Detlev Riesner and Gerhard Steger

24.1
Introduction

As expected from the name of the method, temperature-gradient gel electrophoresis (TGGE) is a combination of a thermodynamic method and an analytical separation technique. The electrophoretic mobility of one or several RNA molecules is analyzed in dependence upon the temperature. Quite different parameters of an RNA affect this temperature-dependent mobility, i.e. size, sequence, secondary and tertiary structure, structural stability, hydrodynamic flexibility, and electrical properties like counterion condensation. Since different electrophoretic mobilities result in well-separated bands in a gel, RNA molecules of different sizes and/or with differences in the other parameters mentioned above can be analyzed in one and the same experiment. Although this feature might appear trivial, one should keep in mind that in other well-established physical methods like spectroscopy, in particular optical melting curves or most hydrodynamic methods, a superimposition of the parameters of different molecules or different conformations is always measured, and can hardly be deconvoluted into the individual parameters.

TGGE can be applied to a wide variety of nucleic acids, DNA and RNA, single-stranded and double-stranded, from oligonucleotides to the size limit of polyacrylamide gel electrophoretic resolution, i.e. a few thousand bases; the most relevant range is between 100 and 1000 bases. Also, different staining protocols can be used for detection of the nucleic acid; the most common is silver staining, but very specific methods like hybridization of a particular sequence in a crude nucleic acid preparation may also be used. The present chapter is restricted to RNA analysis, predominantly single-stranded RNA, and a few more specialized examples of RNA–RNA complexes and RNA–protein complexes. For a detailed description of other examples and application of TGGE to DNA and protein analysis, the reader may refer to several chapters in textbooks [1–3].

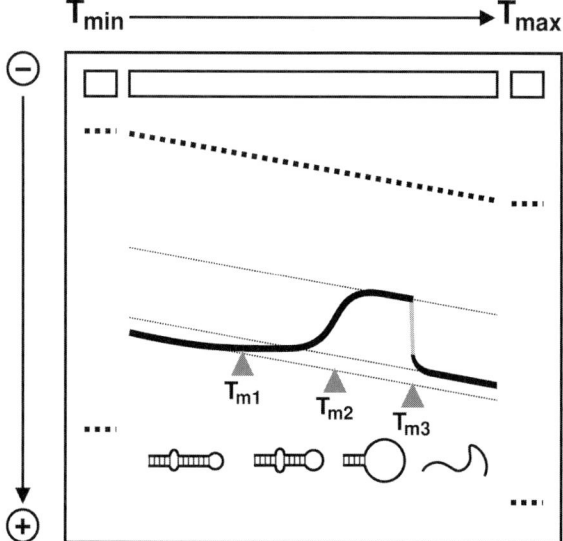

Fig. 24.1. Principle of TGGE. A linear temperature gradient is applied perpendicular to the electric field. The sample is applied to the long slot; small marker slots are on the left and right side of the gel. The mobility of an RNA with secondary structure is slightly decreased after a cooperative, reversible transition at T_{m1}, it is drastically decreased after a further transition at T_{m2} and it is increased after the irreversible transition at T_{m3} from a partially denatured molecule to a state without any base pairs. The secondary structure drawings below the transition curve symbolize the major structures before and after the different transitions. The dashed lines show the migration behavior of double-stranded nucleic acids that do not undergo any transition in the chosen temperature range; their increase in mobility is due to a decrease in viscosity with increasing temperature.

24.2 Method

24.2.1 Principle

As shown in Fig. 24.1, a nucleic acid sample is applied to a slab polyacrylamide gel in a broad slot that extends over nearly the whole width of the gel at the side of the negative electrode. A linear temperature gradient is established perpendicular to the direction of the electric field. The molecules at the left side of the gel migrate at low temperatures, the molecules at the right side at high temperatures and those at positions in between at corresponding intermediate temperatures. Each individual molecule, however, migrates during the whole electrophoretic run at constant temperature. In the example of Fig. 24.1, an RNA is analyzed which undergoes three cooperative conformational transitions: one from a native, highly base-paired state to an altered conformation, a second to a partially denatured state and a third transition to the totally denatured state. The transitions occur at well-defined denaturation temperatures $T_{m1} < T_{m2} < T_{m3}$. The molecules at $T > T_{m1}$ migrate slower than those at $T < T_{m1}$; the molecules at $T > T_{m2}$ migrate much slower than those at $T < T_{m2}$. In the narrow temperature range of the first transition the molecules switch between two states reversibly and assume a mobility averaged according to the degree of transition. The same holds for the second transition. Therefore, after staining the nucleic acid the band represents a transition curve. The third transition is an example for an irreversible step, i.e. there is no thermodynamic equilibrium between the paired structure below T_{m3} with the completely denatured state above T_{m3} under the low salt condition of the gel electrophoresis. The RNA denatures at T_{m3}, but renaturation is not possible, which leads to the jump in migration behavior without a continuous band in between the two states.

In addition to the gradient perpendicular to the electric field, one in parallel to the electric field can also be applied. In that case, the samples are applied to narrow slots and run from low to high temperature. This mode of TGGE is applied mostly to mutant analysis, when heteroduplices of wild-type and mutant sequences form mismatches, and are retarded due to partial melting at a lower temperature as compared to homoduplices [1, 2].

24.2.2
Instruments

The original instruments were home-made instruments, in which the gradient was established by thermostating the two edges of a thermostating block by liquids from two thermostats [4]. Although these instruments worked quite well and were available for some years commercially, they are no longer available. Therefore, the presently available system from Biometra (Göttingen, Germany) will be described.

The Biometra TGGE system is a well-constructed commercial instrument, which is developed for routine use as well as for research use. The microprocessor-driven gradient block based on Peltier elements allows well-defined temperature gradients with good resolution and reproducibility. The Biometra TGGE system is available in two formats. The standard TGGE "mini" system operates small gels and is therefore ideally suited for fast, "first-slot" experiments. The TGGE "maxi" system provides a large separation distance, allows high electrophoretic resolution and is well suited for systematic TGGE analysis of RNA. High parallel sample throughput can be achieved for DNA mutation analysis.

24.2.3
Handling

Handling and protocols are intimately connected with the TGGE instruments. Thus, we refer to the detailed manual (Version 3.02 TGGE MAXI System) from Biometra. The single steps are:

(1) Casting of gels (assembly of the gel cuvette, preparing gel and buffer solutions, pouring gels).

(2) Electrophoresis (electrophoresis conditions, sample loading, setup electrophoresis unit, gel electrophoresis run).
(3) Running conditions (voltage, temperature gradient, electrophoresis time).
(4) Staining.

24.3
Optimization of Experimental Conditions

TGGE relies on the fact that the electrophoretic mobility of a biopolymer is altered due to a conformational change. The extent of the alteration critically depends on several parameters. For a systematic study, it is essential to optimize these parameters first in order to obtain a large alteration. Furthermore, reversibility of a transition may be achieved only by choosing appropriate conditions. Therefore, the variables outlined below should be optimized, which has to be done primarily according to empirical rules.

24.3.1
Attribution of Secondary Structures to Transition Curves in TGGE

In order to attribute secondary structures to transition curves from TGGE, thermodynamic features as well as gel-electrophoretic mobilities have to be taken into account. Whereas the transition temperatures may be calculated quite accurately, the interpretation of gel-electrophoretic mobilities has to rely more on qualitative arguments.

(1) Branched structures migrate slower than extended structures. This effect is known from the denaturation of double-stranded nucleic acid, which leads to drastic retardation as long as the denaturation is incomplete [5]. The effect has also been described with dimeric transcripts of potato spindle tuber viroid (PSTVd) RNA [6].
(2) Structures with large loops migrate extremely slowly. The low mobility of denatured circular viroids and the lower mobility of plasmids in the form of relaxed circles as compared to supercoils are examples of this tendency.
(3) Because of their higher molecular weight and their usually high degree of bifurcations, bimolecular complexes migrate much slower than the corresponding uncomplexed molecules.
(4) Most RNA transitions appear as smooth curves; this is based on the reversible nature of the transitions. Irreversible transitions are recognized by their stepwise behavior; irreversibility is mostly due to a low chance of renaturation in low ionic strengths conditions of gel electrophoresis.

Attempts have been made to derive quantitative relationships from the general rules (see [7]; Mundt and Steger, unpublished). The mobility of partly denatured double-stranded nucleic acid has been calculated with fair success [8].

24.3.2
Pore Size of the Gel Matrix

The pore size of the gel matrix can be varied by the concentration of polyacrylamide and the ratio acrylamide-N,N'-methylenebisacrylamide (Bis). A compromise has to be found between a large change in electrophoretic mobility and an acceptable migration velocity. For example, a concentration of 6% polyacrylamide might be too low because of only small changes, whereas an optimum pore size might be reached at 8% acrylamide. The polyacrylamide concentration may be varied between 4% for larger RNA and up to 20% for short oligonucleotides, and the ratio of acrylamide:bisacrylamide between 20:1 and 40:1, respectively.

24.3.3
Electric Field

The electric field tends to stretch the molecules. These effects are very sensitive to the charge distribution, the conformation and the flexibility of the molecule. The mobility changes of two conformational transitions in one and the same molecule can exhibit different dependencies upon the electric field. Evidently, such a dependence cannot be predicted. An excessive electric field, although inducing large changes, may be disadvantageous because of the large electric current.

24.3.4
Ionic Strength and Urea

A variation of the ionic strength is always connected with a variation of T_m of the conformational transition. Therefore, this parameter cannot be varied independently from other features of TGGE. For nucleic acids one may keep in mind as a general rule that changes in electrophoretic mobility are larger in low ionic strength, which means, however, also at lower temperature. The effect was particularly large with circular RNA such as viroids (*cf.* Fig. 24.2), where the change in mobility was reduced to 40% if the ionic strength was raised from 8.9 to 89 mM Tris–borate. With nucleic acids, high ionic strength always improves the reversibility of a transition, and reversible transitions may be evaluated more easily and more accurately than irreversible or discontinuous transitions. Some examples shown below will prove this. The increase in the transition temperature due to high ionic strength may be compensated by addition of urea. Thus, addition of 5–10 mM NaCl and 4–6 M urea to the standard buffer (8.9 mM Tris, 8.9 mM boric acid) was found advantageous.

24.4
Examples

We will describe four different examples that should show the potency of the TGGE method. All examples were produced with an apparatus which was commercially available from Qiagen (Hilden, Germany) up to about 1995 and which was very similar to the home-made instrument mentioned above [4]; the gel length in this apparatus was 190 mm. However, after adaptation of the applied voltage to the length of a different gel system, the examples should be reproducible on any TGGE apparatus.

Most of the following examples were obtained with PSTVd RNA (for reviews, see [9, 10]). PSTVd is a covalently closed RNA molecule of 359 nt, has no protein-coding capacity, but is able to infect certain plants. It is replicated by DNA-dependent RNA polymerase of the host plant in an asymmetric rolling circle; multimeric (+)-stranded intermediates are processed by plant RNases/ligases to the mature circular molecule.

24.4.1
Analysis of Different RNA Molecules in a Single TGGE

Optical melting curves of a nucleic acid are based on the structure-dependent extinction coefficients; consequently, any impurity of the sample as well as different conformations of the sample at the same temperature add up to the signal and end in a non-interpretable transition curve. In TGGE, molecules are separated according to their hydrodynamic shape, as in any standard gel-electrophoretic method, i.e. nucleic acid molecules of different length are separated according to their size and molecules of identical length are separated according to their different conformation. Thus, sample impurities like abortive transcripts from T7 transcription or DNA templates are not prohibitive for analysis of denaturations in TGGE. For visualization, all gel-staining methods might be applied; of course, silver staining detects all nucleic acids, whereas after blotting, specific RNAs can be detected by hybridization or double-stranded RNAs by double-strand-specific immunostaining.

An example with silver staining is shown in Fig. 24.2. A crude RNA extract, consisting of at least circular and linear PSTVd (359 nt), 7S RNAs (about 305 nt), 5S rRNA (120 nt), and tRNAs, is separated on a TGGE, and the RNA bands are stained by silver. At the lowest temperature of the gel, circular (cPSTVd) as well as linear PSTVd (lPSTVd) have a rod-shaped secondary structure, which leads to a relatively high mobility. At a certain temperature, about 57 °C under the gel-electrophoretic conditions (see arrow in Fig. 24.2), the native structure denatures completely and three extra-stable hairpins are formed; the nucleotide regions forming these hairpins are given in bold characters in the schematic drawing of PSTVd's structure (see Fig. 24.2). The main transition is highly cooperative, leading to the sharp change in mobility at this temperature – the mobility is decreasing because the branched hairpin-containing structure is much bulkier than the rod-like native structure. At higher temperature the most stable hairpin II (HPII) denatures (see arrow in Fig. 24.2); because less base pairs are involved, this transition takes place over a broader temperature range. The completely denatured structure is an expanded circle, which is strongly retarded. Linear PSTVd molecules, which are either replication intermediates or are created by RNases during preparation of the RNA extract, migrate proportional to their length after full denaturation,

404 | 24 Temperature-Gradient Gel Electrophoresis of RNA

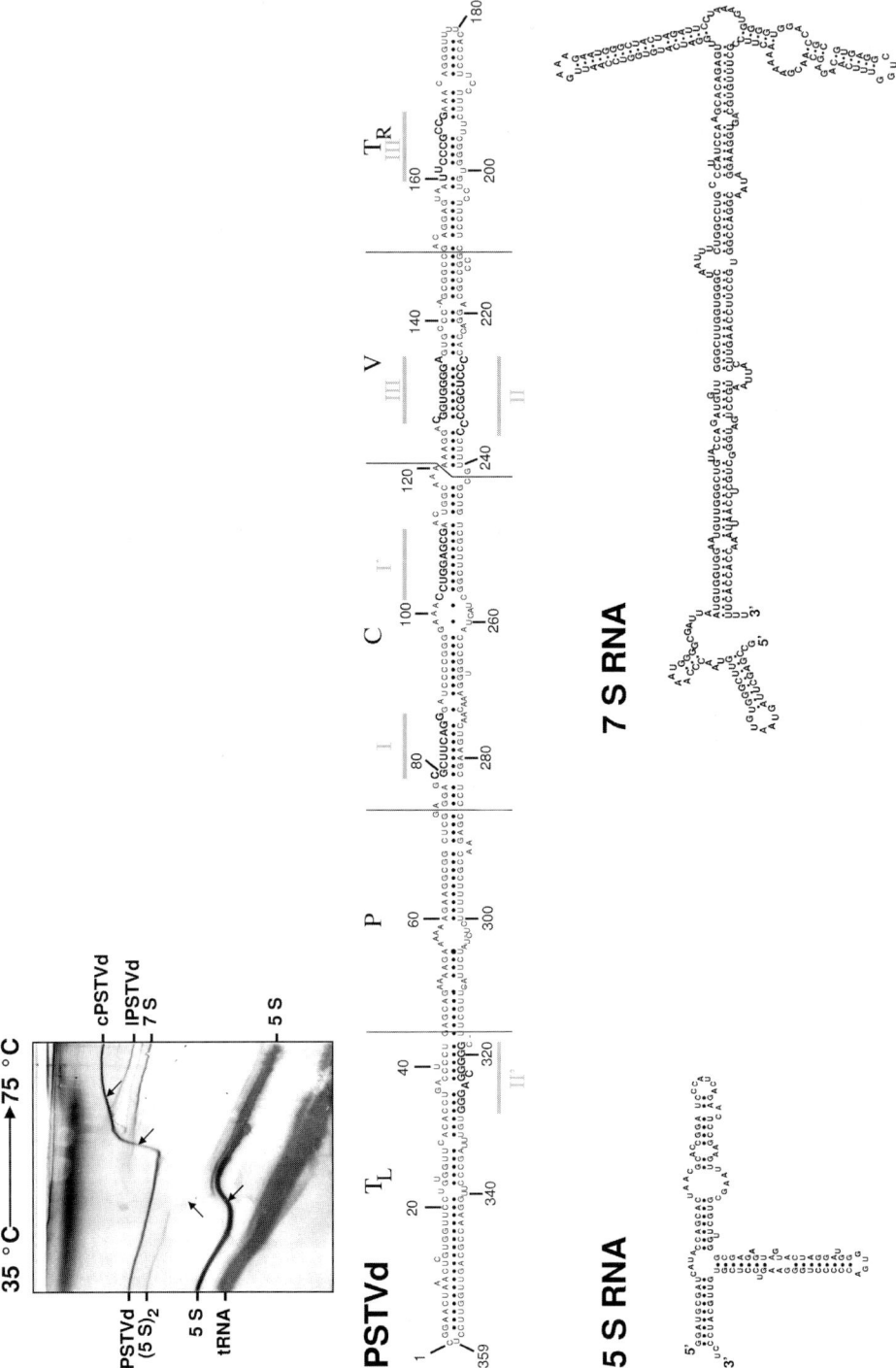

Fig. 24.2

i.e. lPSTVd migrates faster than cPSTVd after full denaturation. The temperature of the main transition of lPSTVd depends on the point of linearization; due to the low concentration of individual lPSTVd molecules, these transitions are not visible on the gel.

Similarly, at low temperature, a band for the 7S RNAs is not visible; after full denaturation, all 7S RNAs co-migrate. The 5S rRNA shows a single transition (see arrow). At the denaturation temperature of 5S RNA, the dimeric complex of 5S RNA also dissociates. At high temperature, the bands derived from monomeric and dimeric 5S RNA do not co-migrate; the small migration difference corresponds to the migration distance of dimeric and monomeric 5S RNA during the pre-electrophoresis step at low temperature before the temperature-gradient was applied.

Note that all bands show an increase in mobility with increasing temperature. This general effect is not based on structural rearrangements, but on the decreasing viscosity with increasing temperature.

24.4.2
Analysis of Structure Distributions of a Single RNA – Detection of Specific Structures by Oligonucleotide Labeling

The bases for structure formation of nucleic acids are hydrogen bonds between bases (Watson–Crick and wobble base pairs) and stacking interactions between neighboring bases and base pairs, which are thermodynamically favorable, i.e. structure formation is a chemical reaction and many different structures are in a thermodynamic equilibrium. Whereas only highly specialized sequences favor a dominant structure, most sequences exhibit a structure distribution. In the case of "RNA switches", an RNA molecule might even be evolutionarily optimized to exhibit more than a single structure. A further possibility for structure distributions is not based on thermodynamics, but on the kinetics of folding, and is called "sequential folding": in most cases the formation of RNA structural elements is faster than the rate of synthesis; therefore, the partial, still elongating RNA chain already

◀

Fig. 24.2. Analysis of a crude RNA extract from tomato plants infected by PSTVd. Conditions of electrophoresis: 0.2 × TAE (8 mM Tris–HCl, 20 mM NaOAc, 0.2 mM EDTA, pH 8.4), 5% (w/v) acrylamide, 0.125% (w/v) bisacrylamide (acrylamide/bisacrylamide 40:1), 0.1% (v/v) N,N,N',N'-tetra-methylethylenediamine (TEMED) and 0.04% (w/v) ammonium peroxodisulfate for starting the polymerization; 10 min pre-electrophoresis at 25 °C, 10 min equilibration for the 35–75 °C temperature gradient, 300 V for 90 min; gel size: 180 × 190 × 0.9 mm; slot size: 130 × 4 mm; silver staining. cPSTVd and lPSTVd: circular and linear forms of PSTVd, respectively; 5 S: 5S RNA; (5 S)$_2$: dimeric complex of 5S RNA; 7 S: 7S RNA. Transitions described in the text are marked by arrows. In the schematic drawing of the PSTVd structure, the following regions are marked: T_L and T_R, terminal left and terminal right region, respectively (cf. Fig. 24.4); P, pathogenicity-related region; C, central conserved region; V, variable region; grey lines marked I/I' and II/II' form extra-stable hairpins I and II, respectively, at temperatures above the main transition (cf. Fig. 24.3).

folds during synthesis into structures which might be thermodynamically suboptimal for the full-length molecule. After finishing the synthesis, the structures generated during synthesis can rearrange if sufficient activation energy is available. Such structure distributions or co-existing structures, either based on thermodynamics or on kinetics, are one of the many problems of chemical and enzymatic mapping for structure determination [11]. TGGE allows for separation of structure distributions, at least when the different structures do not interconvert fast or co-migrate.

Figure 24.3 shows the analysis of sequential folding during synthesis of a (−)-stranded PSTVd transcript [11, 12]. Under thermodynamic equilibrium conditions, which can be established by complete denaturation and slow renaturation under low salt conditions to avoid formation of bimolecular complexes, the transcript forms a single dominant structure (see Fig. 24.3). The native conformation is rod-

Fig. 24.3. Detection of specific structures of a synthetic PSTVd transcript (linearized at a *Sty*I site, nt 337) by oligonucleotide labeling. (A) TGGE of the transcript under equilibrium conditions. The transcript in 0.2 × TBE (17.8 mM Tris, 17.8 mM borate, pH 8.4, 2.5 mM EDTA) was denatured at 70 °C for 15 min and slowly renatured to room temperature at about 0.1 °C/min. Conditions of electrophoresis: 0.2 × TBE, 4% (w/v) acrylamide, 0.13% (w/v) bisacrylamide (acrylamide/bisacrylamide 30:1), 0.1% (v/v) TEMED and 0.04% (w/v) ammonium peroxodisulfate for starting the polymerization; 15 min pre-electrophoresis at 15 °C with 30 V/cm, 10 min equilibration for the 20–55 °C temperature gradient, 30 V/cm for 120 min; gel size: 180 × 190 × 0.2 mm; slot size: 130 × 4 mm; gel was stained with silver. (B and C) TGGE analysis of the transcript after different times of transcription and incubation. T7 transcription assays with cNTP = 1 mM at 25 °C, corresponding to an elongation rate of about 130 nt/s, were stopped after 1 h (B) or after 30 s (C). Conditions of electrophoresis as in (A); detection of transcripts by staining with NBT/BCIP (B) or by autoradiography (C). In (B), the uppermost band is the DNA template. For further details, see [12]. (D) After transcription, (−)-stranded PSTVd exists in different structural conformations due to sequential folding; one of these conformations, which is metastable, contains a long G:C-rich hairpin (HPII) that is thought to be critical for transcription to (+)-strands *in vivo*. In the native conformation the two halves of the helix-stem of HPII are involved in base pairings at two distant positions. The oligonucleotide 27AB (5′-CUUACUUGCUUCCUUUGCGCUGUCGCU-3′), complementary to PSTVd (−)-sequence 318–307/237–251, is designed to hybridize with its full length to the HPII loop of the HPII-containing conformation, whereas binding to the native conformation is possible only with a part of its sequence. (E) Complexes were formed by incubating 200 ng of the *in vitro* transcript for 20 min in buffer (500 mM NaCl, 4 M urea, 1 mM sodium cacodylate, 0.1 mM EDTA, pH 7.0) with 105 c.p.m. oligonucleotide 27AB. After subsequent dialysis against 0.2 × TBE buffer, TGGE analysis was performed. Conditions of electrophoresis: 0.2 × TBE, 5% (w/v) acrylamide, 0.17% (w/v) bisacrylamide (acrylamide/bisacrylamide 30:1), 0.1% (v/v) TEMED and 0.04% (w/v) ammonium peroxodisulfate for starting the polymerization; 15 min pre-electrophoresis at 15 °C with 30 V/cm, 10 min equilibration for the 20–55 °C temperature gradient, 30 V/cm for 90 min; gel size: 180 × 190 × 0.2 mm; slot size: 130 × 4 mm; gel was stained with silver and exposed to X-ray film (Kodak Xomat AR) for detection of the radioactive transcript. On the silver-stained gel (left) several bands are detectable (S, M, Q/P, R), which represent different conformations of the same transcript; conformation S behaves identical to the transcript in the native conformation. On the autoradiograph of the silver-stained gel (right) only the RNA species Q/P and R are visible due to complexing with RNA oligonucleotide 27AB. For further details, see [11].

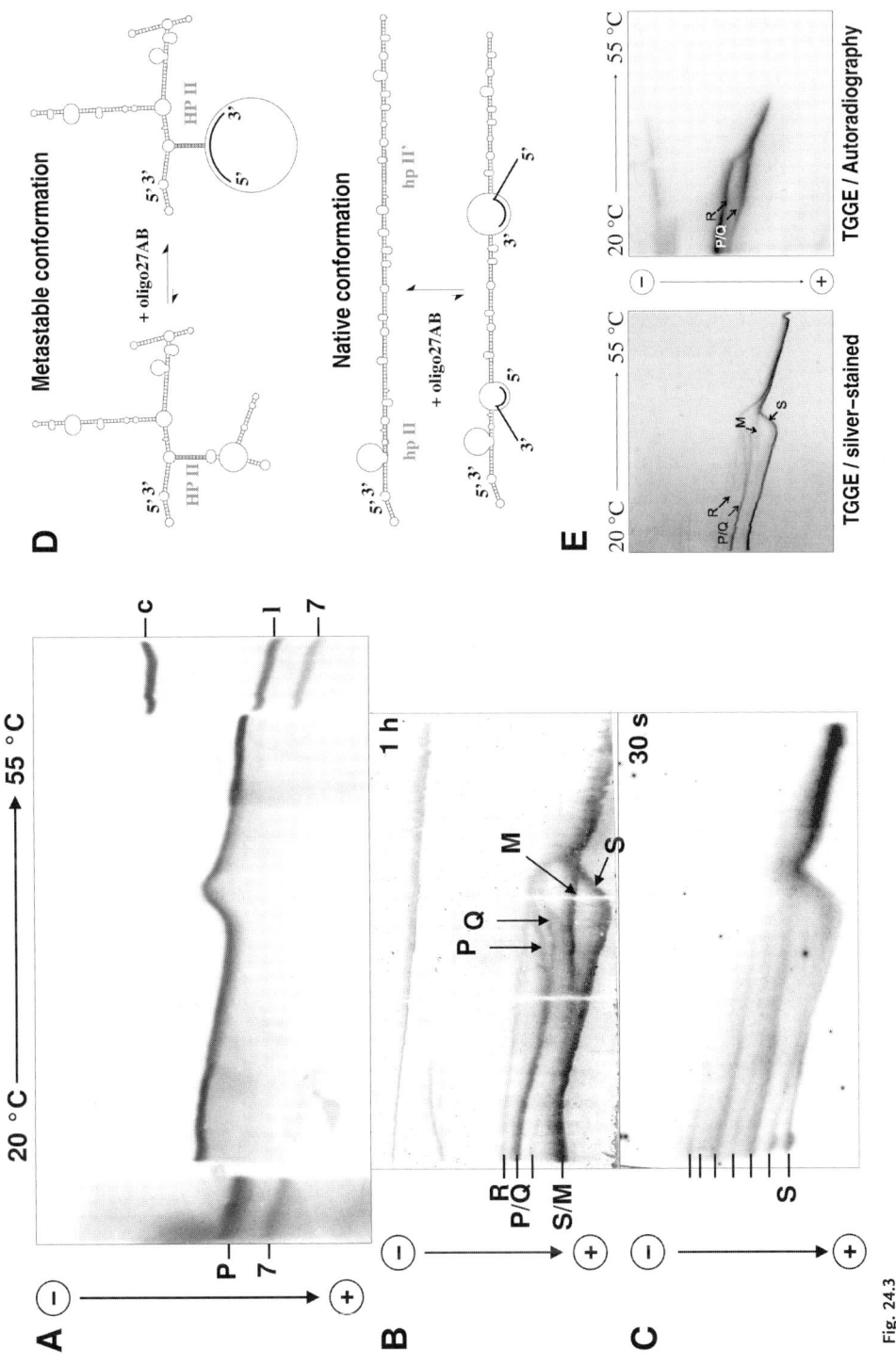

Fig. 24.3

like (see Fig. 24.3D, bottom), and exhibits a single main transition in which the rod-like structure denatures and a particularly stable hairpin HPII is formed. The structure containing HPII is bulky; denaturation of additional helices and finally HPII leads to an increase in mobility at high temperature.

Transcripts from *in vitro* T7 polymerization can also exhibit several different structures (see Fig. 24.3B and C). The band of highest mobility, marked by "S", resembles the band from thermodynamic equilibrium conditions as shown in Fig. 24.3A. If analyzed directly after 30 s of transcription (*cf.* Fig. 24.3C), however, the thermodynamic equilibrium is not established and the rod-like structure is only a minor one. The other bands show lower mobility and therefore represent bulkier structures; they contain many thermodynamically instable helices and the slower bands from the more bulky structures denature at lower temperatures irreversibly. After slower synthesis and incubation under the high salt conditions of synthesis, most structures are of higher mobility and stability in comparison to those after a fast synthesis (*cf.* Fig. 24.3B with C). Note that a single band in TGGE does not ensure a single structure; the bands marked "P/Q" and "S/M" co-migrate at low temperature, but separate prior to the temperature of the main transition.

The migration behavior of certain bands, i.e. steepness of transitions, acceleration or retardation transitions, temperature of transitions, might be correlated with the results from structural calculations. The programs mfold [13] or rnafold [14], which are based on thermodynamic equilibrium calculations, or programs, which take into account kinetics and sequential folding [15, 16], were applied; in both cases the ionic strength difference in gel electrophoresis and calculations had been corrected for. A more direct, experimental confirmation of a structure model is possible, e.g. by "oligonucleotide mapping" – an oligonucleotide designed to hybridize only to a certain structural element is used to mark bands containing that structural element. An example is given in Fig. 24.3(D and E). Those oligonucleotides were used to detect structures of (−)-stranded PSTV transcripts which contain HPII; it should be noted that HPII is thought to be critical for synthesis of (+)-stranded replication intermediates *in vivo*. An oligonucleotide was designed that pairs to both regions neighboring the HPII helix (see Fig. 24.3D, top). With structures containing HPII, the oligonucleotide pairs over its full length; whereas with structures without HPII, e.g. a rod-like structure (see Fig. 24.3D, bottom), the oligonucleotide is able to pair only with half its length. Furthermore, the length of the oligonucleotide was carefully chosen not to shift the concentration ratio between existing structures towards HPII-containing structures. In Fig. 24.3(E), a hybridization analysis of the structure distribution of the transcript (as in Fig. 24.3B) with the radioactively labeled oligonucleotide is shown. By comparison of the silver-stained gel and its autoradiograph, it is obvious that only structures in bands "P/Q" and "R" are marked by the oligonucleotide; these bands are visible in the autoradiograph up to a temperature at which the oligonucleotide dissociates. These bands, however, are not visible in the silver-stained gel due to the low concentration of the oligonucleotide–transcript complex; the assignment is made on the nearly identical migration behavior of the respective complexes from the autoradiograph and the uncomplexed bands after silver staining. Note that the migration behavior of a complex may not be identical to the uncomplexed structure; the molecular weight of the complex is increased and the hydrodynamic shape might be altered.

24.4.3
Analysis of Mutants

Single nucleotide mutations might alter the structure and the structural stability of an RNA [18]. Visualization of those alterations in TGGE can be done either by analysis of a mixture of a mutant and the wild-type RNA or by analysis of the different RNAs after loading them sequentially to the gel. The first method allows for detection of even subtle differences in hydrodynamic migration and/or thermodynamic stability, whereas the second makes it easier to identify the individual species.

An analysis with sequential loading of three slightly different RNAs is shown in Fig. 24.4. The terminal left region (T_L) of PSTVd contains two repeats (marked in light and dark grey, respectively, in Fig. 24.4, right), which are partially complementary to each other. This allows for two different structural arrangements of the T_L region: a rod-like conformation, which is by far the dominant conformation according to calculations, and a branched conformation. For an easy comparison of the different structures, the wild-type "native" structure was modified by 2-bp changes that favor drastically either the rod-like or the branched structure, respectively. In the "rod" mutant, an A_{344}:U_{18} pair is changed to a G:C pair; as seen from Fig. 24.4, this change stabilizes the rod conformation and destabilizes the branched conformation. In the "branched" mutant, an A_5:U_{18} pair is changed to a G:C pair which stabilizes the branched conformation and destabilizes the rod-like conformation. These "designer mutants" could be analyzed experimentally in the silver-stained TGGE (Fig. 24.4, left), as well as in optical melting curves and NMR [17]. The stability of the "rod" structure is increased in comparison to the "native" structure; note the increase in melting temperature and the slight decrease in half-width of the transition, which is based on the increase in $\Delta H°$. In contrast, the "branched" structure shows a much broader transition at relatively low temperature, which might be due to opening of one of the branch helices, overlaid with a second transition at higher temperature (see open arrowhead in Fig. 24.4, left). From all experiments – TGGE, optical melting curves and NMR – it was concluded that the conformations of "native" and "rod" coincide, so that the native conformation of PSTVd is indeed rod-like.

24.4.4
Retardation Gel Electrophoresis in a Temperature Gradient for Detection of Protein–RNA Complexes

A relatively simple analysis of RNA–protein interaction is possible via gel-retardation experiments. In contrast to DNA–protein interaction, RNA–protein interaction is quite often not sequence dependent, but a function of the RNA struc-

24 Temperature-Gradient Gel Electrophoresis of RNA

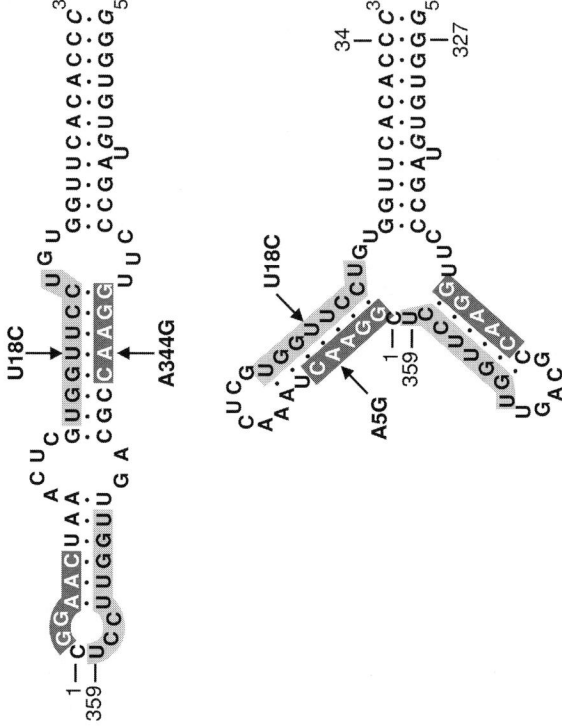
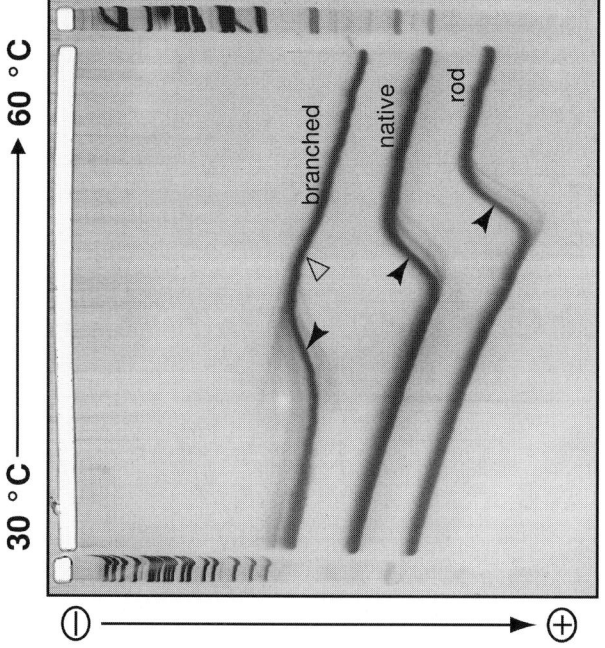

Fig. 24.4

ture. Thus, a combination of gel-retardation for analysis of the interaction and of TGGE for the simultaneous analysis of RNA structure might be helpful.

A corresponding analysis [7] of the interaction of the 5′-untranslated region of spinach chloroplast *psbA* mRNA, which encodes the D1 protein of photosystem II, and a stromal protein extract is shown in Fig. 24.5. To establish complex formation, samples were incubated at 25 °C in the presence of heparin and tRNA to avoid unspecific binding. In a gel-shift assay at 25 °C with 0.5 mg/ml protein extract, 50% of the radioactively labeled RNA (0.5 pmol/ml) is retarded and in the presence of 1 mg/ml protein extract, the complete RNA is shifted into the complexed form (see Fig. 24.5A). Because the fraction of binding proteins in the extract is not known, no binding constant could be estimated. The percentages of bound RNA, however, are drastically dependent on the temperature of incubation and electrophoresis, as shown in Fig. 24.5(B): incubation and analysis of 1 mg/ml protein extract with 0.5 pmol RNA under the same conditions as above, but at 10 °C did not result in any RNA retardation. To analyze the temperature dependence of complex formation in more detail, 67 fmol RNA was incubated with 260 µg protein in 400 µl directly in the long slot of a gel in the presence of a temperature gradient from 10 to 40 °C, then separated by electrophoresis and the RNA visualized by autoradiography (see Fig. 24.5C). The gel shows a distinct temperature range in which retardation of the RNA and thereby complex formation with proteins can be observed. Retardation is observed between about 18 °C up to about 35 °C with a maximum at 22–25 °C. With additional experiments it could be verified that neither the pH gradient, produced by the temperature gradient, nor a

Fig. 24.4. Analysis of transcripts with a sequence derived from the T_L domain of PSTVd. The two possible structural conformations of the "native" sequence are shown on the right; numbering is according to circular PSTVd; the terminal helix was stabilized by addition of a terminal G:C base pair and a mutation $U_{332}G$ in comparison to PSTVd's T_L domain (*cf.* Fig. 24.2). In the mutant transcript "rod", the rod-like conformation was stabilized by the two mutations $U_{18}C$ and $A_{344}G$; in the mutant transcript "branched", the branched conformation was stabilized by the two mutations A_5G and $U_{18}C$. Samples were loaded sequentially onto the gel in the order of elongated rod, native and branched oligomers. The very weak additional bands observed during the transitions are due to $n + 1$ transcripts that could not be removed completely during the purification procedure. The black arrowheads denote the main melting transitions; the open arrowhead denotes an additional transition in the "branched" mutant. Conditions of electrophoresis: 0.2 × (v/v) TBE (17.8 mM Tris, 17.8 mM boric acid, 0.4 mM EDTA, pH 7.5), 5% (w/v) polyacrylamide, 0.17% (w/v) bisacrylamide (acrylamide/bisacrylamide 30:1), 0.1% (v/v) TEMED and 0.04% (w/v) ammonium peroxodisulfate for starting the polymerization. The RNA samples (300 ng) were applied to the broad sample slot (130 × 4 mm), while the small slots (5 × 4 mm) at both sides were used for appropriate size marker DNA (pBR322 digested with *Msp*I). Upon applying 500 V at a uniform temperature of 20 °C for 25 min, the RNA sample migrates a few millimeters into the gel matrix. This step was repeated for the loading of the second and third RNA samples, except the third sample electrophoresis period was only 10 min. Size marker DNA was loaded at the same point as the second RNA loading. Electrophoresis was paused for 15 min for the equilibration of the 30–60 °C gradient and the electrophoresis continued for 75 min at 500 V. The gel was stained by silver. For further details, see [17].

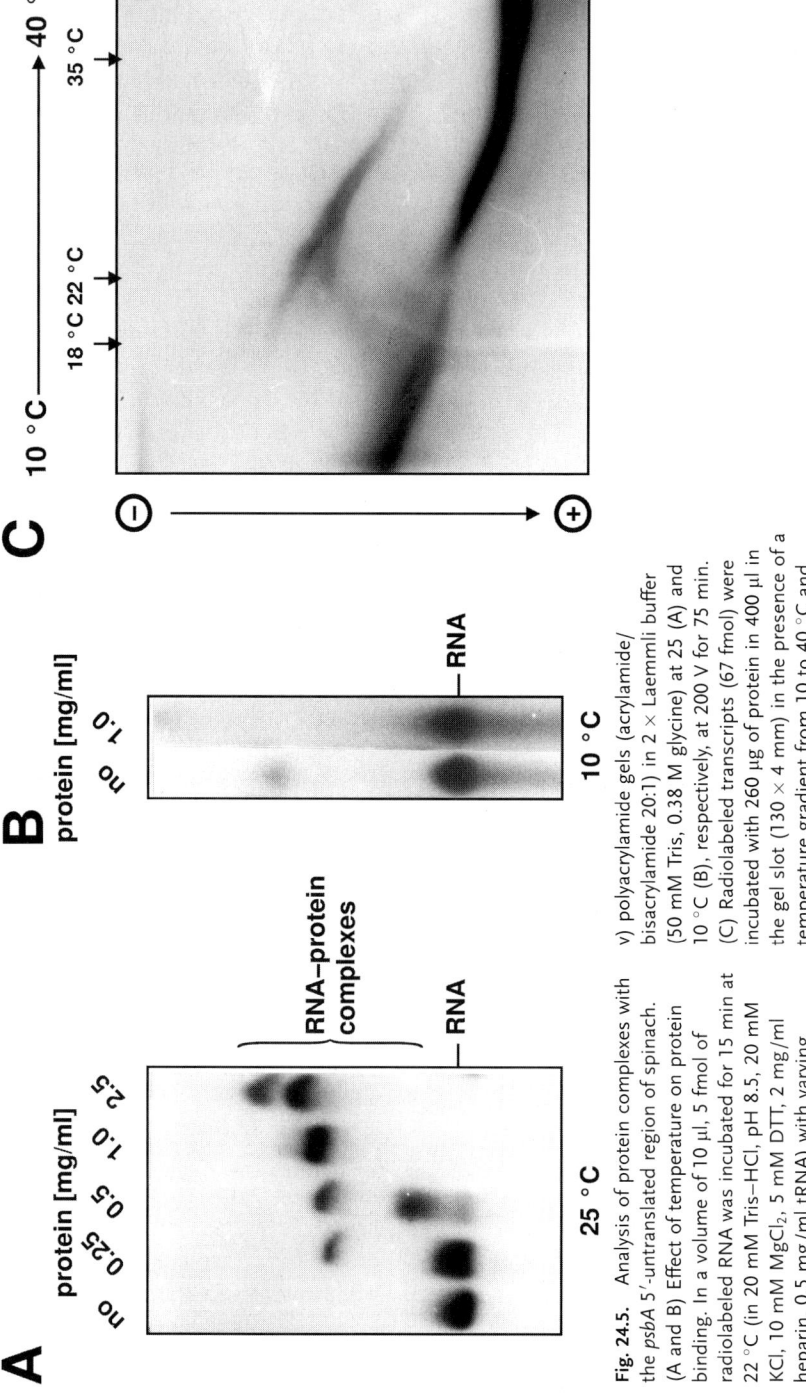

Fig. 24.5. Analysis of protein complexes with the *psbA* 5′-untranslated region of spinach. (A and B) Effect of temperature on protein binding. In a volume of 10 µl, 5 fmol of radiolabeled RNA was incubated for 15 min at 22 °C (in 20 mM Tris–HCl, pH 8.5, 20 mM KCl, 10 mM MgCl$_2$, 5 mM DTT, 2 mg/ml heparin, 0.5 mg/ml tRNA) with varying amounts of stromal protein as indicated. Free and complexed RNA was separated on 8% (w/v) polyacrylamide gels (acrylamide/bisacrylamide 20:1) in 2 × Laemmli buffer (50 mM Tris, 0.38 M glycine) at 25 (A) and 10 °C (B), respectively, at 200 V for 75 min. (C) Radiolabeled transcripts (67 fmol) were incubated with 260 µg of protein in 400 µl in the gel slot (130 × 4 mm) in the presence of a temperature gradient from 10 to 40 °C and then separated on a gel (other conditions as in A and B). For further details, see [7].

protein conformational change (aggregation, precipitation) in the low temperature range, but RNA conformational changes near 22 and 33 °C form basis for the temperature-dependent complex formation. One has to conclude that only the RNA conformation dominant between 22 and 33 °C is responsible for protein binding, whereas the structures dominant below 22 and above 33 °C, respectively, are not adapted for protein binding [7]. In the temperature range above 20 °C, binding of protein to the RNA decreases with increasing temperature; thus, the binding seems to be driven by enthalpy.

24.4.5
Outlook

In the examples we have concentrated on experiments with viroid RNA from our own work. TGGE is not limited, however, to the analysis of viroids, but has to be viewed more generally as a replacement or at least a supporting method for thermodynamic analysis of RNA structure by UV melting curves with the additional advantage to allow for analysis of RNA mixtures.

The group of Bevilacqua [19, 20] has used TGGE to select for and isolate thermodynamically more stable loop variants from combinatorial libraries of small RNA hairpins. They have mainly used a version of TGGE with the temperature gradient in parallel to the direction of RNA migration. This allowed for an easy excision of molecules with structures more stable than the bulk of molecules, because the already denatured molecules run slower than the still structured hairpins at a certain temperature.

TGGE may be used also for the analysis of tertiary structures. Take note, however, that the use of Mg^{2+} ions is necessary for stabilization of tertiary interactions in most cases, but this also catalyzes the degradation of RNA at elevated temperatures similar to basic conditions. The groups of R. Schroeder and E. Westhof monitored the tertiary structure transitions of the *td* intron of bacteriophage T4 and several mutants of this group I intron by TGGE [21]. With two mutant RNAs applied to one gel, similar to the experiment shown in Fig. 24.4, small stability differences of the mutants could be detected while simultaneously checking for deviating conformations and formation of intermolecular dimers.

Guo and Cech [22] searched for *Tetrahymena* ribozymes with an enhanced activity at elevated temperature by *in vitro* evolution. They selected for the thermodynamically most stable tertiary structure variants in the first step and for activity in the second step. In contrast to Bevilacqua's selection procedure [19, 20], they used temperature gradients perpendicular to the migration direction and excised small rectangular regions near to the unfolding transition from the fully folded state including tertiary interactions into a state with mainly secondary structure of the RNA. During eight rounds of selection, the temperature of the tertiary structure unfolding increased from 45 to 52 °C in a buffer containing about 0.4 mM free Mg^{2+} ions and in five subsequent rounds from 35 to 40 °C in about 0.1 mM free Mg^{2+} ions. Indeed the final variants contained up to 11 mutations, which mainly strengthened the active conformation through tertiary interactions and had an increase of the maximum active temperature by 10 °C in comparison to the starting variant.

The last examples demonstrate that TGGE is a favorable method to test for mutations and altered conformation and/or stability in the same analysis. We expect more of these applications in the future.

References

1 D. Riesner, K. Henco, G. Steger, in: *Advances in Electrophoresis 4*, A. Chrambach, M. J. Dunn, B. J. Radola (eds), VCH, Weinheim, **1991**.
2 K. Henco, H. Harders, U. Wiese, D. Riesner, in: *Methods in Molecular Biology*, J. M. Walker (ed.), Humana Press, Clifton, NJ, **1994**.
3 D. Riesner, in: *Antisense Technology: A Practical Approach*, C. Lichtenstein, W. Nellen (eds), Oxford University Press, Oxford, **1997**.
4 V. Rosenbaum, D. Riesner, *Biophys. Chem.* **1987**, *26*, 235–246.
5 G. Steger, P. Tien, J. Kaper, D. Riesner, *Nucleic Acids Res.* **1987**, *15*, 5085–5103.
6 R. Hecker, W. Zhi-min, G. Steger, D. Riesner, *Gene* **1988**, *72*, 59–74.
7 P. Klaff, S. Mundt, G. Steger, *RNA* **1997**, *3*, 1480–1485.
8 L. S. Lerman, S. G. Fischer, I. Hurley, K. Silverstein, N. Lumelsky, *Annu. Rev. Biophys. Bioeng.* **1984**, *13*, 399–423.
9 M. Tabler, M. Tsagris, *Trends Plant Sci.* **2004**, *9*, 339–348.
10 A. Hadidi, R. Flores, J. W. Randles, J. S. Semancik (eds), *Viroids*. CSIRO, Melbourne, **2003**.
11 A. R. W. Schröder, D. Riesner, *Nucleic Acids Res.* **2002**, *30*, 3349–3359.
12 D. Repsilber, U. Wiese, M. Rachen, A. R. Schröder, D. Riesner, G. Steger, *RNA* **1999**, *5*, 574–584
13 M. Zuker, *Nucleic Acids Res.* **2003**, *31*, 3406–3425.
14 I. L. Hofacker, *Nucleic Acids Res.* **2003**, *31*, 3429–3431.
15 M. Schmitz, G. Steger, *J. Mol. Biol.* **1996**, *255*, 254–266.
16 C. Flamm, W. Fontana, I. L. Hofacker, P. Schuster, *RNA* **2000**, *6*, 325–338.
17 A. J. Dingley, G. Steger, B. Esters, D. Riesner, S. Grzesiek, *J. Mol. Biol.* **2003**, *334*, 751–767.
18 P. Schuster, W. Fontana, P. F. Stadler, I. L. Hofacker, *Proc. R. Soc. Lond. B Biol. Sci.* **1994**, *255*, 279–284.
19 J. M. Bevilacqua, P. C. Bevilacqua, *Biochemistry* **1998**, *37*, 15877–15884.
20 D. J. Proctor, J. E. Schaak, J. M. Bevilacqua, C. J. Falzone, P. C. Bevilacqua, *Biochemistry* **2002**, *41*, 12062–12075.
21 P. Brion, F. Michel, R. Schroeder, E. Westhof, *Nucleic Acids Res.* **1999**, *27*, 2494–2502.
22 F. Guo, T. R. Cech, *Nat. Struct. Biol.* **2002**, *9*, 855–861.

25
UV Melting, Native Gels and RNA Conformation

Andreas Werner

25.1
Monitoring RNA Folding in Solution

Due to its conformational versatility, RNA is a performant catalyst and involved in specific recognition. Therefore, it is essential for any biochemical or structural study to monitor the correct formation of its three-dimensional shape. The larger the RNA, the more rugged the free energy landscape and thus the possibility that folding intermediates or alternative conformations persist [1]. Moreover, depending on renaturation protocol and RNA concentration, intermolecular annealing may be preferred over intramolecular folding, as in the case of hairpin versus duplex [2]. UV melting curves cannot only provide a means of controlling these phenomena, but also allow determination of equilibrium constants and energy parameters.

By heating and cooling under quasi-stationary conditions, the RNA can be reversibly forced through the folding process between an ordered, native and a disordered, denatured state. UV absorbance provides a convenient means of monitoring this transition in solution and with little material. This is due to the phenomenon of hypochromicity of stacked bases, which provides a exquisitely sensitive signal as temperature increases and RNA unfolds (Fig. 25.1A). The temperature at the midpoint of this transition is called T_m. When plotting the derivative of the absorbance as a function of temperature (a melting profile), each dA/dT peak corresponds to a partial transition (Fig. 25.1B).

Non-covalent interactions in RNA are commonly classified as secondary or tertiary structure. Secondary structure consists of Watson–Crick base pairs forming helical stacks linked by bulges, loops and junctions. Tertiary structure subsumes all other intra- or intermolecular interactions based on canoncial and non-canonical base pairs, pseudoknots, water bridges, and interactions with metal ions. Translated into melting curves, the first peak in the low temperature range usually corresponds to the opening of the tertiary structure. Higher temperatures are required to unzip stacked bases and thus unfold secondary structure helices. Given a largely modular RNA structure, each independently folding element called a thermosome has its own T_m [3]. The derivative melting curve will be a convolution of normally

Handbook of RNA Biochemistry. Edited by R. K. Hartmann, A. Bindereif, A. Schön, E. Westhof
Copyright © 2005 WILEY-VCH Verlag GmbH & Co. KGaA, Weinheim
ISBN: 3-527-30826-1

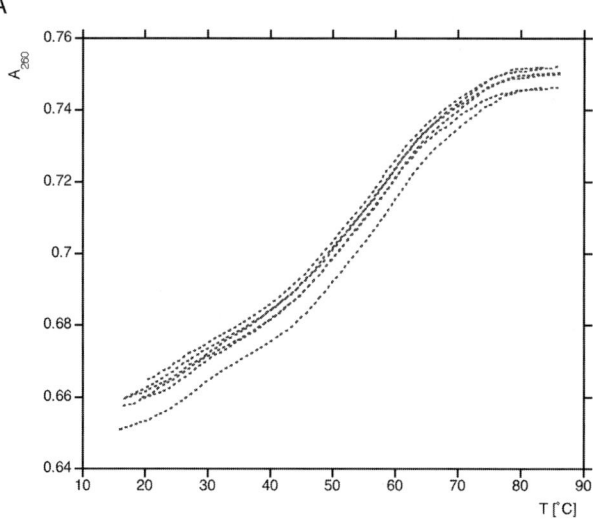

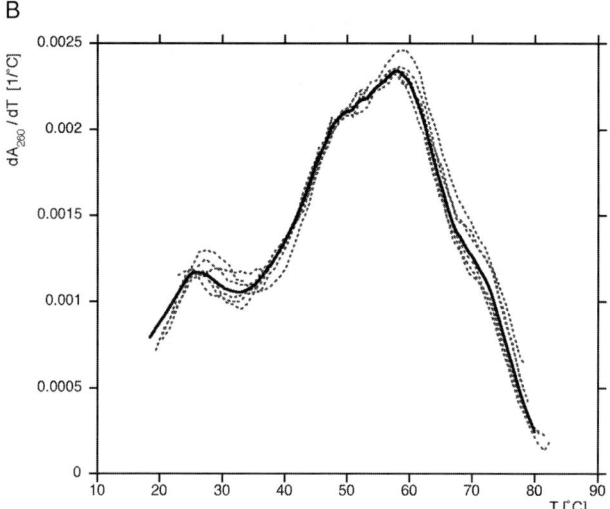

Fig. 25.1. UV melting profile of an RNase P RNA annealed in 10 mM sodium cacodylate pH 7, 10 mM NaCl and 0.5 mM $MgCl_2$. (A) Absorbance melting curve, three subsequent runs. (B) Derivative dA_{260}/dT of (A), the curve showing evaporation has been excluded. Black: average. Transitions visible at 26, 49, 58 and 70 °C.

distributed peaks around these T_ms. Many closely spaced thermosomes give rise to a complex melting profile. In simpler systems (less than 100 nt), it has been possible to deconvolute these thermosomes and assign them to individual transitions [4]. However, interpretation requires additional data through corroborating calorimetric and biochemical methods. For example, native polyacrylamide gels can provide complementary data to understand how the RNA collapses to the native state. In particular cases, a combination of melting and native gels could be applied even to thermostable RNA–protein complexes [5].

25.2 Methods

Most standard UV spectrophotometers can be equipped to measure RNA melting curves. We have used a double-beam Uvikon XL spectrophotometer attached to a Biotek temperature-control unit and a microcomputer running the Biotek Life-Power T_M Junior software (Fig. 25.2A). The instrument offers the advantage that it can hold up to 12 cuvettes in an automatic cell changer. Therefore, up to nine samples plus two standards and a temperature reference can be measured during one experimental run (Fig. 25.2B). Both racks are thermoregulated with a Peltier element and a liquid cooling circuit that allows precise heating/cooling rates of up to 5 °C/min. However, experimental heating rates will be typically in the range of 0.1–0.5 °C/min. On the one hand, a slow heating rate increases the risk of sample degradation by depurination and hydrolysis at high temperatures. On the other hand, the heating ramp should be slow enough to exclude kinetic effects in the processes being observed. For example, the hairpin–duplex transition can be quite

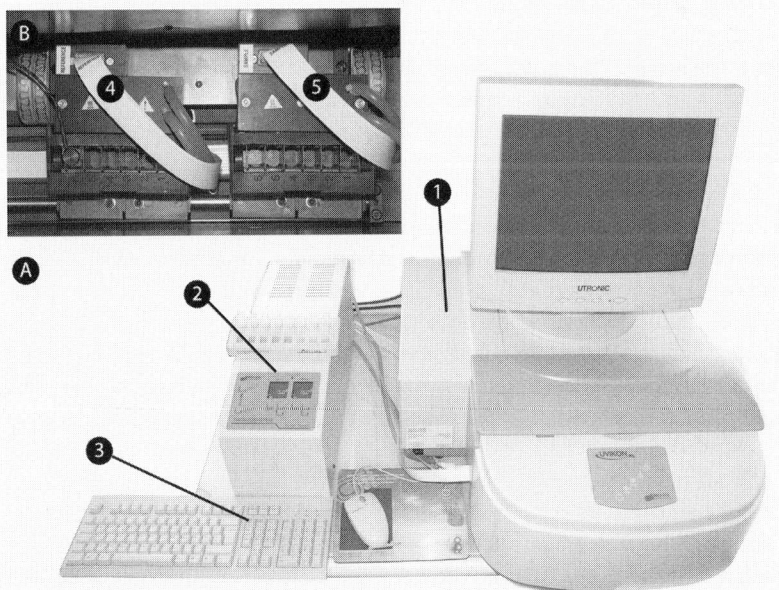

Fig. 25.2. (A) Setup for UV melting studies: UV spectrophotometer Uvikon XL (1), Temperature control unit (2) and computer interface (3). (B) Two independent cuvette racks (4,5).

slow and some group I introns fold on a timescale of minutes. RNA can be annealed prior to the melting run (5 min at 80 °C, then slow cooling to ambient temperature), but preferably in a more controlled way directly in the cuvettes, using the first heating/cooling cycle. Three or more heating/cooling cycles can be run overnight to ensure that results are reproducible. Data is collected from 10 to 90 °C, with data points every 0.2–0.5 °C. Samples would start boiling at higher temperatures and atmospheric pressure, and lower temperatures result in problems with condensation.

RNA can be prepared by transcription or chemical synthesis. A check of purity is advisable prior to spectroscopy. For annealing involving two or more strands, it is important to determine the concentration of each monomer precisely. Concentration is usually determined based on nearest-neighbor calculations of nucleotide extinction coefficients ε. However, depending on base composition, the calculated ε may deviate from the empirical value by up to 20% [6]. If necessary, the ε for any native structure can be determined experimentally by hydrolyzing the RNA enzymatically and measuring the A_{260} of the resulting nucleotides (see Appendix). Different cell path lengths can be used to allow measurements at different concentrations. We have used a set of custom-made 1-cm path length cuvettes (cat.-no. 115; Hellma, Germany) holding 500 µl of sample volume. Shorter path lengths are desirable for low sample concentrations and allow faster heat transfer.

Only little material is required – the hypochromic RNA must be adjusted to an absorbance of 0.1–0.2, corresponding to concentrations of 0.1–100 µM. To remain within the linearity range of the instrument, the maximum signal should not exceed an absorbance of 1.5. Measurements are normally done at the UV absorption maximum of AT base pairs at 260 nm. However, depending on base composition [7] and helix structure, measurements at a different wavelength may yield a better signal difference [8]. This suggests an initial wavelength run at lowest and highest temperature to determine the wavelength with the greatest difference in hypochromicity. Alternatively, the transition can also be monitored by observing circular dichroism as a function of temperature [5]. For model fitting, it may be necessary to obtain absorbance data at two different wavelengths simultaneously. This is because the hypochromicity of A-U and G-C basepairs is the same at 260 nm, but different at 280 nm, which can help to deconvolute two overlapping transitions.

Any ideal buffer for melting curves should not absorb any UV light and its pK_a should not vary with temperature. The standard buffer typically used is sodium cacodylate pH 7. Phosphate buffers, while being temperature insensitive, have the drawback of a high binding constant to divalents. Tris and HEPES show a too high pH variation with temperature. The pH of a Tris-buffered solution drops considerably when increasing the temperature from 35 to 90 °C (Fig. 25.3). For salt ions subject to deprotonation, the buffering capacity should also be considered: 100 mM NH_4^+ buffered in 50 mM Tris leads to a considerable drop in pH by about 1 unit, while the same solution in 10 mM sodium cacodylate (pH 7.0) buffer shows slight acidification (0.3 pH units at 100 mM NH_4). Depending on the salt, a higher buffer concentration may be advisable.

Choice of salt is also critical. Because divalent ions can cause hydrolysis of RNA at high temperatures, many experimenters include 0.1–1 mM EDTA. The major

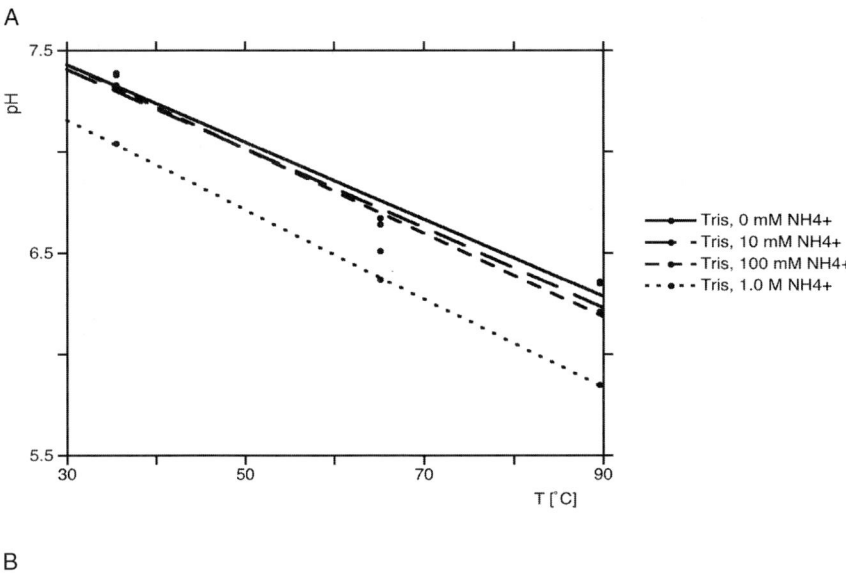

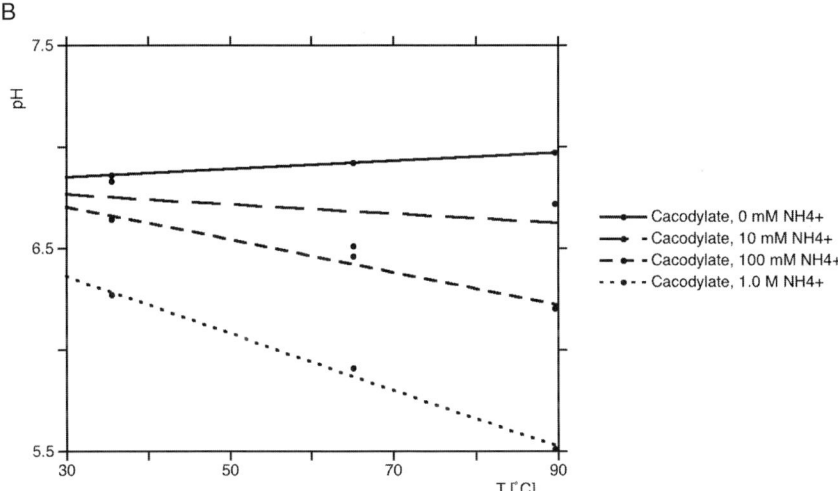

Fig. 25.3. (A) Temperature dependency of pH in Tris- or cacodylate-buffered solutions of NH_4Cl. Conditions: 50 mM Tris, pH 7.5 or 10 mM sodium cacodylate, pH 7.0 (100 mM NaCl, 0.5 mM EDTA).

role of Mg^{2+} is electrostatic, minimizing electrostatic repulsion between nucleic acid strands. The collapse to a compact, native-like state can be induced alternatively by low concentrations of divalents or high concentrations of monovalents, thereby avoiding the problem of hydrolysis. Roughly, each additional charge on the counterion decreases the required ion concentration by 2 log units [9]. The ion concentration should be varied to determine which transitions are visible in the experimentally accessible range (10–80 °C).

After each completed melting/annealing cycle, comparison between initial and final absorbance allows us to determine whether any evaporation or hydrolysis has occurred. This is a frequently encountered problem and some precautions can help to avoid it. Small path lengths seem to favor bubble formation, in this case the solution in each cuvette should be degassed prior to use by bubbling through an inert gas (N_2 or argon) or by applying a vacuum. The surface is sealed with 5–6 drops of PCR-quality mineral oil (Acros Organics, NJ). When applying the oil, care has to be taken not to touch the etched glass surface that should make a tight contact with the Teflon stopper or else the oil will spread and the sample will evaporate. The stopper is sealed with plumber's Teflon tape. For determination of derivative melting curves, minor evaporation losses are not critical.

It is possible to recover and precipitate the RNA after the measurements and separate it on a polyacrylamide gel to check for any degradation. This may be especially important if hydrolysis in the presence of a large concentration of divalent ions is a concern. Before precipitation, the oil can be removed from the aqueous phase by rolling the solution on a Teflon dish or by freezing-out. Separation of our RNA on denaturing 5% polyacrylamide gels containing 8 M urea followed by staining with toluidine blue O failed to detect any RNA degradation after three heating cycles.

25.3
Data Analysis

As stacked bases are hypochromic, thermal unfolding of RNA is accompanied by an increase in absorbance. By assuming a linear temperature dependence for absorbance coefficients of both folded and unfolded RNA, the transition between the two indicates the fraction of unfolded molecules in equilibrium at any given temperature (Fig. 25.4A). A lower (B_0) and upper (B_1) baseline has to be fit to each unfolding transition. In a graphical determination of T_m, the bisector between both baselines intersects the absorbance curve at the midpoint of transition where $\Delta G = 0$. A normalized representation as the fraction of folded molecules θ takes into account the baselines (Fig. 25.4B):

$$\theta = (B_0 - A_{260})/(B_0 - B_1)$$

This is the only way to determine T_m precisely. The data can be fitted with a sigmoidal function, where the peak of the derivative $d\theta/dT$ (or $d\theta/dT^{-1}$) indicates T_m. For intramolecular folding, it can be shown that $d\theta/dT^{-1}$ gives the precise value of the midpoint, while systematically overestimating the T_m of bimolecular annealing. However, most melting profiles with multiple transitions do not allow fitting of baselines with precision. Instead, we use the locally weighted least-square error algorithm provided with the data analysis software to obtain the derivative dA_{260}/dT (Fig. 25.1B). The estimated difference to the real T_m is in most cases no larger than ± 1–$2\ °C$.

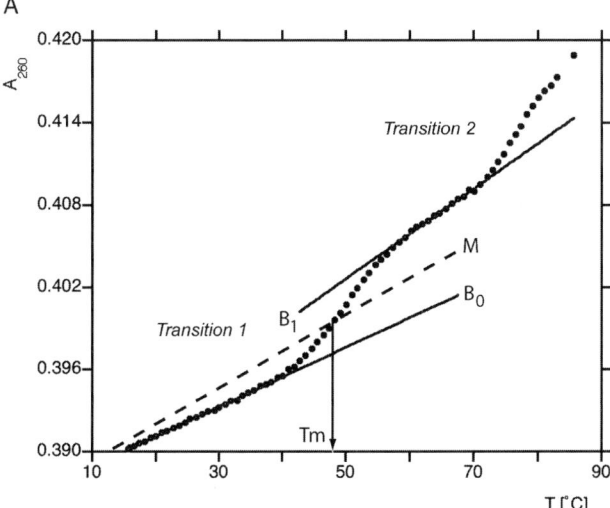

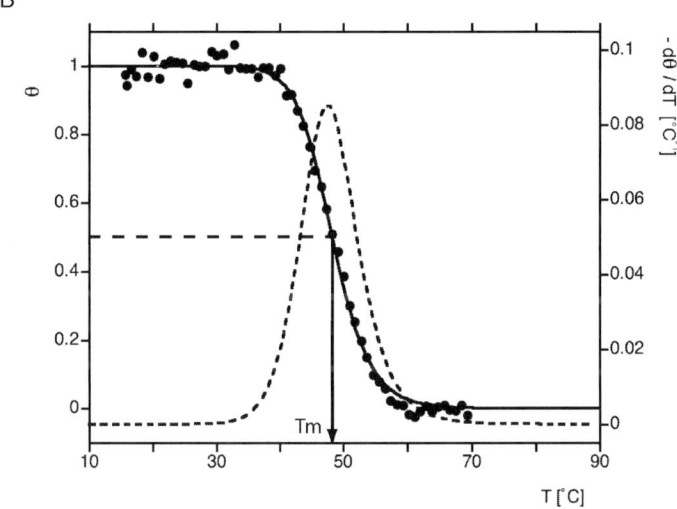

Fig. 25.4. (A) Absorbance data of RNase P RNA annealed in 10 mM sodium cacodylate, pH 7, 100 mM NaCl and 0.5 mM EDTA. The midpoint of transition T_m is defined by the bisector M between upper B_1 and lower tangents B_0. (B) Fraction of dissociated molecules θ calculated from baselines of transition 1 in (A) (left ordinate) and derivative $-d\theta/dT$ (right ordinate, dotted line).

A different approach is based on statistical mechanics. Using the tabulated nearest-neighbor enthalpies for Watson-Crick and non-canonical basepairs ("Turner rules"), the sequential unfolding of every basepair is considered independently to determine the corresponding partition function [4]. Additional fitting parameters are then added to account for the effect of tertiary interactions and make the theoretical curve coincide with the experimental one. Measuring at two different wave-

Tab. 25.1. Calculation of K_a.

Equation	$U \Leftrightarrow N$	$U_1 + U_2 \Leftrightarrow N$	$2U \Leftrightarrow N$
Reaction	monomolecular	bimolecular	bimolecular
Equilibrium constant $K_a =$	$\dfrac{\theta}{(1-\theta)^2}$	$\dfrac{\theta}{[U]_0(1-\theta)^2}$	$\dfrac{\theta}{2[U]_0(1-\theta)^2}$
Examples	Hairpin	Complementary oligos ($[U_1]=[U_2]$)	Autocomplementary oligos

U: unfolded RNA, N: folded RNA. Similar calculations are possible for equilibria involving n molecules [10].

lengths simultaneously allows to determine the minimum number of transitions that have to be considered [13]. Since secondary structures can also be predicted using nearest-neighbor free energy calculations, the peaks can be deconvoluted and, in the case of simple systems, assigned to secondary versus tertiary structure transitions. A number of software packages are available to help performing this kind of analysis, e.g. MeltWin (Windows, http://www.meltwin.com/), and Global-MeltFit (Macintosh, [4]).

25.4
Energy Calculations and Limitations

The spectrophotometric data for each sample consists of a single column of absorbance versus temperature. Sometimes, hysteresis is observed between the denaturation and the annealing curve. This is normally a kinetic effect and an indicator that the heating rate was too high to allow complete equilibration. Using a script, one can extract ascending and descending parts and put them into a table separately. Next, the file is imported into KaleidaGraph (Synergy Software, PA). A plot of absorbance versus temperature (Fig. 25.1A) is fit with the built-in smoothing function (window size 5). Data points of the smoothing function are then differentiated with respect to T and averaged over several runs. Since the derivative is calculated over a small data window (3–5 °C), very sharp transitions are more accurately reproduced by second-order polynomials, as performed by the Savitsky-Golay algorithm included in some software packages. The resulting derivative melting curve is plotted versus temperature (Fig. 25.1B) and contains a peak for each partial transition.

If T_m has been accurately determined, it can be used to calculate the energy parameters [10]. The first step consists in determining if the system can be approximated by any two-state model and accordingly which expression from mass action law is used to calculate the affinity constant K_a (Table 25.1). Next, the natural logarithm of K_a is plotted versus the inverse temperature T^{-1} (in Kelvin) following:

$$\Delta G° = -RT \ln K_a = \Delta H° - T\Delta S°$$

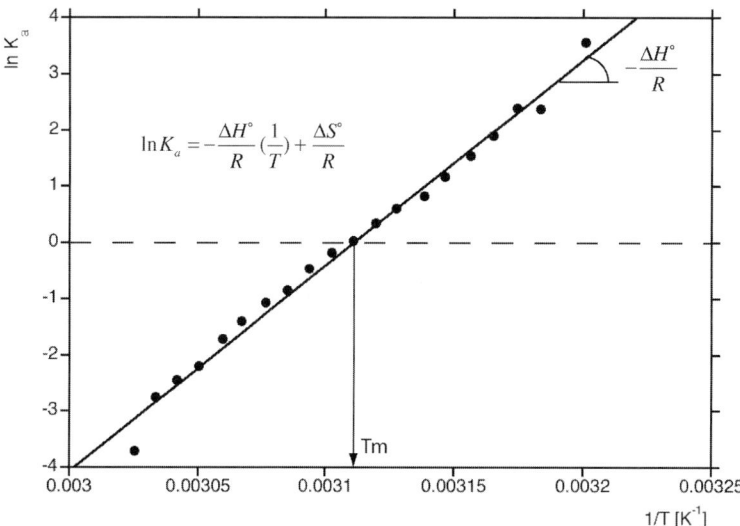

Fig. 25.5. Van t'Hoff plot of data in Fig. 25.4 (see text). K_a: affinity constant, $\Delta H°$: standard enthalpy, $\Delta S°$: entropy, R: molar gas constant.

where $\Delta G°$ is the free Gibbs enthalpy, $\Delta H°$ is the standard enthalpy, $\Delta S°$ is the entropy and R is the molar gas constant. Here, $\Delta H°$ also describes the temperature dependence of the equilibrium. By rearranging, one obtains the van t'Hoff expression:

$$\ln K_a = -\frac{\Delta H°}{R}\left(\frac{1}{T}\right) + \frac{\Delta S°}{R}$$

which allows us to determine $\Delta H°$ from the slope and $\Delta S°$ from the ordinate intersection (Fig. 25.5). In general, the energies determined by this method are less reliable than those determined by calorimetric methods, because they depend on several assumptions: (1) baselines have been determined correctly, (2) a two-state model is valid, (3) the system is in perfect equilibrium at all temperatures (absence of kinetic effects), and (4) $\Delta H°$ is temperature-independent and therefore $\Delta C_p = 0$ [11]. For example, for DNA duplex formation, ΔC_p has been shown to be quite negative [12].

It can be important to determine ΔC_p, which gives information about the exposed surface in folded and unfolded states. In that case, or if precise energy parameters are to be known, it is preferable to rely upon differential scanning calorimetry (DSC) or isothermal calorimetry (ITC) at varied temperatures. These methods provide necessary additional constraints to fit parameters to sequential models consisting of a succession of two-state transitions [4]. Measuring at two different wavelengths simultaneously allows to us determine the minimum number of transitions that have to be considered [13]. Since secondary structures can also be predicted using nearest-neighbor free energy calculations, the peaks can be deconvoluted and, in the case of simple systems, assigned to secondary versus tertiary structure transitions. Because calorimetric methods require large quantities of RNA, it can be useful to narrow down the number of buffer and salt conditions initially by UV melting, before carrying out a small number of calorimetric experiments [14].

25.5
RNA Concentration

For intramolecular annealing only, the T_m is independent of RNA concentration. In all other cases involving secondary and tertiary interactions between molecules, an increase in RNA concentration will facilitate association. Therefore, finding a concentration dependence of the melting profile is a clear indicator of dimerization, as is often observed with small hairpins. The dimerization will raise the T_m, which allows us to determine the enthalpy $\Delta H°$ of the association reaction. Data from a series of melting curves at different RNA concentrations is plotted as $1/T_m$ versus $\ln[RNA]_0$. Again provided that $\Delta C_p = 0$, the reaction enthalpy for the association of n molecules can be directly determined from the slope, $(n-1)R/\Delta H°$. In the case of a bimolecular association:

$$\frac{1}{T_m} = \frac{\Delta S°}{\Delta H°} + \frac{R}{\Delta H°} \ln\left(\frac{1}{2}[RNA]_0\right)$$

which follows from the equilibrium constant given in Table 25.1. Any precise determination supposes that $[RNA]_0$ can be varied over a large range while the absorbance remains within the linearity limits of the instrument.

25.6
Salt and pH Dependence

Increasing the ionic strength of the solution directly increases the melting temperature of the RNA (Fig. 25.6A). From polyelectrolyte theory it has been predicted that the T_m of a secondary structure transition will follow a linear dependence on the logarithm of ion concentration [15]. As for monovalent ions, the effect of divalent ions in solution can be explained by purely electrostatic effects. Through charge screening of the backbone phosphates, an increase in concentration allows the compaction of the structure. Taking into account specific Mg^{2+}-binding sites is only necessary at high background concentrations of monovalents [16] and in very specific cases of large RNAs, where Mg^{2+} locks the already compact, folded structure into the native state.

For helix melting, the T_m will vary by 16–17 °C for every log unit over a wide range of salt concentrations (Fig. 25.6B), but levels off at higher ionic strength. Thus the slope of salt dependence of T_m provides a clue to distinguish secondary from tertiary structure transitions. Moreover, the melting of a tertiary structure is

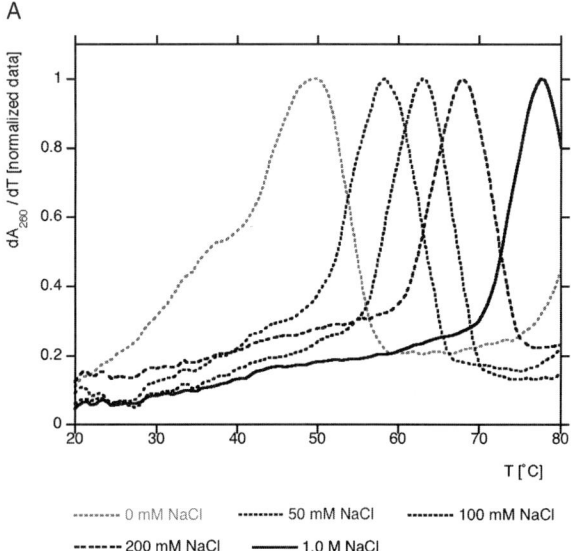
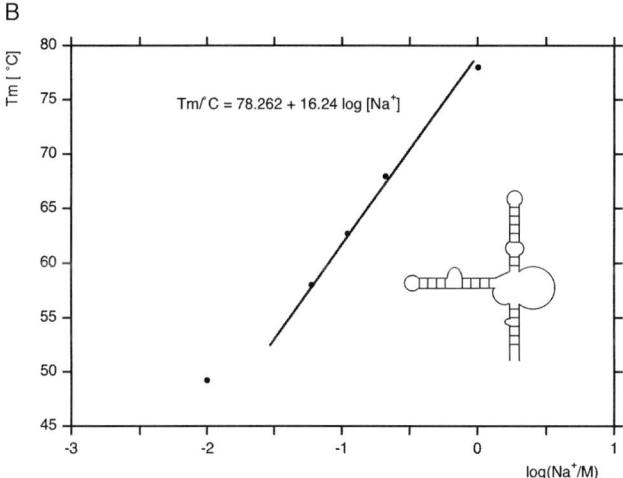

Fig. 25.6. Salt dependence of UV melting curves. (A) RNA annealed in 10 mM sodium cacodylate, pH 7, 0.5 mM EDTA and 0–1.0 M NaCl added. (B) plot of T_m versus log[Na$^+$].

much more dependent on the ionic radius than secondary structure [17]. Therefore, it can be useful to measure melting profiles in the presence of a series of cations with different radii. The pH can also greatly affect the melting temperature, depending on base composition. This is especially true for certain cytosines protonated on N^3. Extensive studies have been performed on the pH dependence of the T_m of pseudoknot motifs [18].

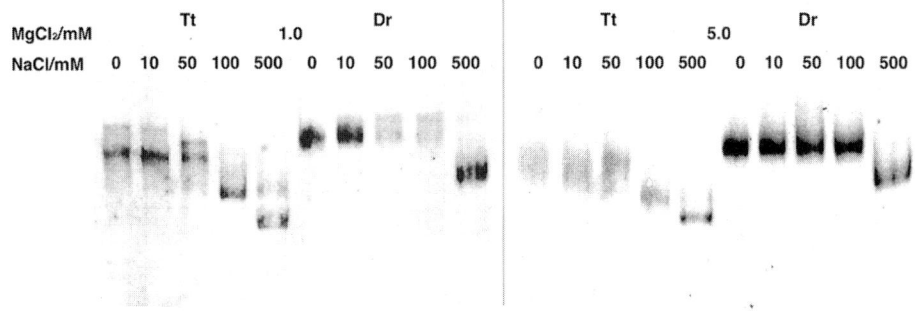

Fig. 25.7. Native gel electrophoresis (5% polyacrylamide). Tt and Dr RNase P RNA has been annealed in 10 mM sodium cacodylate, pH 7, 0–500 mM NaCl before addition of 1.0 (left) or 5.0 mM (right) MgCl$_2$.

25.7
Native Gels

A useful visual control of RNA folding is provided by native gels (Fig. 25.7). Consistent with what has been said for melting curves, the compaction of RNA structure is globally achieved by electrostatic effects. A 100× increase in concentration approximately compensates for the loss of one charge when passing from trivalent to divalent to monovalent ions [9]. Only final rearrangements may depend on the presence of Mg^{2+} at specific binding sites. Here we provide a protocol that can be adapted to the particular conditions of the RNA under investigation. Working under standard RNase-free conditions, RNA is heated to 90 °C for 5 min and annealed by slow cooling in the same buffer used for UV melting studies (10 mM sodium cacodylate, pH 7.0, 10–300 mM NaCl and/or 0–5 mM MgCl$_2$). MgCl$_2$ is added after cooling down to approximately 40 °C. RNA is either radiolabeled to approximately 30 000 c.p.m./μl or stained with the toluidine blue method.

Polyacrylamide gels are adjusted to correct separation range for the given RNA [19] and cast in 1 × TB running buffer (89 mM Tris, pH 8.3, 89 mM sodium borate). It is also possible to include 5% glycerol or low concentrations of Mg^{2+} in the gel matrix. For example, for a 5% gel, we mix 8.33 ml 30% acrylamide:bisacrylamide solution (Roth), 10 ml 5 × buffer, 31.5 ml H$_2$O, 150 μl 1 M MgCl$_2$, 500 μl 10% ammonium persulfate and 50 μl TEMED. To keep the RNA in its conformation, it is important to work quickly and in the cold room. The samples are mixed with 6 × loading buffer directly prior to charging (50% glycerol, 6 × TB buffer, 0.3% bromophenol blue, 0.3% xylene cyanol FF). The gel is run at 15 W and 4 °C. Alternatively, gel units for room temperature exist that are constantly cooled to 4 °C by a circulating water bath.

Finally, for concentrated RNA deposits (below 100 ng per band) the gel stain can be stained for 1 h with toluidine blue solution [0.5 g toluidine blue O (Sigma, MO) in 200 ml EtOH, 295 ml H_2O, 5 ml glacial acetic acid], followed by destaining (H_2O) and drying on a gel dryer. Alternatively, radioactively marked RNA is revealed by autoradiography or phosphoimaging.

Acknowledgments

A. W. acknowledges support from the European contract HPRN-CT2002-00190 (CARBONA).

References

1 WOODSON, S. A. *Nat. Struct. Biol.* **2000**, *7*, 349–352.
2 HOLBROOK, S. R., C. CHEONG, I. TINOCO, JR, S. H. KIM. *Nature* **1991**, *353*, 579–581.
3 GABARRO, J. *Anal. Biochem.* **1978**, *91*, 309–322.
4 DRAPER, D. E., T. C. GLUICK. *Methods Enzymol.* **1995**, *259*, 281–305.
5 BRESCIA, C. C., P. J. MIKULECKY, A. L. FEIG, D. D. SLEDJESKI. *RNA* **2003**, *9*, 33–43.
6 KALLANSRUD, G., B. WARD. *Anal. Biochem.* **1996**, *236*, 134–138.
7 PUGLISI, J. D., I. TINOCO JR. *Methods. Enzymol.* **1989**, *180*, 304–325.
8 MERGNY, J. L., A. T. PHAN, L. LACROIX. *FEBS Lett.* **1998**, *435*, 74–78.
9 HEILMAN-MILLER, S. L., D. THIRUMALAI, S. A. WOODSON. *J. Mol. Biol.* **2001**, *306*, 1157–1166.
10 MARKY, L. A., K. J. BRESLAUER. *Biopolymers* **1987**, *26*, 1601–1620.
11 CHAIRES, J. B. *Biophys. Chem.* **1997**, *64*, 15–23.
12 ROUZINA, I., V. A. BLOOMFIELD. *Biophys. J.* **1999**, *77*, 3242–3251.
13 THEIMER, C. A., Y. WANG, D. W. HOFFMAN, H. M. KRISCH, D. P. GIEDROC. *J. Mol. Biol.* **1998**, *279*, 545–564.
14 GLUICK, T. C., N. M. WILLS, R. F. GESTELAND, D. E. DRAPER. *Biochemistry* **1997**, *36*, 16173–16186.
15 RECORD, M. T., JR. *Biopolymers* **1967**, *5*, 975–992.
16 MANNING, G. S. *Biophys. Chem.* **1977**, *7*, 141–145.
17 HEERSCHAP, A., J. A. WALTERS, C. W. HILBERS. *Biophys. Chem.* **1985**, *22*, 205–217.
18 NIXON, P. L., D. P. GIEDROC. *J. Mol. Biol.* **2000**, *296*, 659–671.
19 SAMBROOK, J., D. W. RUSSELL, J. SAMBROOK. *Molecular Cloning: A Laboratory Manual*, Cold Spring Harbor Laboratory Press, Cold Spring Harbor, NY, **2001**.

26
Sedimentation Analysis of Ribonucleoprotein Complexes

Jan Medenbach, Andrey Damianov, Silke Schreiner and Albrecht Bindereif

26.1
Introduction

Several essential cellular processes such as translation or pre-mRNA splicing are catalyzed by large multimeric complexes that contain essential RNA components. The ordered and stepwise assembly of these complexes often proceeds through complicated maturation pathways. Classical examples for this are provided by ribosomes, heterogeneous nuclear ribonucleoprotein (hnRNP) particles and the spliceosomal small nuclear RNPs (snRNPs) U1, U2 and U4/U6 · U5. The biogenesis of snRNPs, which involves in addition trafficking between different cellular compartments, and the assembly of snRNPs to the active spliceosome are particularly well studied, and are used in the following as a specific example (reviewed in [1]).

Different techniques such as affinity purification have been used to study the assembly and composition of the spliceosomal snRNPs. In particular, the fractionation of snRNPs by density gradient ultracentrifugation has proven a powerful tool for the separation and enrichment of individual snRNPs in their native state. Taking advantage of the stability of the snRNPs under high salt conditions in the presence of Mg^{2+} ions, they could be purified from HeLa nuclei as early as 1980 applying only a series of different cesium chloride gradients [2]. Depending on the conditions, either intact snRNPs or core snRNPs containing the Sm proteins, but lacking the specific protein components, could be obtained [3]. Earlier, the same approaches had been applied to the study of ribosomes and hnRNP particles (see, e.g. [4]). In addition it has been shown for several complexes that catalytic activity is retained after ultracentrifugation (e.g. RNase P [5]), demonstrating that the particles remain in a native and functional state.

A further advantage of the density gradient centrifugation is that it can be applied to crude cellular extracts, such as nuclear extracts [6], as well as to purified fractions or complexes, such as samples obtained by immunoaffinity selection. Thus density gradient ultracentrifugation can be combined with other fractionation techniques. For example, this has recently resulted in the characterization of spliceosomal subcomplexes [7].

In this article we describe the techniques of glycerol and cesium chloride density gradient ultracentrifugation, using as an example the fractionation of the spliceosomal snRNPs present in HeLa cell nuclear extract. The protocols given can also be used to fractionate other samples such as unfractionated splicing reactions or eluates from affinity selections. These two methods separate the samples according to different physical properties. In zonal glycerol gradient density centrifugation the sample is separated in a preformed gradient of a viscous component (glycerol, sucrose) for a defined time resulting in fractionation due to differences in the sedimentation constant (Svedberg, S), which depends on the weight, volume, density and shape of the particle. In contrast, during isopycnic ultracentrifugation in cesium chloride or cesium sulfate the gradient is formed during the run. Centrifugation is continued until an equilibrium of forces is achieved, resulting in the sedimentation of the particles at their isopycnic positions and thereby separating them according to their different densities.

26.2
Glycerol Gradient Centrifugation

As an example, we describe here the fractionation of nuclear extract in a linear 10–30% glycerol gradient. Depending on the particles to be separated, gradients with different glycerol concentrations can be used. Instead of preparing RNA from the fractions (see Section 26.2.3.3), they can also be subjected to immunoaffinity selections or other purification methods. In parallel to RNA, proteins can also be easily prepared from the fractions for further analysis (see Section 26.2.3.4).

26.2.1
Equipment

- SW-40 Ti rotor with polyallomer centrifugation tubes.
- Gradient Mixer (Hoefer Scientific Instruments SG Series or Gradient Master, see Section 26.2.3.6).
- Sterile, RNase-free 1.5-ml Eppendorf tubes.
- Cooling microcentrifuge.
- Glass capillaries or disposable micropipettes (e.g. ringcaps®; Hirschmann).

26.2.2
Reagents

- DEPC-water (1 ml DEPC per 1 l dd-H_2O, stir for 1 h, autoclave twice).
- 1 M DTT in DEPC-H_2O.
- 0.1 M phenylmethyl sulfonyl flouride (PMSF) in ethanol.
- 3 M sodium acetate, pH 5.2, in DEPC-H_2O.
- 20 mg/ml glycogen in DEPC-H_2O.
- 100, 70, 80 and 50% ethanol.

- Phenol/chloroform/isoamylalcohol (25:24:1), saturated with TE buffer (10 mM Tris–HCl, pH 8.0, 1 mM EDTA).
- Buffer D (20 mM HEPES–KOH, pH 8.0, 100 mM KCl, 0.2 mM EDTA, pH 8.0, 1.5 mM $MgCl_2$, 20% glycerol, in DEPC-H_2O, add fresh 0.5 mM PMSF and 0.5 mM DTT).
- 10 × buffer G (200 mM HEPES–KOH, pH 8.0, 1.5 M KCl, 15 mM $MgCl_2$).
- 10% glycerol solution (1 × buffer G, 10% glycerol, in DEPC-H_2O, add fresh per 100 ml 100 μl leupeptin, 50 μl 1 M DTT and 500 μl 0.1 M PMSF solution, filter through a 0.45-μm filter).
- 30% glycerol solution (same as 10% glycerol solution, but with 30% glycerol).
- SDS–PAGE protein sample buffer (2% SDS, 10% glycerol, 50 mM Tris–HCl, pH 6.8, 0.005% bromophenol blue).
- 5 × agarose gel loading buffer (5 × TBE, 10% glycerol, 0.025% bromophenol blue).
- Leupeptin (4 mg/ml in DEPC-H_2O).
- 16S and 23S rRNAs from *Escherichia coli* (Roche; 0206938).

26.2.3
Method

26.2.3.1 Preparation of the Glycerol Gradient

- Clean the gradient mixer with 100 and 50% ethanol, close valves, and rinse the chambers with 30 and 10% glycerol solution, respectively; make sure that no air bubbles become trapped in the connections.
- Place the gradient mixer on a magnetic stirrer, fill the chambers (valves closed) with 5.5 ml of 10 (chamber with the outlet pipe) and 30% glycerol solution without trapping air bubbles, add sterile magnetic stir bar to the chamber with the outlet pipe (Fig. 26.1).
- Place centrifugation tube on ice, fix sterile glass capillary to the pipe of the mixer and place it in the tube so that it touches the bottom.
- Start magnetic stirrer and open the valves, so that the centrifugation tube is slowly filled with the glycerol solution, the 10% solution being underlayed with the denser solution.
- Before pouring another gradient rinse the mixer again with 30% and 10% glycerol solution, respectively.
- Place the gradients at 4 °C for 1 h for equilibration.

26.2.3.2 Sample Preparation and Centrifugation

- Thaw 400 μl of HeLa nuclear extract slowly on ice, clear it of aggregates by a short spin, and mix it with 1.1 ml of 1 × buffer G freshly supplemented with 2 mM DTT and 0.5 mM PMSF.
- Carefully overlay the prepared gradients with the solution and balance the tubes together with the centrifugation buckets with 1 × buffer G.

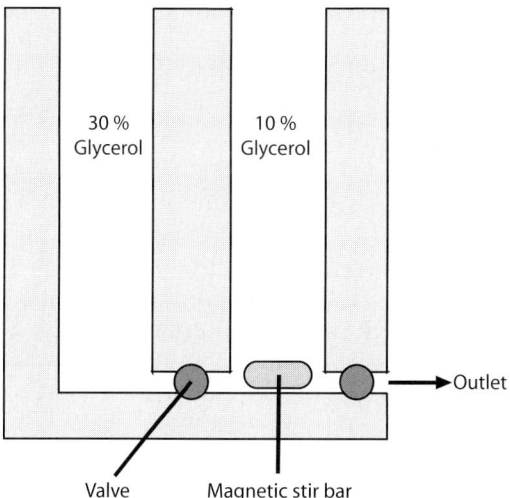

Fig. 26.1. Schematic display of the Hoefer Gradient Mixer for formation of glycerol gradients. For each chamber the concentration of the glycerol stock solution for preparation of a linear 10–30% gradient is given; the outlet pipe is indicated by an arrow.

- Centrifuge for 17 h at 32 000 r.p.m. (corresponding to 130 000 g) and at 4 °C in a precooled SW-40 rotor.
- Carefully fractionate the gradient from top to bottom by taking off 25 fractions of 500 µl each and resuspend the pellet in the last fraction.

26.2.3.3 Preparation of RNA from Gradient Fractions

- Split the fraction in two for easier handling during preparation of the RNA.
- Add 250 µl of phenol/chloroform/isoamylalcohol to each of the fractions, vortex thoroughly and centrifuge for 10 min at 14 000 r.p.m.
- Transfer upper aqueous phase to fresh Eppendorf tubes, and add 20 µl 3 M sodium acetate, 2 µl glycogen (20 mg/ml) and 550 µl 100% ethanol to each of the tubes and mix by inversion.
- Precipitate nucleic acids for 10 min at −70 °C, centrifuge for 20 min at 14 000 r.p.m. and 4 °C, remove supernatant from pellets, and wash at room temperature with 500 µl 70% ethanol.
- Take up pellets in DEPC-H_2O and combine the two aliquots from one fraction.
- Analyze RNA by denaturing PAGE.

26.2.3.4 Simultaneous Preparation of RNA and Proteins

In parallel to the isolation of RNA, proteins can be prepared from the same fractions. Because of the high concentration of proteins in some fractions an initial dilution step of the gradient fractions (1:1 with DEPC-H_2O) is recommended, when crude cellular extracts are fractionated.

- After phenolization and removal of the upper aqueous phase, add 5 volumes of acetone to the phenol phase for precipitating proteins.
- Mix and store samples for at least 1 h at $-20\,°C$.
- Centrifuge for 30 min at 14 000 r.p.m., remove supernatant from pellet and wash with 500 µl of 80% ethanol.
- Dissolve pellet in a small amount (10–20 µl) of SDS–PAGE protein sample buffer, boil for 10 min and analyze by SDS–PAGE.

26.2.3.5 Control Gradient with Sedimentation Markers

Since 5S ribosomal RNA is no longer commercially available, it is recommended to analyze a silver-stained denaturing polyacrylamide gel of a gradient containing nuclear extract to detect the peak of free 5S rRNA (see Fig. 26.2). For the 16S and 23S sedimentation markers proceed as follows:

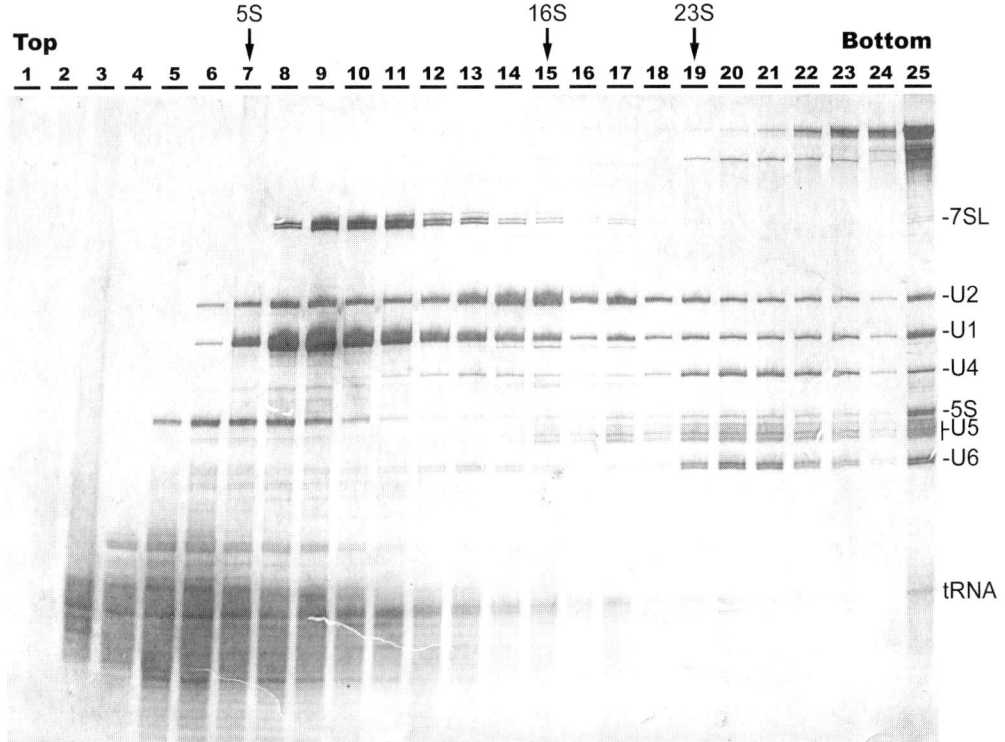

Fig. 26.2. Fractionation of nuclear extract by glycerol gradient centrifugation: RNA analysis. After ultracentrifugation of 200 µl of HeLa nuclear extract through an 11-ml linear 10–30% glycerol gradient (see Section 26.2.3.2 for conditions), 25 fractions of 500 µl each were taken from the top to the bottom of the gradient. RNA from 50 µl of each fraction was isolated and analyzed by denaturing PAGE and silver staining. Marker sizes (in nucleotides, Roche DIG V) are shown on the left, the identities of the RNAs are indicated on the right (U1, U2, U4, U5 and U6 snRNAs, 5S rRNA, 7SL RNA, and tRNAs). The positions of the 5S, 16S and 23S sedimentation markers are given on the top.

- Prepare gradient as described above.
- Mix 10 μl 16S/23S rRNA (4 μg/μl, Roche; 0206938) with 390 μl buffer D (or the buffer corresponding to the sample loaded on the analytical gradients).
- Add 1.1 ml 1 × buffer G and mix.
- Load onto gradient and run as described above.
- Fractionate gradient.
- Mix 12 μl of every second fraction with 3 μl agarose gel sample buffer and analyze on a 0.6% agarose gel in 0.5 × TBE.
- Visualize RNAs by ethidium bromide staining.

26.2.3.6 Notes and Troubleshooting

(1) Depending on the size range and properties of the RNPs to be separated, the running time (17 h), rotor type (SW-40) and conditions (32 000 r.p.m.; 4 °C) described in the example above (see Section 26.2.3.2) may be adjusted. Run the control gradient (see Section 26.2.3.5) under exactly the same conditions.

(2) Instead of glycerol, as described here, gradient sedimentation can also be performed in sucrose gradients (e.g. 15–30% [4] and 15–45%, [8]).

(3) Handle gradient fractions and prepared RNAs always on ice to minimize degradation. During RNA preparation make sure that only the aqueous phase is transferred to a fresh Eppendorf tube; do not touch the interface or the phenol phase which will lead to poor results upon analysis of the RNA due to contaminations. Before taking up the RNA in DEPC-H_2O make sure the pellet is completely dried and that no residual ethanol is left inside the Eppendorf tube. On the other hand, do not overdry the pellet (3–5 min at room temperature are normally sufficient), since otherwise dissolving it later in DEPC-H_2O may become difficult.

(4) Instead of using the Hoefer Gradient Mixer to prepare the gradients (see Section 26.2.3.1), the BioComp Gradient Master, a programmable, gradient-forming instrument (Frederickton, NB) is recommended for higher reproducibility and for faster preparation of the gradients.

(5) Make sure that the samples to be subjected to glycerol gradient centrifugation do not contain too much glycerol – here more than 10% – so that they do not sink below the surface of the prepared gradient upon loading. If necessary, dilute or dialyze against a suitable buffer to reduce the glycerol concentration.

(6) If samples obtained from the glycerol gradient centrifugation are to be subjected to immunoaffinity purification or other methods, we also recommend dilution or dialysis (e.g. with NET-100 buffer: 50 mM Tris–HCl, pH 8.0, 100 mM NaCl, 0.05% Nonidet P-40, 0.5 mM DTT) in order to reduce the high glycerol concentration.

(7) For visualization of RNAs in agarose gels by ethidium bromide staining (see Section 26.2.3.5) the ethidium bromide should be added directly to the agarose solution (final concentration of 500 ng/ml) before casting the gel; if the gel is stained after the run, degradation of the RNA may occur.

26.3
Fractionation of RNPs by Cesium Chloride Density Gradient Centrifugation

When applying isopycnic ultracentrifugation for the analysis of proteins or RNA–protein complexes, one has to consider that the high ionic strength may destabilize and dissociate the complexes, resulting in denaturation and precipitation of the proteins. Spliceosomal core snRNPs, however, are stable under high-salt conditions in the presence of 15 mM $MgCl_2$. This allows us to separate the snRNPs from free proteins that stay at the top of the gradient and from free RNAs that are pelleted. The sedimentation behavior of the snRNPs yields additional information on their protein:RNA ratio, since that determines their buoyant density ρ (see Section 26.3.3.4). As a typical example we describe here the fractionation of nuclear extract by cesium chloride density gradient ultracentrifugation.

26.3.1
Equipment

- Beckman Optima TLX benchtop ultracentrifuge.
- Beckman TLA 120.2 rotor with thick-walled polycarbonate centrifugation tubes (11 × 34 mM, Beckman 343778).
- Sterile, RNase-free 1.5-ml Eppendorf tubes.
- Cooling microcentrifuge.

26.3.2
Reagents

- DEPC-H_2O (see Section 26.2.2).
- 1 M $MgCl_2$ in DEPC-H_2O.
- 3 M sodium acetate, pH 5.2, in DEPC-H_2O.
- 20 mg/ml glycogen in DEPC-H_2O.
- 1 M DTT in DEPC-H_2O.
- 0.1 M PMSF in ethanol.
- Phenol/chloroform/isoamylalcohol (25:24:1) saturated with TE buffer (10 mM Tris–HCl, pH 8.0, 1mM EDTA).
- 10% SDS.
- 70 and 100% ethanol.
- Buffer D/$MgCl_2$ (see Section 26.2.2, but with 15 mM $MgCl_2$).
- CsCl stock solution (dissolve CsCl in buffer D/$MgCl_2$ to a final density of 1.55 g/ml, the easiest way to achieve this is to mix two buffer D/$MgCl_2$ solutions, one containing no CsCl, the other containing approximately 1.8 g/ml CsCl; adjust the CsCl density of 1.55 g/ml precisely, since this is critical for the reproducibility of the gradients. The amount x of CsCl (expressed in grams) needed to prepare 1 ml of a solution with the density ρ can also be calculated using the following formula [9]: $x = (\rho - 1)/0.92$ (ρ indicating the numerical value of the density expressed in g/ml). Note that it is not necessary to add PMSF and DTT to buffer D/$MgCl_2$ at this stage, store at room temperature, see Section 26.3.3.4).

26.3.3 Method

26.3.3.1 Preparation of the Gradient and Ultracentrifugation

- Thaw nuclear extract carefully on ice, clear it of precipitates by a short spin (1 min at 14 000 r.p.m. and 4 °C) and add $MgCl_2$ solution to a final concentration of 15 mM.
- Take an aliquot of the extract as input control.
- Supplement 3 ml of the CsCl stock solution freshly with 0.5 mM DTT and 0.5 mM PMSF.
- Mix 200 µl of the extract with 300 µl of the prepared CsCl solution.
- Pipette 500 µl CsCl stock solution supplemented with PMSF and DTT into a precooled 1-ml polycarbonate tube (11 × 34 mM; part no. 343778) and overlay it carefully with the prepared extract CsCl mixture.
- Balance the tubes carefully with buffer D.
- Centrifuge at 90 000 r.p.m. for 20 h at 4 °C in a Beckman TLX tabletop ultracentrifuge, using a precooled TLA 120.2 rotor.
- Carefully fractionate the gradient from top to the bottom by taking off 10 100-µl fractions; in the 10th tube collect the pellet in the residual gradient solution by resuspension.

26.3.3.2 Preparation of RNA from the Gradient Fractions

- Add 300 µl of DEPC-H_2O and 40 µl of 10% SDS to each fraction, then add 400 µl phenol/chloroform/isoamylalcohol, vortex thoroughly and separate phases by centrifugation (10 min, 4 °C, 14 000 r.p.m.).
- Transfer upper aqueous phase to a new Eppendorf tube and add 40 µl of 3 M sodium acetate, pH 5.2, 2 µl glycogen (20 mg/ml) and 1 ml ethanol.
- Mix solution by inversion of the tube and incubate for 10 min at −70 °C.
- Precipitate nucleic acids by centrifugation for 20 min at 14 000 r.p.m. and 4 °C, remove supernatant from pellets and wash with 500 µl of 70% ethanol (room temperature).
- Analyze RNA by denaturing PAGE (Fig. 26.3).

26.3.3.3 Control Gradient for Density Calculation

- Prepare a gradient as described, replacing the nuclear extract by buffer D.
- Run gradient and fractionate as described above.
- Precisely weigh an aliquot of each fraction (e.g. 50 µl), which gives the density distribution across the gradient.

26.3.3.4 Notes and Troubleshooting

(1) By measuring the buoyant density of a particle the approximate percentage of protein mass in the complex can be calculated using the following empirical

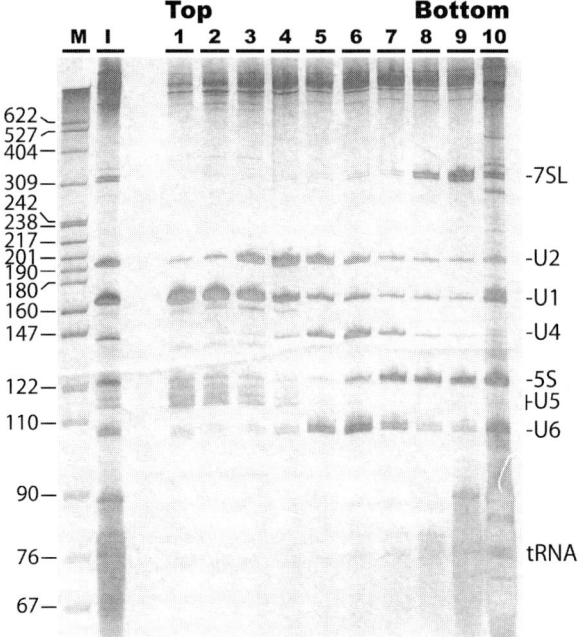

Fig. 26.3. Fractionation of nuclear extract by CsCl density gradient ultracentrifugation: RNA analysis. After ultracentrifugation of 200 µl of HeLa nuclear extract in a 1-ml CsCl density gradient (see Section 26.3.3.1 for conditions), 10 fractions of 100 µl each were taken from the top to the bottom of the gradient. RNA was prepared from each fraction (1–10, 10 including the pellet) and analyzed by denaturing polyacrylamide gel electrophoresis, followed by silver staining. Marker sizes (pBR322 DNA digested with *Hpa*II) are shown on the left, RNAs are indicated on the right. For example, the U4/U6 snRNP with base-paired U4 and U6 snRNAs peaks in fractions 5–7, free tRNAs are found in the pellet fraction (10). For comparison, lane I contains RNA prepared from 10 µl of HeLa nuclear extract corresponding to 10% of the input.

formula [9, 10]: % protein = $(1.85 - \rho)/0.006$ (ρ indicating the numerical value of the density expressed in g/ml). For example, densities of 1.36 and 1.51 g/ml, respectively, were observed for the *Trypanosoma brucei* U5 and U4/U6 core snRNPs; these values correspond to protein ratios of 82 (U5 core snRNP) and 57% (U4/U6 core snRNP). Taking the known masses of the RNA components into account, this results in total protein masses of 93 kDa per U5 core complex and 89 kDa per U4/U6 core complex [11].

(2) Instead of the standard CsCl, Cs_2SO_4 has also been used in isopycnic density gradient centrifugation (see, e.g. [4, 12, 13]); note that this appears to result in different stringencies for the RNPs (as discussed for the trypanosomal U2 snRNP [14]).

(3) For the preparation of the CsCl stock solution it is not recommended to add PMSF and DTT to buffer D/$MgCl_2$, since they are degraded during storage. In-

stead, PMSF and DTT should be added to the CsCl stock solution immediately prior to gradient preparation (see Section 26.3.3.1).

(4) Due to the high ionic strength of the fractions prepared from the gradient it is recommended to dilute or to dialyze the fractions (e.g. against buffer D) before subjecting them to immunoaffinity purification or other methods (see, e.g. [15]). The high ionic strength also interferes with degradation of the RNAs, nevertheless it is recommended to handle the obtained samples on ice, especially after removal of the salts by precipitation and washing. During phenol/chloroform extraction make sure that only the upper, aqueous phase is transferred to a new Eppendorf tube without touching the interface or the phenol phase, as this may lead to poor results during analysis of the RNA due to contaminations.

(5) Always completely remove ethanol from the RNA pellet by drying (which normally takes 3–5 min at room temperature) before taking it up in water. Avoid overdrying the pellet, since this may make dissolving it in DEPC-H_2O difficult.

Acknowledgments

This work was supported by grants from the Deutsche Forschungsgemeinschaft (Bi 316/9-2 and 10-3).

References

1 C. L. WILL, R. LÜHRMANN, *Curr. Opin. Cell Biol.* **2001**, *13*, 290–301.
2 C. BRUNEL, J. S. WIDADA, M. N. LELAY, P. JEANTEUR, J. P. LIAUTARD, *Nucleic Acids Res.* **1981**, *9*, 815–830.
3 C. BRUNEL, G. CATHALA, *Methods Enzymol.* **1990**, *181*, 264–273.
4 T. PEDERSON, *J. Mol. Biol.* **1974**, *83*, 163–183.
5 C. HEUBECK, A. SCHÖN, *Methods Enzymol.* **2001**, *342*, 118–134.
6 J. D. DIGNAM, R. M. LEBOVITZ, R. G. ROEDER, *Nucleic Acids Res.* **1983**, *11*, 1475–1489.
7 E. M. MAKAROV, O. V. MAKAROVA, H. URLAUB, M. GENTZEL, C. L. WILL, M. WILM, R. LÜHRMANN, *Science* **2002**, *298*, 2205–2208.
8 G. AST, D. GOLDBLATT, D. OFFEN, J. SPERLING, R. SPERLING, *EMBO J.* **1991**, *10*, 425–432.
9 M. G. HAMILTON, *Methods Enzymol.* **1971**, *20*, 512–521.
10 A. S. SPIRIN, *Eur. J. Biochem.* **1969**, *10*, 20–35.
11 S. LÜCKE, T. KLÖCKNER, Z. PALFI, M. BOSHART, A. BINDEREIF, *EMBO J.* **1997**, *16*, 4433–4440.
12 S. MICHAELI, T. G. ROBERTS, K. P. WATKINS, N. AGABIAN, *J. Biol. Chem.* **1990**, *265*, 10582–10588.
13 W. SZYBALSKI, *Methods Enzymol* **1968**, *12*, 330–360.
14 M. CROSS, A. GÜNZL, Z. PALFI, A. BINDEREIF, *Mol. Cell. Biol.* **1991**, *11*, 5516–5526.
15 Z. PALFI, G.-L. XU, A. BINDEREIF, *J. Biol. Chem.* **1994**, *269*, 30620–30625.

27
Preparation and Handling of RNA Crystals

Boris François, Aurélie Lescoute-Phillips, Andreas Werner and Benoît Masquida

27.1
Introduction

The crystal structures of prokaryotic ribosomal subunits [1–3], as well as sequence analysis coupled to molecular modeling, have demonstrated that RNA structure is modular [4, 5]; in other words, it can be decomposed in individual building blocks (modules), recurrently found in various RNA molecules, that are assembled together to form the overall RNA fold.

Because ribosomal subunits are the most abundant native particles of a growing cell, there is no need to use *in vitro* transcription or chemical synthesis to overproduce them. This is crucial since it 'reduces' molecular handling to biochemical purification of native particles. However, in most cases, overproduction techniques are needed and the biochemist faces subsequent RNA folding or protein–ligand/RNA association problems, unless the RNA is rather short. This partly explains why small RNA structures can be solved fairly quickly, whereas RNAs beyond 100 nt require time-consuming biochemical characterization before successful crystallization. In fact, except ribosomal subunits, only three RNA structures over 100 nt have been solved to date [6–8]. Thus, the folding of individual RNA motifs is apparently easier to control than the whole assemblies they are part of. Consequently, the structure of these motifs can be studied individually. Here, we focus on intermediate-size RNA motifs either produced by *in vitro* transcription or chemical synthesis (below 50 nt) [9–12].

The motifs of interest are extracted from a large RNA assembly and sequences are designed in order to favor a unique secondary structure. This can be assessed by using *in silico* folding programs [13–15] and native PAGE techniques (see Chapter 25). When the RNA has a substrate, further biochemical characterization might be required. Then the fragments are either cloned and *in vitro* transcribed or chemically synthesized. In this chapter we describe experimental procedures routinely used in the laboratory to obtain highly pure RNA molecules suitable for crystallization studies. Once the RNA has been synthesized, it has to be purified and concentrated. The correct length product is separated from contaminants by gel elec-

Handbook of RNA Biochemistry. Edited by R. K. Hartmann, A. Bindereif, A. Schön, E. Westhof
Copyright © 2005 WILEY-VCH Verlag GmbH & Co. KGaA, Weinheim
ISBN: 3-527-30826-1

trophoresis and/or chromatography. The RNA is then eluted, concentrated and desalted. Folding assays are performed prior to crystal screening.

27.2
Design of Short RNA Constructs

RNA structures can be seen as assemblies built from a construction set consisting of building blocks of various shape and complexity which obey conservation rules at the level of sequence and structure. In order to understand RNA architecture, it is therefore necessary to elucidate the structure of these building blocks. To achieve this goal, the motifs are analyzed in their wild-type contexts and, after alignment of the sequences, designed so as to ensure that they will conserve their original structure. It is worth to note that the best situation is when the secondary structure is well supported by biochemical data and that the edges of the motif (5′ and 3′ ends) are well located. Attention should be given to the design process in order to increase the probability for the structure to adopt the wild-type conformation. The stability of the constructs can be evaluated using UV-melting techniques under various ionic strengths (see Chapter 25).

Design of short RNA constructs is greatly helped by *in silico* folding programs [13–16]. They allow for testing modifications of the secondary structure upon modifications of the various base pairs. This step is crucial because apparently insignificant events such as the reversal of a GC pair can sometimes have major implications. A second point to think about is that the RNA motif may not reveal any propensity to pack and consequently to grow crystals. To address this problem it is advised to design constructs exhibiting various edges producing different situations regarding length and sequence that would eventually be more favorable for crystal growth.

27.3
RNA Purification

RNA molecules can be purified either by PAGE or liquid chromatography (HPLC or FPLC). These methods can even be coupled to improve the results. While PAGE is applicable to any RNA length, HPLC is dedicated to RNA up to about 35 nt long, but the latter method is always useful to cleanse RNA preparations purified on gels. Routine techniques mentioned in this chapter are described in more detail in [17].

27.3.1
HPLC Purification

When the RNA oligonucleotide is shorter than around 35 nt, it can be purified using FPLC or HPLC techniques. The best results are obtained using salt gradients

Tab. 27.1.

	Buffer A	Buffer B
MES	20 mM pH 6.2	20 mM pH 6.2
Urea	4 M	4 M
NaClO$_4$	1 mM	400 mM

A sodium perchlorate gradient is run over 70 min from 15 to 70% buffer B with a 1 ml/min flow rate [MES: 2-(N-morpholino)ethanesulfonic acid]. In the lab we run these gradient on a Dionex system equipped with a Nucleopac-PA-100 (0.45 × 25 × π mm^3) column. After a wash step at 90% buffer B, the column is re-equilibrated in 15% of buffer A to prepare the next run.

on anion exchange columns bearing quaternary amines like mono-Q matrices. HPLC presents the advantage that the column can be heated in an oven to temperatures up to 90 °C, thus promoting the unfolding of the RNA and increasing the retention time on the column for a better separation. The addition of chaotropic agents like urea or formamide enhances the effect of heating the sample. However, formamide should be used with caution in the presence of RNA. Heat leads to formamide decomposition into carbon monoxide and ammonia – the latter can very quickly hydrolyze the RNA preparation. In such denaturing conditions, the RNA mix is fractionated according to the size of the present species, close to what can be achieved using gel electrophoresis. A typical protocol is described in Table 27.1. Other HPLC purification procedures for RNA have been described elsewhere [18].

Several pitfalls should nonetheless be avoided. If the RNA has been produced by *in vitro* transcription, proteins should be removed by phenol/chloroform extraction. Otherwise the column bed may get coated with proteins and the column will lose its loading capacity over time. The sample should be assayed for precipitation by mixing with the highest salt buffer that is going to be used for separation to avoid clogging the HPLC. Since the sample is going to be heated, divalent ion contamination should be avoided. Hence, the pK_a value of the buffer should be in the acidic range so as to minimize spontaneous hydrolysis of phosphodiester bonds in the case of a contamination with divalent cations. To achieve this, we recommend to use PEEK (poly-ether-ether-ketone)-coated pumps as well as peek tubing.

27.3.2
Gel Electrophoresis

RNAs of any size (up to 500 nt) can be purified efficiently using PAGE under denaturing, semi-denaturing or native conditions. Various urea concentrations can be tried at an analytical scale before going to preparative scale in order to identify the most appropriate protocol. A sequencing electrophoresis apparatus with an aluminum back to homogenize the glass plates' temperature allowing the use of 30 × 40 cm^2 gel plates is recommended. The running temperature is usually set between

50 and 60 °C. Of course, the gel thickness has to be significantly increased when going to preparative scale (use at least 1.5-mm thick spacers) to separate precisely the RNAs of different length in the sample. The volume of gel to be prepared is thus around 250 ml.

The following equipment is needed to set up the experimental procedure:

- PAGE equipment: electrophoresis apparatus, siliconized glass plates (around 30×40 cm^2), comb and spacers (at least 1.5-mm thick).
- Acrylamide 20%/urea 8 M stock solution (made from 500 ml acrylamide-bisacrylamide 38% (w/v) acrylamide, 2% (w/v) N,N′-methylene bisacrylamide mixed with 480.5 g urea and Millipore water to give a volume of 1 l).
- $10 \times$ TBE (Tris–borate–EDTA buffer [17]).
- 8 M Urea solution.

Once a gel of the appropriate acrylamide percentage has been prepared (see [17]), the RNA solution to be fractionated is mixed with one volume of 8 M urea and then loaded onto the gel. The preparative gel electrophoresis usually requires a power value around 25 W. Progress of the migration is followed by the course of the bromophenol and xylene cyanol dyes loaded in a lane containing no RNA. At the end of the migration, glass plates are removed, the gel is wrapped in plastic film, and the RNA bands are visualized by UV shadowing using a UV lamp and a silica plate as a screen (see Chapter 3 for details).

27.3.3
RNA Recovery

The RNA contained in the visualized bands has to be eluted from the gel and concentrated before subsequent experiments. To achieve this goal, the bands are first delineated with an indelible marker on the plastic wrap. Then, they are cut out of the gel using a sterile scalpel blade.

27.3.3.1 Elution of the RNA from the Gel
The oligoribonucleotides are recovered by passive elution at 4 °C in Millipore water. The gel is crushed in a mill (A11 basic analysis mill; IKA) and poured in a 50 ml polypropylene conical tube with water (around 30 ml of gel under preparative conditions). The RNA-containing tube is placed in a "rock & roll" stirrer at 4 °C overnight. Finally, the eluted RNA solution is filtered on a 0.22-µm sterile filtration unit (Nalgene) to get rid of the acrylamide particles.

27.3.3.2 Concentration and Desalting
Whatever the technique employed to purify the RNA, it is necessary to desalt and concentrate it prior to use in crystallization trials. A very efficient way of achieving this is to use reverse-phase Sep-Pak columns that can be used on the bench (Waters Sep-Pak C_{18} Classic short-body). These are operated by gravity or using a syringe.

A classical protocol consists of the following steps:

(1) Plug the column inlet to the luer of a 10 ml syringe and fix it to a bench stand.
(2) Equilibrate the column using 10 ml of methanol.
(3) Pass through 10 ml of Millipore water.
(4) Load the sample.
(5) Wash the sample with 10 ml of Millipore water.
(6) Elute the sample with 5 ml of water/acetonitrile (1:4) in 1 ml fractions.

Three facts should be kept in mind when using Sep-Pak cartridges. The pH of the sample should not exceed 7 to guarantee efficient binding to the column bed. The loading step should not exceed 10 min to minimize loss of material due to driving by the mobile phase. If loading would take longer, the sample should be fractionated on more than one column. The column should never run dry to prevent the loss of the sample. Hence, the syringe luer should be removed with caution in intermediate steps. The next solution should be added when there is still a small volume (100 μl) of the previous phase in the syringe.

The RNA-containing water/acetonitrile solution is then evaporated to dryness in a Speed Vac. The pellet can be resuspended in the solution of choice for further studies.

27.4
Setting Crystal Screens for RNA

After purification of a sufficient amount of RNA, conditions for crystallization have to be found. An economical screening method should use the least possible amount of RNA. Therefore, it is recommended to start by screening a large number of combinations, and then switch to other methods to optimize crystal shape and size. For a broad general screen, specific crystallization sparse matrices for proteins or nucleic acids have been published [19–22] and some are commercially available (www.hamptonresearch.com, www.decode.com, www.nextalbiotech.com). These have been designed based on extensive mining of previously published crystal-yielding conditions. Although the general considerations for crystal screens of proteins equally apply to RNA, some particularities have been identified. If no crystals have been obtained during the first trials, it is often more promising to vary the sequence instead of sampling a larger variety of combinations. In the crystallization process, sequence and shape of the molecules will drive the nucleation and subsequent crystal growth through a network of packing interactions mediated among symmetry-related molecules. Considering RNA, these factors have even more drastic effect than for proteins since the former are usually less globular in shape than the latter. Thus, various RNA constructs with different sequences and helix length, in other words with different shapes, should be tried when no crystals appear for a given construct [21, 23]. Chemical synthesis makes this process relatively straightforward for small RNAs (below 30 nt). Higher crystallization temper-

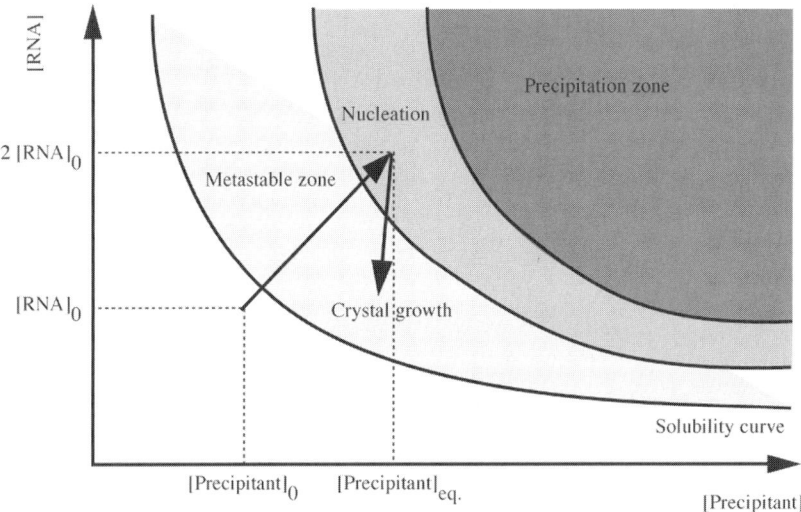

Fig. 27.1. Solubility diagram. During equilibration, the concentration of both precipitant and macromolecule increase until precipitation occurs. The formation of crystal nuclei reduces the amount of solvated macromolecule and allows the system to remain in the metastable zone where crystals can grow.

atures (37 °C) also seem to favor formation of RNA crystals. Finally, the choice of crystallization method (vapor diffusion or batch crystallization) may also influence the success.

Once suitable starting conditions have been found, the strategy consists of a rational variation of conditions. The crystallization process, as visualized in the phase diagram (Fig. 27.1), is influenced by numerous variables $x_1, x_2, \ldots$ (called *factors*), i.e. RNA sequence, crystallization temperature, buffer and pH, and kind and percentage of precipitant and salts. Each of these factors can be adjusted to different *levels* (for example, factor [LiCl] to 150 mM and factor [MPD] to 25%). In order to quantitate each observation (clear drop, precipitate, spherulites, microcrystals or crystals), an arbitrary score (*response*) is assigned to it and represented on a multi-dimensional *response surface* $f(x_1, x_2, \ldots)$. The aim of the crystallization screen is to explore this surface (Fig. 27.2) where the expected summit would yield the optimal result (best crystals). Yet, in practice, only a limited number of all possible factor combinations can be tested. The simplest approach would be varying a single factor at a time, while keeping all others constant. While it is possible to reach the optimum by sheer luck, the response surface shows that the score will more likely converge to a plateau or local maximum. Furthermore, the results of each testing series cannot be generalized and interactions between different factors are neglected.

These problems are avoided using *experimental design*, where multiple factors are varied simultaneously between different crystallization trials. Each experiment n represents a combination $Cn = (x_1, x_2, \ldots, x_k)$ of k factors. For each combination,

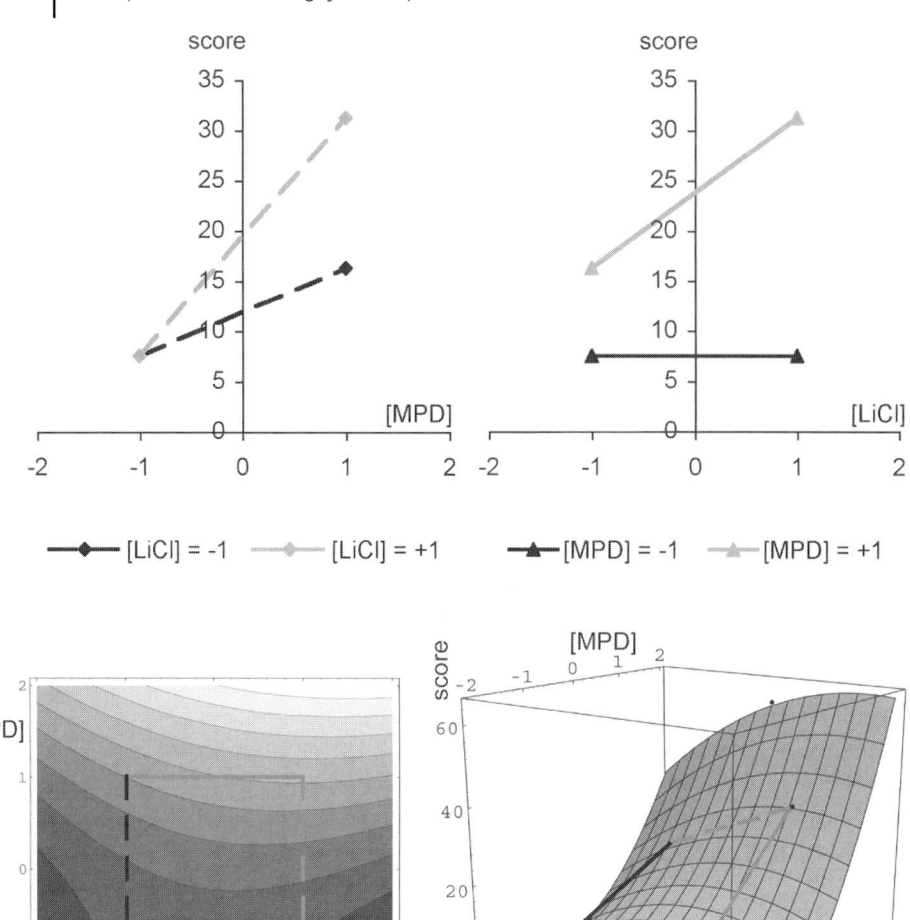

Fig. 27.2. (Top) Scoring of influence of two factors on crystal growth (see text, score in arbitrary units). (Bottom) response surface fitted to result, top view (left) and three-dimensional view (right).

multiple factors are changed simultaneously according to a predefined plan. A detailed exploration of this powerful technique is beyond the scope of this text and instead can be found in books on engineering statistics, or see references [24] and [25] for its application to crystallization. In general, at least two levels are defined for each factor at a chosen, equal distance around a central point, staking out a well-balanced experimental space in which the response of the system (the quality of crystals) is noted. Rather than just trying out some extreme values, it is important to choose reasonable values within a well-known range to avoid veering towards some jagged region, or close to an asymptote. The different levels of each factor are coded in coefficients. In our two-factor example ($k = 2$), both 50 mM/250 mM [LiCl] and 10%/40% MPD would be listed in an experimental design matrix and coded as $+1/-1$. Following scoring, the response surface spread out over all combinations indicates the direction where the best score is to be expected (Fig. 27.2). If k is not too large, one can also take into account possible interactions between factors. If factor x influences factor y, then the response surface $f(x_1, x_2)$ is not only a polynomial of $c_1 \cdot x_1$ and $c_2 \cdot y_2$, respectively, but the additional interaction term $c_3 \cdot x_1 x_2$ also has to be considered. To help with the design process, several computer programs have been made available [26]. A number of different designs have been coined with the common goal of reducing the number of experiments without compromising the well-balanced exploration of the experimental space.

Initial screens can be distinguished between those used to determine what factors are most important, and follow-up screens that allow optimization and improvement of crystal quality (Table 27.2). In experimental design, this is known

Tab. 27.2. Application of various experimental designs in crystallization.

Field of application	Experimental design	Factors and levels	No. of experiments	Comments
Initial screen: which factors are most important?	full 2-level factorial design	k factors at 2 levels	2^k	accounts for interactions between factors, but too many experiments necessary if $k > 5$ factors
	incomplete 2-level factorial design		2^{k-p}	p factors are confounded, effect of interactions cannot be evaluated, but less experiments necessary
	Plackett–Burman design		$k + 1$	greatly reduced number of experiments; interaction bias neglected
Optimization	steepest ascent	k factors at 2 levels		follow-up on 2-level design: reduction of step size when approaching maximum
	central composite Box–Behnken; Hardin–Sloane	k factors at 5 levels		follow-up on steepest ascent design. quadratic model, approach to optimum
	Randomized block designs	3–5 factors at 3–4 levels		one single factor of primary interest interaction bias neglected
	Simplex matrix	k factors at 3 levels	initially $k + 1$	iterative triangulation towards optimum

as the "Box–Wilson strategy" [27]. The first group of screens is generally based on a so-called *factorial plan* which determines the polynomial coefficients of a function with k variables (factors) fitted to the response surface. It can be shown that the number of necessary experiments n increases with 2^k if all interactions are taken into account. Instead of running an unrealistic large number of initial experiments, the full factorial matrix can be advantageously replaced by a fractional factorial matrix or a Plackett–Burman design [28]. Here, interactions between factors are partially or completely neglected. For example, if a multiplicative effect of salt concentration and MPD can be ruled out, the interaction between these factors can be neglected, thereby reducing the number of necessary experiments.

Based on the response surface obtained, a second round of optimization follows, using the *steepest ascent method* where the direction of the steepest slope indicates the position of the optimum. Alternatively, a quadratic model can be fitted around a region known to contain the optimum somewhere in the middle. This so-called *central composite design* contains an imbedded factorial design with center points, points at $-1/+1$ and an additional group of outlying "star points" α as upper and lower limits, which allows an estimation of the curvature (see Fig. 27.3 for an example with combinations of three factors). There are alternative designs, if the number of factors is small and optimization is the main goal. *Randomized block designs* (Latin Squares and Greco-Latin Squares) are useful if there is one main factor to consider. The design helps to separate it from the influence of *nuisance factors* that may affect the measured result, but are not of primary interest.

Finally, the simplex design has also been adopted for crystallization purposes [29]. This is an iterative approach starting with one more combination than factors under investigation. In an example with three factors at three equally spaced levels 0, p and q, the first set consists of combinations C0 (0, 0, 0), C1 (p, q, q), C2 (q, p, q) and C3 (q, q, p) (Fig. 27.3). Combination C2 giving the worst result, it is replaced in the following round by combination C4 with coordinates exactly opposite to C2, where the mirror plane is defined by C0, C1 and C3. Comparing these

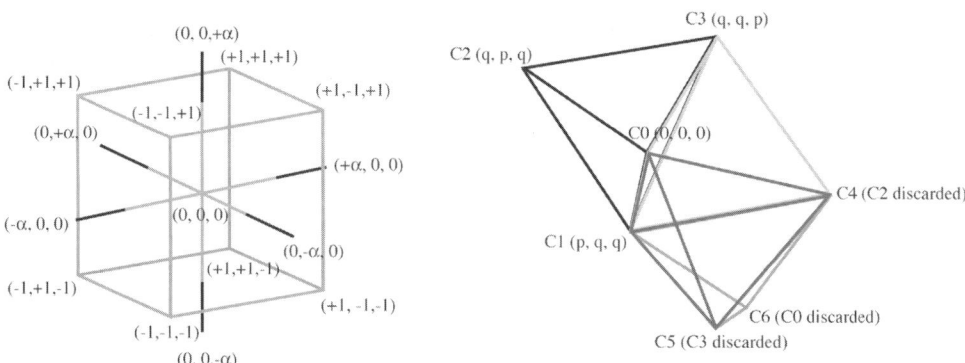

Fig. 27.3. Optimization designs. (Left) Centered composite design with three factors. (Right) Three factors optimized in four steps using a simplex design.

points with C4, the worst result is now C3 and therefore replaced by mirror point C5, and so forth. After several rounds of triangulation in the experimental space, the optimum is reached when no further improvements are observed. While multiple rounds of optimization are required, this extremely economical approach is especially useful when too little sample is available for extended factorial plans.

27.4.1
Renaturing the RNA

Prior to setting up crystallization experiments, the concentrated RNA has to be properly folded in the native state. This is performed by a heating step in a heating-block for 1 min at 70–85 °C (depending on the melting temperature) in the presence of monovalent salts only. Then, the solution is left in the switched-off heating-block to cool down slowly until room temperature is reached. In order to avoid self-cleavage of the RNA, the pH is usually chosen slightly acidic and divalent cations are added only around 35 °C.

27.4.2
Setting-up Crystal Screens

The main technique employed to setup crystal screens is the vapor diffusion method either in the hanging drop or sitting drop setup. This method is based on slowly concentrating the droplet solution against a reservoir solution of infinite volume (milliliter scale) compared to the volume of the droplet (microlitre scale, see Fig. 27.1). The choice between the various plastic-ware commercially available will be driven by the amount of RNA sample and the number of crystallization conditions to be tested. Nowadays, more and more laboratories have the opportunity to use crystallization robots that permit to decrease the drop volume to hundreds of nanoliters and, with the same amount of material, to set up thousands of trays on very short time scales.

The different crystallization screens are set up by adding the biomolecule solution to the crystallization solution. Once a first hit has been obtained, one needs to optimize the conditions. In the example of the aminoglycoside/ribosomal A site complexes (see below), crystals were optimized using different crystallization solutions to test various glycerol/MPD ratios (see Table 27.3). All trials are done at 37 °C using the hanging drop vapor diffusion method: 1 μl of RNA-antibiotic complex solution is added to 1 μl crystallization solution and equilibrated over a reservoir containing 500 μl of 40% MPD.

27.4.3
Forming Complexes with Organic Ligands: The Example of Aminoglycosides

RNA molecules bind various organic ligands. Different RNA fragments based on the *Escherichia coli* A site located in the penultimate helix of the 16S ribosomal RNA [10–12, 30] have been tested in the presence of their natural ligands, anti-

Tab. 27.3. Crystallization conditions testing various glycerol/MPD ratios.

Crystallization condition (reservoir)		1	2	3	4	5	6
	Stock						
MPD (%)	60	1	2	2	2	1.5	2
Glycerol (%)	100	5	4	2	1.5	1	1
Na cacodylate pH 6.4 (M)	0.85	0.05					
KCl (M)	3	0.15					
Glycerol/MPD ratio		5	2	1	0.75	0.67	0.5

biotics of the aminoglycoside family. The RNA construct was designed as a self-complementary oligonucleotide so as to incorporate two A sites in a head-to-head manner (Fig. 27.4). This choice eliminates two drawbacks. First, since the internal loop is asymmetric, one would otherwise need to synthesize, purify and mix 1:1 two different RNA strands in order to obtain a single site. Second, one could also use a single site capped by a stable hairpin of the GNRA family, for example. However, in such cases, it is frequently observed that the crystallized structure reveals a

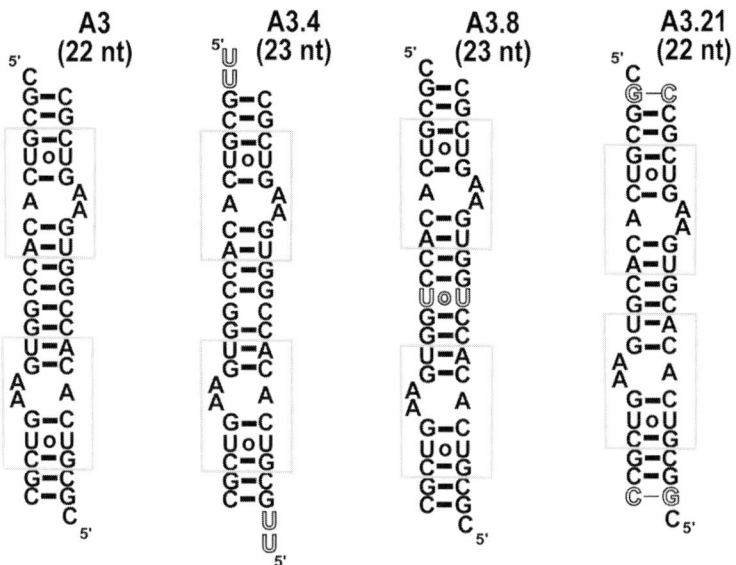

Fig. 27.4. Four self-complementary RNA fragments containing a tandem array of two *E. coli* 16S ribosomal A site modules [10–12].

full duplex with several non-Watson–Crick pairs [31]. In order to monitor the effect of sequence variations for the crystallization of these complexes, various modifications have been performed such as addition of a 5′ UU overhang, insertion of a UU pair or moving the two A sites closer to one another (Fig. 27.4).

Routinely, the purified oligoribonucleotides are solubilized in a solution containing 2 mM RNA, 25 mM NaCl, 5 mM $MgSO_4$ and 100 mM sodium cacodylate buffer, pH 6.4. This solution is first heated to 85 °C for 2 min and then slowly cooled until a temperature of 37 °C is reached. One volume of a 4 mM aminoglycoside solution is added to the RNA solution and incubated for 2 h at 21 °C. The two solutions should be at the same temperature. Since the RNA fragments contain two antibiotic-binding sites, the aminoglycoside concentration is twice the RNA concentration. A general rule is that the organic ligand concentration should be 100 times higher than its dissociation constant (K_D) to ensure binding site saturation which is usually easily achieved in the millimolar range.

In the example of the aminoglycoside/A Site complexes, different crystallization solutions were prepared to test various glycerol/MPD ratios: 5, 2, 1, 0.75, 0.67 and 0.5 (Table 27.3). All trials are performed at the optimal temperature of 37 °C using the vapor diffusion method in the hanging drop setup: 1 μl RNA-antibiotic complex solution is added to 1 μl crystallization solution and equilibrated over a 40% MPD reservoir.

27.4.4
Evaluate Screening Results

Since first crystallization attempts will not automatically yield crystals or they may be of too poor quality for X-ray diffraction experiments, evaluating screening results is required prior to proceed to crystallization optimization. This is performed by using a binocular microscope hooked up to a digital camera to record observations. A numerical scoring value describing the content of the droplet (Fig. 27.5) is reported on a paper scoring sheet (Table 27.4).

Two weeks are enough for droplets of about 3 μl to equilibrate under any conditions [32]. During this period, droplets should be inspected daily to follow up the appearance of crystals. Crystals may still form after 2 weeks, but this is less likely in the case of oligonucleotides. Crystals can then be cryo-protected and frozen or capillary mounted to be tested.

27.4.5
The Optimization Process

Here are provided non-exhaustive guidelines to interpret the droplet content of crystallization screenings (Table 27.4) and possible ways to optimize positive hits. See also [33] for more details.

- *Clear drops* – indicates that the RNA supersaturation state has not been reached, the RNA concentration is outside the nucleation zone (Fig. 27.1). These experi-

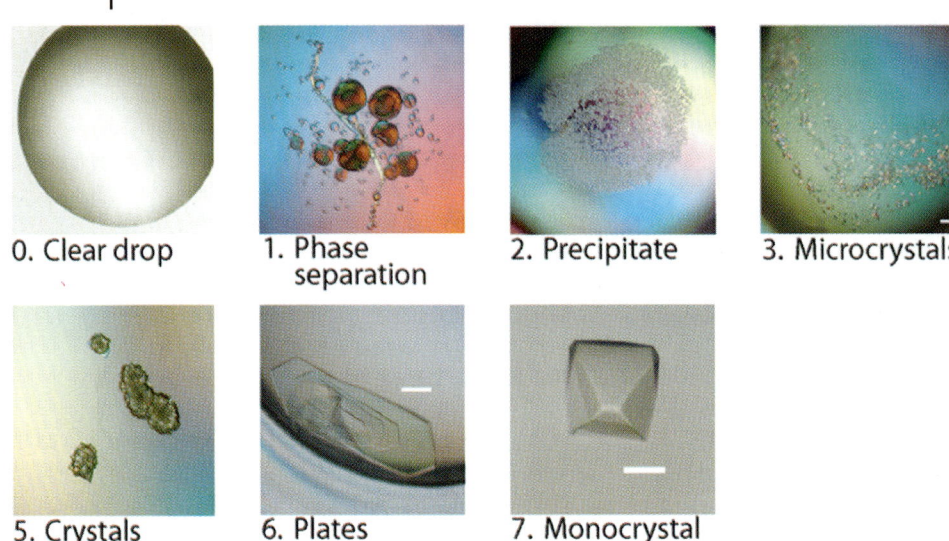

Fig. 27.5. Numerical scoring terms.

Tab. 27.4. Scoring sheet for rows A and B from a 24-well LINBRO crystallization plate.

| CRYSTALLIZATION PLATE NUMBER: _____ |
| SAMPLE & ADDITIVES: _____ |
| DATE OF SETTING: _____ |
| TEMPERATURE: |

	DATE								
A	1								
	2								
	3								
	4								
	5								
	6								
B	1								
	2								
	3								
	4								
	5								
	6								

ments must be repeated with higher sample and/or salt concentrations. The temperature can also be lowered.
- *Phase separation* – indicates a need to increase the monovalent salt concentration and/or to test a smaller precipitant concentration (MPD, PEG) to make the RNA sample more soluble.
- *Light precipitates* – indicates that the relative supersaturation between sample and reagent is too high. Prepare new tests with a decreased RNA and/or precipitant concentration or dilute the droplet by vapor diffusion by adding water into the reservoir.
- *Strong precipitates* – indicates that the sample has been partially denatured. The sample must be tested at a lower concentration or less salt should be used. Note that a fresh test should be prepared in this case.
- *Small precipitates* – must be carefully inspected using polarized light because it may contain a microcrystalline shower. A microscope with a magnification factor greater than 100-fold can be useful in this case.
- *Cluster of homogenous crystals* – try to slow down the nucleation; test the conditions at lower temperature, cover the reservoir solution with oil to slow down the water diffusion.

Different parameters can be tested to optimize the growth of single monocrystals: salt type, additives, temperature. Finally, if no crystal can be obtained with a given construct, new RNA sequences have to be designed so as to provide new potential scaffoldings to help crystal packing.

27.5 Conclusions

Protocols to purify and concentrate large amounts of RNA under controlled buffer and salt conditions for crystallization experiments have been described. The crucial points to preserve the RNA sample are: the use of slightly acidic pH and the avoidance of divalent cations. Usually RNases, feared by most RNA scientists, are introduced into the solution by an upstream experiment such as plasmid and protein preparations (e.g. T7 RNA polymerase). It is, thus, strictly recommended to assess the RNase activity of a solution before using it on the whole RNA sample by incubating an aliquot in the presence of the RNA for few hours and check the extent of the digestion by PAGE. Then strategies to crystallize RNA oligonucleotides in the presence or absence of ligands have been presented. A peculiarity of crystallization experiments is the absence of a negative control. To circumvent this, we advise to attempt to crystallize simultaneously several related RNA sequences in the same well. Only one or few RNAs, if any, will crystallize if any, leading to the conclusion that RNA crystals instead of salt crystals have been obtained. In cases where no crystal is observed, it is recommended to design a new set of oligonucleotides bearing slight sequence changes in order to enhance interactions between symmetry-related molecules that lead to regular crystal packing interactions.

References

1. N. Ban, P. Nissen, J. Hansen, P. B. Moore, T. A. Steitz, *Science* **2000**, *289*, 905–920.
2. W. M. Clemons, Jr, D. E. Brodersen, J. P. McCutcheon, J. L. May, A. P. Carter, R. J. Morgan-Warren, B. T. Wimberly, V. Ramakrishnan, *J. Mol. Biol.* **2001**, *310*, 827–843.
3. M. M. Yusupov, G. Z. Yusupova, A. Baucom, K. Lieberman, T. N. Earnest, J. H. Cate, H. F. Noller, *Science* **2001**, *292*, 883–896.
4. N. B. Leontis, J. Stombaugh, E. Westhof, *Nucleic Acids Res.* **2002**, *30*, 3497–3531.
5. N. B. Leontis, J. Stombaugh, E. Westhof, *Biochimie* **2002**, *84*, 961–973.
6. J. H. Cate, A. R. Gooding, E. Podell, K. Zhou, B. L. Golden, C. E. Kundrot, T. R. Cech, J. A. Doudna, *Science* **1996**, *273*, 1678–1684.
7. B. L. Golden, A. R. Gooding, E. R. Podell, T. R. Cech, *Science* **1998**, *282*, 259–264.
8. A. S. Krasilnikov, X. Yang, T. Pan, A. Mondragon, *Nature* **2003**, *421*, 760–764.
9. B. Masquida, C. Sauter, E. Westhof, *RNA* **1999**, *5*, 1384–1395.
10. Q. Vicens, E. Westhof, *Structure* **2001**, *9*, 647–658.
11. Q. Vicens, E. Westhof, *Chem. Biol.* **2002**, *9*, 747–755.
12. Q. Vicens, E. Westhof, *J. Mol. Biol.* **2003**, *326*, 1175–1188.
13. M. Zuker, A. B. Jacobson, *RNA* **1998**, *4*, 669–679.
14. H. Isambert, E. D. Siggia, *Proc. Natl Acad. Sci. USA* **2000**, *97*, 6515–6520.
15. Y. Ding, C. E. Lawrence, *Nucleic Acids Res.* **2003**, *31*, 7280–7301.
16. M. Zuker, *Nucleic Acids Res.* **2003**, *31*, 3406–3415.
17. J. Sambrook, D. W. Russell, *Molecular Cloning: A Laboratory Manual*, Cold Spring Harbor Laboratory Press, Cold Spring Harbor, NY, **2001**.
18. A. C. Anderson, S. A. Scaringe, B. E. Earp, C. A. Frederick, *RNA* **1996**, *2*, 110–117.
19. J. A. Doudna, C. Grosshans, A. Gooding, C. E. Kundrot, *Proc. Natl Acad. Sci. USA* **1993**, *90*, 7829–7833.
20. I. Berger, C. H. Kang, N. Sinha, M. Wolters, A. Rich, *Acta. Crystallogr. D* **1996**, *52*, 465–468.
21. W. G. Scott, J. T. Finch, R. Grenfell, J. Fogg, T. Smith, M. J. Gait, A. Klug, *J. Mol. Biol.* **1995**, *250*, 327–332.
22. J. H. Cate, J. A. Doudna, *Methods Mol. Biol.* **1997**, *74*, 379–386.
23. A. C. Anderson, B. E. Earp, C. A. Frederick, *J. Mol. Biol.* **1996**, *259*, 696–703.
24. H. Petersen, *Grundlagen der Statistik und der statistischen Versuchsplanung*. Ecomed, Landsberg/Lech, **1991**.
25. C. W. J. Carter, in: *Crystallization of Nucleic Acids and Proteins: A Practical Approach*, 2nd edn, A. Ducruix, R. Giegé (eds), IRL Press, Oxford, **1999**, pp. 75–120.
26. C. D. Potter, *The Scientist* **1994**, *8*, 18.
27. G. E. P. Box, W. G. Hunter, J. S. Hunter, *Statistics for Experimenters. An Introduction to Design, Data Analysis, and Model Building*, Wiley, New York, **1978**.
28. R. L. Plackett, J. P. Burman, *Biometrika* **1946**, *33*, 305–325.
29. B. D. Prater, S. C. Tuller, L. J. Wilson, *J. Crystal Growth* **1999**, *196*, 674–684.
30. F. Walter, Q. Vicens, E. Westhof, *Curr. Opin. Chem. Biol.* **1999**, *3*, 694–704.
31. V. Kacer, S. A. Scaringe, J. N. Scarsdale, J. P. Rife, *Acta. Crystallogr. D* **2003**, *59*, 423–432.
32. V. Mikol, J.-L. Rodeau, R. Giegé, *Anal. Biochem.* **1990**, *186*, 332–339.
33. A. Ducruix, R. Giegé, *Crystallization of Nucleic Acids and Proteins: A Practical Approach*, 2nd edn, IRL Press, Oxford, **1999**.

II.3
Fluorescence and Single Molecule Studies

28
Fluorescence Labeling of RNA for Single Molecule Studies

Filipp Oesterhelt, Enno Schweinberger and Claus Seidel

28.1
Introduction

Visualizing single molecules is no mystery. The human eye is sensitive enough to detect single photons in non-color vision. Certain fluorophores can give up to 100 000 fluorescence photons per second when excited intensively – certainly sufficient to create a colored visual impression when looking through a microscope. However, developing technical devices that match that sensitivity of the human eye was a major task. Thus it took until the late 1980s before the first single molecule fluorescence measurement was performed [1, 2].

Single molecule measurements allow a much more detailed investigation of structural and dynamic characteristics at the molecular level than ensemble measurements. In ensemble measurements, only average values for the measured fluorescence parameters are obtained. In contrast, single molecule experiments allow the direct observation of the structure or folding pathways of the individual molecules.

When analyzing samples that contain mixtures of molecules or even identical molecules that can be in different states, in ensemble measurements many details would get lost due to averaging or could even give mean values of no relevance. However, when measuring one molecule after the other, different species can be easily distinguished. Subsequently, average values can be calculated for each species alone as well as for the whole sample.

To measure the dynamics of folding, in ensemble measurements the process needs to be triggered for synchronization. However, in which case, multiple pathways are averaged out and fluctuations between different states as well as short-living intermediates cannot be observed. In contrast, single molecule measurements allow us to measure the full individual dynamic behavior over different time ranges from milliseconds to minutes without triggering.

Due to the high fluorescence intensities of most fluorophores, the decisive factor for effective single molecule detection is the reduction of the unspecific background signal than the detection sensitivity. Additionally, one has to ensure that

Handbook of RNA Biochemistry. Edited by R. K. Hartmann, A. Bindereif, A. Schön, E. Westhof
Copyright © 2005 WILEY-VCH Verlag GmbH & Co. KGaA, Weinheim
ISBN: 3-527-30826-1

the detected fluorescence originates from only one molecule. In order to fulfill both requirements, the measuring volume has to be minimized by shrinking the excitation and/or the detection volume. Depending on the goal of the study, different techniques are applied to realize small measuring volumes: optical near-field excitation through pointed fibers (scanning near-field microscope), total internal reflection (TIR) at a glass–water interface and the confocal microscopy technique (CM) where a laser beam is focused into the sample and a special optical arrangement restricts the excitation as well as the detection volume.

Applying TIR, the molecules under investigation are usually fixed to the surface, which has the advantage of long observation times, so that even slow dynamics are accessible. However, care has to be taken regarding the possibility that the biomolecule can be strongly influenced by interaction with the surface. To observe single molecules on a surface, their density should not be more than one per 10 μm^2 to avoid an overlap of the single fluorophores images.

The CM setup gives detection volumes in the femtoliter range, which corresponds to the size of bacteria. Compared to the TIF technique, the CM setup has an inferior signal-to-noise ratio, but measurements on freely diffusing molecules are possible, avoiding the risk of surface artifacts. To guarantee that mostly only one fluorophore will diffuse through the focus of the CM setup at a time, the fluorophore concentration has to be in the range of 100 pM, leaving approximately one fluorophore in the detection volume.

A major point which has to be considered in single molecule detection is photodestruction of the fluorophore. A fluorophore can typically do 10^6-10^7 absorption–emission cycles on average before it is likely to be destroyed by chemically reacting in the excited state, mainly with oxygen. Thus, the total number of photons that can be detected from one fluorophore is limited. They can be detected either in a short time, when exciting the fluorophore with a higher intensity, getting a high time resolution to observe fast dynamics, but only for a short observation time, or, when exciting with a lower intensity, one can achieve a long observation times up to minutes, but will miss fast dynamic events.

A prerequisite for single molecule fluorescence detection is the existence of a suitable fluorophore. In biomolecules, however, only a few intrinsic fluorophores like the flavin-adenine dinucleotide (FAD) are suitable for single molecule spectroscopy [3]. In nucleic acids, one can substitute bases by some fluorescent analogs, e.g. 2-aminopurine or ethenoadenosine. Unfortunately they all suffer from low photostability and are therefore not suitable for single molecule measurements. Therefore, in most cases efficient fluorescent dyes are covalently coupled to the sample. These fluorophores are used as probes to test the molecular properties of the biomolecules.

In fluorescence microscopy, the static distribution of fluorescently labeled molecules in cells is directly observed. Their dynamic transport between different cell compartments can only be observed when it is triggered either by the injection of labeled molecules or by bleaching them in certain areas and observing their redistribution, called fluorescence recovery after photobleaching (FRAP). From these methods only average diffusion coefficients can be calculated and the mobility of the single molecules is still hidden behind the average.

The full dynamic behavior of the molecules is only accessible when the movement of individuals is observed. Today, digital cameras are available that are sensitive and fast enough to take video sequences of single fluorescent molecules. This allows tracking the pathway of individual molecules through the cell [4], to observe their diffusion in membranes [5] or their directed transport in cells.

The motion of a labeled molecule depends on its mass and, thus, the analysis of dynamic behavior gives information about the change of mass due to molecular complexation. Another technique able to analyze the mobility of molecules is fluorescence correlation spectroscopy (FCS). In this technique, the intensity fluctuations in the fluorescence signal are observed that are caused by single molecules diffusing in and out of a confocal detection volume. The analysis of the signal fluctuations gives the typical diffusion times at the defined position of the detection volume [6–8].

Several physical properties of the fluorophores are influenced by their surroundings and by that provide functional information about the sample. Close contact of the fluorophore to certain molecules can lead to quenching and thus to the emission of less fluorescence photons. In ensemble assays this decrease of fluorescence intensity can be used to monitor binding events. In addition to the fluorescence intensity, the fluorescence lifetime changes when the fluorophore is quenched and thus also can indicate molecular interaction.

A further important parameter of fluorescence is anisotropy. It can give information about the mobility of the fluorophore. When exciting fluorophores are in solution with linear polarized light, the fluorescence emission is not isotropic, i.e. has an anisotropy, even if the fluorophores have a random orientation. If the fluorophore rotates within its lifetime, i.e. in the time between excitation and fluorescence emission, this leads to a decrease of anisotropic emission. This can be used to determine the molecular rotational time [9] and, thereby, the binding events due to the change in mass of the molecular complexes.

In addition to these one-label techniques that allow the analysis of position and mobility of one molecule or molecular complex, techniques using two different labels allow us to analyze the proximity of different molecules.

With a mixture of molecules labeled with two different fluorophores, one can visualize the two fluorophores separately by taking images in the spectral ranges of only one or the other fluorophore. Superposition of those two images allows one to determine the co-localization with an accuracy of up to tenths of nanometers, far below the wavelength of light, depending only on the number of photons one can detect per molecule [10, 11].

Using two different labels in fluorescence correlation analysis, interactions can be monitored much better than via the diffusion times. So-called two-color cross-correlation analysis detects only fluctuations that occur simultaneously in two different detection channels, each sensitive for only one of the two fluorophores.

Thus, only that part of the intensity fluctuations is detected which originates from both fluorophores moving together bound to one complex [12–14].

28.2
Fluorescence Resonance Energy Transfer (FRET)

Co-localization studies can only measure distances down to tenths of nanometers. To measure inter- or intramolecular distances below 10 nm, one can take advantage of a process called FRET. Instead of emitting a photon, an excited fluorophore (called a donor) can transfer its energy to another fluorophore (called an acceptor) which instead emits a photon. The acceptor has to be in a vicinity of 1–10 nm and its absorption spectrum has to overlap with the emission spectrum of the donor. From this decrease of the donor fluorescence and the increase of acceptor fluorescence the efficiency of the energy transfer and the distance between the two fluorophores can be calculated. In addition to the donor and the acceptor fluorescence intensity, the donor lifetime is also reduced if energy transfer occurs and thus the measured change in lifetime can also be used to calculate the distance.

FRET is probably the most versatile fluorescence tool for single molecule measurements. It allows one to get both dynamic and structural information at the single molecule level. If the donor and acceptor are bound to two different segments of an RNA structure, all structural changes that affect the distance between the segments will be visible as a change in the energy transfer. Conformational changes can be observed at a time range from milliseconds up to minutes. While dwell times or time constants of fluctuations can be directly calculated from the FRET signal, the measurement of exact distances is more complicated. Different parameters have to be taken into account when calculating distances from the measured fluorescence intensities or the measured lifetime. Here, single molecule measurements are a great advantage, since many artifacts that occur in ensemble measurements due to averaging can be eliminated, which allows a more accurate calculation of distances.

28.2.1
Measurement of Distances via FRET

According to Förster's theory [15, 16] the efficiency E of energy transfer from the donor fluorophore to the acceptor fluorophore depends on their distance R:

$$E = R_0^6/(R_0^6 + R^6)$$

Here R_0 is called the Förster distance. At that distance the transfer efficiency is 50%. R_0 is a typical parameter for each donor–acceptor pair.

R_0 depends on several parameters:

$$R_0^6 = 8.8 \times 10^{-28} J\kappa^2 \Phi_D n^{-4}$$

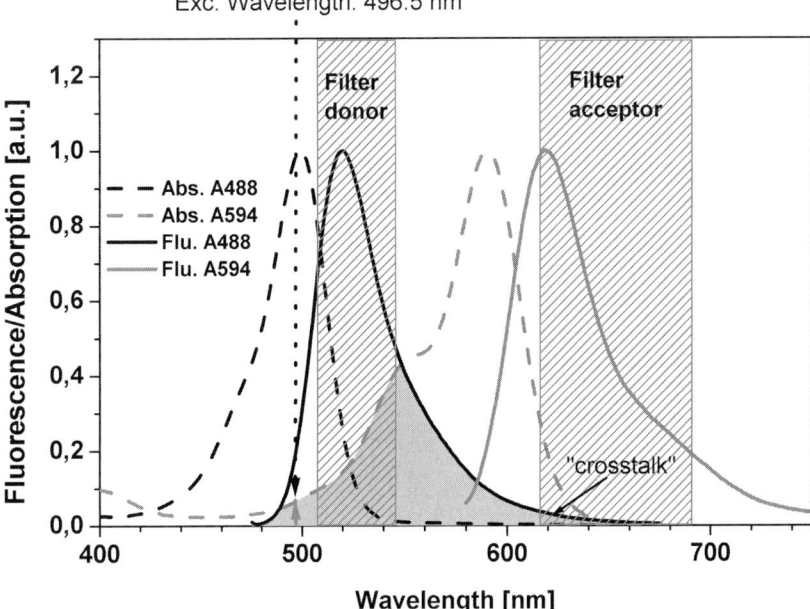

Fig. 28.1. Donor (Alexa 488) and acceptor (Alexa 594) spectra are shown. FRET occurs if the donor emission overlaps with the acceptor absorption; illustrated here by the grey area. Fluorescence detection with different filters allows detection of donor and acceptor fluorescence separately. The fluorescence intensities have to be corrected for direct excitation (indicated by the grey arrow at 496.5 nm) of the acceptor and for crosstalk of the donor fluorescence into the acceptor channel.

Here J is the spectral overlap between donor emission and acceptor absorption. κ^2 accounts for the relative orientation of the two dyes. If both dyes are free to rotate on a time scale faster than their fluorescence lifetimes, an averaging of all orientations results in $\kappa^2 = 2/3$. Φ_D is the fluorescence quantum yield of the donor fluorophore and n the optical refraction index of the medium. For the most used donor–acceptor pairs, R_0 is in the range between 2 and 6 nm. See Fig. 28.1.

The transfer efficiency needed to calculate distances can be determined in two independent ways. Since the lifetime of the donor fluorophore is shortened by the energy transfer, the ratio of the donor lifetimes in the presence ($\tau_{D(A)}$) and absence (τ_D) of the acceptor is thus a measure of the transfer efficiency:

$$E = 1 - (\tau_{D(A)}/\tau_D)$$

On the other hand, the transfer efficiency can also be determined by measuring the fluorescence intensities of the donor (F_D) and the acceptor (F_A):

$$E = (1 + F_D \Phi_A / F_A \Phi_D)^{-1}$$

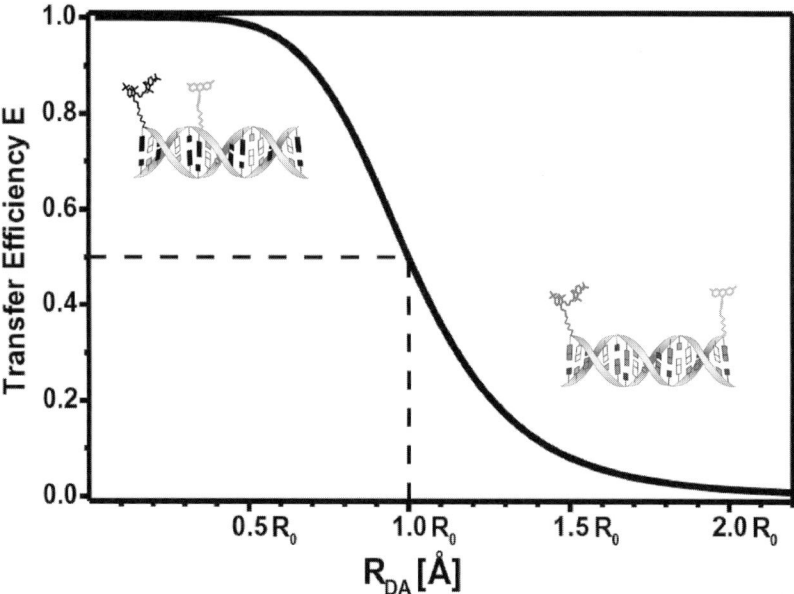

Fig. 28.2. FRET efficiency according to the Förster theory. At small distances the energy transfer approaches 100% and decreases for longer distances. At R_0 the FRET efficiency is 50%. It is there that FRET is most sensitive to distance changes.

Here Φ_D and Φ_A are the fluorescence quantum yields for donor and acceptor. The simultaneous measurement of the fluorescence lifetime as well as the fluorescence intensity for the donor and acceptor fluorescence reduces statistical errors and helps to identify systematic ones.

To test the accuracy of different single molecule or ensemble techniques that measure distances via FRET, several groups used the DNA helix with its well-known structure as a molecular ruler [9, 17–19]. See Fig. 28.2.

28.3
Questions that can be Addressed by Single Molecule Fluorescence

Very good work is published about the dynamics of RNA structures measured in detail by single molecule fluorescence. Conformational fluctuations of three- and four-way junctions as well as protein-induced structural changes have been investigated *in vitro*. Folding pathways of complex RNA structures like ribozymes were analyzed [20–24]. Single molecule measurements in living cells promise detailed insight in molecular distribution and dynamic processes. The simultaneous acquisition of lifetime information, FRET efficiencies and anisotropies can reveal molecular conformation and its interaction with other molecules. Thus, one could get a functional image of specific molecules in a cell. The autofluorescence of cells makes single molecule detection difficult, but single molecule imaging in cells has nonetheless already been shown.

Meanwhile, the advantages of single molecule detection are also used in different techniques for nucleic acid analysis. Single molecule detection is applied especially in sequencing and fragment sizing in order to increase the throughput and reduce the amount of sample needed.

28.3.1
RNA Structure and Dynamics

The FRET technique was used by Kim et al. [20] to study Mg^{2+}-facilitated conformational changes in a three-helix junction – a ribosomal junction that initiates the folding of the 30S ribosomal subunit. The junctions were bound to a surface and labeled with donor and acceptor at the ends of two helical arms. Structural changes of the junction caused a change in the angle between the helical arms and thus could be observed as a change in the energy transfer. The authors analyzed the fluctuation of the single molecule signals visible in the measured time traces by correlation analysis. In contrast to solution FCS measurements, fluorescence correlation of surface-bound single molecules is not limited by diffusion times. The experiments showed that for this particular junction the structural fluctuations were Mg^{2+} and Na^+ dependent, but did not result from binding and unbinding of the ions. Just the intrinsic conformational fluctuations were altered by the uptake of ions.

Ha et al. used FRET to measure structural changes induced by protein binding [21]. They observed the conformational change of the 16S rRNA three-way junction induced by the binding of ribosomal protein S15. In ribosomal assembly the binding of the S15 protein nucleates the assembly of the central domain of the 30S ribosomal subunit by folding the RNA such that distant sites are brought together, allowing the binding of subsequent proteins in the assembly cascade. This distance change was monitored by FRET. In two-color images of surface-bound junctions, those with no protein bound could easily be distinguished from the ones that formed a complex with a S15 protein. The protein-binding-induced structural change reduced the distance between donor and acceptor, leading to high energy transfer and high red acceptor fluorescence. In the same image, protein-free junctions are distinguished by their high green donor fluorescence.

Zhuang et al. demonstrated the potential of FRET for the study of folding of RNA structures. They measured the complex dynamics of the hairpin ribozyme in its minimal form [22]. There it consists of two helix–loop–helix segments. These associate non-coaxially in the active folded structure in a way that brings catalytically important loop nucleotides into close proximity. Donor and acceptor bound to the ends of the two segments allowed direct observation of the enzyme opening and closing. The enzyme–substrate complex exists in either docked (active) or undocked (inactive) conformations. The authors found complex structural dynamics with several docked states of distinct stabilities. With the complex structural dynamics they could quantitatively explain the heterogeneous cleavage kinetics common to many catalytic RNAs.

However, in the natural form where the ribozyme assembles in the context of a four-way helical junction, the folding pathway includes an additional intermediate state. Tan et al. [23] found that this intermediate step originates from the four-way junction and is obligatory for the folding process. This intermediate step, which could not be discovered by ensemble measurements, brings the two loop elements in close vicinity. It increases the probability of their interaction and accelerates the folding by nearly three orders of magnitude, allowing the ribozyme to fold rapidly in physiological conditions.

Zhuang et al. [24] used the single molecule FRET technique to analyze much bigger RNA structures. They observed folding steps of individual *Tetrahymena* group I intron ribozymes. The analysis of time trajectories allowed them to identify a rarely populated state, which was not measured by ensemble measurements before. Intermediate folding states and multiple pathways were observed. As well as previously established pathways, a new folding pathway could also be observed.

28.3.2
Single Molecule Fluorescence in Cells

The spatial and temporal distribution of RNA species in living cells has been studied successfully in many systems. The mechanisms of RNA localization and pathways of cellular transport in transcription, splicing and translation processes have been investigated.

When observing single molecules in cells one has to take into account the autofluorescence background. To reduce the background fluorescence, one should select the fluorophores emission spectrum according to a minimal overlap with the autofluorescence spectrum. Also, several reagents have been described that reduce autofluorescence, but they cannot remove it completely [25].

28.3.2.1 Techniques used for Fluorescent Labeling RNA in Cells
Several techniques have been established to introduce fluorophores into living cells. Direct microinjection of labeled RNA allows us to use photostable fluorophores with high quantum efficiency [26–28]. After microinjection one can easily follow the temporal and spatial distribution by live cell microscopy. Transport between cytoplasm and nucleus and the assembly of RNA in different foci was observed by several groups [29–31]. Microinjection also makes it possible to control the number of molecules injected to a single cell. Knemeyer et al. showed that one can count the number of injected molecules by positioning a confocal detection volume at the end of the micropipette [28]. However, it must be taken in account that the injected labeled nucleotides may follow different routes and show different kinetics due to the fluorescent modifications. When working with excess RNA, cellular transport and processing routes may also become saturated.

Another approach uses RNA-binding domains that are fused to Green Fluorescent Protein (GFP) [32]. This approach is useful for organisms that cannot be microinjected or do not allow the penetration of macromolecules. Due to unfavorable photophysical properties and low photostability of all fluorescent proteins, the detection of single RNA molecules is only achieved if multiple binding is possible.

Instead of directly labeling RNA, the hybridization of labeled oligonucleotide probes to endogenous RNA, called fluorescence *in vivo* hybridization (FIVH), is often used to label RNA specifically in live cells [33, 34]. Different groups used nucleotide probes complementary to polyadenylated RNA, spliceosomal small nuclear RNA or ribosomal rRNA to detect their *in vivo* distribution dynamics [28, 35]. In the cytoplasm ribonucleic probes are degraded by RNases. To circumvent this, nuclease resistant 2′-*O*-methyl oligoribonucleotides with high affinity to complementary RNA or DNA strands are used. A major problem in FIVH is to distinguish between specifically hybridized probes and background due to non-bound fluorescent probes.

The concept of molecular beacons, developed for the detection of nucleic acid amplification products, is used to localize specific nucleic acids in fixed cells. They can also be applied to live cells. Molecular beacons are double-labeled oligonucleotides with a fluorophore at one end and a quencher at the opposite end. The molecular beacons are designed to have a target-specific probe sequence positioned centrally between two short self-complementary segments. These special sequences form a structure that consists of a loop and a stem, so that the dye and quencher are in close vicinity. Upon hybridization to the target, the stem opens and the fluorophore starts to fluoresce. This technique has been used to detect different mRNAs in the cytoplasm of living cells [36, 37]. Unfortunately it is reported that molecular beacons, especially in the nucleus, often do not show better results than linear oligonucleotide probes, meaning that molecular beacons often already open before hybridization to the target sequence due to non-specific interaction with nuclear proteins or RNAs.

Another promising approach uses FRET to detect only specific hybridized probes. In this case the acceptor and donor are bound to two different nucleotide probes, which hybridize with the target sequence such that donor and acceptor are placed close enough for energy transfer to occur [37–40]. FRET occurs only when donor and acceptor probes are hybridized to the target sequence simultaneously. However, the probability to assemble a FRET probe is much smaller than to hybridize one single fluorophore probe alone.

The detection of singly fluorescently labeled oligo(dT) probes hybridized to mRNA was demonstrated [28, 41]. This was possible due to the use of spectrally resolved fluorescent lifetime microscopy. The combination of spectral and lifetime information allows one to distinguish clearly between autofluorescence background and the used probes. Oligo(dT) nucleotides were microinjected in 3T3 mouse fibroblast cells and hybridized to polyadenylated mRNA. From single molecule imaging Knemeyer et al. estimated the fraction of immobile and mobile mRNA. This technique showed that imaging single molecules in living cells is possible. Thus, the toolbox of single molecule multiparameter fluorescence detection allows us to analyze structural and dynamic properties of single molecules and molecular complexes in living cells.

28.3.2.2 Intracellular Mobility

Several attempts have been made to detect the detailed individual mobility of single particles or even single molecules. Different groups use fast CCD cameras that are sensitive enough to detect few or single fluorophores. Time series of those single particle images reveal particle movement in cells (time-lapse microscopy). The trajectories of single viruses and spliceosomal particles have been analyzed, revealing diffusive movement and active transport [4, 42, 43]. A different approach is under development; it uses a confocal scanning system, which allows tracking the movement of single particles [44, 45]. There the detection focus is moved repetitively in a circular path, allowing us to determine the position of a fluorophore in the plane of movement by the intensity distribution along the circular scanning path.

28.3.3
Single Molecule Detection in Nucleic Acid Analysis

In nucleic acid analysis, single molecule detection is an advantage due to the small amount of sample needed. It also allows the development of fast assay formats with high throughput [46].

28.3.3.1 Fragment Sizing

To determine the size of restriction fragments, one can take advantage of nucleic acid staining fluorophores that intercalate or bind in the helical grooves. Fluorophores are available for single-stranded as well as for double-stranded oligonucleotides. Several fluorophores can bind to a single DNA fragment. The number of fluorophores depends on the length of the fragment and thus the fluorescent brightness of the labeled fragment can be used as a measure of its length. Since single molecule detection counts the number of fragments, an accurate calculation of sample concentration is easily possible. This is a big advantage over gel electrophoresis, where the brightness of the bands depends on the number of molecules as well as on the length of the fragments. To analyze the labeled fragments individually, microfluidic systems similar to flow cytometric instruments are used. Capillary systems lead the molecules to an excitation and detection volume. For the detection of the individual fluorescent molecules, imaging with a sensitive CCD camera or detection in a confocal setup is usually used [47, 48].

28.3.3.2 Single Molecule Sequencing

With the availability of biological sequences new applications emerge and the need for sequence information grows exponentially. Single molecule sequencing has the potential to produce faster and cheaper sequences by facilitating a high parallelization to reduce the time and amount of material needed. Sequencing then enables also fast analytical purposes such as single nucleotide polymorphism (SNP) detection or microbial typing [49, 50].

In the search for a robust technique for single molecule sequencing, two different approaches are under development – the sequential degrading of fluorescently labeled nucleic acid strands and the successive incorporation of labeled bases.

To observe sequential degrading, the labeled nucleic acid is placed in an ultrasensitive flow cytometric setup. When the exonuclease degrades the nucleotide, the flow drags the cleaved nucleotides to the detection volume. There the labeled nucleotides are excited with a focused laser beam and detected mostly by a CCD camera. If all four bases are labeled with different fluorophores, the sequence can be read directly. Since the efficiency of enzymatic processing of nucleic acids is reduced by labeled nucleotides, it is useful not to label all nucleotides. Different groups recently demonstrated this technique for DNA sequences containing labeled nucleotides at each T position [51], and two distinguishable fluorophores at the U and C positions, respectively [52]. Instead of labeling all nucleotides, it is enough to label only two different base types at a time. That will result in a two-base sequence. Measuring all six possible two-base sequences allows the assembly of the full sequence.

The other approach uses the possibility to image single fluorophore labeled molecules on a surface with a CCD camera. Therefore, primers are hybridized to nucleic acid strands that are bound to the surface. Then a primer extension is performed by a DNA polymerase that incorporates a fluorescently labeled nucleotide. In principle, when using distinguishable fluorophores for the four base types, this would allow us to read the sequence while the extension takes place. Mainly due to unspecific adsorption to the surface and high background fluorescence caused by the high amount of labeled nucleotides needed for the polymerase reaction, the construction of an experimental setup is more complicated. It was demonstrated that with the use of FRET between a fluorescently labeled primer and the newly incorporated fluorophore, it is possible to detect clearly specific incorporation and to identify the sequence [53].

This technology is still under development and recent progress, especially in the field of background reduction [54], allows to hope that in near future it will lead to fast sequencing that can be highly parallelized, and needs only the tiniest amounts of material and probe. Direct sequencing of DNA as well as RNA without previous amplification should be possible. mRNA sequencing can then be performed by using either the reverse transcriptase or DNA polymerase after a DNA strand is synthesized.

28.4
Equipment for Single Molecule FRET Measurements

28.4.1.1 Excitation of the Fluorophores
For any fluorescence experiment an excitation source is mandatory. Modern single molecule FRET measurements are performed using a laser with suitable wave-

length to excite the donor dye, usually in the range from 450 to 550 nm. For most applications, a diode laser operated in the in the continuous wave (c.w.) mode is sufficient. For experiments were fluorescence lifetime information is needed, the laser has to shoot short flashes of light (pulsed excitation).

A special excitation technique is the two-photon excitation. Here, excitation is induced by simultaneous absorption of two photons that have twice the wavelength used in one-photon excitation. Since the two photons have to hit the dye at almost the same time, two-photon excitation requires a high instantaneous photon flux and therefore bleaching is a major problem. The advantage of this technique is a higher spatial resolution compared to one-photon excitation, since the intensity and thus the probability that two photons are absorbed at the same time is sufficiently high only in a very small area. The long wavelength of the exciting laser reduces autofluorescence significantly, which is especially useful for measurements in cells.

The light is mostly focused on the sample via an objective. Total internal reflection fluorescence (TIRF) experiments use oil immersion objectives with a high numerical aperture ($NA = 1.4$) to create an evanescent field at the glass surface where the sample is bound. For experiments where the sample molecules diffuse free in solution, water immersion objectives ($NA = 1.2$) are used.

In particular, y for experiments on immobilized molecules in combination with confocal detection, it is necessary to scan the sample. Therefore a piezo-driven scanning stage is needed, which allows precise movement of the e sample precisely the x-, y- and z-directions.

In many setups, the laser excitation light as well as the fluorescence passes the same objective. In this case a dichroic mirror has to be implemented to separate the fluorescence from excitation light (laser is reflected and fluorescence passes or *vice versa*, see Fig. 28.3).

28.4.1.2 Fluorescence Detection

In a setup for a single molecule FRET measurement, the fluorescence originated by the donor dye is separated from acceptor fluorescence either by another dichroic and/or by suitable band pass filters. The fluorescence is detected by an avalanche photodiode or a photomultiplier tube to collect single photons. To get an image, the sample is scanned or a CCD camera is used for wide-field detection.

Thus, a laser, an objective, two dichroics, two lenses and two detectors are needed for the most simple setup which allows single molecule FRET experiments. This "minimal" setup allows the registration of the intensities for the donor and acceptor fluorescence, and permits the calculation of the FRET efficiency for every single molecule event. If a polarizing beam splitter is implemented and the donor fluorophore is excited with pulsed excitation, the fluorescence lifetime as well as the anisotropy for the donor and acceptor dye becomes additionally accessible. Employing this supplementary information, the precision of single molecule FRET measurements can be enhanced and possible artifacts, like restricted motion (κ^2 effects) or local quenching of the dyes, can be recognized and taken into account for further analysis [55, 56]. Therefore this kind of setup, shown in

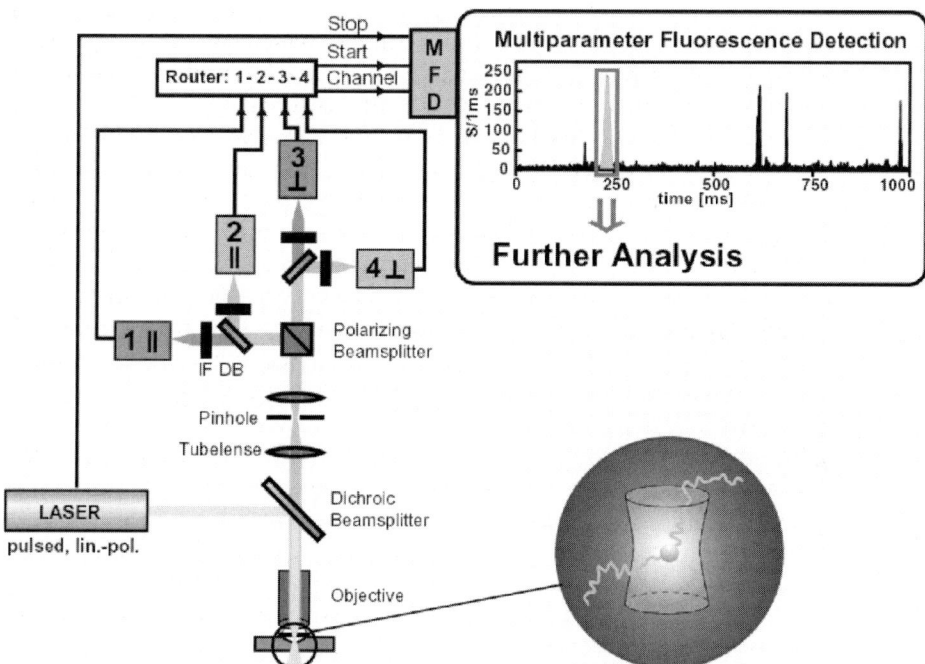

Fig. 28.3. A confocal microscopy arrangement for multiparameter fluorescence detection (MFD). Freely diffusing molecules are excited by a pulsed, linear polarized laser. After reflection by a dichroic mirror the laser beam is focused by a microscope objective. The fluorescence is collected and then refocused by the same objective to the image plane, where a pinhole is placed, which ensures that only fluorescence from the focal plane can pass. After the collected light is separated according to its polarization by a polarizing beam splitter, the fluorescence is divided by a dichroic mirror into the fractions originating from the acceptor and donor fluorophores. Subsequently, band pass filters are used to discriminate fluorescence from scattered laser light. Finally, the fluorescence light is detected by four detectors: two for the donor $(2 + 4)$ and two the acceptor fluorescence $(1 + 3)$.

Fig. 28.3, is especially useful to investigate processes where the expected FRET changes are small or structural information should be gained.

28.4.1.3 Data Analysis

Even more challenging than building a setup capable of detecting single fluorophores, is the data processing of single molecule experiments. Using a confocal setup with time-correlated single photon counting, the amount of data gained in a single molecule experiment is tremendous. However, since commercial computers with fast processors and a memory with a high capacity are available, data collection is rarely a problem anymore.

Most of the collected data originate from the background due to dark counts of the detector and remaining Raman, respectively, Rayleigh scattering. After a single molecule solution experiment all intensity peaks, called fluorescence bursts, that originate from single fluorophores diffusing through the focus can be identified and further evaluations of the spectroscopic properties can be restricted to the selected events.

28.5
Sample Preparation

28.5.1
Fluorophore–Nucleic Acid Interaction

The interaction between the fluorophore and the biomolecule under investigation should be minimized or well known. The fluorophores should not introduce any perturbation to the biomolecules' structure or dynamics. When labeling RNA, care has to be taken so that the interaction of the bases is not disturbed, especially in complex non-helical structures. Also, the fluorophores' characteristics should not be changed by the biomolecule [57].

For FRET measurements it is desirable that the dyes rotate fast and freely, so one can assume an average κ^2 of 2/3. Therefore, great care has to be taken so that the reporter dyes do not stick to the nucleic acid under investigation. Steric hindrance of the fluorophores can introduce a reasonable uncertainty [58].

On the other hand, a sticky fluorophore can be used to calculate the characteristic rotation of the oligonucleotide from its fluorescence anisotropy.

However, the interaction with the bases can lead to quenching of the fluorophores, mainly due to electron transfer from the fluorophore to guanosine bases [59, 60]. Then the fluorophore is dark in the bound state and bright if it is free. Thus fluctuations between the bound and unbound state become visible as intensity fluctuations that could be misinterpreted as dynamics of the nucleic acid structure. Fluorophore–nucleic acid interactions can be reduced if the fluorophores are negatively charged and thus rejected by the negatively charged backbone.

28.5.2
RNA Labeling

28.5.2.1 Fluorophores for Single Molecule Fluorescence Detection
Fluorophores used for single molecule fluorescence have to fulfill several criteria.

- They need high photostability to allow long observation times, a high probability to absorb photons for excitation (high extinction coefficient) and a high fluorescence quantum yield.
- They have to be excitable with one of the commercially available lasers in the visible range.
- They have to be water soluble and should not be pH insensitive.

Several fluorophores have been developed that fulfill these conditions. Many of them belong to the rhodamine or cyanine dye family. They are available under trade names like Alexa, Cy, Atto and others.

28.5.2.2 Fluorophores used for FRET Experiments

For FRET applications the selected donor–acceptor pair should have large spectral separation to minimize leakage of donor fluorescence into the acceptor channels and direct excitation of the acceptor, but an overlap of the donor emission and acceptor absorption spectrum still big enough to guarantee energy transfer.

Most publications report the use of Cy3–Cy5 [23, 61] or Alexa488–Cy5 [9, 56] as FRET pairs for single molecule FRET studies. With their Förster distances R_0 of 5.8 nm (Cy3–Cy5) [62] and 5.1 nm (Alexa488–Cy5) [63] they allow distance measurements between 2 and 8 nm with highest sensitivity around 5 nm.

In spite of the fact that Cy5 is widely used, it has some drawbacks. Often up to 50% of the fluorophores are found to be inactive after coupling to biomolecules, probably due to pre-bleaching. However, in single molecule experiments this is not a severe problem, since these can be separated from the ones carrying active acceptors. Cy5 also undergoes photo-induced isomerization between a highly fluorescent *trans* and a weakly fluorescent *cis* state. This leads to a loss of almost 50% of the fluorophores' capacity and to intensity fluctuations that interfere with correlation analysis of other fluctuations [64]. Alternatively, the Alexa594 dye is sometimes used as acceptor. This dye from the rhodamine family has no dark states, but direct excitation of the acceptor is higher due to its blue-shifted fluorescence spectrum.

The constant need for brighter and more photostable fluorophores ensure the continuing rapid development of fluorophores. Thus, for single molecule experiments it is always worth looking for the best fluorophore available at the time.

28.5.2.3 Attaching Fluorophores to RNA

Intercalating or groove-binding dyes can be used to stain DNA, but these dyes do not bind in a sequence-specific manner to the DNA. For structural investigations on DNA or RNA, one has to attach the dyes to specific positions within the oligonucleotide.

To attach fluorophores to specific positions in RNA, modified bases can be incorporated in oligo synthesis to which then fluorophores are attached in a second step. Mostly amino modifications are used. All standard fluorophores are available as activated amine-reactive derivatives. There are four major classes of commonly used reagents to label amines: succinimidyl esters, isothiocyanates, sulfonylchlorides and tetrafluorophenyl esters. Of the four, tetrafluorophenyl esters are the preferred chemistry for conjugations. Similar to the succinimidyl esters, they produce stable carboxamide bonds. They are less susceptible to hydrolysis than succinimidyl esters and therefore can provide more reaction time in aqueous-based reactions. However, the succinimidyl ester is still by far the most commonly used amine-reactive group. Oligos with 3'- or 5'-end amino-modifications or end-labeled with fluorophores are commercially available almost everywhere. Here, the dyes are usually attached as phosphoramidites during synthesis.

For internal labeling the modification should be located at the 5 position of the base U or C. The fluorophore is then positioned in the major groove of the double strand. This position minimizes interference to protein–RNA interactions which mostly take place in the minor groove. RNA oligos with internal modification are available from a few companies.

5-Aminoallyl U and 5-aminoallyl C are commercially available as triphosphates. Some companies also offer fluorophores already bound to UTP. Since this modification is compatible with enzymatic processing, primer extension and ligation can be used to incorporate the modifications at the desired position. For enzymatic incorporation or successive attachment of fluorophores we refer to the methods recommended by the respective provider of the modified bases or fluorophores.

For FRET experiments, donors and acceptors are mostly bound to the complementary strands. Double labeling of one RNA strand is much more complicated and mostly leads to loss of sample. Ligation of different labeled strands might introduce pre-bleaching while handling. Using different RNA modifications and coupling strategies for donor and acceptor is often limited by the availability of the respective modifications. However, some commercial suppliers do offer double-labeled oligonucleotides.

28.5.2.4 Linkers

The linker used to couple the fluorophore to the RNA has to be considered carefully. To avoid orientational effects the linker has to be long enough to give the fluorophore enough mobility to test all orientations within its fluorescence lifetime. Usually a carbon C_6 linker serves this purpose well enough. Care has to be taken when calculating absolute distances for shorter linkers. The value of the orientation factor κ^2 might deviate from its average value of $2/3$, introducing uncertainties in the calculated distances. When using C_6 linkers it always has to be taken into account that the mean position of the fluorophores due to the linker length is not at the base where it is attached to. The distance between the Base and the mean position of the fluorophore mostly is between 1 and 2 nm. The position of the fluorophore relative to the anchoring point either has to be determined experimentally, e.g. from systematic distance measurements on a linear nucleic acid helix (then being valid only in systems where the fluorophore is bound to a helical segment) or has to be estimated by molecular modeling.

28.5.3
Fluorescence Background

28.5.3.1 Raman Scattered Light

The laser light used for excitation produces Raman light that is shifted to longer wavelengths and thus can interfere with the fluorescence. Intensity and spectral distribution depend on the solvent used and the excitation wavelength. This can be efficiently suppressed with the right selection of excitation wavelength, filters and fluorophores.

28.5.3.2 Cleaning Buffers

Using picomolar concentrations of host molecules, background fluorescence of the solvents becomes a big issue and particles scattering the excitation light as well as background fluorescence of ingredients of the used buffer have to be carefully avoided. Impurities in the sample are less critical, since they also get diluted when preparing the right sample dilution for single molecule experiments. Fluorescent impurities of the buffers can be efficiently removed by adding activated charcoal granula, mixing and subsequently filtration with standard sterile filters [65].

28.5.3.3 Clean Surfaces

In single molecule measurements at a surface, two sources of background signal always have to be taken into account – impurities adsorbed to the surface, which can be reduced by cleaning, and autofluorescence of the surfaces material itself.

Two procedures are commonly used for cleaning – sonicating in a series of different solvents [66] or flaming of the surface [20].

The following protocol for sonication is given in [66]:

(1) 30% detergent solution 1 h
(2) Distilled water 5 min
(3) Acetone 15 min
(4) Distilled water 5 min
(5) 1 M KOH 15 min
(6) Ethanol 15 min
(7) 1 M KOH 15 min
(8) Distilled water 15 min

Additionally, or alternatively, the surface can be flamed for a few seconds in a propane torch.

As well cleaning, the sonication procedure serves the purpose of providing a chemically uniform surface for subsequent functionalization. The autofluorescence of the surface substrate cannot be removed, but instead of glass, a quartz glass can be used which has significantly less autofluorescence.

28.5.4
Surface Modification

Surface modification serves two purposes – it has to provide specific attachment of the molecules and at the same time prevents unspecific adsorption.

28.5.4.1 Coupling Single Molecules to Surfaces

The biotin–streptavidin system is mostly used for specific coupling of RNA to surfaces [22, 66]. Biotinylated bovine serum albumin (BSA) is adsorbed to the surface and subsequently streptavidin is attached. Then biotinylated RNA is anchored to the surface. The following protocol for RNA coupling is given in [20]:

(1) Biotinylated BSA 1 mg/ml
(2) Streptavidin 0.2 mg/ml
(3) 50 pM biotinylated RNA

Each step lasts 5 min followed by washing with buffer (10 mM Tris/50 mM NaCl, pH 8).

For covalent surface modification, aminosilanization with subsequent binding of PEG is often used in DNA chip production. For most single molecule experiments there is no need for covalent modification since no long-term stability is needed, but quick and simple sample preparation is desired.

28.5.4.2 Surface Passivation

Unspecific adsorption of the sample to the surface has to be prevented since it produces strong artifacts or leads to loss of sample. BSA provides already good passivation against RNA adsorption. Unfortunately, this procedure does not apply to proteins. A dense layer of polyethylene glycol (PEG) is usually used to prevent reject protein adsorption [66].

28.5.5
Preventing Photodestruction

Photobleaching of the fluorophores is an issue in any single molecule experiment. Even the most photostable organic fluorophores like rhodamine or cyanine dyes bleach sooner or later. Since the presence of oxygen is known to reduce the photostabiltity of almost all dyes used in single molecule experiments, the reduction of the oxygen content helps to prevent photobleaching and longer observation times are achievable. So far, most groups use an enzymatic oxygen-scavenging system, where oxygen is consumed in an enzymatic (glucose oxidase/catalase) catalyzed oxidation of glucose. The use of approximately 400 µM ascorbic acid is another possibility to remove the oxygen from the sample [11]. However, one has to keep in mind that oxygen also serves as a triplet quencher for many dyes. Consequently, if the oxygen is removed completely, the fluorescence intensity drops because the dyes are in the triplet state most of the time and do not fluoresce [66].

Thus, there are no universally valid conditions to avoid photobleaching in a single molecule experiment; in fact, the excitation power as well as the oxygen content has to be adjusted for every experiment individually.

28.6
Troubleshooting

Here we want to discuss some typical problems that can occur in sample preparation or in the measurement.

28.6.1.1 Orientation Effects

After labeling of the sample, unbound fluorophores have to be removed. This is mostly done by PAGE. If the fluorophores could not be removed thoroughly, they have to be distinguished from the host molecules that carry only the donor (e.g. a

RNA single strand) in the single molecule measurements. They, understandably, reduce the relative amount of host molecules. In solution experiments free fluorophores can be distinguished by their diffusion times either in single molecule events or in FCS.

Since the orientation factor κ^2 is important for the calculation of absolute distances, it has to be ensured that the fluorophore is freely mobile so that the average value for κ^2 may be taken. Also, in experiments that concentrate on dynamic aspects, signal fluctuations might arise from changes in the way the fluorophore interacts with the host molecule. As a control, the measurement of the anisotropy is recommended either in an ensemble or single molecule experiment. If the fluorophore is nearly freely mobile, the measured anisotropy should be close to zero, if the host molecules rotation is also slow.

If single molecules bound to a glass surface are observed, interaction with this surface may also reduce the mobility of the fluorophore. Here the polarization response of the immobilized molecules should be analyzed. Again, if the fluorophore is nearly freely mobile, the polarization should be close to zero.

28.6.1.2 Dissociation of Molecular Complexes

To measure single molecules in solution, the sample has to be diluted to sub-nanomolar concentrations. Thereby experiments on molecular complexes may be limited due to their low affinity constant, leading to dissociation upon dilution. To prevent dissociation, some components may be added in higher concentrations. Normally the component carrying the acceptor can be used in up to 10-fold higher concentrations compared to the donor, before the background gets to high due to the increased direct excitation. If the experiment is designed such that the donor and acceptor are bound to the same molecule, the other unlabeled binding partners can be added in much higher concentrations as long as no unspecific complexation is induced or their autofluorescence becomes significant.

28.6.1.3 Adsorption to the Surface

In solution experiments, the low concentration of host molecules makes the sample highly sensitive to loss due to adsorption to the surface. If the measured mean count rate drops during the measurement, this is mainly due to adsorption. Bleaching also leads to depopulation of the sample, but this can be neglected in typical sample volumes around 50 μl and gets important only in sub-microliter volumes.

28.6.1.4 Diffusion Limited Observation Times

Diffusion normally limits observation times of biomolecules of around 100 kDa to a few milliseconds. For longer observation times, molecules are mostly bound to surfaces at the cost of possible sample–surface interaction. A good alternative is the reduction of diffusion time by attaching the host molecule to a bigger mass that has a low diffusion time. In nucleic acid analysis, a long DNA sequence fragment with a sticky end can easily serve this purpose. Another alternative sometimes proposed is changing the viscosity of the used solvent. Great care has to be taken when using organic solvents, which can produce strong Raman light that interferes with the detected fluorescence.

28.6.1.5 Intensity Fluctuations

Since most dyes are not ideal emitters, when analyzing fluctuations in FRET signals one has to take into account that some fluorophores show intrinsic fluctuations. Fluctuations of the donor between the bright and dark state can be disregarded, since it results in total annihilation of donor and acceptor fluorescence, and thus just no information is available during dark states. In contrast, dark states of the acceptor lead to a reduction of the energy transfer and thus to an increase of the donor fluorescence. This can be misinterpreted as big conformational changes where the dyes are brought totally out of the range of energy transfer. Photo-induced transitions between the dark and bright state, as found in the case of Cy5, are intensity dependent. Thus, an intensity series can decide whether the fluctuations are due to conformational changes of the host molecule or due to intrinsic properties of the fluorophore.

References

1 MOERNER, W. E., KADOR, L., *Phys. Rev. Lett.* **1989**, *62*, 2535–2538.
2 ORRIT, M., BERNARD, J., *Phys. Rev. Lett.* **1990**, *65*, 2716–2719.
3 LU, H. P., XUN, L., XIE, X. S., *Science* **1998**, *282*, 1877–1882.
4 KUES, T., DICKMANNS, A., LUHRMANN, R., PETERS, R., KUBITSCHECK, U., *Proc. Natl. Acad. Sci. USA* **2001**, *98*, 12021–12026.
5 SAKO, Y., MINOGUCHI, S., YANAGIDA, T., *Nat. Cell Biol.* **2000**, *2*, 168–172.
6 SCHWILLE, P., KORLACH, J., WEBB, W. W., *Cytometry* **1999**, *36*, 176–182.
7 SCHWILLE, P., HAUPTS, U., MAITI, S., WEBB, W. W., *Biophys. J.* **1999**, *77*, 2251–2265.
8 KÖHLER, R. H., SCHWILLE, P., WEBB, W. W., HANSON, M. R., *J. Cell Sci.* **2000**, *113*, 3921–3930.
9 WIDENGREN, J., SCHWEINBERGER, E., BERGER, S., SEIDEL, C. A. M., *J. Phys. Chem. A* **2001**, *105*, 6851–6866.
10 ENDERLE, T., HA, T., OGLETREE, D. F., CHEMLA, D. S., MAGOWAN, C., WEISS, S., *Proc. Natl. Acad. Sci. USA* **1997**, *94*, 520–525.
11 YILDIZ, A., FORKEY, J. N., MCKINNEY, S. A., HA, T., GOLDMAN, Y. E., SELVIN, P. R., *Science* **2003**, *300*, 2061–2065.
12 SCHWILLE, P., MEYER-ALMES, F. J., RIGLER, R., *Biophys. J.* **1997**, *72*, 1878–1886.
13 KETTLING, U., KOLTERMANN, A., SCHWILLE, P., EIGEN, M., *Proc. Natl. Acad. Sci. USA* **1998**, *95*, 1416–1420.
14 HEINZE, K. G., KOLTERMANN, A., SCHWILLE, P., *Proc. Natl. Acad. Sci. USA* **2000**, *97*, 10377–10382.
15 FÖRSTER, T., *Ann. Phys.* **1948**, *2*, 55–75.
16 STRYER, L., HAUGLAND, R. P., *Proc. Natl. Acad. Sci. USA* **1967**, *58*, 719–726.
17 DENIZ, A. A., DAHAN, M., GRUNWELL, J. R., HA, T. J., FAULHABER, A. E., CHEMLA, D. S., WEISS, S., SCHULTZ, P. G., *Proc. Natl. Acad. Sci. USA* **1999**, *96*, 3670–3675.
18 JARES-ERIJMAN, E. A., JOVIN, T. M., *J. Mol. Biol.* **1996**, *257*, 597–617.
19 CLEGG, R. M., MURCHIE, A. I. H., ZECHEL, A., LILLEY, D. M., *Proc. Natl. Acad. Sci. USA* **1993**, *90*, 2294–2298.
20 KIM, H. D., NIENHAUS, G. U., HA, T., ORR, J. W., WILLIAMSON, J. R., CHU,

S., *Proc. Natl. Acad. Sci. USA* **2002**, *99*, 4284–4289.
21 HA, T., ZHUANG, X. W., KIM, H. D., ORR, J. W., WILLIAMSON, J. R., CHU, S., *Proc. Natl. Acad. Sci. USA* **1999**, *96*, 9077–9082.
22 ZHUANG, X., KIM, H., PEREIRA, M. J. B., BABCOCK, H. P., WALTER, N. G., CHU, S., *Science* **2002**, *296*, 1473–1476.
23 TAN, E., WILSON, T. J., NAHAS, M. K., CLEGG, R. M., LILLEY, D. M., HA, T., *Proc. Natl. Acad. Sci.* **2003**, *100*, 9308–9313.
24 ZHUANG, X., BARTLEY, L. E., BABCOCK, H. P., RUSSELL, R., HA, T., HERSCHLAG, D., CHU, S., *Science* **2000**, *288*, 2048–2051.
25 ANDERSSON, H., BAECHI, T., HOECHL, M., RICHTER, C., *J. Microsc.* **1998**, *191*, 1–7.
26 AINGER, K., AVOSSA, D., DIANA, A. S., BARRY, C., BARBARESE, E., CARSON, J. H., *J. Cell Biol.* **1997**, *138*, 1077–1087.
27 WANG, J., CAO, L. G., WANG, Y. L., PEDERSON, T., *Proc. Natl. Acad. Sci. USA* **1991**, *88*, 7391–7395.
28 KNEMEYER, J.-P., HERTEN, D.-P., SAUER, M., *Anal. Chem.* **2003**, *75*, 2147–2153.
29 JACOBSON, M. R., PEDERSON, T., *Proc. Natl. Acad. Sci. USA* **1998**, *95*, 7981–7986.
30 JACOBSON, M. R., CAO, L. G., WANG, Y. L., PEDERSON, T., *J. Cell Biol.* **1995**, *131*, 1649–1658.
31 LANGE, T. S., GERBI, S. A., *Mol. Biol. Cell* **2000**, *11*, 2419–2428.
32 BERTRAND, E., CHARTRAND, P., SCHAEFER, M., SHENOY, S. M., SINGER, R. H., LONG, R. M., *Mol. Cell* **1998**, *2*, 437–445.
33 POLITZ, J. C., TUFT, R. A., PEDERSON, T., SINGER, R. H., *Curr. Biol.* **1999**, *9*, 285–291.
34 POLITZ, J. C., BROWNE, E. S., WOLF, D. E., PEDERSON, T., *Proc. Natl. Acad. Sci. USA* **1998**, *95*, 6043–6048.
35 DIRKS, R. W., MOLENAAR, C., TANKE, H. J., *Histochem. Cell Biol.* **2001**, *115*, 3–11.
36 PERLETTE, J., TAN, W., *Anal. Chem.* **2001**, *73*, 5544–5550.
37 SOKOL, D. L., ZHANG, X., LU, P., GEWIRTZ, A. M., *Proc. Natl. Acad. Sci. USA* **1998**, *95*, 11538–11543.
38 TSUJI, A., SATO, Y., HIRANO, M., SUGA, T., KOSHIMOTO, H., TAGUCHI, T., OHSUKA, S., *Biophys. J.* **2001**, *81*, 501–515.
39 TSUJI, A., KOSHIMOTO, H., SATO, Y., HIRANO, M., SEI–IIDA, Y., KONDO, S., ISHIBASHI, K., *Biophys. J.* **2000**, *78*, 3260–3274.
40 MATSUO, T., *Biochim. Biophys. Acta* **1998**, *1379*, 178–184.
41 SAKO, Y., HIBINO, K., MIYAUCHI, T., MIYAMOTO, Y., UEDA, M., YANAGIDA, T., *Single Mol.* **2000**, *1*, 159–163.
42 SEISENBERGER, G., RIED, M. U., ENDREß, T., BÜNING, H., HALLEK, M., BRÄUCHLE, C., *Science* **2001**, *294*, 1929–1932.
43 KUES, T., PETERS, R., KUBITSCHECK, U., *Biophs. J.* **2001**, *80*, 2954–2967.
44 DAHAN, M., LEVI, S., LUCCARDINI, C., ROSTAING, P., RIVEAU, B., TRILLER, A., *Science* **2003**, *302*, 442–445.
45 LEVI, V., RUAN, Q., KIS-PETIKOVA, K., GRATTON, E., *Biochem. Soc. Trans.* **2003**, *31*, 997–1000.
46 KELLER, R. A., AMBROSE, W. P., ARIAS, A. A., CAI, H., EMORY, S. R., GOODWIN, P. M., JETT, J. H., *Anal. Chem.* **2002**, *74*, 316A–324A.
47 VAN ORDEN, A., KELLER, R. A., AMBROSE, W. P., *Anal. Chem.* **2000**, *72*, 37–41.
48 GAO, Q., SHI, Y., LIU, S., *Fresenius. J. Anal. Chem.* **2001**, *371*, 137–145.
49 RONAGHI, M., ELAHI, E., *J. Chromatogr. B* **2002**, *782*, 67–72.
50 POURMAND, N., ELAHI, E., DAVIS, R. W., RONAGHI, M., *Nucleic Acids Res.* **2002**, *30*, e31.
51 WERNER, J. H., CAI, H., JETT, J. H., REHA-KRANTZ, L., KELLER, R. A., GOODWIN, P. M., *J. Biotechnol.* **2003**, *102*, 1–14.
52 SAUER, M., ANGERER, B., ANKENBAUER, W., FOLDES-PAPP, Z., GOBEL, F., HAN, K. T., RIGLER, R., SCHULZ, A., WOLFRUM, J., ZANDER, C., *J. Biotechnol.* **2001**, *86*, 181–201.
53 BRASLAVSKY, I., HEBERT, B., KARTALOV, E., QUAKE, S. R., *Proc. Natl. Acad. Sci. USA* **2003**, *100*, 3960–3964.

54 Levene, M. J., Korlach, J., Turner, S. W., Foquet, M., Craighead, H. G., Webb, W. W., *Science* **2003**, *299*, 682–686.
55 Margittai, M., Widengren, J., Schweinberger, E., Schroder, G. F., Felekyan, S., Haustein, E., Konig, M., Fasshauer, D., Grubmuller, H., Jahn, R. et al., *Proc. Natl. Acad. Sci. USA* **2003**, *100*, 15516–15521.
56 Rothwell, P. J., Berger, S., Kensch, O., Felekyan, S., Antonik, M., Wöhrl, B. M., Restle, T., Goody, R. S., Seidel, C. A. M., *Proc. Natl. Acad. Sci. USA* **2003**, *100*, 1655–1660.
57 Gaiko, N., Hillisch, A., Berger, S., Seidel, C. A. M., Diekmann, S., Griesinger, C. **2002**, in preparation.
58 Ha, T., Glass, J., Enderle, T., Chemla, D. S., Weiss, S., *Phys. Rev. Lett.* **1998**, *80*, 2093–2096.
59 Seidel, C. A. M., Schulz, A., Sauer, M. H. M., *J. Phys. Chem.* **1996**, *100*, 5541–5553.
60 Seidel, C., *Proc. SPIE Int. Soc. Opt. Eng.* **1991**, *1432*, 91–104.
61 McKinney, S. A., Declais, A. C., Lilley, D. M., Ha, T., *Nat. Struct. Biol.* **2003**, *10*, 93–97.
62 Ishii, Y., Yoshida, T., Funatsu, T., Wazawa, T., Yanagida, T., *Chem. Phys.* **1999**, *247*, 163–173.
63 Widengren, J., Schweinberger, E., Berger, S., Seidel, C. A. M., *J. Phys. Chem. A* **2001**, *105*, 6851–6866.
64 Widengren, J., Schwille, P., *J. Phys. Chem. A* **2000**, *104*, 6416–6428.
65 Borsch, M., Diez, M., Zimmermann, B., Reuter, R., Graber, P., *FEBS Lett.* **2002**, *527*, 147–152.
66 Ha, T., *Methods* **2001**, *25*, 78–86.

29
Scanning Force Microscopy and Scanning Force Spectroscopy of RNA

Wolfgang Nellen

29.1
Introduction

RNA research has significantly advanced by using sophisticated methods to investigate secondary and tertiary structures, and RNA–RNA as well as RNA–protein interactions. Most biochemical and biophysical methods rely on large numbers of molecules – the average reaction or deduced structures are interpreted to reconstruct the behavior or shape of a single molecule. This "top-down approach" can answer many fundamental questions to understand RNA function. It is contrasted by the "bottom-up approach" where single molecules are investigated and a large number of individual observations are used to statistically determine the average behavior or structure of the molecule(s). (It should be kept in mind that biochemists and biophysicists, on the one hand, and single molecule investigators, on the other, are approximately 17 orders of magnitude apart when they speak of "large numbers of molecules"!).

One of the highly promising methods in single molecule research is scanning probe microscopy (SPM), based on the scanning force microscopy (SFM) developed by Binnig et al. [1]. This allows us to record three-dimensional (3-D) topographic maps of biological samples with a resolution of a few nanometers.

The basic principle of SFM (Fig. 29.1) is scanning of a sample with a tip a few nanometers in diameter. The tip is attached to a flexible cantilever arm fixed at one end in a holder. A laser beam is focused on the cantilever and reflected to the center of a four-quadrant diode. The sample is mounted on a stage that can be moved by piezo elements in the x-, y- and z-directions. When, during scanning in the x-direction, the tip hits an obstacle, i.e. the molecule of interest, the cantilever arm will bend and the laser beam is deflected off the center of the diode. An electronic feedback will re-adjust the stage in the z-direction and the signal is recorded. Every scan line thus presents a height profile of the sample. Usually, 512 x-profiles adjacent to each other in the y-direction create a topographic image.

In contrast to electron microscopy, images are taken directly from "live" molecules and not indirectly from metal or carbon coatings of the molecules. Since

Handbook of RNA Biochemistry. Edited by R. K. Hartmann, A. Bindereif, A. Schön, E. Westhof
Copyright © 2005 WILEY-VCH Verlag GmbH & Co. KGaA, Weinheim
ISBN: 3-527-30826-1

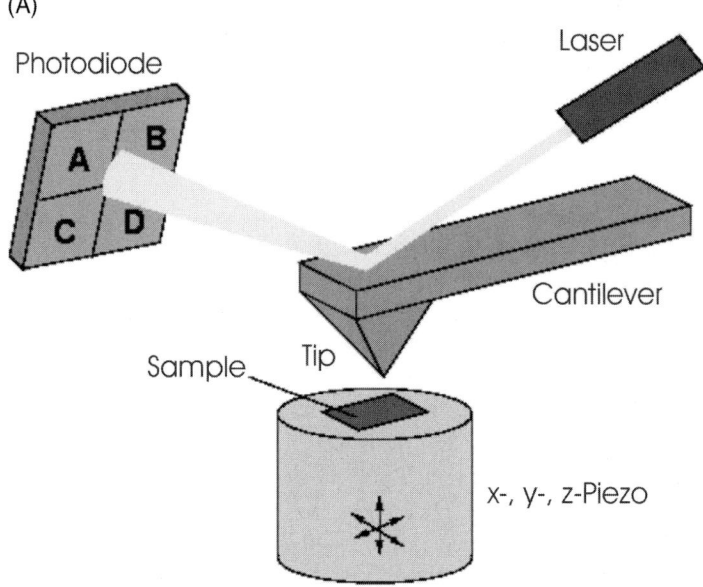

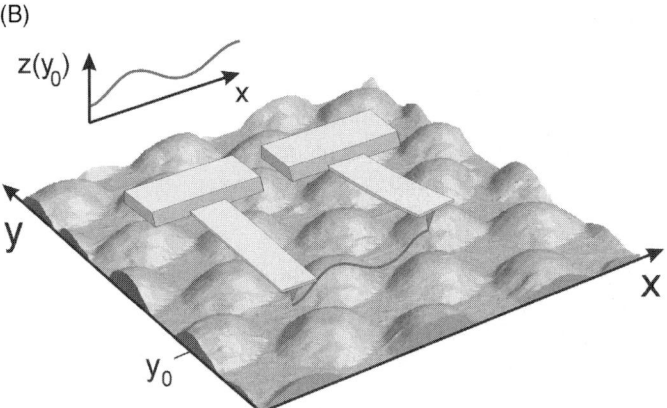

Fig. 29.1. (A) The principle of the scanning force microscope: see text for details. (B) Scanning of a 3-D surface. The cantilever is moving in the x-direction across the sample, and is shown in the start (left) and end (right) position based on the diagram above. Note the bending of the cantilever arm (right position) when it encounters an obstacle.

samples are physically scanned under ambient conditions, biomolecules may also be investigated under near physiological conditions, i.e. in aqueous solutions.

In the "contact mode", the rigid cantilevers are in direct contact with the sample and "scrape" the surface. Thus, shearing forces are exerted on the sample and soft material may be scratched or cut. Therefore, the "tapping mode" is usually employed to minimize contact with the sample. The cantilever oscillates and only briefly touches the sample and lateral forces are therefore minimized. Cantilevers used for these measurements usually oscillate with a resonance frequency of 250–400 kHz and a free amplitude of about 100 nm. The set point is adjusted to 10–30% below the free amplitude to obtain clear images with minimal damage to the sample.

29.2
Questions that could be Addressed by SFM

At present, SPM is mainly within the realm of physicists who, with the support of biologists and biochemists, developed methods to provide a "proof of principle". The number of publications presenting new, additional information in biology and biochemistry is still limited. However, the relative simplicity of sample preparation and use of the instrument should result in a rapid increase in applications in the near future. RNA research by SFM started out with the visualization of molecules [2, 3]. Imaging of mRNA circularization by interaction of poly(A)-binding proteins with EIF4E was one of the major contributions of SFM to the understanding of translation complexes [4]. The detection of preferential binding sites of an antibody directed against double-stranded RNA provided evidence that RNA–protein interactions could be demonstrated that had previously escaped biochemical analysis [5] (Fig. 29.2).

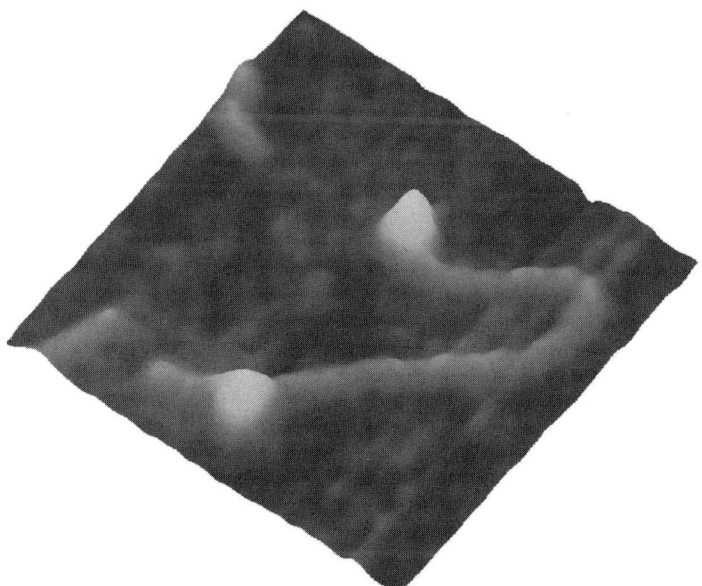

Fig. 29.2. SFM: surface plot image of a 700-bp double-stranded RNA with two proteins bound. Image size is 125 × 125 nm.

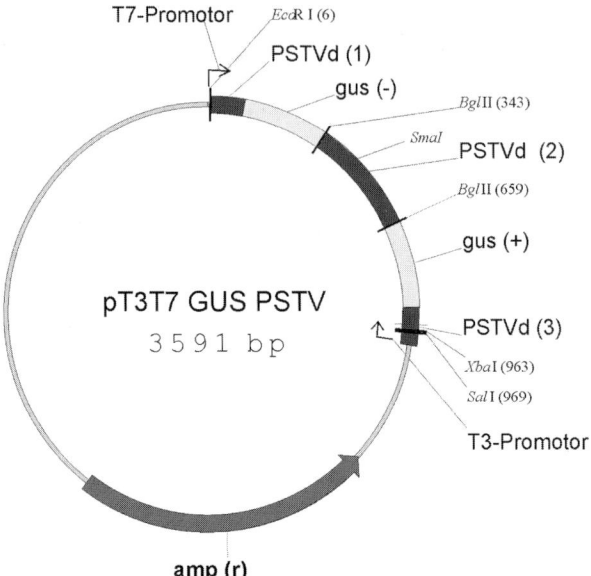

Fig. 29.3. Plasmid for the generation of double-stranded RNA with inserted protein binding sites. The fragment to be transcribed by T7 or T3 RNA polymerase consists of two gus segments that are complementary and form a complete double helix; in addition, three PstVd segments form imperfect double strands flanking the gus hybrid. The single SmaI site within PstVd (2) is used to insert a sequence of interest.

A more general approach to study RNA–protein interactions was initiated by the construction of a versatile vector that could be used to generate a fold-back RNA into which any potential interaction site of interest could be inserted (Fig. 29.3). The backbone RNA contained one completely double-stranded and two partially double-stranded regions derived from a coding mRNA and the potato spindle tuber viroid (PstVd), respectively. The RNA of 980 nt that forms a rigid rod-like structure thus contains, in addition to the interaction site of interest, a multitude of non-specific competitor sequences and structures. This backbone with the appropriate sequences inserted allows for the determination of protein-binding specificity in comparison to other sites and for large-scale structural changes (like bending and kinking) upon binding of a partner. With the insertion of a second site for binding of a different molecule, protein–protein interactions could be investigated as long as the distance between both sites is large enough to allow for sufficient flexibility [6].

RNA-modifying enzymes like adenosine deaminase that acts on RNA (ADAR) have a general, promiscuous binding affinity to and modifying activity on double-stranded RNA *in vitro*, but modify only specific sites *in vivo*. By SFM, the binding frequency to specific and non-specific sites could be determined with a precision of approximately 40 bp (the sequence covered by one protein molecule). In parallel biochemical experiments, editing efficiency was measured at the respective sites and could be directly compared to binding. These experiments provided new, unexpected insights into the mechanisms of RNA editing [7]. It should, however, be noted that the SFM experiments cannot be correlated to association constants, but rather give quantitative values of a less-well-defined "binding frequency" in comparison to competing sites.

SFM measurements as described above can be easily done on dried samples in air – they visualize the situation after a reaction has been carried out in solution and stopped by spreading the molecules on the mica surface. More sophisticated experiments can be done in liquid using a "fluid cell" that allows for imaging under close to physiological conditions. While measurements in air are usually done at ambient temperature, reaction conditions in a fluid cell can be controlled and reaction partners can be sequentially added by a delivery system. Technical problems arise because of insufficient adherence of molecules to the surface and consequently displacement by the lateral forces of the cantilever during scanning. Between 1 and 3 mM $NiCl_2$ has been used to improve binding of RNA to the substrate. Fay et al. [8] have used these conditions to visualize hairpin ribozymes in solution and, by using mutants, were able to distinguish between different conformational states during the self-cleaving process. Obviously, adjusting salt conditions for optimal binding of nucleic acids to the surface may interfere with molecular interactions with other reaction partners.

The resolution of imaging depends on the variable scanning field (usually 10 × 10 µm) and the scanning speed (0.3–2 Hz). Images of 512 × 512 pixels are set up within 4–25 min. More rapid scanning may damage the sample and decrease the resolution. Even though some attempts have been made to monitor dynamic changes in biological samples [9–11], the compromise between resolution and speed of commercial cantilevers is not yet favorable for real-time recording of molecular movements. However, new, very small cantilevers of 10–20 µm length, 2–5 µm width and 100–140 nm thickness with resonance frequencies up to 650 kHz (in water) have now been developed [12] and permit rapid scanning. Images of 100 × 100 pixels can be captured within 80 ms [13], thus allowing for the observation of biomolecules in motion.

"Pseudo-dynamic" measurements may be done when a reaction or interaction can be started, for example, by adding ATP. By spreading and drying the sample in the mica surface, the reaction is stopped at different times and the evaluation of the reaction intermediates could provide insight into consecutive structural alterations with time [14].

SFM of biological samples is limited by several parameters. To avoid background structures, extremely flat surfaces are required for loading the sample. Freshly cleaved mica (Goodfellow, Cambridge) has proven functional in most cases of nucleic acid and protein analysis. For the specific, oriented attachment of molecules (see below), gold-coated chips have frequently been used.

Even though the tip forces on the sample have been strongly reduced with the development of the "tapping mode", molecules are still subjected to mechanical insult that may change the integrity of a molecule or a complex. In fact, cantilever forces can be adjusted so that single molecules may be "nano-dissected" in a defined way [15, 16]. Lateral forces exerted by the cantilever can dislocate the sample during the scanning process and result in blurred images. On the other hand, strong binding of the sample to the surface could interfere with structural flexibility and thus biochemical properties. The "non-specific" interaction forces of nucleic acids and proteins with mica are not defined, but they are sufficiently strong in air. In liquid, however, nucleic acid molecules are more loosely attached or are only bound at a few sites along the chain. Therefore, specific precautions have to be taken to avoid displacement by the cantilever forces (see above).

Tip diameters present another limitation of SFM: commercially available tips have a radius of 1–10 nm and thus result in an apparent broadening of the sample. Double-stranded RNA with a width of 2.5 nm therefore appears 5–10 nm wide depending on the quality of the tip. In contrast, the height of a sample is frequently underestimated because the mechanical contact of the tip with the soft material causes flattening. Heights of 1.2–1.7 nm are usually measured instead of the expected 2.5 nm for double-stranded RNA. Due to these errors and to the fact that scanning only visualizes the top view of an object (but not possible indentations at the bottom), volumes of a sample can hardly be determined and the analysis of 3-D fine structures is also limited.

Double-stranded DNA and double-stranded RNA are easily visualized since they form rather rigid rods. The investigation of, for example, protein-binding sites is, however, hampered by the fact that the orientation of the nucleic acid cannot be determined. For double-stranded RNA, the constructs to produce transcripts can be modified by including defined asymmetric secondary structures of at least 30 bp (equivalent to 9 nm). Branches of this size are easily visible by SFM, and unambiguously identify the left and right of the molecule.

Loops, bulges and longer stretches of unpaired bases may contribute to the length of the molecules but so far in a rather unpredictable way. An unpaired loop of 60 nt appeared as a bulge of approximately 10 nm length and 15 nm width; in addition, the height of this structure was significantly increased in comparison to completely base-paired double-stranded RNA. Presumably, a predominant 3-D structure was adopted by the single-stranded RNA, but the resolution of the method was insufficient to obtain any more detailed information (Bonin and Nellen, unpublished).

Many RNA-binding proteins display high affinities to double-stranded RNA ends. Substrates generated by *in vitro* transcription of sense and antisense strands followed by hybridization generate open termini that are "sticky" for some proteins. Limited digestion with RNase A to remove single-stranded overhangs did not abolish the problem. A general backbone construct largely reduced the problem of non-specific end binding: *in vitro* transcripts are generated from an inverted repeat that folds back to a double-stranded and partially double-stranded molecule with the 5' end, and the 3' end embedded in a secondary structure [6].

Visualization of single-stranded RNA has only been possible in some special cases. This is due to extensive inter- and intramolecular secondary structures that are rapidly formed in solution or even in the remaining liquid on mica before drying. Single strands may become visible when they are fixed by a protein on one side and stabilized by a double-stranded region on the other side [14]. For some applications, single-stranded RNA can be incubated with the interacting protein and subsequently parts of the RNA are "stretched out" by annealing DNA oligos [4].

29.3
Statistics

Biochemical analysis provides an average of the reactivity of 10^{17}–10^{20} molecules, but does not address the detailed behavior of single molecules. On the other hand, SPM, observes single molecules and every non-typical reactivity gains a much higher weight than in bulk analysis. Non-typical behavior could be due to rare but significant reactions, but also to damaged molecules, mistakes in synthesis of the components or other artifacts. To distinguish between artificial and rare, but specific, interactions, a large number of molecules have to be inspected. Overview images rapidly supply a first general, although subjective, impression of molecular interactions. To measure sizes, distances and heights of several 100 molecules in a reasonable time and to gather sufficient data for solid statistics, the appropriate software is usually supplied with the instrument. With measuring 100–500 events, even minor interactions that may escape biochemical analysis could be detected.

29.4
Scanning Force Spectroscopy (SFS)

The scanning force microscope can also be employed to determine inter- and intramolecular binding forces. The basic principle is that one binding partner is attached to the cantilever and the other to the surface. Upon approaching the cantilever to the surface in the z direction, two molecules eventually interact. When the cantilever is then retracted from the sample, binding forces bend the arm until the retracting force equals and finally exceeds the binding force and the interaction is disrupted. The force required for a conformational change or disruption of binding is calculated from the recorded bending of the cantilever and the spring constant – a value describing the flexibility of an individual cantilever.

Conventional scanning force microscopes may be used for SFS but a separate z-piezo has to be integrated to individually control the approach and retract movements. Specific force microscopes are available that do not allow for imaging, but provide optimal hardware and software for interaction measurements. Hybrid instruments have now been developed and become especially interesting for biological applications: a sample can be scanned to provide a topological image and then a specific molecule for force measurements can be approached. It has to be noted that a functionalized cantilever, i.e. a tip with a biomolecule attached, may well be used for scanning and subsequent force measurements.

Figure 29.4 displays a typical force–distance curve taken from double-stranded DNA. In this case, non-specific binding forces between the nucleic acid and the mica surface, on one side, and between the DNA and the silicon nitrate tip, on

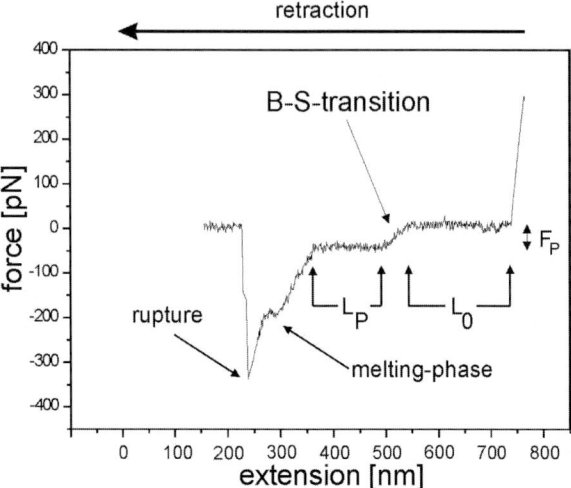

Fig. 29.4. SFS: force–distance curve taken from double-stranded DNA. See text for explanation.

the other side, are sufficiently strong to measure binding forces within the molecule. The graph shows the bending of the cantilever while the z-piezo moves away from the tip: on the far right, tip and substrate are in contact. Overcoming nonspecific adhesion to the substrate, the tip jumps to the zero line. A DNA molecule attached to the surface and the tip is lifted up without applying additional force until it is extended to its entire free length (here approximately 200 nm = L_0). The plateau force (F_p) of 65 pN is then required to convert the B-form DNA to the stretched S-form. After the B–S transition, S-form DNA can be further extended without additional force by a factor of 0.6 (S factor) of the original B-form length. This is shown by the plateau length L_p. Then, a strong force depending on various parameters (attachment site to tip and surface, G + C content, length of molecule) has to be applied to reach the melting phase and a further force finally leads to either disruption of the two strands, disruption of the bond to the tip or disruption of the bond to the surface. After this rupture, the cantilever swings back to the zero line [17, 18].

This example demonstrates the principle of SFS – pulling at a complex of two molecules may lead to one or more defined structural conversions that require defined forces and a final disruption of the interaction caused by an additional force. As for SFM, multiple measurements are made to obtain statistically solid data. When the x–y position of the cantilever is not changed, the same complex may be recorded multiple times. This is useful to demonstrate reversibility and reproducibility, but different complexes at other locations should also be measured. Attempts have been made to detect secondary structures in RNA molecules by SFS, but the results are so far difficult to interpret [19] or have only confirmed relatively simple structures that were already known.

DNA and double-stranded RNA are easy samples since they make strong, non-specific bonds to the mica surface. The exact nature of these bonds is not known and also the position of attachment to the surface within a long nucleic acid is variable. With a 1000-bp double-stranded RNA (or DNA) molecule L_0 values from close to zero up to a maximum of 300 nm can be obtained. In the latter case, the attachment to the mica surface would be close to one end while the molecule is picked up by the cantilever close to the other end. For most other experiments, both or at least one of the interacting partners should be covalently attached to the substrate or the cantilever. The most common method for RNA is to use thiolated oligos that can be bound to a gold surface. Alternatively, thiolated DNA oligos may be used as an anchor and the RNA of interest is transcribed *in vitro* with a tail complementary to the oligo. The hydrogen bonds of the hybrid are usually stronger than, for example, the protein–RNA interactions to be measured. However, one has to consider that the B–S transition of the hybrid may be superimposed on the forces disrupting the interaction of interest.

The measured unbinding forces depend on the velocity of cantilever retraction (loading rate = retraction velocity × elasticity of molecule). With low loading rates, thermal fluctuation contributes significantly to overcoming the energy barrier to separate the molecules and therefore lower forces are measured. With higher loading rates, higher forces are usually required to achieve unbinding. Different protein binding characteristics to target nucleic acids may only be detected in the high loading rate regime [20].

In many cases, it is advisable to attach RNA and/or protein via a flexible linker either to the surface or the cantilever or both. Polyethylene glycol (PEG) linkers of defined length have been successfully used [21].

29.5
Questions that may be Addressed by SFS

SFS can address binding forces between different molecules as well as intramolecular forces like secondary structures of proteins [22] or the forces that hold a protein in a membrane [23]. In combination with molecular genetics, the influence of defined mutations on interactions can be determined. Although association kinetics cannot so far be determined, thermal off-rates can be derived from experiments using different loading rate regimes [20]. At least in some cases this has provided insights into binding mechanisms that could not be obtained by conventional molecular biology approaches.

29.6
Protocols

Protocol: SFM studies on RNA–protein interactions

- To provide internal controls and to reduce the frequently observed preferential end binding of proteins to RNA, cloning of the (putative) protein interaction site into a vector like pT3T7–gus–PstVd is recommended. A single *Sma*I site within the PstVd segment (Fig. 29.3) is used to insert the sequence of interest (usually 30–100 bp). mfold analysis [24] of the expected transcript is carried out to confirm that the inserted sequence is exposed and does not disturb the structure of the backbone molecule. RNA is synthesized by *in vitro* transcription with T7 or T3 RNA polymerase, the reaction mixture is extracted with phenol/chloroform, precipitated with ethanol and washed with 70% ethanol. RNA is taken up in an appropriate volume of buffer (see below) and the concentration is determined by $OD_{260/280}$. It is recommended to evaluate the RNA by native as well as by denaturing gel electrophoresis. In native agarose gels, the approximately 1000-nt gus–PstVd backbone transcripts display a mobility similar to a 500-bp DNA fragment. Recombinant protein is purified by affinity chromatography under native conditions.

- Mica plates are mounted with double-adhesive tape to 1-cm diameter metal disks for easier handling in the SFM. Mica is freshly cleaved by removing the top layer with adhesive tape and is then activated by exposure for 1 min to an air plasma at 0.2 mbar, 600 V and 20 kHz.
- Reactions are carried out under the appropriate buffer conditions in solution. Buffers should contain 3–15 mM $MgCl_2$. Tris and KCl should be avoided since they form salt precipitates on the mica when dried. Instead, HEPES and NH_4Cl are recommended. The concentrations of the reaction partners have to be tested to identify the optimal density for molecules in the microscope. Depending on intrinsic properties of the reactants (like affinity to the surface) and the properties of the mica, concentrations in the range of 3–30 nM are appropriate. Too high concentrations may result in the formation of aggregates during or after drying. Overcrowding of the reaction mixture may not be directly obvious since large aggregates may be preferentially washed off in the following step leaving only a few molecules on the surface. Aliquots of 10 µl of the reaction are placed on the mica surface for 2–10 min and then washed off with 1 ml of water. The surface is blow-dried with nitrogen or argon and ready for microscopy. If the reaction conditions are not compatible with SFM (high salt), a more concentrated assay may be diluted 10-fold in 5 mM $MgCl_2$ and then spread on mica.
- All water and buffers used for SFM are set up with MilliQ purified water, autoclaved and passed through a 0.45-µm sterile filter unit.
- The scanning force microscope is usually placed on a heavy stone slab supported by bungee cords. Alternatively, a dynamic vibration isolation system is used to grant stable measurements.
- SFM conditions depend very much on the microscope that is used. Scanners with a range of $125 \times 125 \times 5$ µm (x, y, z) or $10 \times 10 \times 2.5$ µm are employed. Silicon cantilevers with a nominal resonance frequency of 200–400 kHz (in air), a spring constant of 10–50 N/m and a point diameter of 1–5 nm provide good resolution in tapping mode scanning. High-quality images are taken at 0.5–1 Hz (5–10 min per image). Appropriate software to process the images is provided by the suppliers of the microscope. Additional software for length measurements may be required. Height profiles can be calculated by integrated software for any straight line drawn across the sample.

Protocol: SFS studies on protein–nucleic acids interactions

To our knowledge, measurements of RNA–protein interaction forces have so far not been reported, but can be carried out according to established protocols for DNA–protein interaction.

Functionalized tips and surfaces

- Si_3N_4 tips may be prepared for binding of ligands by different ways.
  (1) Tips are activated by brief incubation in concentrated nitric acid and then silanized for 2 h in 2% aminopropyltriethoxysilane in dry toluene and extensively washed in toluene. They are then incubated in 0.1 mM potassium phosphate pH 8 containing 1 mM N-hydroxysuccinimide–PEG–maleimide for 30 min. Cantilevers are washed with potassium phosphate buffer and are ready for binding the nucleic acid. 5′-SH modified RNA oligonucleotides (10 ng/µl) are incubated with the prepared tips in 50 mM Tris, 100 mM NaCl, 0.1 mM $NiCl_2$ at pH 8.3 at 4 °C for 10 h. After washing with binding buffer, the tips are ready for spectroscopy measurements or may be stored at 4 °C for about 1 week [20].
  (2) An alternative is the use of defined length PEG spacers with an amine-reactive and a thiol-reactive end. Silicon nitride cantilevers are activated for 10 min in chloroform and then for 30 min in H_2SO_4/H_2O_2 (70:30), washed in sterile water and baked for 2 h at 180 °C. Tips are then incubated in 55% ethanolamine chloride in dimethylsulfoxide overnight at 100 °C together with 0.3-µm molecular sieve beads under vacuum with H_2O trapping [21]. PEG spacers are coupled to the amine groups as described by Haselgrübler et al. [25]. Thiolated RNA oligos are then coupled in an oriented way to the thiol-reactive end as described above.
- Several cantilevers (five to 10) should be derivatized simultaneously. The concentration of protein or nucleic acid to be bound to the tip is so low that statistically only a few molecules will bind to the tip. Cantilevers with no ligand or too many ligands have to be identified experimentally and will be discarded.
- For derivatization of surfaces, mica is freshly cleaved, baked at 180 °C and processed as described under (2) for nucleic acid or protein binding.
- For protein binding, the surface is silanized with aminopropyltriethoxysilane in an exsiccator and then incubated for 1 h at 4 °C with the purified protein of interest (5 µM) in 0.1 M potassium phosphate buffer (pH 7.5) containing 20 µM bis(sulfosuccinimidyl)suberate-sodium salt [20], this couples the N-terminus of the protein covalently to the derivatized surface.
- Another method to functionalize surfaces is using gold chips or gold-coated mica to anchor thiolated oligos or proteins covalently [26].

29.7
Troubleshooting

The most sensitive component for imaging, but also for spectroscopy, are the tips. It is very difficult to predict the lifetime of a cantilever: they may be good for measuring 2–3 days or they may become blunt within one measurement. The manufacturing process and quality control is becoming much more reliable, but 10% of rejects are still to be expected. A broadening of the image indicates that a cantilever is blunt. A frequent observation is a "shadow image", a weaker duplication of an image in the same frame. This indicates a double tip either due to the original cantilever, to some damage during scanning or to contamination picked up during scanning of the sample. The shadow image may appear 100 nm or more from the original image and not be obvious at the first glance. Since double tips usually have a different distance to the substrate, the weaker and often thinner shadow image may be misinterpreted as single-stranded DNA or RNA.

As with electron microscopy, the interpretation of an image may be problematic since contaminations may be similar to the expected shape of the sample. A famous example is the steps in carbon surfaces that resembled in shape structure and dimension DNA double strands.

High concentrations of nucleic acids result in aggregates – they cannot be interpreted and may be regarded as contamination only and a complete failure of the experiment. High concentration of protein can cause the same effect; in addition, non-specific multimers may give the impression that the sample is inhomogenous with many different sizes of protein. Since only a "top view" of the molecules is possible, even specific dimers are frequently not observed as particles with double the volume or double the surface of a monomer. To reduce the concentration of the sample is a simple remedy for this problem. However, especially for RNA–protein interactions measured in air, concentrations in the reaction tube are not necessarily reflected in the image because RNA, protein and complexes may adhere with different efficiency to the surface.

Low contrast could be due to set point adjustment too close to the free amplitude of the cantilever. Samples measured in air are usually robust under ambient conditions. High humidity of the air may, however, decrease contrast because a water film will form on the sample. Repeated drying with argon or nitrogen will temporarily solve the problem but water will re-accumulate after 20–40 min.

To discuss specific problems and applications the SFM/SFS forum at http://spm.di.com/listinfo.html is recommended.

29.8
Conclusions

SFM and SFS have advanced beyond the realm of physicists and of providing proof-of-principle for nucleic acids applications. The current generation of instruments is user friendly and designed for biologists who wish to approach biological problems, and not necessarily improve the technology. In combination with biochemistry and molecular biology, SFM and SFS have proven to provide additional information and substantial insight into molecular mechanisms.

Acknowledgments

I thank F. W. Bartels (University of Bielefeld) and I. Sagi (Weizmann Institute) for making data available prior to publication. Nils Anspach and C. Hammann are acknowledged for critical reading of the manuscript. K. Pross (Veeco) is acknowledged for contributing an extensive compilation of the relevant literature.

References

1. G. Binnig, C. F. Quate, C. Gerber, *Phys. Rev. Lett.* **1986**, *56*, 930–933.
2. Y. Lyubchenko, L. Shlyakhtenko, R. Harrington, P. Oden, S. Lindsay, *Proc. Natl. Acad. Sci. USA* **1993**, *90*, 2137–2140.
3. A. Y. Lushnikov, A. Bogdanov, Y. L. Lyubchenko, *J. Biol. Chem.* **2003**, *278*, 43130–43134.
4. S. E. Wells, P. E. Hillner, R. D. Vale, A. B. Sachs, *Mol. Cell* **1998**, *2*, 135–140.
5. M. Bonin, J. Oberstrass, N. Lukacs, K. Ewert, E. Oesterschulze, R. Kassing, W. Nellen, *RNA* **2000**, *6*, 563–570.
6. M. Bonin, J. Oberstrass, U. Vogt, M. Wassenegger, W. Nellen, *Biol. Chem.* **2001**, *382*, 1157–1162.
7. Y. Klaue, A. M. Kallman, M. Bonin, W. Nellen, M. Ohman, *RNA* **2003**, *9*, 839–846.
8. M. J. Fay, N. G. Walter, J. M. Burke, *RNA*, **2001**, *7*, 887–895.
9. M. Argaman, R. Golan, N. H. Thomson, H.-G. Hansma, *Nucleic Acids Res.* **1997**, *25*, 4379–4384.
10. M. Guthold, M. Bezanilla, D. A. Erie, B. Jenkins, H. G. Hansma, C. Bustamante, *Proc. Natl. Acad. Sci. USA* **1994**, *91*, 12927–12931.
11. M. Guthold, X. Zhu, C. Rivetti, G. Yang, N. H. Thomson, S. Kasas, H. G. Hansma, B. Smith, P. K. Hansma, C. Bustamante, *Biophys. J.* **1999**, *77*, 2284–2294.
12. M. B. Viani, T. E. Schäffer, A. Chand, M. Rief, H. E. Gaub, P. K. Hansma, *J. Appl. Phys.* **1999**, *86*, 2258–2262.
13. T. Ando, N. Kodera, E. Takai, D. Maruyama, K. Saito, A. Toda, *Proc. Natl. Acad. Sci. USA* **2001**, *98*, 12468–12472.
14. A. Henn, O. Medalia, S. P. Shi, M. Steinberg, F. Franceschi, I. Sagi, *Proc. Natl. Acad. Sci. USA* **2001**, *98*, 5007–5012.
15. S. Iwabuchii, T. Mori, K. Ogawa, K. Sato, M. Saito, Y. Morita, T. Ushiki, E. Tamiya, *Arch. Histol. Cytol.* **2002**, *65*, 473–479.
16. S. Thalhammer, R. W. Stark, S. Muller, J. Wienberg, W. M. Heckl, *J. Struct. Biol.* **1997**, *119*, 232–237.
17. H. Clausen-Schaumann, M. Rief, C. Tolksdorf, H. E. Gaub, *Biophys. J.* **2000**, *78*, 1997–2007.
18. M. Rief, H. Clausen-Schaumann, H. E. Gaub, *Nat. Struct. Biol.* **1999**, *6*, 346–349.
19. M. Bonin, R. Zhu, Y. Klaue, J. Oberstrass, E. Oesterschulze, W. Nellen, *Nucleic Acids Res.* **2002**, *30*, e81.
20. F. W. Bartels, B. Baumgarth, D. Anselmetti, R. Ros, A. Becker, *J. Struct. Biol.* **2003**, *143*, 145–152.
21. P. Hinterdorfer, W. Baumgartner, H. J. Gruber, K. Schilcher, H. Schindler, *Proc. Natl. Acad. Sci. USA* **1996**, *93*, 3477–3481.
22. M. Rief, M. Gautel, F. Oesterhelt, J. M. Fernandez, H. E. Gaub, *Science* **1997**, *276*, 1109–1112.
23. D. J. Muller, J. B. Heymann, F. Oesterhelt, C. Moller, H. Gaub, G. Buldt, A. Engel, *Biochim. Biophys. Acta* **2000**, *1460*, 27–38.
24. M. Zuker, *Nucleic Acids Res.* **2003**, *31*, 3406–3415.
25. T. Haselgrubler, A. Amerstorfer, H. Schindler, H. J. Gruber, *Bioconjug. Chem.* **1995**, *6*, 242–248.
26. O. Medalia, J. Englander, R. Guckenberger, J. Sperling, *Ultramicroscopy* **2001**, *90*, 103–112.